ELECTROCHEMICAL BIOMASS CONVERSION

Woodhead Series in Bioenergy

The *Woodhead Series in Bioenergy* publishes titles on the topic of bioenergy as it relates to renewable sources of energy. The series notably covers biological resources, chemical and bio-logical processes, and biomass products for energy production, along with the environmental, social, management, and economic aspects of bioenergy.

Series Editor

Prof. Yves Gagnon, Université de Moncton, Moncton, Canada

Previous Volumes in the Series

Kishor Kumar Sadasivuni and Mithra Geetha, Biofuel Cells and Energy Generation, 2025, 9780443216022

Krishna Kumar Jaiswal, Bhaskar Singh and Amit K. Jaiswal, Microalgal Biofuels: Sustainable Production, Conversion, and Life Cycle Assessment, 2025, 9780443241109

Jeyabalan Sangeetha and Devarajan Thangadurai, Microalgal Biomass for Bioenergy Applica- tions, 2024, 9780443139277

Lina María Grajales Agudelo, Juan Carlos Valdés Serra and Erich Collicchio, Agroenergy, 2024, 9780443214301

Stefano Ferrari Interlenghi, José Luiz de Medeiros and Ofélia de Queiroz Fernandes Araújo, Sustainability of Methylic and Ethylic Biodiesel Production Routes, 2024, 9780443219429

Hwai Chyuan Ong, Islam Md Rizwanul Fattah and Indra Mahlia, Waste Valorization for Bioenergy and Bioproducts, 2024, 9780443191718

Woodhead Series in Bioenergy

ELECTROCHEMICAL BIOMASS CONVERSION

Sustainable Processes and Applications for Bioenergy, Biofuels, and Bioproducts

Edited by

RAJENDER BODDULA
School of Sciences, Woxsen University, Hyderabad, Telangana, India

RAMYAKRISHNA POTHU
School of Physics and Electronics, College of Chemistry and Chemical Engineering, Hunan University, Changsha, Hunan, China

NOORA AL-QAHTANI
Center for Advanced Materials, Qatar University, Doha, Qatar

ELSEVIER

WP
WOODHEAD PUBLISHING
An imprint of Elsevier
elsevier.com/books-and-journals

Woodhead Publishing is an imprint of Elsevier
50 Hampshire Street, 5th Floor, Cambridge, MA 02139, United States

ISBN: 978-0-443-26757-4

For Information on all Woodhead Publishing publications visit our website at https://www.elsevier.com/books-and-journals

Publisher: Priscilla Braglia
Acquisitions Editor: Ruchi Bhardwaj
Editorial Project Manager: Akansha Rathi
Production Project Manager: A. Maria Shalini
Cover Designer: Arumugam Kothandan

Typeset by Aptara, New Delhi, India

Contents

Contributors

Taniya Rose Abraham School of Energy Materials, Mahatma Gandhi University, Kottayam, Kerala, India

Anmol School of Natural Sciences, GNA University, Sri Hargobindgarh, Phagwara, Punjab, India

Isha Arora School of Natural Sciences, GNA University, Sri Hargobindgarh, Phagwara, Punjab, India

Neelesh Ashok Amrita School of Artificial Intelligence, Amrita Vishwa Vidyapeetham, Coimbatore, Tamil Nadu, India

Garima Awasthi School of Basic & Applied Sciences, K.R. Mangalam University, Gurugram, Haryana, India

Kumud Kant Awasthi School of Basic & Applied Sciences, K.R. Mangalam University, Gurugram, Haryana, India

Liya Babu School of Energy Materials, Mahatma Gandhi University, Kottayam, Kerala, India

Rukhsana Badarudeen School of Energy Materials, Mahatma Gandhi University, Kottayam, Kerala, India

Jayeeta Bhaumik Department of Biotechnology (DBT), Laboratory of Bioproduct Chemistry, BRIC-National Agri-Food and Biomanufacturing Institute (BRIC NABI), Government of India, Punjab, India

Priya Bijesh Department of Chemistry, KCG College of Technology, Chennai, Tamil Nadu, India

Rajender Boddula School of Sciences, Woxsen University, Hyderabad, Telangana, India

Nikita Choudhary Vivekananda Global University, Jaipur, Rajasthan, India

Sancler da Costa Vasconcelos Institute of Chemistry, Fluminense Federal University, Niterói, Brazil

Joyeeta Das Department of Chemistry, Center of Forensic Science, Banaras Hindu University, Varanasi, Uttar Pradesh, India

Thiago de Melo Lima Institute of Chemistry, Fluminense Federal University, Niterói, Brazil

Carolina Guimarães de Souza Lima Institute of Chemistry, Fluminense Federal University, Niterói, Brazil

Surbhi Dhadda Vivekananda Global University, Jaipur, Rajasthan, India

Mayane Barbosa dos Santos Institute of Chemistry, Fluminense Federal University, Niterói, Brazil

Manoj Eswaran Department of Electronics and Communication Engineering, KPR Institute of Engineering and Technology, Coimbatore, Tamil Nadu, India

Ramasamy Ganesamoorthy Department of Chemistry, Vinayaka Mission's Kirupananda Variyar Arts and Science College, Vinayaka Mission's Research Foundation (Deemed to be University), Salem, Tamil Nadu, India

Jayanthi Gopal Department of Chemistry, KCG College of Technology, Chennai, Tamil Nadu, India

Vetha Gopal Department of Chemistry, KCG College of Technology, Chennai, Tamil Nadu, India

Shubhi Gupta Department of Fuel, Minerals and Metallurgical Engineering, Indian Institute of Technology (Indian School of Mines), Dhanbad, Jharkhand, India

Keerthi Kumar Kulanthaigounder Jaganathan Department of Electronics and Communication Engineering, KPR Institute of Engineering and Technology, Coimbatore, Tamil Nadu, India

Ganesamoorthy Jeevitha Department of Molecular Physics, Saveetha School of Engineering, Saveetha Institute of Medical and Technical Sciences (SIMATS), Chennai, Tamil Nadu, India

Vijayakrishnan Jeyavani Department of Physics, Rathinam College of Arts and Science, Coimbatore, Tamil Nadu, India

Davis Johny School of Energy Materials, Mahatma Gandhi University, Kottayam, Kerala, India

Suba Kannaiyan Department of Chemistry, Sri Sairam Engineering College, Chennai, Tamil Nadu, India

Ashish Kapoor Department of Chemical Engineering, Harcourt Butler Technical University, Kanpur, Uttar Pradesh, India

Kuljit Kaur School of Natural Sciences, GNA University, Sri Hargobindgarh, Phagwara, Punjab, India

Amit Kumar Department of Chemistry, School of Allied Sciences, Dev Bhoomi Uttarakhand University, Dehradun, Uttarakhand, India

Ponnaiah Sathish Kumar Gyeongbuk Institute of Science & Technology (DGIST), Daegu, Republic of Korea

Poonam Kumari School of Allied Health Sciences, Jaipur National University, Jaipur, Rajasthan, India

Sungwon Lee Gyeongbuk Institute of Science & Technology (DGIST), Daegu, Republic of Korea

Ramakrishna Madaka School of Sciences, Woxsen University, Hyderabad, Telangana, India

Sajay K. Meena Vivekananda Global University, Jaipur, Rajasthan, India

Prasenjit Mondal Department of Chemical Engineering, Indian Institute of Technology Roorkee, Roorkee, Uttarakhand, India

Deepthi R Murali Post Graduate and Research Department of Chemistry, Sree Narayana College for Women, Kollam, Kerala, India

Lakshmana Reddy Nagappagari Catalysis and Spectrochemistry Laboratory, ENSICAEN, CNRS, LCS, University of Caen Normandy, Caen, France

Varad Nagar Department of Chemistry, Center of Forensic Science, Banaras Hindu University, Varanasi, Uttar Pradesh, India

Prathyusha P.V. School of Energy Materials, Mahatma Gandhi University, Kottayam, Kerala, India

Mrutyunjay Padhiary Department of Agricultural Engineering, Assam University, Silchar, Assam, India

Priya Pallavi School of Sciences, Woxsen University, Hyderabad, Telangana, India

Harsh Pandey Department of Biotechnology and Chemical Engineering, Manipal University, Jaipur, Rajasthan, India

Muthu Senthil Pandian Department of Physics, SSN Research Centre, Sri Sivasubramaniya Nadar College of Engineering, Kalavakkam, Chennai, Tamil Nadu, India

Ajmal Parveena Department of Physics, Macquarie University, Sydney, NSW, Australia

Ruchi Pathania Department of Medicinal Chemistry, Center for Natural Products, Drug Discovery and Development, University of Florida, Gainesville, FL, United States

Arun Paulose CSIR–Institute of Genomics and Integrative Biology, New Delhi, India

Ramyakrishna Pothu Centre for Research and Development Cell, Sharda University, Greater Noida, Uttar Pradesh, India; School of Physics and Electronics, College of Chemistry and Chemical Engineering, Hunan University, Changsha, Hunan, China

Monika Kumari Prajapati Department of Forensic Science, SBMS, Galgotias University, Greater Noida, Uttar Pradesh, India

Megha Pramod School of Energy Materials, Mahatma Gandhi University, Kottayam, Kerala, India

Amit Kumar Rajak Department of Energy & Environmental Engineering, CSIR–Indian Institute of Chemical Technology, Hyderabad, Telangana, India

Jagadeeswari Rangaraman Department of Chemistry, KPR Institute of Engineering and Technology, Coimbatore, Tamil Nadu, India

Yeddula Nikhileshwar Reddy Department of Biotechnology (DBT), Laboratory of Bioproduct Chemistry, BRIC-National Agri-Food and Biomanufacturing Institute (BRIC NABI), Government of India, Punjab, India; Department of Biotechnology, School of Bioengineering, SRM Institute of Science and Technology, Kattankulathur, Tamil Nadu, India

Eduardo Magalhães Rodrigues Institute of Chemistry, Fluminense Federal University, Niterói, Brazil

Anupam Singha Roy School of Sciences, Woxsen University, Hyderabad, Telangana, India

Mahipal Singh Sankhla School of Basic & Applied Sciences, K.R. Mangalam University, Gurugram, Haryana, India

Bipul Sarkar Catalytic Depolymerization Area, CSIR-Indian Institute of Petroleum, Dehradun, Uttarakhand, India

Ganesan Kausalya Sasikumar Centre of Research and Development, KPR Institute of Engineering and Technology, Coimbatore, Tamil Nadu, India; School of Sciences, Woxsen University, Hyderabad, Telangana, India

Himanshu Sharma Department of Chemistry, Science College, Udaipur, Rajasthan, India

Mehak Sharma School of Natural Sciences, GNA University, Sri Hargobindgarh, Phagwara, Punjab, India

Sonali Sharma Vivekananda Global University, Jaipur, Rajasthan, India

Vaibhav Sharma Department of Medical Elementology and Toxicology, School of Chemical and Life Sciences, Jamia Hamdard, (Deemed to be University), New Delhi, India

Omvir Singh Department of Sciences and Humanities, Rajiv Gandhi Institute of Petroleum Technology, Jais, Amethi, Uttar Pradesh, India; Catalytic Depolymerization Area, CSIR-Indian Institute of Petroleum, Dehradun, Uttarakhand, India

N. Soujanya School of Technology, Woxsen University, Hyderabad, Telangana, India

Amit Srivastava Centre Algatech, Institute of Microbiology of the Czech Academy of Sciences, Trebon, Czech Republic

Rajesh K. Srivastava Department of Biotechnology, Gandhi Institute of Technology and Management (GITAM), Visakhapatnam, Andhra Pradesh, India

Ashwani Tapwal ICFRE-Himalayan Forest Research Institute, Shimla, Himachal Pradesh, India

Naveen Thakur ICFRE-Himalayan Forest Research Institute, Shimla, Himachal Pradesh, India

Santhosh Kumar Thatikonda School of Technology, Woxsen University, Hyderabad, Telangana, India

Karchiyappan Thirugnanasambandham Nammazhvar Organic Farming Research Centre, Tamil Nadu Agricultural University, P N Pudur, Coimbatore, Tamil Nadu, India

Sabu Thomas School of Energy Materials, Mahatma Gandhi University, Kottayam, Kerala, India

Sneha Thomas CSIR–Institute of Genomics and Integrative Biology, New Delhi, India

Selvaraj Vaithilingam Department of Chemistry, KCG College of Technology, Chennai, Tamil Nadu, India; Department of Chemistry, University College of Engineering, Villupuram, Tamil Nadu, India

Andal Gopal Venugopal Department of Chemistry, KCG College of Technology, Chennai, Tamil Nadu, India

Foreword

The energy sector is undergoing a significant transition to meet growing energy demands, reduce greenhouse gas emissions, support climate change mitigation, and protect the planet for future generations. A major component of this energy transition is the integration of large-scale renewable energy sources, including bioenergy, into the overall energy mix of countries.

Bioenergy refers to the conversion of biomass, derived from recently living organic materials, into electricity, heat, and transportation fuels. Alongside wind and solar energy, bioenergy is poised to play a significant role in the global energy transition.

To support both the research community and the commercial bioenergy sector, the Woodhead Series in Bioenergy publishes research-driven, application-focused titles that explore bioenergy as a renewable source of energy. The series covers biological resources, chemical and biological processes, and biomass-based products for energy production, all framed by the principles of sustainability, encompassing environmental, social, managerial, and economic dimensions.

This book focuses on electrochemical biomass conversion for bioenergy applications, an alternative to thermochemical transformations, which require heat and entail additional environmental impacts. Electrochemical biomass conversion offers enhanced selectivity and operates at lower temperatures and pressures, producing biofuels and bioproducts while minimizing environmental impacts and optimizing resource efficiency.

Building on his experience in scientific publication and book editing, Assoc. Prof. Dr. Rajender Boddula, from Woxsen University, India, teams up with two promising early-career researchers, Ramyakrishna Pothu from Hunan University, China, and Research Assistant Prof. Dr. Noora Al-Qahtani from Qatar University, Qatar, to present this title on electrochemical biomass conversion, focusing on sustainable processes and with applications in bioenergy, biofuels, and bioproducts.

The book presents an extensive coverage of electrochemical biomass conversion, from fundamental mechanisms to practical applications, along with the latest advancements in industrial applications.

This specialized book in the Woodhead Series in Bioenergy covers niche technologies that are of interest to academics, researchers, and industry practitioners working on research and the development of biomass conversion processes and applications in bioenergy and bioproducts.

Prof. Yves Gagnon P.Eng., D.Sc.
Université de Moncton, Canada
Series Editor, Woodhead Series in Bioenergy

CHAPTER

1

Introduction to biomass electrochemistry

Ajmal Parveena[a], *Taniya Rose Abraham*[b], *Davis Johny*[b], *Sneha Thomas*[c], *and Arun Paulose*[c]

[a]Department of Physics, Macquarie University, Sydney, NSW, Australia, [b]School of Energy Materials, Mahatma Gandhi University, Kottayam, Kerala, India, [c]CSIR-Institute of Genomics and Integrative Biology, New Delhi, India

1.1 Introduction

As climate change aggravates, it is evident that we urgently need to switch from fossil fuels to sustainable alternatives. As the production of conventional chemicals heavily depends on nonrenewable carbon sources, environmental degradation and greenhouse gas emissions have become a serious concern. The Intergovernmental Panel on Climate Change (IPCC) highlights the need to limit global warming to 1.5°C by 2030 by emphasizing the significance of cutting carbon emissions by at least 45%. Breakthroughs in green chemistry and low-carbon technology, specifically involving the utilization of renewable resources, are driven by the pressing circumstances. As an abundant and sustainable carbon resource, biomass provides a workable approach to decarbonize the chemical sector while raising the benchmarks for the circular economy established by domestic and international frameworks [1] (Fig. 1.1).

The composition of biomass is largely defined by lignin, hemicellulose, and cellulose, with additional fractions of proteins, lipids, and other organic materials depending on the source. The functionalized carbon molecules comprising these structures provide a rich and intricate matrix that can be transformed into materials, chemicals, and fuels. When

Electrochemical Biomass Conversion: Sustainable Processes and Applications for Bioenergy, Biofuels, and Bioproducts
DOI: https://doi.org/10.1016/B978-0-443-26757-4.00010-6

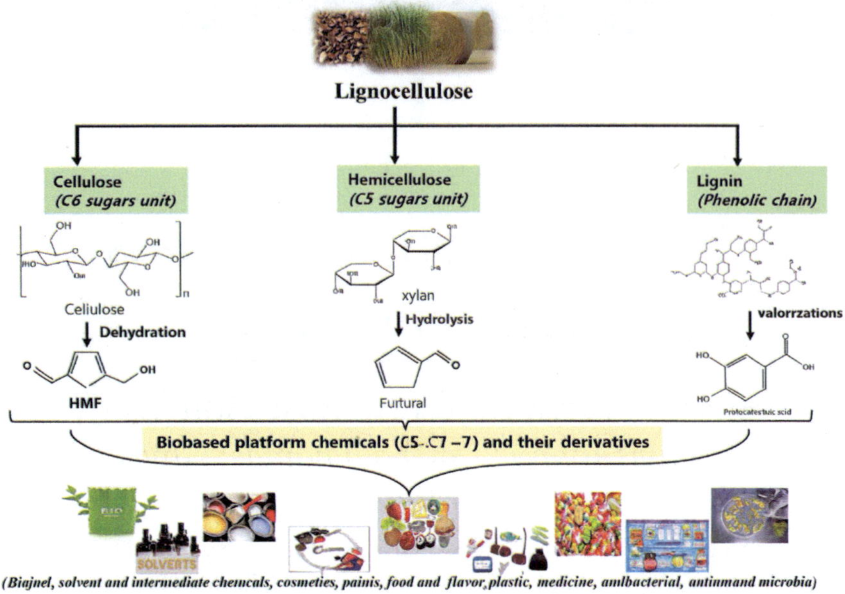

FIGURE 1.1 Utilization of bio-based feedstocks for the production of platform chemicals and their derivative applications. *Source: P.P. Upare, R.E. Clarence, H. Shin, B.G. Park, An overview on production of lignocellulose-derived platform chemicals such as 5-hydroxymethyl furfural, furfural, protocatechuic acid. Processes 11 (2023) 2912, doi:10.3390/pr11102912.*

integrated with sustainable agricultural strategies, biomass can result in carbon-neutral or even carbon-negative results as it is renewable on a human timescale, unlike fossil resources. Furthermore, using lignocellulosic and nonfood biomass offers a more sustainable and ethically sound approach for chemical feedstock as it avoids conflicts between food and fuel and makes use of forestry waste or agricultural residues [2,3].

The electrochemical conversion of biomass-derived compounds under ambient conditions—free from hazardous chemicals, high pressures, or high temperatures—provides a potent and environmentally friendly pathway for green chemical production. This approach supports decentralized and modular chemical production systems, simplifies integration with intermittent renewable electricity, and permits fine control over redox processes through electrochemical potential control. These characteristics are well aligned with the principles of decentralized biorefineries and green chemistry concepts. Due to its low energy requirements, minimal waste generation, and seamless integration with renewable energy sources, the electrochemical oxidation or reduction of biomass platform molecules, such as furfural or 5-hydroxymethylfurfural (HMF), into useful monomers and fuels represents a rapidly emerging area of research [4,5].

Among biomass-derived intermediates, HMF has emerged as a pivotal platform molecule due to its versatility in generating a broad range of value-added compounds, such as bio-based monomers, fuels, and fine chemicals. HMF is typically produced by the acid-catalyzed dehydration of hexose sugars from lignocellulosic biomass, such as fructose and glucose. Due to the presence of both hydroxymethyl and aldehyde functional groups, it can undergo a wide range of chemical reactions, such as etherification, hydrogenation, reduction, and oxidation. 2,5-diformylfuran (DFF), 2,5-furandicarboxylic acid (FDCA), and 2,5-bis(hydroxymethyl)furan (BHMF) are key HMF-derived compounds that serve as important precursors for the synthesis of bioplastics and sustainable polymers [6].

Electrocatalytic approaches provide a greener and more energy-efficient alternative to traditional thermocatalytic routes for converting HMF and related biomass-derived carbohydrates. Electro-oxidation of HMF to FDCA usually occurs via two pH-dependent pathways: (1) oxidation of the hydroxymethyl group (HMF → DFF → FFCA → FDCA) under mildly alkaline conditions, and (2) oxidation of the aldehyde group (HMF → HMFCA → FFCA → FDCA) under strong alkaline conditions. Both noble metals (such as Au, Pt, and Pd) and non-noble substitutes like Ni, Co, and Mn, as well as their corresponding oxides and layered double hydroxides (LDHs), which show encouraging activity, selectivity, and scalability, have been the focus of electrocatalyst development [3,7].

The circular bioeconomy and sustainable chemistry meet at the point of electrochemical biomass conversion. It offers a versatile platform for turning waste or underutilized biomass sources into renewable chemicals, fuels, and materials. Crucially, it makes it possible to combine biomass oxidation with beneficial cathodic reactions like CO_2 reduction or hydrogen evolution, improving system economics and energy efficiency. For example, scalable FDCA production and concurrent green hydrogen generation have been achieved using continuous flow electrolyzers employing bifunctional catalysts, providing a feasible route to integrated electrochemical biorefineries [8] (Fig. 1.2).

1.2 Fundamentals of electrochemical biomass conversion

Electrochemical cell topologies, usually consisting of an anode, a cathode, and an electrolyte, often divided by an ion-conducting membrane, are essential for electrochemical biomass conversion. While the cathode supports reduction reactions like hydrogen evolution (HER) or biomass hydrogenation, the anode promotes oxidation reactions like the conversion of HMF to FDCA. Through effective spatial and functional separation of half-reactions, this setup allows for paired electrolysis, maximizing system

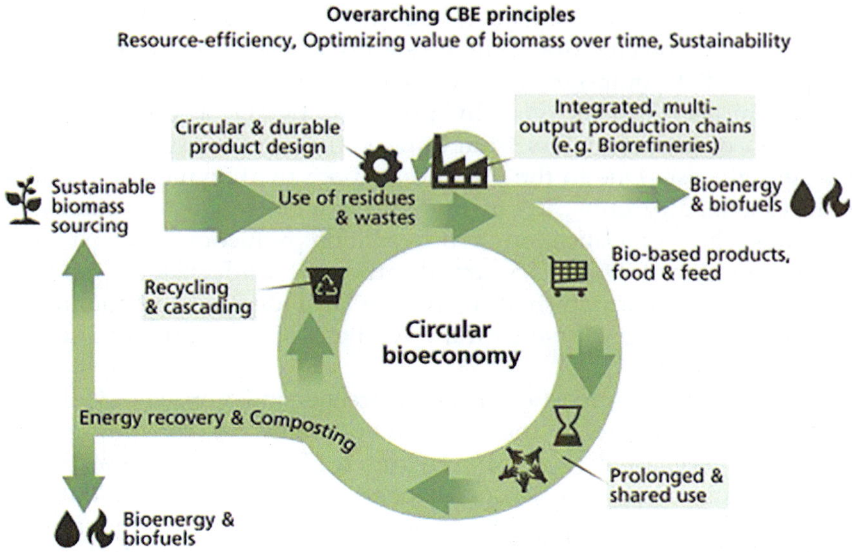

FIGURE 1.2 Core elements of the circular bioeconomy and their interconnections. *Source: P. Stegmann, M. Londo, M. Junginger, The circular bioeconomy: its elements and role in European bioeconomy clusters. Res. Conserv. Recy. 6 (2020) 100029, doi:10.1016/j.rcrx.2019.100029.*

efficiency by enabling both electrodes to generate useful products. Such a configuration improves electron efficiency and overall energy consumption [9] (Fig. 1.3).

The kind and structure of the electrocatalyst have a significant impact on biomass electroconversion performance. The availability of active sites, conductivity, stability in electrochemical settings, and binding strength to important intermediates are important factors. Noble metals with high activity, like Pt, Au, and Pd, have problems with durability and cost. As a result, studies have turned to non-noble catalysts like layered double hydroxides (LDHs) and transition metal-based systems composed of nickel, cobalt, manganese, and their oxides. The performance and scalability of these materials are being further advanced by innovations such as heteroatom-doped nanostructures, single-atom catalysts, and bimetallic sulfides generated from metal-organic framework (MOF) [10].

Electrolyte pH, applied voltage, electrode material, and competing reactions are some of the crucial factors that must be taken into consideration for successful electrochemical biomass valorization. The stability of electrocatalysts and the oxidation route of HMF are both impacted by the electrolyte pH. Strong alkaline media, such as 1 M KOH, are frequently chosen because of their superior conductivity and solubility. However, particularly on specific metal surfaces, side reactions like the hydrogen evolution reaction (HER) may interfere with intended oxidative

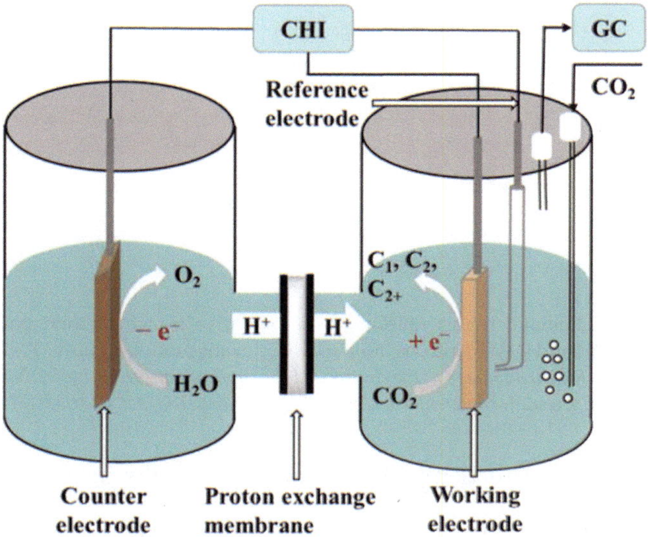

FIGURE 1.3 Diagram of a dual-compartment H-type electrochemical cell employed in the biomass electroconversion process. *Source: L. Wu, L. Wu, C. Guo, Y. Guan, H. Wang, J. Lu, Progress in electroreduction of CO2 to form various fuels based on Zn catalysts. Processes 11 (2023) 1039, doi:10.3390/pr11041039.*

or reductive changes. Therefore, controlling reaction selectivity, energy consumption, and Faradaic efficiency is crucial when developing effective electrochemical systems for biomass conversion [11].

1.3 Biomass-derived platform molecules

Platform molecules are intermediates made from biomass waste streams or lignocellulosic biomass that have functional groups that can be converted into products with additional value later on. Molecular accessibility, functional group variety, availability from non-food biomass, and capacity to substitute monomers obtained from petrochemicals are among the selection factors. Many of these requirements are met by molecules such as lignin-derived aromatics, levulinic acid, glycerol, furfural, and HMF [6,12].

A number of chemicals generated from biomass have become more well-known because of their capacity for conversion. Furfural comes from C5 hemicellulose sugars, whereas HMF is a top contender made from C6 sugars. Cellulose is typically depolymerized into glucose, while hemicellulose yields xylose upon hydrolysis. Trihydroxy functionality is provided by glycerol, a byproduct of the manufacturing of biodiesel. Levulinic acid is a precursor to several acids and ketones. These compounds are prized for the

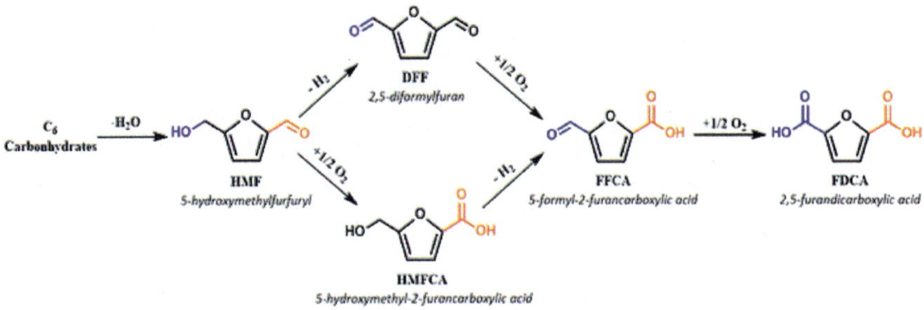

FIGURE 1.4 Schematic representation of the oxidative transformation pathways of 5-hydroxymethylfurfural (HMF) into an industrially relevant product. *Source: Y. Zhao, M. Cai, J. Xian, Y. Sun, G. Li, Recent advances in the electrocatalytic synthesis of 2,5-furandicarboxylic acid from 5-(hydroxymethyl)furfural. J. Mater. Chem. A 9 (2021) 20164–20183, doi:10.1039/D1TA04981J.*

variety of their electrochemical transformation paths in addition to their availability [7,13].

HMF has become the most researched platform molecule in electrochemical valorization because it is derived from fructose or glucose and has two functional groups (aldehyde and alcohol). FDCA, a monomer for polyethylene furanoate (PEF), a sustainable substitute for PET, can be produced by oxidizing HMF. HMF's major significance in research and industry is justified by its ease of electrooxidation in comparison to water oxidation and its close connection to the circular bioeconomy through PEF [6].

1.4 Electrochemical oxidation pathways

1.4.1 Oxidation of 5-hydroxymethylfurfural

The electrochemical oxidation of HMF has been extensively researched due to its potential to generate high-value products under mild and environmentally benign conditions. Depending on the pH and the type of catalyst surface, two main oxidation processes are typically seen: (1) the alcohol-first pathway (HMF → DFF → FFCA → FDCA) and (2) the aldehyde-first pathway (HMF → HMFCA → FFCA → FDCA). The aldehyde route predominates in extremely alkaline settings ($pH > 13$), whereas the alcohol route does so in milder alkaline environments ($pH < 13$). These mechanistic pathways are very pH sensitive [6,14,15] (Fig. 1.4).

1.4.2 Electrooxidation to 2,5-furandicarboxylic acid

FDCA, produced via the electrooxidation of HMF, is a vital platform chemical in the synthesis of renewable plastics such as PEF, driving extensive research into efficient and green conversion methods. In alkaline media, catalysts such $NiCo_2O_4$/Ni foam, NiFe LDH, and MOF-derived CoxNiyS exhibit near-quantitative yields and good selectivity for FDCA [14, 16, 17]. High activity is demonstrated by mediator-free systems based on Ni, Co, and Ru, but superior selectivity is made possible by mediator-assisted systems employing substances like 2,2,6,6-tetramethylpiperidine-1-oxyl (TEMPO) or acetoxylated compounds (ACT) [7,9,18].

1.4.3 Oxidation of furfural and derivatives

Another important platform molecule, furfural, is electrooxidized to provide maleic and furoic acids. These oxidation processes, which involve the transition of aldehyde and furan moieties, are comparable to HMF oxidation. pH and catalyst composition have an impact on furfural-derived electrooxidation pathways. Although noble metals like Au, Pd, and Pt have been used, transition metal oxides and bimetallic catalysts are becoming more reliable and affordable substitutes [8].

1.4.4 Coupled anodic oxidation benefits

Energy savings and the co-production of compounds with added value are two benefits of using biomass oxidation at the anode in place of the oxygen evolution process (OER). When compared to traditional water electrolysis, coupled systems that combine HMF oxidation with cathodic hydrogen evolution (HER) can drastically lower total energy consumption. These arrangements also enhance Faradaic efficiency and electron economy, which is in line with the objectives of sustainable chemical manufacture [9].

1.5 Electrochemical reduction pathways

1.5.1 Hydrogenation of 5-hydroxymethylfurfural

An array of reduced products, such as BHMF, DMF, and other hydrogenolysis derivatives, is produced when HMF is electrochemically hydrogenated (ECH). These products are necessary for applications involving monomers and biofuels. In order to maximize product selectivity and reaction efficiency, catalyst development is essential. Pd single-atom catalysts are being investigated for DMF synthesis, whilst Ag/C and Cu-based catalysts have been used to facilitate the reduction of HMF to BHMF. In order to prevent adverse reactions like hydrodimerization, cathodic potential control is essential [9,17].

1.5.2 Sugar hydrogenation

Sugar alcohols like sorbitol and xylitol can be produced by electrocatalytically hydrogenating sugars originating from biomass, such as glucose and xylose. An environmentally friendly substitute for conventional thermal hydrogenation is offered by this method. Performance is strongly impacted by catalyst surface morphology, particularly electrochemical roughening. Among the investigated metals, such as Ag, Au, Pt, and Zn, rough Cu electrodes showed the best selectivity and conversion. Potential tuning is necessary to manage the competitive HER, which is still difficult [11,16].

1.5.3 Paired electrochemical processes

By combining oxidation and reduction reactions in a single electrochemical cell, paired electrochemical systems significantly increase energy efficiency. In one instance, HMF undergoes simultaneous reduction to BHMF and oxidation to FDCA, resulting in product efficiencies of 98% and 85%, respectively. When compared to single-path systems, this method allows for an electron utilization efficiency of >180%. In order to suppress undesirable side reactions and meet prospective requirements, catalysts must be carefully selected. This combination demonstrates the possibility for integrated electrochemical biorefineries while avoiding energy-intensive oxygen evolution [9,19] (Fig. 1.5).

1.6 Catalyst and reactor design

The key to effective electrochemical biomass conversion is creative catalyst design. To increase active site density and reaction kinetics, techniques include morphological control, heteroatom doping, surface roughening, and defect engineering. As evidenced by CoxNiyS electrocatalysts that produced an 84.5% HMF conversion and a 54% FDCA yield [10], MOF-derived catalysts have become more and more popular because of their customizable porosity and bimetallic compositions. By improving the adsorption of intermediates and promoting electron transfer, single-atom catalysts and spinel oxides also demonstrate encouraging performance [8,20].

Scalability and reaction efficiency are strongly influenced by reactor architecture. While continuous flow systems have shown industry relevance with output rates reaching 40 g h^{-1} FDCA and stable operation over 100 h, traditional batch reactors are appropriate for preliminary experiments. Long-term operation and improved mass transport are made possible by flow electrolyzers and membrane electrode assemblies. To reduce ohmic

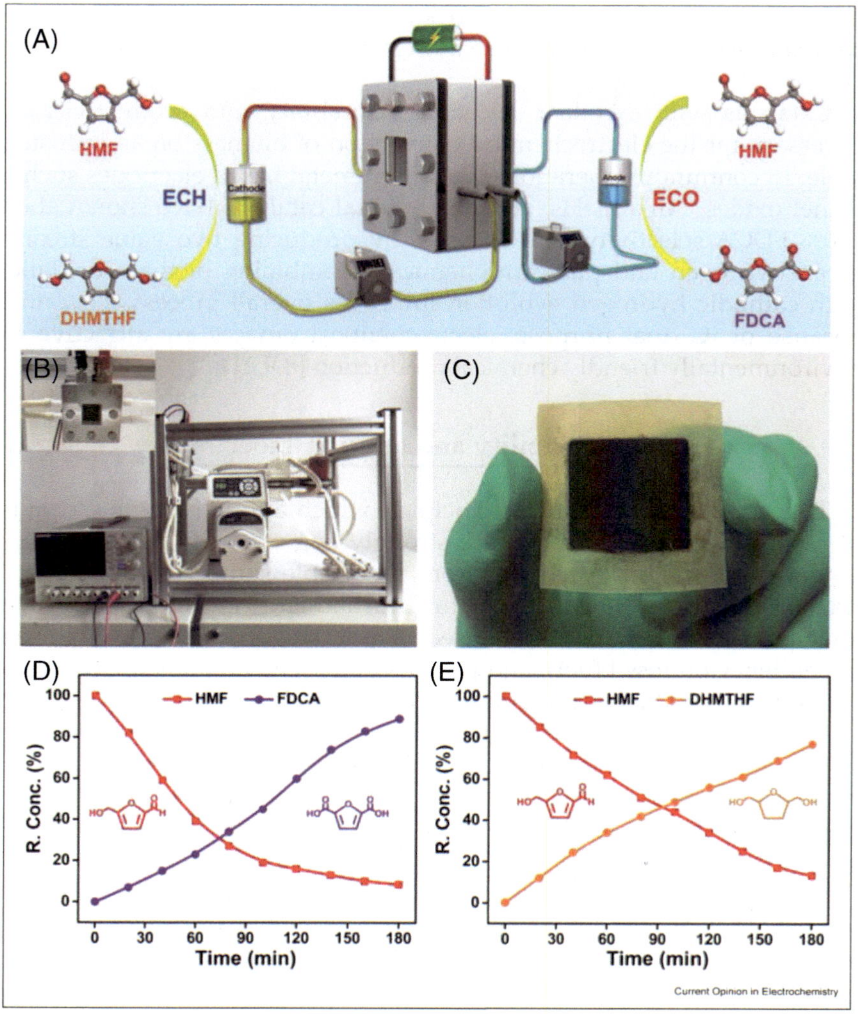

FIGURE 1.5 Integrated paired electrolysis system for coupled oxidative and reductive electrochemical pathways (ECO and ECH) for HMF valorization. (A) Schematic illustration showing the use of vanadium nitride (VN) and palladium-loaded VN (Pd/VN) as electrode materials for ECO and ECH, respectively. (B) Experimental photograph of the paired electrolyzer setup used for the simultaneous reactions. (C) Bipolar membrane cell incorporating VN and Pd/VN catalyst-coated electrodes. (D) Relative concentration profiles of HMF and its oxidation product, 2,5-furandicarboxylic acid (FDCA), over time. (E) Relative concentration profiles of HMF and its hydrogenation product, 2,5-bishydroxymethyl-tetrahydrofuran (DHMTHF), highlighting parallel reduction performance. *Source: S. Li, X. Sun, Z. Yao, X. Zhong, Y. Cao, Y. Liang, et al. Biomass valorization via paired electrosynthesis over vanadium nitride-based electrocatalysts. Adv. Funct. Mater. 29 (2019) 1904780, doi:10.1002/adfm.201904780.*

losses and increase conversion, design parameters such as membrane selection, electrode surface area, and electrolyte flow rate must be tuned [7].

Catalysts with excellent selectivity, durability, and productivity are necessary for the electrochemical conversion of biomass on an industrial scale. In continuous operations, transition metal-based electrodes such as spinel oxides, NiFe LDHs, and bifunctional catalysts have shown about 100% FDCA selectivity. Furthermore, by producing two value streams, green hydrogen and platform chemicals, combining biomass oxidation with cathodic hydrogen evolution improves overall process economics. Because of its dual purpose, electrochemical devices are attractive for environmentally friendly chemical production [8,9,21].

1.7 Sustainability and circular bioeconomy

The circular bioeconomy's principles, which aim to cascade the value of renewable carbon sources, are strongly aligned with electrochemical biomass conversion. Electricity from renewable sources can be used to transform platform molecules like furfural and HMF into bioplastics, fuels, and specialty chemicals. This closed-loop strategy lowers waste, lessens dependency on fossil fuels, and permits integrated biorefinery systems to generate several high-value products at once [1,22].

In contrast to thermal catalytic processes, electrochemical methods do not require stoichiometric oxidants, high-pressure hydrogen gas, or high reaction temperatures. By functioning in mild or ambient settings, they improve safety and lower greenhouse gas emissions. These systems are also flexible enough to be used for decentralized manufacturing because they can be colocated with renewable energy sources. When electrochemical oxidation and hydrogen evolution are combined, economic feasibility is further enhanced, and two revenue streams are created [7,8].

Electrochemical biomass valorization holds potential, but there are still a number of challenges to overcome. Industrial application is hampered by difficult separation procedures, poor product selectivity under scaled-up conditions, and catalyst degradation. Economic evaluations also need to take reactor maintenance, membrane and catalyst durability, and electricity costs into consideration. It will take strong reactor design, system integration, and lifecycle analysis to close the performance gap between lab-scale efficiency and industrial-scale performance [9,23]

1.8 Future outlook

The future of sustainable chemical processes will increasingly depend on the integration of electrochemical biomass conversion with renewable

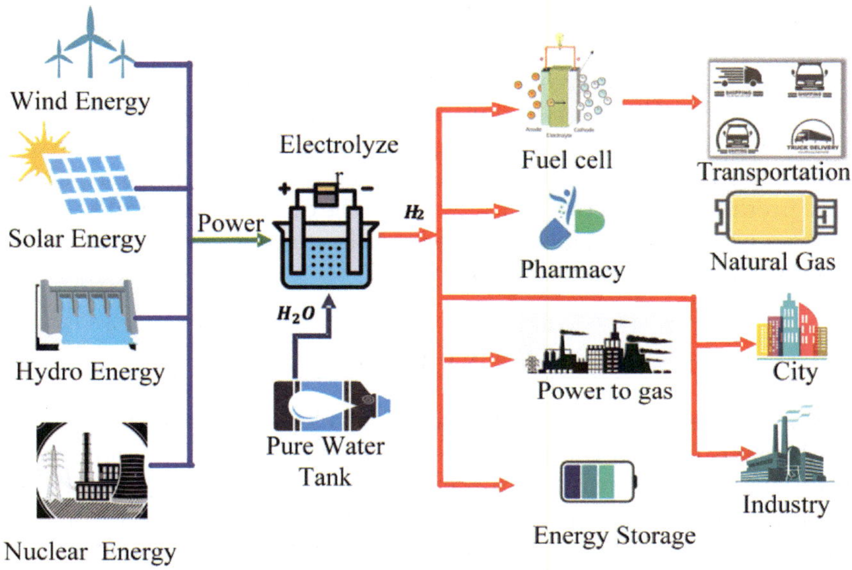

FIGURE 1.6 Schematic of a hybrid renewable energy system where intermittent renewable power (e.g., wind, solar) drives an electrolyzer to produce hydrogen, which is then used in fuel cells and other chemical conversion pathways. *Source: A. Alzahrani, S.K. Ramu, G. Devarajan, I. Vairavasundaram, S. Vairavasundaram, A review on hydrogen-based hybrid microgrid system: topologies for hydrogen energy storage, integration, and energy management with solar and wind energy. Energies 15 (2022) 7979, doi:10.3390/en15217979.*

power sources, including solar and wind energy. Because electrochemical systems are naturally adaptable and modular, they can be integrated with smart grids and used to provide demand-responsive electricity. Such integration could improve overall grid stability and allow for low-cost chemical manufacturing during times of excess energy [8].

New hybrid approaches for upgrading biomass are made possible by combining electrochemical techniques with biocatalysis. For example, electrochemical processes carry out oxidation or reduction, whereas enzymatic processes can be utilized for selective bond cleavage or functionalization. Compared to strictly thermal or chemical techniques, these synergies can lower energy input and achieve improved selectivity. Such integration has been proposed for cascade reactions and multistep biomass conversion routes [11] (Fig. 1.6).

Next-generation electrocatalyst development is anticipated to be accelerated by machine learning, high-throughput screening, and digital technologies. Predicting degradation pathways, determining the best catalyst compositions, and directing synthetic techniques are all made possible by artificial intelligence. These resources will lessen experimental trial-and-error and facilitate the advancement of catalyst systems with enhanced stability and performance [10, 19].

The commercialization of electrochemical biorefineries requires scalable reactors, standardized operating procedures, and integration with upstream and downstream processing. Promising advancements have been made with the demonstration of continuous flow systems that achieve great productivity and product selectivity. The commercial viability of such systems will also be made possible by policy incentives, carbon pricing, and investments in green infrastructure [7, 9].

1.9 Conclusions

Biomass electrochemistry has quickly developed into a sustainable and feasible substitute for traditional thermocatalytic methods of biomass valorization. It is now feasible to transform platform molecules like HMF, furfural, and glucose into economically and environmentally beneficial compounds, like bioplastics, fine chemical intermediates, and energy-dense fuels, by strategically using renewable electricity and electrocatalysis. Ambient operating conditions, integration with sustainable energy technologies, and the possibility of decentralized, modular deployment are only a few benefits of the electrochemical approach.

The development of catalysts has advanced significantly, with materials generated from MOFs, transition metal oxides, and doped sulfides providing high activity and selectivity for both oxidation and reduction reactions. Scalable demonstration of these technologies has been made possible by reactor developments, including membrane electrode assembly and continuous flow systems. Additionally, process energy efficiency and economic viability are improved by combining electrochemical biomass conversion with paired redox processes and green hydrogen production.

In spite of these developments, catalyst stability, product separation, techno-economic analysis, and lifecycle evaluation still face challenges. It will take teamwork in engineering, digital design, and supportive regulations to overcome the challenges of scaling from research to industrial production. It is anticipated that future developments like biocatalyst integration, AI-guided catalyst discovery, and smart grid coupling would influence the development of economically feasible electrochemical biorefineries. As a result, this subject has enormous potential to further the global shift toward a circular and sustainable chemical economy.

References

[1] P. Stegmann, M. Londo, M. Junginger, The circular bioeconomy: its elements and role in European bioeconomy clusters, Res. Conserv. Recy. 6 (2020) 100029, doi:10.1016/j.rcrx.2019.100029.

[2] M. Ghasemlou, C.J. Barrow, B. Adhikari, The future of bioplastics in food packaging: an industrial perspective, Food Packag. Shelf Life 43 (2024) 101279, doi:10.1016/j.fpsl.2024.101279.

[3] T.P. Kainulainen, P. Erkkilä, T.I. Hukka, J.A. Sirviö, J.P. Heiskanen, Application of Furan-based dicarboxylic acids in bio-derived dimethacrylate resins, ACS. Appl. Polym. Mater 2 (2020) 3215–3225, doi:10.1021/acsapm.0c00367.

[4] L. Guo, X. Zhang, L. Gan, L. Pan, C. Shi, Z. Huang, et al., Advances in selective electrochemical oxidation of 5-hydroxymethylfurfural to produce high-value chemicals, Adv. Sci. 10 (2023) 2205540, doi:10.1002/advs.202205540.

[5] O. Simoska, Z. Rhodes, S. Weliwatte, J.R. Cabrera-Pardo, E.M. Gaffney, K. Lim, et al., Advances in electrochemical modification strategies of 5-hydroxymethylfurfural, ChemSusChem 14 (2021) 1674–1686, doi:10.1002/cssc.202100139.

[6] S. Jung, K.S. Kim, G.H. Park, H.G. Cha, H. Jeong, M.J. Kang, Strategies on utilizing biomass derived 5-hydroxymethylfufural by catalytic reactions: pathways and mechanisms, Mater. Today Sustain. 29 (2025) 101058, doi:10.1016/j.mtsust.2024.101058.

[7] S. Prasad, A.J. Khalid, V. Narishetty, V. Kumar, S. Dutta, E. Ahmad, Recent advances in the production of 2,5-furandicarboxylic acid from biorenewable resources, Mater. Sci. Energy Technol. 6 (2023) 502–521, doi:10.1016/j.mset.2023.04.005.

[8] L. Gidi, J. Amalraj, C. Tenreiro, G. Ramírez, Recent progress, trends, and new challenges in the electrochemical production of green hydrogen coupled to selective electrooxidation of 5-hydroxymethylfurfural (HMF), RSC Adv. 13 (2023) 28307–28336, doi:10.1039/D3RA05623F.

[9] X.H. Chadderdon, D.J. Chadderdon, T. Pfennig, B.H. Shanks, W. Li, Paired electrocatalytic hydrogenation and oxidation of 5-(hydroxymethyl) furfural for efficient production of biomass-derived monomers, Green Chem. 21 (2019) 6210–6219, doi:10.1039/C9GC02264C.

[10] C. Guo, Y. Huo, Q. Zhang, K. Wan, G. Yang, Z. Liu, et al., MOF material-derived bimetallic sulfide CoxNiyS for electrocatalytic oxidation of 5-hydroxymethylfurfural, Nanomaterials 13 (2023) 2318, doi:10.3390/nano13162318.

[11] J.P. Oña, R.M. Latonen, N. Kumar, J.-E. Eriksson, I. Angervo, H. Grénman, Electrocatalytic hydrogenation of glucose and xylose on electrochemically roughened metal catalysts, ACS Catal. 13 (2023) 14300–14313, doi:10.1021/acscatal.3c04043.

[12] T. Liu, H. Yabu, Biomass-derived electrocatalysts: low-cost, robust materials for sustainable electrochemical energy conversion, Adv. Energy Sustain. Res. 5 (2024) 2470001, doi:10.1002/aesr.202470001.

[13] S. Li, S. Wang, Y. Wang, J. He, K. Li, Y. Xu, et al., Doped Mn enhanced NiS electrooxidation performance of HMF into FDCA at industrial-level current density, Adv. Funct. Mater. 33 (2023) 2214488, doi:10.1002/adfm.202214488.

[14] V. Chaerusani, A.C.A. Zahra, A. Anniwaer, P. Zhang, N. Chaihad, J. Rizkiana, et al., Catalytic upgrading of bio-oils derived from terrestrial and marine biomass over various types of zeolites, J. Anal. Appl. Pyrolysis 168 (2022) 105735, doi:10.1016/j.jaap.2022.105735.

[15] F.W.S. Lucas, R.G. Grim, S.A. Tacey, C.A. Downes, J. Hasse, A.M. Roman, et al., Electrochemical routes for the valorization of biomass-derived feedstocks: from chemistry to application, ACS Energy Lett. (2021) 1205–1270, doi:10.1021/acsenergylett.0c02692.

[16] X. Xu, X. Song, X. Liu, H. Wang, Y. Hu, J. Xia, et al., A highly efficient nickel phosphate electrocatalyst for the oxidation of 5-hydroxymethylfurfural to 2,5-furandicarboxylic acid, ACS Sustainable Chem. Eng. 10 (2022) 5538–5547, doi:10.1021/acssuschemeng.2c00121.

[17] Q. Zhang, C. Yang, A. Guan, M. Kan, G. Zheng, Photocatalytic CO_2 conversion: from C1 products to multi-carbon oxygenates, Nanoscale 14 (2022) 10268–10285, doi:10.1039/D2NR02588D.

[18] T. Rafaïdeen, S. Baranton, C. Coutanceau, Electroreforming of glucose/xylose mixtures On PdAu-based nanocatalysts, ChemElectroChem 9 (2022) e202101575, doi:10.1002/celc.202101575.

[19] J.-S. Kim, Y.-S. Lee, Effect of reaction pH on enolization and racemization reactions of glucose and fructose on heating with amino acid enantiomers and formation of melanoidins as result of the Maillard reaction, Food Chem. 108 (2008) 582–592, doi:10.1016/j.foodchem.2007.11.014.

[20] A.T. Governo, L. Proença, P. Parpot, M.I.S. Lopes, I.T.E. Fonseca, Electro-oxidation of d-xylose on platinum and gold electrodes in alkaline medium, Electrochim. Acta. 49 (2004) 1535–1545, doi:10.1016/j.electacta.2003.11.013.

[21] X. Liu, J. Wang, H. Li, X. Qu, Y. Xu, L.-L. Shen, et al., Ionic liquids modified electrochemical interfaces for enhanced biomass valorization, J. Catal. 450 (2025) 116330, doi:10.1016/j.jcat.2025.116330.

[22] T. Rafaïdeen, S. Baranton, C. Coutanceau, Highly efficient and selective electrooxidation of glucose and xylose in alkaline medium at carbon-supported alloyed PdAu nanocatalysts, Appl. Catal. B 243 (2019) 641–656, doi:10.1016/j.apcatb.2018.11.006.

[23] J. Bi, Q. Zhu, W. Guo, P. Li, S. Jia, J. Liu, et al., Simultaneous CO_2 reduction and 5-hydroxymethylfurfural oxidation to value-added products by electrocatalysis, ACS Sustainable Chem. Eng. 10 (2022) 8043–8050, doi:10.1021/acssuschemeng.2c0211720.

CHAPTER

2

Fundamentals and processes for electrochemical biomass conversion

N. Soujanya[a], *Pallavi Saxena*[b]

[a] School of Technology, Woxsen University, Hyderabad, Telangana, India, [b] School of Sciences, Woxsen University, Hyderabad, Telangana, India

2.1 Introduction

Biomass electroconversion is the process of converting feedstocks produced from biomass into materials, fuels, and chemicals with added value using electrochemical methods. Because it directly tackles the dual problems of environmental deterioration and the depletion of fossil fuels, this strategy has become a potential avenue within the larger subject of sustainable energy and green chemistry. Since biomass is the sole renewable source of carbon, it is a flexible and plentiful feedstock that may be obtained from forestry waste, industrial byproducts, agricultural residues, and energy crops.

The ability to function in benign circumstances (ambient temperature and pressure) with renewable electricity as the source of power is the main reason for the electrochemical conversion of biomass. As a result, the process's overall carbon footprint decreases because much less external thermal energy needs to be generated. The possibility of scaling up such systems for industrial applications has been additionally enhanced by recent developments in reactor engineering, catalyst design, and in-situ characterization methods [1–3].

Traditional biomass conversion routes, such as thermochemical (e.g., pyrolysis, gasification) and biochemical (e.g., fermentation, enzymatic hydrolysis) methods, have played a dominant role in bio-based production

Electrochemical Biomass Conversion: Sustainable Processes and Applications for Bioenergy, Biofuels, and Bioproducts
DOI: https://doi.org/10.1016/B978-0-443-26757-4.00012-X

for decades [4,5]. While these techniques are mature and commercially deployed, they present notable limitations in terms of energy efficiency, selectivity, and environmental impact [6].

Thermochemical processes often require high temperatures (>400°C) and pressures, leading to substantial energy consumption and the formation of complex product mixtures that necessitate costly downstream separation. They may also generate undesired byproducts such as tars or char, which can complicate reactor operation and reduce yield. Biochemical methods, although operating under milder conditions, typically demand long reaction times, sensitive microorganisms, and pretreatment steps to break down recalcitrant biomass structures such as lignin [7–9].

In contrast, electrochemical biomass conversion offers a number of distinct advantages, such as energy source flexibility, reaction selectivity, milder operational conditions, and process coupling. Moreover, electrochemical processes inherently separate oxidation and reduction reactions into two compartments, allowing independent optimization and facilitating coproduction strategies. While challenges remain in terms of catalyst durability, feedstock variability, and scale-up, the tunability and modularity of electrochemical methods position them as a compelling alternative to conventional technologies [10,11].

2.2 Biomass-derived feedstocks for electro conversion

2.2.1 Monosaccharides and polysaccharides

Electrochemical mineralization of glucose to carbon dioxide is a multi-electron process (24 electrons per glucose) and proceeds via many intermediate steps:

$$C_6H_{12}O_6 + 6H_2O \longrightarrow 6CO_2 + 24H^+ + 24e^-$$

Between these two extremes lies an array of partial oxidation products (gluconic acid, glucaric acid, formic acid, and smaller C2–C3 acids) and reduction products (sugar alcohols such as sorbitol from catalytic hydrogenation). Product distribution depends strongly on the applied potential, catalyst surface chemistry, electrolyte pH, and mass transport conditions (Fig. 2.1).

Polysaccharides require different handling. Because cellulose and many hemicelluloses are insoluble in water, they are typically depolymerized into soluble oligomers or monomers before selective electrochemical upgrading. This depolymerization is commonly achieved by thermochemical pretreatments (acid or base hydrolysis, steam explosion), enzymatic hydrolysis (cellulases), or emerging solvents (ionic liquids, deep eutectic solvents) that disrupt crystalline hydrogen bonding [11]. Only after sufficient

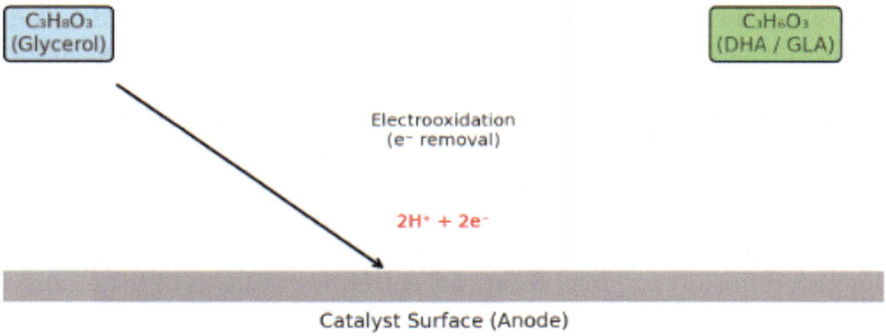

FIGURE 2.1 Figure depicting electrooxidation of glycerol.

hydrolysis do the resulting monosaccharides become amenable to conventional electrochemical oxidation or reduction at electrode surfaces. Recent research has also explored direct electrochemical depolymerization, where anodic oxidation or electrochemically generated reactive oxygen species (e.g., hydroxyl radicals) break glycosidic linkages; however, controlling selectivity in direct electro-depolymerization remains challenging because radical-mediated pathways tend to produce complex product mixtures and can rapidly foul electrode surfaces.

The choice of electrode material and catalyst largely determines activity and selectivity for carbohydrate electroconversion. Noble metals such as gold and platinum are historically important for alcohol and sugar oxidation because they provide well-defined adsorption sites and good electrical conductivity; gold electrodes in particular show favorable selectivity for the oxidation of glucose to gluconic acid in near-neutral conditions [12]. Earth-abundant transition metals and their oxides or hydroxides (e.g., nickel-, copper-, and cobalt-based materials) have attracted attention because they form catalytically active oxyhydroxide phases (e.g., NiOOH in alkaline media) that mediate alcohol and polyol oxidations efficiently and at lower cost. Carbon-based supports and heteroatom-doped carbons (N- or S-doping) enhance dispersion and electronic coupling of metal nanoparticles and can also show intrinsic catalytic activity for certain transformations [13]. In parallel, bioelectrocatalysis using immobilized enzymes, such as glucose oxidase, provides unmatched selectivity—the enzyme converts glucose to glucono-δ-lactone while transferring electrons via mediators or direct electron transfer—but enzyme-based systems face stability and scale-up limitations compared to inorganic catalysts.

Reaction environment, specifically pH and electrolyte composition, strongly influences reaction pathways. Alkaline electrolytes favor the deprotonation of hydroxyl groups and the formation of metal oxyhydroxide active phases, which often lowers the overpotential for the oxidation of sugars and their alcohol derivatives. By contrast, neutral or mildly acidic media may be preferred when product stability or downstream separation is a concern [14–16].

Mass transportation and electrode design are realistic limitations when working with carbohydrate feedstocks. Monosaccharide reactions in solution are often diffusion-limited at typical lab electrodes; rotating-disk electrode or flow-cell experiments are therefore used to decouple kinetics from mass-transport limitations. Polysaccharides or viscous slurries exhibit sluggish diffusion and may form concentration gradients at the electrode surface, which complicates both kinetic analysis and product distribution. Electrode fouling is another recurring problem: oligomeric byproducts, polymerized tars, or adsorption of strongly bound intermediates can block active sites and decrease performance. Mitigation strategies include pulsed electrolysis, periodic oxidative cleaning cycles, surface functionalization to reduce nonspecific adsorption, and the use of mediators that shuttle electrons without producing strongly adsorbing species.

Selectivity control is a central research goal. Two complementary strategies are widely used to steer carbohydrate electrochemistry toward desired products. The first is catalyst design [17]: tailoring surface composition, oxidation state, and nanostructure to favor adsorption geometries and intermediate stabilization that lead to a particular product. For example, catalysts that stabilize adsorption through the aldehyde (C1) position favor gluconic acid formation, whereas surfaces that permit C–C bond scission will produce smaller acids. The second strategy is reaction engineering: using redox mediators such as nitroxyl radicals (e.g., TEMPO) [18] to effect selective oxidation of primary alcohol groups, applying controlled potentials to avoid overoxidation, and pairing anodic sugar oxidation with a valuable cathodic reaction (such as hydrogen evolution) so that energy input is efficiently used while producing two useful outputs. TEMPO-mediated electrooxidation has been successfully used in the selective oxidation of primary alcohols on polysaccharide surfaces, enabling, for instance, the oxidation of cellulose primary hydroxyls to carboxylates with minimal backbone cleavage.

Comprehensive product analysis and mechanistic probing are essential for advancing carbohydrate electroconversion. Electrochemical methods such as cyclic voltammetry and linear sweep voltammetry provide information on onset potentials, currents, and kinetic features; electrochemical impedance spectroscopy reveals charge-transfer and mass-transport resistances; rotating-disk voltammetry helps decouple diffusion effects; and long-term chronoamperometry assesses stability. Chemical analysis of liquid-phase products is typically performed with HPLC (for sugars and organic acids), GC or GC–MS (for volatile fragments), NMR (for structural elucidation of complex mixtures), and FTIR (for functional-group identification). Together, these diagnostic tools allow the calculation of conversion, selectivity, and Faradaic efficiency (FE). For a given product, FE is calculated as

$$FE = (n.F.M_{product}/Q_{total}) \times 100$$

where n is the number of electrons exchanged per mole of product, F is Faraday's constant, and $M_{product}$ is the moles of product formed.

2.2.2 Lignin and aromatic compounds

Lignin is the principal aromatic biopolymer in terrestrial biomass and, after cellulose, is the second most abundant organic polymer on Earth. Structurally, lignin is a highly irregular, three-dimensional polymer built from three primary monolignol units—p-coumaryl, coniferyl, and sinapyl alcohols—which give rise, after incorporation, to p-hydroxyphenyl (H), guaiacyl (G), and syringyl (S) structural motifs. These monomers are linked by a spectrum of carbon–oxygen and carbon–carbon bonds formed through radical coupling during lignification; among these linkages, the β-O-4 aryl ether is the most common, and other motifs such as β-β, β-5, and 5-5 create a dense, cross-linked network. The intrinsic heterogeneity of linkage types, variable substitution patterns on the aromatic rings, and differences in monomer composition across biomass sources (softwoods are generally G-rich, hardwoods contain both G and S units, and grasses include significant H units) produce a material that is chemically complex and resistant to straightforward conversion.

Nevertheless, lignin is an attractive feedstock because it contains aromatic rings that, if liberated as monomeric phenolics, could serve as renewable platform chemicals—replacing petrochemical aromatics in polymers, resins, vanillin production, and fine chemicals. The challenge for chemical and electrochemical valorization is therefore twofold: first, to depolymerize lignin selectively into defined low-molecular-weight aromatics; and second, to avoid deep oxidation or uncontrolled repolymerization that produces low-value tars and insoluble char.

Electrochemical approaches to lignin valorization aim to address these challenges by exploiting controlled electron transfer rather than high temperatures or stoichiometric chemical oxidants. Anodic strategies typically seek to oxidize lignin fragments or phenolic moieties to generate reactive cationic or radical intermediates that can undergo selective bond cleavage (e.g., aryl ether scission at β-O-4 linkages) or functionalization (oxidation to aldehydes/carboxylates). Mechanistically, anodic oxidation can proceed either by direct electron transfer from lignin units adsorbed on the electrode surface or indirectly via electrochemically generated oxidants or redox mediators. Direct single-electron oxidation of a phenolic subunit produces a phenoxy radical that can rearrange or fragment, while mediated routes (using stable radical mediators or transition-metal redox couples) can lower required potentials and promote selectivity. Conversely, reductive electrochemical routes exploit electrogenerated hydrogen atoms or cathodic hydrogenation at metal cathodes to cleave ether bonds (hydrogenolysis) under milder conditions than thermal hydrogenolysis; cathodic approaches are particularly attractive when paired with hydrogen evolution or other valuable cathodic processes, so overall energy use is optimized.

The choice of electrochemical cell configuration, electrode material, and solvent/electrolyte environment strongly influences outcomes. Boron-doped diamond (BDD) and certain metal oxide anodes generate highly oxidative surface species or hydroxyl radicals and are effective for deep oxidative depolymerization, but such aggressive conditions favor overoxidation and fragmentation. Noble metal and transition metal electrodes (e.g., Pd, Ru, Ni, and Cu) supported on conductive carbons can catalyze more controlled oxidative or reductive pathways, while carbon-based electrodes (graphitic, nanotextured, or heteroatom-doped carbon materials) promote adsorption and electron transfer with tunable surface chemistry. Solvent choice is equally important: lignin solubility and reactivity are strongly affected by solvent polarity and hydrogen-bonding ability, so electrochemical studies often use organic cosolvents, ionic liquids, or deep eutectic solvents to dissolve lignin and mitigate mass transport limitations. Electrolyte pH regulates the oxidation state and formation of active catalyst phases (e.g., metal oxyhydroxides), and it also affects the stability of target aromatic products.

Several practical and mechanistic obstacles are characteristic of electrochemical lignin processing. The first is repolymerization: intermediates such as phenoxy radicals readily couple to form new C–C linkages, leading to increased molecular weight and char formation; controlling radical lifetime via mediators, electrode potential windows, or rapid extraction of monomers is therefore critical. Second electrode fouling and blockage of pores due to high density and then heterogeneity of lignin [19–22].

Analytical characterization is essential to chase depolymerization and to identify valuable products. Techniques commonly deployed include spectroscopies (FTIR, UV–Vis, and Raman) and total phenolic assays provide rapid screening, while operando electrochemical techniques (in-situ FTIR, Raman, or electrochemical mass spectrometry) are increasingly used to follow intermediate formation and electrode surface chemistry in real time. From a process perspective, maximizing the economic value of lignin conversion requires integrating electrochemical depolymerization with downstream separations, product upgrading, and valorization of coproducts [23]. Promising directions include paired electrolysis (oxidation of lignin at the anode coupled with cathodic hydrogen generation or CO_2 reduction), use of selective redox mediators to improve monomer yields, continuous-flow electrochemical reactors that improve mass transfer and mitigate fouling, and hybrid strategies that combine mild thermochemical pretreatment with electrochemical upgrading. Life-cycle and techno-economic analyses indicate that feedstock pretreatment, solvent choice, and energy efficiency are decisive factors for commercial feasibility; therefore research continues to focus on low-energy, selective processes that are compatible with existing biorefinery streams.

2.2.3 Glycerol, furfural, and other platform molecules

Glycerol and furfural occupy central roles among platform molecules derived from biomass because each is simultaneously abundant, chemically versatile, and amenable to selective electrochemical transformation. Glycerol emerges as a major byproduct of the transesterification of vegetable oils during biodiesel manufacture; its high availability and low cost make it an attractive feedstock for producing fine chemicals [24]. Structurally, glycerol is a triol (1,2,3-propanetriol) whose multiple hydroxyl groups offer several distinct reactive sites: the secondary (central) hydroxyl can be selectively oxidized to give a ketone (dihydroxyacetone, DHA), while the primary hydroxyls can be oxidized to aldehydes and then to carboxylic acids (glyceric acid [GLA], tartronic acid, and eventually smaller acids or CO_2 under deep oxidation). Electrochemical oxidation provides a route to these products under much milder conditions than conventional thermochemical oxidation: for example, oxidation of the central secondary alcohol to the ketone typically involves removal of two hydrogen equivalents (a two-electron process), whereas oxidation of a terminal alcohol to the corresponding carboxylic acid proceeds in a stepwise fashion via aldehyde intermediates and involves multiple electrons. The applied potential, electrolyte pH, and the nature of the electrode surface therefore exert decisive control over selectivity; alkaline media frequently favor the formation of metal oxyhydroxide active phases (e.g., NiOOH) that promote polyol oxidation at lower overpotentials, while neutral or mildly acidic electrolytes can stabilize specific products and simplify downstream separation [25].

Strategy to enhance selectivity is mediated by electrochemistry—redox mediators such as TEMPO derivatives selectively oxidize primary alcohols to aldehydes/carboxylates at low overpotentials while leaving secondary centers intact. Paired electrolysis—where glycerol oxidation at the anode is coupled to a valuable cathodic reaction such as hydrogen evolution or CO_2 reduction—improves the overall energy balance and enables coproduction of hydrogen while upgrading glycerol to a higher-value chemical [26–29] (Fig. 2.2).

Furfural and related furanic compounds arise from the dehydration of pentose sugars in hemicellulose and constitute another class of high-value platform molecules. Furfural (2-furaldehyde) can be electrochemically

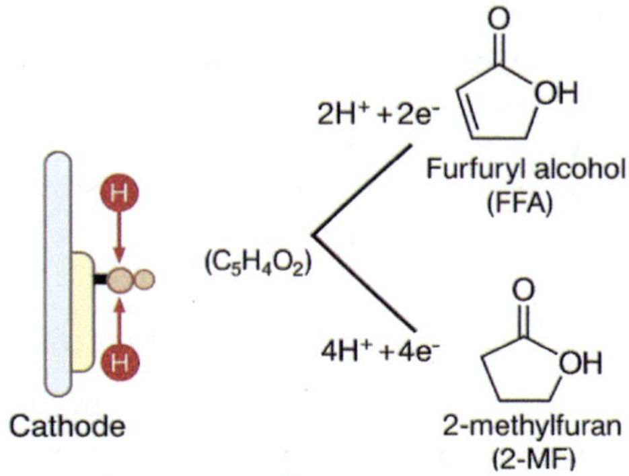

FIGURE 2.2 Reduction of furfural to furfuryl alcohol and 2-methylfuran.

reduced at the cathode to furfuryl alcohol, a feedstock for resins and polymers, or oxidized at the anode to furoic acid and other oxygenated derivatives [30,31]. The cathodic hydrogenation of furfural is typically a two-electron, two-proton process producing the alcohol; catalysts that favor hydrogenation pathways (e.g., Cu, Ni, Pd-modified surfaces) can achieve good selectivity to furfuryl alcohol at modest overpotentials. By contrast, anodic oxidation of the aldehyde functionality to the corresponding carboxylic acid follows a different mechanistic sequence but is also typically a two-electron process if stopped at the carboxylate stage; more aggressive anodic conditions drive ring opening and further oxidation to low-molecular-weight acids or CO_2. Furfural's relative structural simplicity (compared with lignin) and the commercial demand for its derivatives make it a useful model for demonstrating selective electrochemical upgrading. However, control of hydrogenation versus polymerization (which can form humins) remains a practical concern during scale-up [32,33] (Fig. 2.3).

Beyond glycerol and furfural, a suite of other platform molecules—notably hydroxymethylfurfural (HMF), levulinic acid, sorbitol, succinic acid, and others—are readily accessed from carbohydrate feedstocks and are promising targets for electrochemical conversion. HMF, derived from hexose dehydration, is especially important because its oxidation to 2,5-furandicarboxylic acid (FDCA) yields a bio-based monomer that can replace terephthalic acid in polyester production. The HMF→FDCA transformation is a multielectron oxidation involving sequential oxidation of both the hydroxymethyl and aldehyde moieties; achieving high selectivity while avoiding ring degradation requires careful catalyst and potential control [34]. In glycerol electro-oxidation, if the intermediate aldehyde is rapidly hydrated and further oxidized, the product stream will shift toward carboxylic acids [35,36]; if instead the intermediate desorbs or is selectively hydrogenated at the cathode, partial oxidation products accumulate.

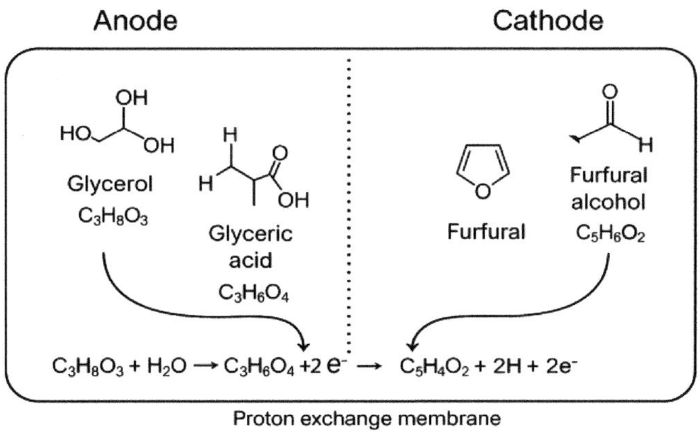

FIGURE 2.3 Coupled redox reaction of furfural and glycerol.

From an engineering perspective, the practical implementation of electrochemical upgrading relies on cell design, mass transport management, and downstream separations. Flow cells and divided membrane reactors are preferred for continuous operation and for preventing undesirable cross-reactions between anodic and cathodic products. Membranes (anion or cation exchange) allow independent optimization of anode and cathode environments, which is valuable when, for instance, glycerol oxidation performs best in alkaline media while cathodic CO_2 reduction requires a different pH. Mass transport, especially for viscous glycerol streams or concentrated HMF solutions, becomes rate-limiting unless addressed by turbulence promoters, rotating electrodes, or forced flow. Product separation is also nontrivial because many partially oxidized products are water-soluble and chemically similar; integration of electrosynthesis with selective extraction, membrane separations, or reactive crystallization is therefore an important area of development.

Analytical characterization and performance metrics focus on conversion, selectivity, and FE. Because multiple parallel pathways compete, reporting product distributions by molar yield and quantifying the fraction of charge that produces each desired product (FE) are standard practices. Chromatography (HPLC, GC–MS), NMR, and electrochemical techniques (CV, LSV, and EIS) are used in tandem to correlate catalyst behavior with observed product suites. Stability is also a crucial metric: many organic substrates and intermediates can foul electrodes or poison active sites, so long-term chronoamperometry and periodic regeneration studies are necessary to demonstrate technological viability.

Economically and environmentally, glycerol and furfural offer complementary opportunities. The oversupply of crude glycerol from biodiesel operations motivates value addition to improve refinery economics, while furfural and HMF are closer to conventional chemical value chains that

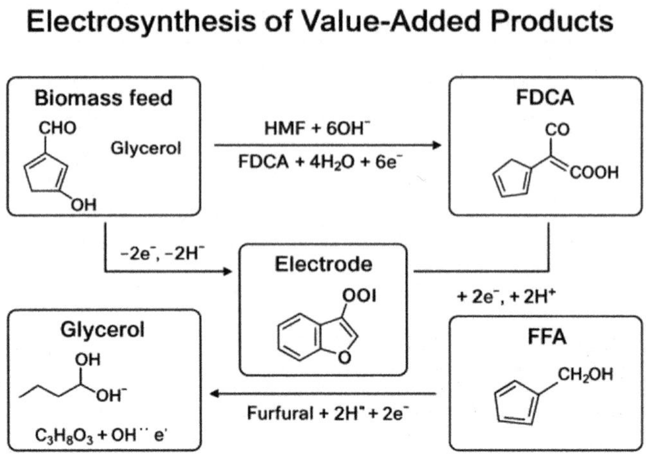

FIGURE 2.4 Electrosynthesis of value-added products.

can supply polymer precursors and solvents. Pairing anodic oxidation of platform molecules with cathodic hydrogen generation or CO_2 electroreduction creates integrated processes that improve energy utilization and coproduce valuable streams. Nevertheless, scaling these routes requires addressing catalyst cost and durability, electrolyte management, and product purification energy costs to achieve broader commercial adoption.

2.2.4 Waste biomass and biorefinery residues

Waste biomass and biorefinery residues constitute a diverse and abundant class of feedstocks that hold great promise for electrochemical valorization. Unlike purified platform molecules, these materials arrive as heterogeneous, often moist, mixtures containing solids, dissolved organics, ash-forming inorganics, and traces of lipids, proteins, or salts. Typical sources include agricultural residues (straw, corn stover, rice husks), forestry residues (branches, bark), process streams from pulp and paper mills (black liquor, lignosulfonates), crude glycerol from biodiesel production, food and municipal organic wastes; algal slurries from cultivation ponds, and aqueous fractions from hydrothermal liquefaction or pyrolysis. Each pretreatment option carries trade-offs in energy consumption, additional reagents, and downstream separation burden; choosing the optimal pathway depends on the targeted product slate and overall process integration strategy [37–41] (Fig. 2.4).

Once a processable intermediate is obtained, whether it is an aqueous carbohydrate-rich stream, a lignin-containing solution, or a bio-oil aqueous fraction, electrochemical routes can be molded to the chemistry of that

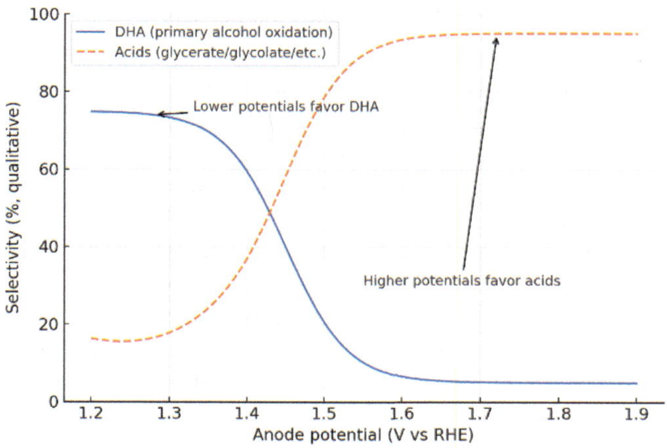

FIGURE 2.5 Selectivity of the catalyst vs potential graph.

stream. For carbohydrate-rich wastewaters, anodic oxidation can selectively convert sugars to organic acids or value-added derivatives, often in alkaline media where many nonprecious metal catalysts are active. For lignin-rich residues and technical lignins from pulping, electrochemical depolymerization or functionalization aims to break ether linkages (β-O-4) and liberate phenolic monomers with retained aromaticity. For bio-oils and aqueous fractions from thermal processing, electrocatalytic hydrogenation and oxidation can upgrade oxygenated species to more valuable chemicals or platform molecules, and paired electrolysis can couple the anodic valorization of organics with cathodic hydrogen production or CO_2 reduction, thereby improving energy utilization.

The reactor configuration and electrode strategy must account for the solid or slurry nature of many residues. Flow-through electrochemical reactors, filter-press cells, and slurry-fed flow cells are common options that enable continuous operation and mitigate fouling. In designs that process particulate slurries, robust mixing, controlled residence time, and hydrodynamic shear are necessary to maintain mass transfer and prevent sedimentation. Membrane-separated cells (anion or cation exchange) allow independent optimization of anode and cathode chemistries—important when, for instance, anodic oxidation favors alkaline conditions while the cathode process (e.g., CO_2 reduction) requires neutral pH. Solid–liquid interfaces in such systems are prone to fouling by high-molecular-weight tars, polymeric residues, or inorganic precipitates, so electrode materials and surface treatments that resist adsorption (or allow facile regeneration) are advantageous. Periodic in-situ electrochemical cleaning, sacrificial anodes, or continuous leaching strategies can be incorporated to maintain performance (Fig. 2.5).

Catalyst selection for residues emphasizes robustness and tolerance to impurities as much as intrinsic activity. While noble metals deliver high activity and defined selectivity for many model substrates, their susceptibility to poisoning and cost limit widespread use with dirty feedstocks. Transition-metal-based catalysts (Ni, Cu, Co, and Fe), metal oxides, doped carbons, and BDD electrodes have been investigated for their relative resilience and capacity to effect desirable transformations in complex matrices. Redox mediators and bio-electrocatalysts (enzymes or microbial consortia) offer alternative selectivity levers: stable mediators can enable selective oxidation at lower potentials while enzymes provide specificity for particular functional groups, although enzyme stability in waste streams is a concern. Designing catalyst layers with hierarchical porosity and hydrophilic/hydrophobic balance helps manage wetting and mass transport for viscous or multiphase streams.

Analytical and separation challenges are amplified with residues because products form complex mixtures that can include soluble organics, oligomers, and inorganic salts. Reliable product quantification thus requires integrated analytical workflows—chromatography (HPLC, GC–MS) for small molecules, size-exclusion for oligomers, NMR for structural elucidation, and ICP or ion chromatography for inorganic species.

2.3 Integrated electrochemical conversion and process coupling

2.3.1 Electrooxidation and electroreduction pathways of biomass-derived molecules

Glycerol oxidation can proceed via partial oxidation to form DHA, glyceraldehyde (GLAD), and GLA, or via deeper oxidation to tartronic acid and mesoxalic acid. As discussed in Section 2, selectivity is strongly influenced by catalyst composition and reaction potential. Noble metals (Pt, Au, and Pd) often favor partial oxidation due to high surface reactivity, whereas transition metal oxides (Ni, Co, and Cu) can enable more complete oxidation with lower noble metal content.

A general anodic oxidation reaction can be expressed as in Eq. (2.1).

$$\text{R-OH} + \text{H}_2\text{O} \longrightarrow \text{R=O} + 2\text{H}^+ + 2\text{e}^- \quad (2.1)$$

where

- R–OH is a hydroxyl-containing biomass derivative (e.g., glycerol, sorbitol).
- R=O represents the oxidized product (aldehyde, ketone, or acid).

- H^+ and e^- are generated simultaneously, often feeding into a hydrogen evolution reaction (HER) at the cathode.

Eq. (2.2) exhibits a simplified glycerol oxidation mechanism.

$$C_3H_8O_3 + H_2O \longrightarrow C_3H_6O_3 + 2H^+ + 2e^- \tag{2.2}$$

Here the selectivity depends on:

- Applied potential, where higher potentials yield deeper oxidation.
- Catalyst composition (Au favors partial oxidation; Pt tends toward complete oxidation).
- Electrolyte pH where an alkaline medium improves the reaction kinetics.

At the cathode, the electrons generated from electrooxidation are consumed in **electroreduction** reactions. The most direct cathodic reaction is the hydrogen evolution reaction (HER) as in Eq. (2.3).

$$2H^+ + 2e^- \longrightarrow H_2 \tag{2.3}$$

Also if CO_2 is electrochemically reduced to various carbon-based products such as HCOOH, HCHO, and CH_3OH as shown in Eqs. (2.4–2.6).

$$CO_2 + 2H^+ + 2e^- \longrightarrow HCOOH \tag{2.4}$$

$$CO_2 + 4H^+ + 4e^- \longrightarrow HCHO + H_2O \tag{2.5}$$

$$CO_2 + 6H^+ + 6e^- \longrightarrow CH_3OH + H_2O \tag{2.6}$$

Some of the intermediates from biomass (e.g., furfural, $C_5H_4O_2$) undergo targeted reductions to produce products like furfuryl alcohol or 2-methylfuran [42], allowing additional value creation as given in Eqs. (2.7) and (2.8).

$$C_5H_4O_2 + 2H^+ + 2e^- \longrightarrow C_5H_6O_2 \text{ (Furfuryl Alcohol)} \tag{2.7}$$

$$C_5H_4O_2 + 4H^+ + 4e^- \longrightarrow C_5H_8O \text{ (2-methylfuran)} + H_2O \tag{2.8}$$

2.3.2 Coupled redox reactions and electrosynthesis of value-added products

Coupled redox reactions in biomass electro-conversion is the process where oxidation and reduction reactions occur simultaneously at the anode and cathode, enabling efficient conversion of biomass-derived molecules into value-added chemicals while generating coproducts such as hydrogen or other fuels. This approach maximizes resource utilization, minimizes energy losses, and enables a broad product spectrum.

2.3.2.1 *Principle of coupled redox reactions*

In an electrochemical cell, the anodic reaction means the oxidation of biomass-derived molecules into intermediates or valuable chemicals, while the cathodic reaction can be adjusted to generate complementary products such as hydrogen, alcohols, or hydrocarbons. By combining both of them, the electrodes contribute to product generation instead of wasting energy on side reactions like oxygen evolution.

2.3.2.2 *General cell representation*

At the anode, the biomass-derived compound generates an oxidized product and electrons, and at the cathode, the generated electron combines with the substrate to reduce it.

Finally, Biomass-derived compound + substrate ⟶ reduced product.

As an example, to understand with consider oxidation of glycerol coupled with Furfural reduction. At the anode, the glycerol combined with H_2O yields GLA, and at the cathode, furfural combined with the generated protons and electrons yields furfuryl alcohol. The anodic, cathodic, and overall reactions are as depicted below.

$$C_3H_8O_3 + H_2O \longrightarrow C_3H_6O_4 + 2H^+ + 2e^- \tag{2.9}$$

$$C_5H_4O_2 + 2H^+ + 2e^- \longrightarrow C_5H_6O_2 \tag{2.10}$$

$$C_3H_8O_3 + C_5H_4O_2 + H_2O \longrightarrow C_3H_6O_4 + C_5H_6O_2 \tag{2.11}$$

This coupling not only reduces the cell voltage compared with HER-OER, but also generates two value-added products—GLA and furfuryl alcohol—used in pharmaceuticals, resins, and polymer industries [43].

2.3.2.3 *Electrosynthesis of value-added products*

Electrosynthesis is a process that harnesses electrical potential to change biomass-derived molecules into higher-value chemicals under relatively mild conditions. Compared with thermochemical routes that rely on high temperatures and external oxidants or hydrogen, electrosynthesis uses the electrode itself as the redox reagent. This decouples activation energy from heat, permits precise control over reaction driving force through the applied potential, and allows direct pairing with renewable electricity. In biomass valorization, electrosynthesis is a powerful technique because many feed molecules like sugars, polyols, furanics, and lignin fragments are oxygen-rich and already poised for selective oxidation or reduction to marketable products (monomers, solvents, flavors/fragrances, polymer intermediates, and fuel-range molecules).

At the anode, electrosynthetic oxidation converts alcohols and aldehydes into aldehydes, ketones, or carboxylates with high atom economy. In alkaline media, the chemistry can be outlined as a two-electron step in which adsorbed alkoxides are dehydrogenated on a metal-oxyhydroxide surface, as shown in Eqs. (2.12) and (2.13).

$$\text{R-CH}_2\text{OH} + 2\text{OH}^- \longrightarrow \text{R-CHO} + \text{H}_2\text{O} + 2e^- \quad (2.12)$$

$$\text{R-CHO} + 2\text{OH}^- \longrightarrow \text{R-COO}^- + \text{H}_2\text{O} + 2e^- \quad (2.13)$$

A canonical example is the oxidation of **5-HMF** to **2,5-FDCA** as a bio-derived replacement for terephthalic acid, given in Eq. (2.14) [44], also depicted in Fig. 2.2.

$$\text{HMF} + 6\text{OH}^- \longrightarrow \text{FDCA} + 4\text{H}_2\text{O} + 6e^- \quad (2.14)$$

As both an aldehyde and a hydroxymethyl group are present in HMF, reaction selectivity mostly depends on the catalyst and potential window, as already discussed; along with oxidizing both side chains, suppression of the ring opening is useful to yield FDCA. Also DHA, GLAD, glyceric, and tartronic acids can be formed from glycerol by changing the surface chemistry (e.g., Au vs. Pt vs. NiOOH) and pH [45]. The formation is given in Eq. (2.15).

$$C_3H_8O_3 \longrightarrow C_3H_6O_3 + 2H^+ + 2e^- \quad (2.15)$$

where $C_3H_8O_3$ denotes a partially oxidized C_3 product such as DHA. At the cathode, reductive electrosynthesis exploits the electrons generated at the anode to hydrogenate, hydrodeoxygenate, or carboxylate biomass-derived substrates. The reduction of furfural to furfuryl alcohol (FFA) is a standard transformation that proceeds via a two-electron, two-proton addition to the aldehyde, as shown in Eq. (2.16).

$$C_5H_4O_2 + 2H^+ + 2e^- \longrightarrow C_5H_6O_2 \quad (2.16)$$

FFA can be further elevated to 2-methylfuran, a high-energy-density fuel component, with more sturdy hydrogenating surfaces or stepwise deoxygenation. Carbonyl-rich platform molecules follow analogous pathways. Like levulinic acid, which accepts two electrons and two protons to form 4-hydroxypentanoic acid, which cyclizes to gamma (γ) Valero lactone (GVL) that is a versatile green solvent and fuel precursor, the equation is given in Eq. (2.17) [46].

$$C_5H_8O_3 + 2H^+ + 2e^- \longrightarrow C_5H_{10}O_3 \longrightarrow \text{GVL} + H_2O \quad (2.17)$$

whereas hydrogen often originates as hydrogen atoms formed at the cathode side from water reduction, avoiding pressurized H_2. Electrosynthesis emerges as a unique system when paired, that means when an oxidation reaction, which creates a valued product [47] (e.g., glycerol →GLA) is run in parallel with a reduction reaction that also makes a valued product (e.g., furfural →FFA) in a cell. As the biomass oxidation mostly occurs at a lower potential than the oxygen evolution reaction, the overall cell voltage drops, current efficiency rises, and both the generated products are value-added. The example equation is shown in Eq. (2.18).

$$C_3H_8O_3\ (\text{anode}) + C_5H_4O_2\ (\text{cathode}) + H_2O \longrightarrow C_3H_6O_3 + C_5H_6O_2 \quad (2.18)$$

A similar kinetics at both electrodes must be maintained to obtain high selectivity. Also maintaining compatible electrolytes (often aided by ion-exchange membranes) and designing flow architectures that minimize concentration polarization and product crossover are essential.

The catalyst used and the oxidation state change the adsorption geometry and stability of key intermediates; also nano-structuring and composite formation can shift binding energies to favor partial oxidation (e.g., Au for DHA/FDCA) over deep mineralization. Electrolyte pH directs the formation of active phases such as NiOH, the speciation of carboxylates and aldehydes, and proton availability for hydrogenation. The applied potential chooses the elementary steps for activation. When the potential is too low, the electron transfer is slow, and conversely, when it is too high, the ring-opening or C–C scission pathways dominate. With viscous or concentrated feeds, mass transfer is often rate-limiting, so flow cells, porous electrodes, and gas-diffusion layers (for CO_2-involving cathodes) are commonly used to increase the current densities, maintaining a high FE.

Persistent organic radicals such as nitroxyls (e.g., TEMPO) can be electro-regenerated at low overpotential and selectively oxidize primary alcohols to aldehydes or acids without strong adsorption on the anode, reducing fouling and enabling homogeneous contact with bulky substrates, Eqs. (2.19) and (2.20).

$$RCH_2OH + TEMPO^+ \longrightarrow RCHO + TEMPOH \quad (2.19)$$

$$TEMPOH \longrightarrow TEMPO^+ + H^+ + e^- \quad (2.20)$$

Conversely soluble metal redox couples (e.g., Co, Cu complexes) can shuttle electrons between electrode and substrate, broadening the operational window and sometimes improving chemo selectivity.

FDCA (polyester monomer), glyceric/tartronic acids (fine-chemical intermediates), and vanillic acid or vanillin relatives from lignin streams for flavors, fragrances, and resins are delivered through an oxidation process. Reduction processes generate furfuryl alcohol (furan resins), GVL (solvent and fuel precursor), BHMF, and 2,5-dimethylfuran from HMF derivatives (polymer and fuel domains), and formate/methanol with CO_2 as a substrate.

Translating laboratory electrosynthesis to practice is still pending. As divided cells have anion- or cation-exchange membranes, these prevent product crossover and allow different pH at different electrodes, whereas undivided cells can be opted for when cross-reactions are acceptable and ohmic losses must be minimized. Shorter diffusion paths and higher space-time yields can be obtained from continuous cells with porous catalysts. As many of the biomass streams are impure, catalyst layers are selected in such a way to decrease resistance, and easy regeneration and operating points are chosen to avoid polymerization and tar formation. FE (fraction of charge going to the target), energy efficiency (product enthalpy per electrical input), and space time yield are the main parameters that quantify the overall success of the process, all of which guide scale-up and techno-economic analysis (TEA).

2.3.3 Coupling biomass oxidation with hydrogen production

The concept of coupling biomass oxidation with hydrogen production builds directly on the anodic and cathodic reaction principles discussed in Sections 3.1 and 3.2. Instead of performing water oxidation at the anode, the electrolyzer utilizes biomass-derived molecules as anodic feedstock, while hydrogen evolution occurs at the cathode. This integrated process reduces the energy requirement for electrolysis and generates hydrogen and an oxidized chemical intermediate or final product.

2.3.3.1 Concept and operating principle

A thermo-electrochemical system integrates two energy inputs:

1. *Thermal energy*: It increases the temperature of the electrolyte and electrode surfaces, reducing the activation energy of reactions, accelerating charge transfer, and improving mass transport.
2. *Electrical energy*: Supplies the necessary driving energy for oxidation or reduction reactions, ensuring high selectivity and controllability of product formation.

The two energy forms can be applied simultaneously or in a sequential manner:

1. In simultaneous mode, both heat and electricity are applied continuously, maintaining optimal temperature and potential conditions.
2. Whereas, in the sequential mode, the thermal activation is used for pretreatment or intermediate reaction steps, followed by electrochemical finishing.

2.3.3.2 Integration with renewable energy

Integration of renewable energy improves both the sustainability and the economic feasibility of these systems. Numerous configurations are possible:

- *Solar hybrid systems*: Solar thermal collectors deliver heat, while photovoltaic (PV) panels provide electrical energy to the electrochemical cell.
- *Wind/electric hybrid*: Wind turbines generate electrical energy that powers the cell, with supplementary heat from resistive heating or waste-heat recovery.
- *Biomass-fuelled heat + electrolysis:* Combustion or gasification of biomass generates thermal energy and, in some cases, electrochemically active species for upgrading.
- *Geothermal–electrochemical coupling:* Stable geothermal heat maintains elevated cell temperatures for year-round operation.

This integration is particularly attractive for remote or off-grid sites, where local renewable resources can sustain continuous operation without reliance on centralized power infrastructure.

2.3.3.3 *Thermo-electrochemical synergy*

There are several advantages, especially at elevated temperatures (typically 60°C–200°C, depending on electrolyte stability), as given later: reduced overpotential, enhanced kinetics, pathway selectivity, and improved electrode wettability.

2.3.4 Life cycle assessment

Life cycle assessment (LCA) is a method for quantifying the environmental impacts associated with all the stages of a product or a process, from resource extraction to manufacturing, operation, and, as a final stage, disposal or recycling. In the context of biomass electro-oxidation, LCA allows researchers and engineers to identify stages where energy consumption, greenhouse gas (GHG) emissions, or other environmental burdens are highest.

2.3.4.1 *Goal and scope definition*

- To define the purpose of assessment, such as comparing a biomass-electrolysis hybrid system with conventional water electrolysis or fossil-based hydrogen production.
- To identify the system boundaries (e.g., "cradle-to-gate" for production phases or "cradle-to-grave" including disposal).

2.3.4.2 *Special considerations for biomass systems*

- The sustainability of biomass sourcing helps in avoiding the usage of land and food crop displacement.
- Carbon neutrality claims are only valid when biomass regrowth offsets CO_2 emissions.

- Integration with waste biomass streams (e.g., agricultural residues) aids in environmental sustainability.

2.3.4.3 *Techno-economic analysis*

TEA evaluates the economic feasibility of a technology by integrating engineering design parameters with cost data. For biomass electrochemical systems, TEA helps determine whether the process can compete with established industrial methods.

2.3.4.4 *Core components of techno-economic analysis*

- Process design and simulation
  - Ø Develop a process flow diagram covering biomass pretreatment, electrochemical conversion, separation, and product purification.
  - Ø Use simulation tools to analyze energy consumption, material requirements, and output yields.
- Capital expenditure
  - Ø Account for costs of electrochemical reactors, catalysts, renewable energy systems, and auxiliary units.
- Operating expenditure
  - Ø Includes feedstock costs, electricity consumption, maintenance, labor, and replacement parts.
- Economic performance indicators
  - Ø **Levelized cost of hydrogen (**LCOH**)** for hydrogen coproduction.
  - Ø **Payback period** and **net present value** for the entire system.
  - Ø **Internal rate of return** to assess investment attractiveness.
- Sensitivity analysis
  - Ø Examines how fluctuations in electricity price, feedstock cost, or product market price affect profitability.
- Scenario comparison
  - Ø Compare stand-alone electrolysis, biomass-only oxidation, and integrated biomass–electrolysis processes.

2.3.4.5 *Cost reduction opportunities*

Ø Use of inexpensive, earth-abundant catalysts instead of noble metals.
Ø Scaling up to reduce per-unit equipment cost.
Ø Integration with industrial waste heat or renewable energy surplus.

2.3.4.6 *Integrating life cycle assessment and techno-economic analysis for decision-making*

Combining LCA and TEA provides a complete view. A process that is economically feasible and environmentally harmful may fail to gain approval and public acceptance (Fig. 2.6). But an environmentally benign but

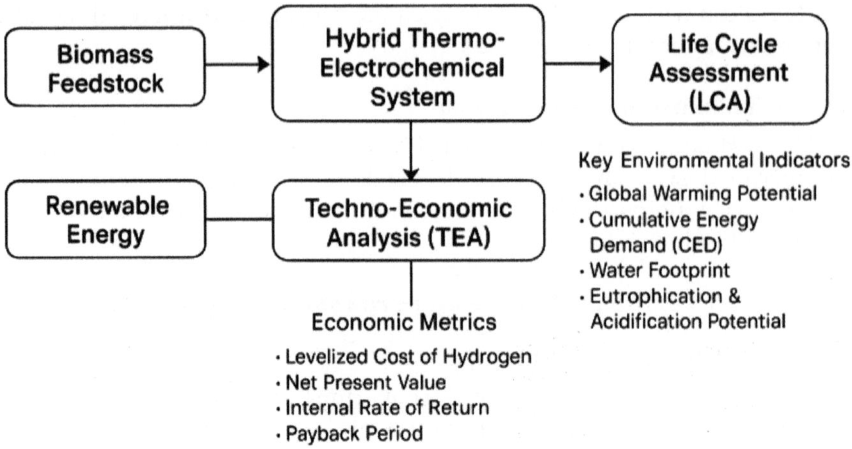

FIGURE 2.6 Lifecycle and techno-economic aspects.

cost-inefficient process will not attract investment. Integrated assessment enables:

- **Trade-off analysis** between cost and environmental impact.
- **Policy guidance** for incentives such as carbon credits or feed-in tariffs.
- **Strategic investment planning** for industrial partners.

2.3.4.7 *Case example*

For a biomass-based electrolysis hybrid plant producing both hydrogen and lactic acid:

- LCA reveals that renewable electricity integration reduces GHG emissions by 65% compared with a grid-powered baseline.
- TEA shows that LCOH can be reduced from 6.5 to 4.2 USD/kg with improved electrode stability and feedstock cost reduction.

2.4 Challenges and mitigation strategies

One of the sustainable paths for generation of value added chemicals and fuels is electrochemical biomass conversion, but still several challenges are suppressing its large scale expansion. These processes have low reaction selectivity and also a variety of by products are formed. Their efficiency is mostly affected by electrode and catalyst degradations, mass transfer limitations and less conductivity of some of the biomass derived intermediates. To overcome these challenges a proper selection of catalysts which are nano structured and multifunctional must be done. Appropriate electrode materials must be opted to optimise the reaction kinetics and also reduce electrode fouling. Also developing robust electro chemical reactors and integrating these with the renewable energy sources could enhance the efficiency, performance and scalability of the electrochemical biomass conversion techniques.

Electrochemical conversion of biomass (EBC) has emerged as a transformative technology for producing both energy carriers and specialty chemicals, including hydrogen and 2,5-furandicarboxylic acid. Unlike conventional thermal conversion processes, EBC is highly regarded for its efficiency and its environmentally benign characteristics, presenting a compelling alternative for sustainable production pathways. Despite these advantages, several obstacles impede the broader acceptance and commercialization of EBC. Central among these are the intricate reaction mechanisms, inherent inefficiencies in current systems, elevated operational costs, and persistent material limitations. These multifaceted challenges collectively hinder the ability of EBC technologies to scale and compete with well-established alternatives. A major contributor to operational inefficiency is the complexity of electrochemical reactions and the substantial energy needed for processes such as external reforming of biomass feedstocks particularly under low and medium temperature conditions. This often results in actual system performances falling short of their theoretical potential. Furthermore, high capital expenditures, especially the investment associated with electrolysis systems and concerns over their durability, significantly challenge commercial viability. The quest for advanced, robust catalysts and materials remains ongoing, as these are needed to ensure reliable performance under demanding operating conditions. Another significant dimension shaping the evolution of EBC is its potential synergy with renewable energy sources. While this integration could offer solutions to the intermittent issues of renewables, practical realization is constrained by the present infrastructure and associated economics. Policy landscapes and regulatory demands introduce further complexity, as balancing environmental compliance with economic practicality requires carefully engineered solutions. In response, research has intensified on optimizations in reaction processes, development of novel electrocatalysts, and innovative system designs. Such advances are intended not only to boost performance and cost-effectiveness but also to establish EBC as a cornerstone technology for reducing reliance on fossil-derived resources in the chemical and energy sectors.

2.4.1 Challenges

2.4.1.1 Complexity of reaction mechanisms and kinetics

One of the foremost challenges lies in the intricate nature of the electrochemical reactions involved. Comprehensive elucidation of the underlying reaction mechanisms and formulation of general kinetic models are fundamental steps necessary to optimize the conversion process and improve overall efficiency. Current scientific efforts focus on deepening insights into these mechanisms to refine system design and operation parameters.

2.4.1.2 System efficiency and performance

While electrochemical approaches generally offer superior efficiency compared to traditional thermal methods, substantial difficulties remain in achieving high electrical efficiency and power densities. Fuel cell systems fueled by biomass-derived materials frequently operate below their theoretical maxima, primarily due to complex reaction pathways and the substantial energy demands of external fuel reforming, especially at low to intermediate temperatures. These factors collectively limit the attainable performance of current systems.

2.4.1.3 Economic barriers and commercial viability

Financial considerations present significant hurdles for the commercial deployment of electrochemical biomass conversion technologies. The initial investment required for electrolysis equipment is high, compounded by concerns regarding durability and reduced net efficiency over time, making market entry challenging. Furthermore, scaling these technologies involves confronting technical issues related to system longevity and feedstock variability, all of which increase operational risks and costs.

2.4.1.4 Material and technical constraints

The development of robust materials, including catalysts and electrolytes, is vital for enhancing the functionality of electrochemical reactors. Current materials often lack the durability to withstand rigorous operational environments, limiting system lifetime and performance consistency. Additionally, the use of both solid and liquid electrolytes introduces stability and efficiency issues that must be surmounted for sustained application.

2.4.1.5 Renewable energy integration

Integrating electrochemical biomass conversion with renewable energy sources has the potential to mitigate challenges related to the intermittent nature of renewables such as solar and wind. However, practical deployment is hindered by limitations in existing infrastructure and the high costs associated with energy storage and management. Enhancing system flexibility and responsiveness is critical to maximizing this synergy within renewable energy-dominant grids.

2.4.1.6 Regulatory and environmental challenges

Compliance with environmental regulations and management of the ecological impact of biomass sourcing and processing add layers of complexity. Balancing economic viability with environmental stewardship requires carefully calibrated policies and innovative technological solutions to ensure sustainable operation within regulatory frameworks.

2.4.2 Mitigation strategies

To address the numerous challenges facing electrochemical biomass conversion, a multi-faceted approach targeting production optimization, catalyst development, and system design is essential. These strategies aim to improve process efficiency, reduce costs, and support sustainable energy transitions by enabling the cost-effective conversion of biomass into hydrogen and 2,5-furandicarboxylic acid (FDCA).

2.4.2.1 *Optimizing production processes*

Enhancing the electrochemical production of hydrogen and FDCA begins with fine-tuning reaction conditions and parameters. Rigorous evaluation of required reaction potentials ensures an optimized electrochemical environment that facilitates efficient interaction between reactive species. Such optimization not only boosts the selectivity towards preferred products but also minimizes energy consumption. Employing sophisticated electrochemical techniques like linear and cyclic voltammetry assists in pinpointing ideal operational parameters, thus maximizing yield while maintaining energy efficiency.

2.4.2.2 *Development of advanced electrocatalysts*

A key pillar of mitigation involves the discovery and engineering of novel electrocatalysts exhibiting high activity, selectivity, and long-term stability. Catalysts based on abundant transition metals such as nickel, iron, and copper have demonstrated promising performance in biomass electro-oxidation processes, pairing cost-effectiveness with catalytic efficiency. Research continues to refine these materials' physical and chemical properties through innovative synthesis approaches, aiming to sustain activity under harsh operational conditions. The advent of bifunctional catalysts capable of facilitating multiple reactions, including hydrogen evolution and FDCA production, offers further simplification of the system design and operational workflow.

2.4.2.2.1 System design and integration

Engineering efforts focused on scalable and robust system design are pivotal to overcoming current limitations. The development of large-area electrodes enhances industrial applicability by facilitating scale-up from laboratory models. Incorporating real-time analytical tools such as gas chromatography (GC) and high-performance liquid chromatography (HPLC) within the production process allows for continuous monitoring and control, facilitating process improvements and ensuring product quality. Coupling energy generation with biomass waste conversion in integrated systems can close resource loops and improve overall sustainability, providing simultaneous environmental and economic benefits.

2.4.2.2.2 Paired electro-oxidation techniques

Advanced paired electro-oxidation strategies have been investigated to enhance the overall reaction efficiency by coupling biomass conversion reactions with organic transformation processes. Notably, the use of indirect electro-oxidation has shown efficacy in mitigating competing side reactions such as oxygen evolution and reduction reactions. This targeted approach improves the selectivity and efficiency of the conversion process, thereby advancing the feasibility of electrochemical systems for biomass valorization.

2.4.2.2.3 Implementation challenges

Despite technological advances, deploying these systems at scale faces hurdles including system complexity, feedstock quality control, and economic viability. The high capital and operational costs, coupled with uncertainties in feedstock supply consistency, limit the scalability of these approaches. Additionally, maintaining efficient performance over prolonged operation remains a demanding task, requiring further refinement of materials and system architecture.

2.4.2.2.4 Carbon capture and utilization innovations

Innovations in carbon capture and utilization (CCU) technologies integrated with biomass power generation present pathways toward achieving negative emissions. By coupling electrochemical conversion processes with CCU, these systems can amplify sustainability benefits, transforming biomass conversion from a carbon-neutral process into a carbon-negative one. Emerging evidence from case studies demonstrates the efficacy of such integrations in reducing overall greenhouse gas emissions while enhancing process efficiency.

2.5 Conclusion and future directions

As the global demand for sustainable energy solutions intensifies, the field of electrochemical biomass conversion (EBC) stands at the threshold of major advancements. Future research efforts are expected to focus on several critical areas to augment both the efficiency and practicability of these emerging technologies.

2.5.1 Enhancing electrochemical performance

A paramount objective is improving the electrochemical efficiency of systems such as microbial fuel cells and other electrochemical reactors. This can be accomplished by developing high-surface-area, catalytically active electrodes that support greater electrolyte flow rates, thereby accelerating reaction kinetics and boosting energy conversion efficiencies. Innovations in reactor designs and membrane technologies will also be vital to minimize undesired species migration and prevent energy losses, ultimately enhancing system stability and output.

2.5.2 Renewable energy integration

Incorporating EBC technologies with renewable energy sources remains a promising avenue, particularly for managing the intermittent nature of solar and wind power. Future strategies will likely include energy storage and shifting techniques designed to equilibrate supply and demand within power grids dominated by renewables. These advances will enable more reliable and efficient use of sustainable electricity to power biomass conversion processes.

2.5.3 Microbial electrosynthesis (MES)

Further exploration of microbial electrosynthesis presents significant potential for producing value-added fuels and chemicals directly from carbon dioxide using microorganisms. MES systems powered by renewables could generate methane, acetate, and other biofuels with high efficiency. Ongoing research is required to optimize microbial strains and operational parameters for industrial-scale implementation.

2.5.4 Economic and commercial scale-up

Addressing the economic challenges and scalability of EBC technologies is as crucial as technical innovation. Developing sustainable business models, cost reduction strategies, and manufacturing frameworks will facilitate the transition from laboratory prototypes to commercial reality. Integration with existing energy infrastructures and supply chains will be essential to maximize impact and market penetration.

2.5.5 Environmental and regulatory considerations

Future research must also account for the environmental implications of biomass production and conversion, particularly regarding land use change and indirect emissions. Ensuring compliance with evolving environmental regulations while maintaining cost-effectiveness poses a complex but necessary challenge to guarantee sustainable deployment.

2.5.6 Policy framework and economic incentives

The role of policy in fostering the adoption of electrochemical biomass conversion is critical. Government incentives such as subsidies for renewables, carbon pricing, and supportive regulations create favorable conditions for investment and innovation. Case studies underscore the importance of integrated policy measures to balance environmental objectives with economic feasibility, thereby accelerating commercial deployment.

References

[1] M. Zhang, J. Zhang, S. Ran, W. Sun, Z. Zhu, Biomass-derived sustainable carbon materials in energy conversion and storage applications: status and opportunities. A mini review, Electrochem. Commun. 138 (2022) 107283.

[2] N.C. Ramos, M.Manye Ibanez, R. Mittal, M.J. Janik, A. Holewinski, Combining renewable electricity and renewable carbon: understanding reaction mechanisms of biomass-derived furanic compounds for design of catalytic nanomaterials, Acc. Chem. Res. 56 (2023) 2631–2641.

[3] P. Prabhu, Y. Wan, J.-M. Lee, Electrochemical conversion of biomass derived products into high-value chemicals, Matter 3 (2020) 1162–1177.

[4] K.K. Pant, P. Mohanty, Biomass, conversion routes and products–an overview, in: Transformation of Biomass: Theory to Practice, Wiley, 2014, pp. 1–30, doi:10.1002/9781118693643.

[5] A. Faaij, Modern biomass conversion technologies, Mitig. Adapt. Strateg. Glob. Chang. 11 (2006) 343–375.

[6] G. Kumar, A.P. Eswari, S. Kavitha, M.D. Kumar, R.Y. Kannah, L.C. How, et al., Thermochemical conversion routes of hydrogen production from organic biomass: processes, challenges and limitations, Biomass. Convers. Biorefin. 13 (2023) 8509–8534.

[7] J. Carneiro, E. Nikolla, Electrochemical conversion of biomass-based oxygenated compounds, Annu. Rev. Chem. Biomol. Eng. 10 (2019) 85–104.

[8] H.P. Ramesh, R.N. Tharanathan, Carbohydrates—the renewable raw materials of high biotechnological value, Crit. Rev. Biotechnol. 23 (2003) 149–173.

[9] S. Chaudhary, V.P. Jain, G. Jaiswar, The composition of polysaccharides: monosaccharides and binding, group decorating, polysaccharides chains, in: Innovation in Nano-Polysaccharides for Eco-Sustainability, Elsevier, 2022, pp. 83–118, doi:10.1016/B978-0-12-823439-6.00005-2.

[10] J.F. Kornecki, D. Carballares, P.W. Tardioli, R.C. Rodrigues, Á. Berenguer-Murcia, A.R. Alcántara, et al., Enzyme production of D-gluconic acid and glucose oxidase: successful tales of cascade reactions, Catal. Sci. Technol. 10 (2020) 5740–5771.

[11] A. Bjelić, B. Hočevar, M. Grilc, U. Novak, B. Likozar, A review of sustainable lignocellulose biorefining applying (natural) deep eutectic solvents (DESs) for separations, catalysis and enzymatic biotransformation processes, Rev. Chem. Eng. 38 (2022) 243–272.

[12] M.P. van der Ham, Towards the Electrocatalytic Oxidation of Sugar Alcohols Saccharides and Polysaccharides, Wageningen University and Research, Wageningen, 2024.

[13] Z.I. Lai, L.Q. Lee, H. Li, Electroreforming of biomass for value-added products, Micromachines 12 (11) (2021) 1405.

[14] W. Zhang, S. Wei, P. Bai, W. Liu, C. Yang, L. Xu, Temperature versus type: Which is the determining factor in biomass-based electrocatalyst performance? Appl. Catal. B 325 (2023) 122391.

[15] R.B. Melati, F.L. Shimizu, G. Oliveira, F.C. Pagnocca, W. de Souza, C. Sant'Anna, et al., Key factors affecting the recalcitrance and conversion process of biomass, BioEnergy Research 12 (1) (2019) 1–20.

[16] M. Yang, Z. Yuan, R. Peng, S. Wang, Y. Zou, Recent progress on electrocatalytic valorization of biomass-derived organics, Energy Environ. Mater. 5 (4) (2022) 1117–1138.

[17] C.Y. Toe, C. Tsounis, J. Zhang, H. Masood, D. Gunawan, J. Scott, et al., Advancing photo reforming of organics: highlights on photocatalyst and system designs for selective oxidation reactions, Energy Environ. Sci. 14 (3) (2021) 1140–1175.

[18] W. Zeng, Y. Wang, C. Peng, Y. Qiu, Organo-mediator enabled electrochemical transformations, Chem. Soc. Rev. (2025), doi:10.1039/D4CS01142B.

[19] D. Shen, G. Liu, J. Zhao, J. Xue, S. Guan, R. Xiao, Thermo-chemical conversion of lignin to aromatic compounds: effect of lignin source and reaction temperature, J. Anal. Appl. Pyrolysis 112 (2015) 56–65.

[20] C.C. Azubuike, M.N. Allemann, J.K. Michener, Microbial assimilation of lignin-derived aromatic compounds and conversion to value-added products, Curr. Opin. Microbiol. 65 (2022) 64–72.

[21] D.D. Laskar, M.P. Tucker, X. Chen, G.L. Helms, B. Yang, Noble-metal catalyzed hydrodeoxygenation of biomass-derived lignin to aromatic hydrocarbons, Green Chem. 16 (2) (2014) 897–910.

[22] B. Zhang, Q. Meng, H. Liu, B. Han, Catalytic conversion of lignin into valuable chemicals: full utilization of aromatic nuclei and side chains, Acc. Chem. Res. 56 (24) (2023) 3558–3671.

[23] Y. KN, M.U. TM, K. S, S. Sachdeva, S. Thakur, A.K. S, R.B. J, Lignocellulosic biorefinery technologies: a perception into recent advances in biomass fractionation, biorefineries, economic hurdles and market outlook, Fermentation 9(3) (2023) 238.

[24] X. Fan, R. Burton, Recent development of biodiesel feedstocks and the applications of glycerol: a review, Open Fuels Energy Sci. J. 2 (2009) 100–109.

[25] M. Braun, C. Andronescu, Deciphering the electrochemical activity and selectivity of earth-abundant transition metal-based catalysts for the alcohol electrooxidation—current status, ChemCatChem 17 (8) (2025) e202402013.

[26] S. Carrettin, P. McMorn, P. Johnston, K. Griffin, C.J. Kiely, G.J. Hutchings, Oxidation of glycerol using supported Pt, Pd and Au catalysts, Phys. Chem. Chem. Phys. 5 (6) (2003) 1329–1336.

[27] A. Villa, N. Dimitratos, C.E. Chan-Thaw, C. Hammond, L. Prati, G.J. Hutchings, Glycerol oxidation using gold-containing catalysts, Acc. Chem. Res. 48 (5) (2015) 1403–1412.

[28] G. Wu, X. Dong, J. Mao, G. Li, C. Zhu, S. Li, et al., Anodic glycerol oxidation to formate facilitating cathodic hydrogen evolution with earth-abundant metal oxide catalysts, Chem. Eng. J. 468 (2023) 143640.

[29] K. Fernández-Caso, M. Molera, T. Andreu, J. Solla-Gullón, V. Montiel, G. Díaz-Sainz, et al., Coupling glycerol oxidation reaction using Ni-Co foam anodes to CO_2 electroreduction in gas-phase for continuous co-valorization, Chem. Eng. J. 480 (2024) 147908.

[30] L. Liu, H. Liu, W. Huang, Y. He, W. Zhang, C. Wang, et al., Mechanism and kinetics of the electrocatalytic hydrogenation of furfural to furfuryl alcohol, J. Electroanal. Chem. 804 (2017) 248–253.

[31] D. Ouyang, D. Gao, J. Hong, Z. Jiang, X. Zhao, Promoting and controlling electron transfer of furfural oxidation efficiently harvest electricity, furoic acid, hydrogen gas and hydrogen peroxide, J. Energy Chem. 79 (2023) 135–147.

[32] X. Li, L. Cong, N. Lin, C. Tang, Efficient electrochemical upgradation strategies for the biomass derivative furfural, J. Mater. Chem. A 11 (43) (2023) 23133–23147.

[33] C. Xu, E. Paone, D. Rodríguez-Padrón, R. Luque, F. Mauriello, Recent catalytic routes for the preparation and the upgrading of biomass derived furfural and 5-hydroxymethylfurfural, Chem. Soc. Rev. 49 (13) (2020) 4273–4306.

[34] J. Zhang, J. Li, Y. Tang, L. Lin, M. Long, Advances in catalytic production of bio-based polyester monomer 2, 5-furandicarboxylic acid derived from lignocellulosic biomass, Carbohydr. Polym. 130 (2015) 420–428.

[35] T. Li, D.A. Harrington, An overview of glycerol electrooxidation mechanisms on Pt, Pd, and Au, ChemSusChem 14 (6) (2021) 1472–1495.
[36] J.F. Gomes, C.A. Martins, M.J. Giz, G. Tremiliosi-Filho, G.A. Camara, Insights into the adsorption and electrooxidation of glycerol: self-inhibition and concentration effects, J. Catal. 301 (2013) 154–161.
[37] C. Arce, L. Kratky, Mechanical pretreatment of lignocellulosic biomass toward enzymatic/fermentative valorization, iScience 25 (7) (2022) 104610.
[38] J.S. Tumuluru, Biomass Preprocessing and Pretreatments for Production of Biofuels: Mechanical, Chemical, and Thermal Methods (2018).
[39] R. Łukajtis, P. Rybarczyk, K. Kucharska, D. Konopacka-Łyskawa, E. Słupek, K. Wychodnik, et al., Optimization of saccharification conditions of lignocellulosic biomass under alkaline pre-treatment and enzymatic hydrolysis, Energies 11 (4) (2018) 886.
[40] E. Scelsi, A. Angelini, C. Pastore, Deep eutectic solvents for the valorization of lignocellulosic biomasses towards fine chemicals, Biomass 1 (1) (2021) 29–59.
[41] M.K. Hrnčič, G. Kravanja, Ž. Knez, Hydrothermal treatment of biomass for energy and chemicals, Energy 116 (2016) 1312–1322.
[42] Y. Nakagawa, M. Tamura, K. Tomishige, Catalytic reduction of biomass-derived furanic compounds with hydrogen, ACS Catal. 3 (12) (2013) 2655–2668.
[43] Y. Yang, T. Mu, Electrochemical oxidation of biomass derived 5-hydroxymethylfurfural (HMF): pathway, mechanism, catalysts and coupling reactions, Green Chem. 23 (12) (2021) 4228–4254.
[44] M. Sajid, X. Zhao, D. Liu, Production of 2, 5-furandicarboxylic acid (FDCA) from 5-hydroxymethylfurfural (HMF): recent progress focusing on the chemical-catalytic routes, Green Chem. 20 (24) (2018) 5427–5453.
[45] T. Li, D.A. Harrington, An overview of glycerol electrooxidation mechanisms on Pt, Pd, and Au, ChemSusChem 14 (6) (2021) 1472–1495.
[46] M. Sajid, U. Farooq, G. Bary, M.M. Azim, X. Zhao, Sustainable production of levulinic acid and its derivatives for fuel additives and chemicals: progress, challenges, and prospects, Green Chem. 23 (23) (2021) 9198–9238.
[47] M.C. Leech, K. Lam, A practical guide to electrosynthesis, Nature Rev. Chem. 6 (4) (2022) 275–286.

CHAPTER

3

Electrochemical techniques for lignocellulosic biomass conversion

Ruchi Pathania[a], *Amit Srivastava*[b], *and Pallavi Saxena*[c]

[a]Department of Medicinal Chemistry, Center for Natural Products, Drug Discovery and Development, University of Florida, Gainesville, FL, United States, [b]Centre Algatech, Institute of Microbiology of the Czech Academy of Sciences, Trebon, Czech Republic, [c]School of Sciences, Woxsen University, Hyderabad, Telangana, India

3.1 Introduction

The growing necessity to reduce greenhouse gas emissions, rising energy demands, and depleting fossil fuel reserves have made the shift toward sustainable energy and bio-based chemical production essential [1]. This global shift has accelerated efforts to replace fossil resources with renewable alternatives, with lignocellulosic biomass emerging as the largest reservoir of renewable organic carbon on Earth, available at scales capable of significantly contributing to the global energy and materials supply chain [2]. The valorization of biomass not only facilitates the energy transition but also addresses broader environmental challenges by reducing reliance on fossil fuels and mitigating pollution [3]. Unlike food crops, lignocellulosic feedstocks are largely nonedible and readily available from agricultural residues, forestry byproducts, and dedicated energy crops, ensuring that their use does not compete with food production. Their carbon neutrality and widespread availability make them central to sustainable and economically viable bioprocesses [4].

Electrochemical Biomass Conversion: Sustainable Processes and Applications for Bioenergy, Biofuels, and Bioproducts
DOI: https://doi.org/10.1016/B978-0-443-26757-4.00007-6

Lignocellulosic biomass typically consists of 20% to 60% cellulose, 10% to 50% hemicellulose, and 5% to 40% lignin [2,5]. Cellulose, a semicrystalline polymer in plant cell walls, is composed of glucose units linked by β-1,4-glycosidic bonds and exhibits a degree of polymerization between 1000 and 15,000 [6]. Hemicellulose, by contrast, is a heterogeneous group of short-branched polysaccharides, including both pentoses and hexoses, connected mainly through β-1,4-glycosidic linkages [7]. Lignin, a rigid heteropolymer primarily composed of phenolic units such as coniferyl, p-coumaryl, and syringyl alcohols and cross-linked by β–O–4, α–O–4, β–β, and β–5 bonds, fills the spaces between cellulose and hemicellulose within the plant's three-dimensional matrix, where it functions as a barrier to both enzymatic and chemical degradation [8]. Together, these components make lignocellulose a valuable carbon reservoir for the production of biofuels, organic acids, solvents, aldehydes, aromatic compounds, and a wide range of other industrially significant chemicals [9]. Despite its great potential, the intrinsic complexity and heterogeneity of lignocellulosic biomass make it highly recalcitrant, posing significant challenges to efficient deconstruction. Consequently, targeted fractionation during pretreatment becomes essential to enhance both the yield and value of biomass-derived products. In 2010, the US Department of Energy updated its list of the top 10 platform chemicals identified as essential building blocks for chemical synthesis. Among those, glycerol, 5-hydroxymethylfurfural (HMF), and levulinic acid (LA) have attracted particular attention for their catalytic conversion into various high-value products with applications in pharmaceuticals, cosmetics, polymers, and solvents [10]. Rising global demand for these platform chemicals, coupled with their broad industrial utility, underscores the need for efficient and selective valorization technologies.

Conventional biomass conversion methods, including thermochemical and biological routes, have demonstrated potential but face major drawbacks. Thermochemical processes require harsh conditions of high temperature and pressure, making them energy-intensive and prone to producing complex mixtures that complicate downstream separation [11]. Biological methods such as enzymatic hydrolysis and fermentation operate under milder conditions but require extensive pretreatment to breakdown lignin, followed by long processing times and costly enzyme inputs [12]. These limitations restrict scalability and economic competitiveness, highlighting the urgent need for alternative approaches that are both efficient and environmentally sustainable.

Electrochemical conversion has recently emerged as a promising strategy to address these challenges [13]. These processes can utilize electricity preferably sourced from renewable energy, such as solar, wind, or hydropower, to drive targeted oxidation or reduction reactions. Unlike conventional methods, electrochemical conversion operates under relatively mild conditions, avoids the use of corrosive or toxic reagents, and enables precise control of reaction pathways through the modulation of applied potential or current density. Fine-tuning these electrochemical parameters allows the selective formation of target bio-based products, thereby enhancing conversion efficiency while minimizing the generation of undesired by-products. A major advantage of electrochemical biomass conversion lies in its compatibility with other electrochemical processes and its potential to reduce the overall energy input [14,15]. For example, coupling the anodic oxidation of biomass or biomass-derived intermediates with cathodic reactions can replace the conventional oxygen evolution reaction (OER), a process known for sluggish kinetics and high overpotential. This replacement not only improves energy efficiency but also generates high-value chemicals at the anode, effectively upgrading biomass while producing hydrogen or other reduction products at the cathode. Similarly, electroreduction strategies enable the conversion of lignocellulose-derived intermediates into advanced fuels and chemicals, offering synergies with CO_2 utilization and hydrogen-based energy systems. This chapter provides a comprehensive overview of biomass conversion methods, with a particular focus on electrochemical approaches, including the underlying principles, electrode and electrolyte design, reaction mechanisms, and key operating conditions. It further highlights the variety of products obtainable from the electrochemical conversion of lignocellulosic biomass. The chapter also addresses critical challenges such as scaling up, enhancing catalyst durability, improving selectivity, and achieving economic feasibility, while outlining opportunities for industrial implementation. Overall, it emphasizes the pivotal role of electrochemical valorization in promoting sustainable and circular bio-based chemical production.

3.2 Biomass conversion methods

Biomass can be converted into fuels, chemicals, and materials through various approaches, broadly categorized as thermochemical, chemical, biological, physical, and electrochemical processes [16,17]. Each method offers distinct advantages and faces specific limitations, which depend on factors such as feedstock type, particle size, moisture content, target products, and operating conditions (as summarized in Table 3.1). The biomass conversion efficiency is also influenced by chemical characteristics, including carbon and nutrient availability, sugar composition, and the presence of acidic or alkaline hydrolysis agents and phenolic inhibitors, as well as process parameters such as pH, temperature, and residence time. Recent studies have provided comparative evaluations of these methods, highlighting differences in efficiency, selectivity, and overall sustainability [18–21].

TABLE 3.1 Comparative advantages and disadvantages of various biomass conversion methods.

Biomass Conversion Methods	Advantages	Disadvantages
Physical	• Enhances the accessibility of biomass • Can be combined with other methods • Environmentally friendly	• Limited direct conversion to chemicals • High energy input • Effectiveness depends on biomass type
Thermochemical	• Rapid conversion of biomass • Can handle diverse feedstocks • Produces bio-oil, syngas, and biochar	• High energy requirement • May generate undesirable by-products • Limited selectivity
Chemical	• High conversion efficiency • Produces specific chemicals • Can be tuned with catalysts	• Often requires harsh conditions • Potential formation of inhibitors • Corrosion and safety issues
Biological	• Mild operating conditions • High product specificity • Uses renewable feedstocks • Environmentally friendly	• Slow conversion rates • Sensitive to inhibitors • Limited to fermentable sugars • Scale-up can be challenging
Electrochemical	• Mild and controllable conditions • High selectivity • Can be integrated with renewable electricity • Simultaneous anodic/cathodic valorization	• Requires specialized electrodes and setups • Mass transfer limitations • Scaling up is challenging • High initial cost

3.2.1 Physical pretreatments

This includes ultrasonication, microwave irradiation, and mechanical milling, which enhance biomass accessibility by disrupting its recalcitrant structure [22]. These methods are often applied before or in combination with other treatments to improve overall conversion efficiency. Ultrasonication generates cavitation bubbles that collapse to produce localized temperatures of 2000–10,000 K, creating intense physical and chemical effects. This process cleaves key lignin linkages, thereby facilitating polysaccharide release and lignin depolymerization [23,24]. Microwave irradiation also disrupts biomass but operates through both thermal and nonthermal mechanisms. The thermal effect weakens hydrogen bonds and damages cell walls, enhancing hydrolysis, while the nonthermal effect arises from interactions between electromagnetic fields and dielectric components, leading to relaxation and polarization that further destabilize biomass. Mechanical milling reduces the particle size of lignocellulosic biomass and increases its surface area, thereby enhancing the efficiency of subsequent conversion processes [25].

3.2.2 Thermochemical approach

This approach relies on heat to break down biomass into simpler molecules [26,27]. Techniques such as pyrolysis, combustion, gasification, liquefaction, and torrefaction enable the production of a wide range of bio-based products, including solid fuels like biochar, liquid fuels such as bio-oil, gaseous fuels such as syngas, and other value-added chemicals. Combustion, though widely applied for domestic waste and landfill disposal, is considered a relatively low-efficiency recycling method in which waste is burned to generate heat, CO_2, and water [28]. Pyrolysis, a thermal conversion process conducted at 400–800°C in the absence or limited amount of oxygen, decomposes biomass into fuels and chemicals. It is generally classified into slow, fast, and flash pyrolysis [29]. Slow pyrolysis, performed at lower heating rates and extended residence times, favors char production. Fast pyrolysis, operating at 450 to 650°C with short residence times (0.5–5 s) and high heating rates, yields predominantly bio-oil. Flash pyrolysis, characterized by extremely high heating rates (~1000°C/s), short residence times (0.1–1 s), and temperatures up to 1000°C, mainly produces gases and aerosols that are rapidly quenched and condensed into useful products. Recently, advanced variants such as microwave-assisted and solar-assisted pyrolysis have gained interest for improving efficiency that depends solely on external heat or electricity. Torrefaction, often described as mild pyrolysis or high-temperature drying, is carried out at 200 to 300°C under oxygen-limited or inert conditions [30]. The resulting torrefied biomass (biochar) exhibits higher energy density, greater heating value, improved hydrophobicity, and enhanced grindability and flowability compared to raw biomass, making it more coal-like. In liquefaction, solvents such as water, ethanol, acetone, glycerol, phenols, or tetralin are commonly used. The process typically produces approximately 40% to 60% bio-oil, 10 to 12% biochar, and 35% to 45% gases. It can be broadly classified into two main routes: hydrothermal liquefaction (HTL) and solvolytic liquefaction. HTL employs water as both solvent and catalyst, facilitating the cleavage of bonds between hemicellulose and cellulose under elevated temperature and pressure, producing liquid fuels with relatively high efficiency [31]. In contrast, solvolytic liquefaction relies on organic solvents and can operate at moderate or even low temperatures, sometimes at atmospheric pressure, making it more energy-efficient but highly solvent-dependent [32]. Gasification is an endothermic thermochemical process in which biomass undergoes partial oxidation at 600 to 1500°C in the presence of gasifying agents such as air, oxygen, steam, or CO_2[33]. At lower temperatures, the process produces a mixture of CO, CO_2, H_2, CH_4, and small hydrocarbons, whereas higher temperatures favor the formation of syngas, primarily composed of H_2, CO, CO_2, and H_2O, with reduced impurities.

3.2.3 Chemical approach

This approach employs acids, bases, or other catalysts to depolymerize biomass and alter its chemical structure [34]. Acid hydrolysis, for instance, converts polysaccharides into monosaccharides, whereas base-catalyzed reactions can break lignin linkages or drive transesterification to produce various bioderivatives. These methods enable tunable reactions under controlled environments but often generate inhibitory by-products and may require additional steps for neutralization and waste management.

3.2.4 Biological conversion

This method relies on enzymes or microorganisms to transform biomass into valuable products [12,35]. These processes operate under mild conditions near-neutral pH, moderate temperatures, and ambient pressure, providing high selectivity but generally requiring longer processing times than thermochemical methods. By utilizing enzymatic and microbial capabilities, this environment- friendly approach converts biomass into products, which are relatively safe and nontoxic. Enzymes, including α-amylase, cellulase, lipase, protease, and xylanase, enable the breakdown of complex biomaterials. The key strategies include fermentation, which converts fermentable carbohydrates into liquid fuels (e.g., ethanol, butanol) or organic acids under mild anaerobic conditions [36], and anaerobic digestion, which produces methane and CO_2 through microbial consortia of bacteria and fungi that enhance biomass degradation and saccharification [37]. Although widely applied, for example, in commercial bioethanol production, biological methods face scalability challenges due to long residence times and the high costs of enzyme inputs.

3.2.5 Electrochemical method

This method utilizes electrical energy to facilitate redox reactions, enabling the selective cleavage and conversion of complex biomass components such as cellulose, hemicellulose, lignin, polyols, and furan derivatives [38]. Both direct and mediated electro-oxidation and electroreduction pathways offer precise control over reaction routes, which can be further optimized by tuning key electrochemical parameters such as applied potential, current density, electrolyte composition, and electrode material [39]. A unique advantage of this approach is its ability to valorize biomass simultaneously at both the anode and cathode, effectively doubling the potential for high-value product generation within a single process. Beyond this versatility, the method offers notable sustainability benefits, including a reduced environmental impact. They typically operate under mild conditions, ambient temperature and pressure, making them safer and less energy-intensive than other methods. In addition, their compatibility with renewable electricity sources such as solar, wind, tidal, and geothermal energy enables low-carbon, flexible, and decentralized operation [40]. This integration not only reduces dependence on fossil-based energy but also supports the development of carbon-neutral chemical manufacturing and circular bioeconomy models.

3.3 Fundamentals of electrochemical biomass conversion

Electrochemical biomass conversion is driven by two complementary electrode processes: electrooxidation and electroreduction [13,41]. Fundamentally, this approach relies on redox reactions at the electrode–electrolyte interface within an electrolytic cell. When an external potential is applied, oxidation occurs at the anode and reduction at the cathode, enabling selective bond cleavage or the formation of target biomass-derived molecules. Electrocatalysts play a central role in this process by facilitating electron transfer, lowering activation energy barriers, and enhancing both reaction kinetics and product selectivity under well-controlled electrochemical conditions. The dual reactivity of anodic and cathodic reactions allows simultaneous valorization, improving overall efficiency and providing a distinct advantage over conventional methods. This enhanced selectivity not only increases the yield of desired products but also minimizes the formation of unwanted by-products and reduces energy consumption, making electrochemical strategies a highly efficient and tunable platform for biomass valorization. A systematic literature search was conducted to ensure comprehensive coverage of different techniques for lignocellulosic biomass conversion, along with their advantages and limitations. Databases including Scopus, Web of Science, Google Scholar, and ScienceDirect were queried using relevant keywords such as electrochemical conversion, lignocellulosic biomass, electro-oxidation, electroreduction, and biomass valorization. Only peer-reviewed original research articles and review papers focusing on biomass conversion approaches applied to lignocellulosic feedstocks were included to ensure a thorough representation of current advances.

Electro-oxidation involves the removal of electrons from biomass or biomass-derived intermediates at the anode [38,42]. This process typically cleaves C–C, C–H, or C–O bonds, yielding oxygenated products such as organic acids, aldehydes, ketones, and other high-value chemicals. Importantly, electro-oxidation provides a sustainable alternative to the OER, thereby reducing the overall energy demand of electrochemical systems. However, several challenges remain. Overoxidation can reduce product selectivity, while catalyst deactivation caused by adsorbed intermediates can impair performance. Additionally, competition with the sluggish OER, which requires high overpotentials, further complicates the process [43]. Electroreduction, in contrast, involves the gain of electrons by biomass-derived compounds at the cathode. Common substrates include furanics, carboxylic acids, and carbonyl-containing intermediates, which can be selectively reduced into alcohols, alkanes, or hydrogenated derivatives [44]. For example, furfural can be electrochemically reduced to furfuryl alcohol, a valuable precursor for polymer and resin production. A major challenge is competition with the hydrogen evolution reaction (HER), which consumes electrons and protons but does not generate desired products, thereby lowering both faradaic efficiency and selectivity [45]. The design of highly selective cathodic catalysts is therefore essential, as they can direct electron transfer toward targeted products while inhibiting unwanted hydrogen formation [46]. Despite its promise, electroreduction still faces obstacles related to catalyst stability, substrate adsorption behavior, and energy efficiency, which must be optimized for industrial scalability.

A promising strategy couples biomass electrooxidation with value-added cathodic reactions, effectively suppressing the competing OER, enhancing energy efficiency, and enabling the co-production of valuable chemicals [47]. Integrated electrooxidation and electroreduction facilitate dual valorization, allowing simultaneous conversion at both electrodes [48]. For instance, anodic oxidation of glycerol to formic acid or tartronic acid can be paired with cathodic reduction of furfural to furfuryl alcohol, maximizing resource utilization while minimizing energy demand [49]. This integrated approach replaces the energy-intensive OER while ensuring that electrical energy is efficiently converted into high-value products, advancing both sustainability and economic viability. In the context of lignocellulosic biomass, the electrochemical method utilizes the cellulose, hemicellulose, and lignin components for conversion. Oxidative reactions at the anode can depolymerize lignin, generating aromatic monomers such as phenols, aldehydes, and organic acids, while reductive reactions at the cathode can target glycosidic linkages in cellulose and hemicellulose, producing fermentable sugars, sugar alcohols, and other platform chemicals [50,51]. The efficiency and selectivity of electrochemical biomass conversion depend upon multiple factors, including electrode potential, electrolyte composition, catalyst properties, electrode architecture, pH, temperature, and reaction duration [52].

3.3.1 Role of electrodes, electrocatalysts, electrolytes, and reaction conditions

In electrochemical biomass conversion, electrodes, electrocatalysts, electrolytes, and reaction conditions collectively determine process efficiency, selectivity, and sustainability, with each component playing a synergistic and complementary role in driving overall performance. Electrodes act as the primary sites for electron transfer, where oxidation and reduction reactions break down complex biomass polymers into smaller molecules [53]. Their material and structure strongly influence conductivity, stability under reaction conditions, and accessibility of active sites, thereby governing reaction rates and product yields. Electrocatalysts, typically integrated with electrode surfaces, enhance these reactions by lowering activation energy barriers and directing pathways toward targeted products [54]. Electrolytes, meanwhile, provide the ionic medium for charge transport, maintaining electrical neutrality and enabling steady current flow [55].

Electrolyte composition and concentration are critical determinants of ionic conductivity, charge balance, reaction kinetics, electrode stability, and product selectivity. Acidic electrolytes (e.g., H_2SO_4 and HCl) and alkaline electrolytes (e.g., NaOH and KOH) adjust local pH and thereby direct oxidation or reduction pathways, while neutral salts (e.g., Na_2SO_4 and KCl) primarily enhance conductivity without significantly altering pH [56]. Adding redox mediators can further enhance electron transfer and inhibit unwanted side reactions. Electrolyte concentration plays a dual role: moderate levels promote ionic conductivity and faster reaction rates, while excessive concentrations may hinder mass transport, block active sites, or alter pathways, ultimately reducing faradaic efficiency [57]. Importantly, the microenvironment at the electrode–electrolyte interface exerts a greater influence on reaction outcomes than the bulk solution conditions, highlighting the need for precise electrolyte engineering [58]. Proper electrolyte design minimizes cell resistance, lowers operating voltages, sustains stable current flow, and maintains charge balance, which are crucial for efficient, selective, and scalable electrochemical biomass valorization. Extremely harsh acidic or alkaline conditions, however, can degrade substrates and pose challenges for industrial applications, emphasizing the importance of controlled conditions. pH plays a central role by regulating proton availability, intermediate stability, and reaction equilibrium. For instance, alkaline media often favor selective oxidation of alcohols to aldehydes or acids, while acidic environments tend to promote hydrogen evolution or alternative reduction pathways [59].

The selection of the solvent also impacts solubility, ionic conductivity, mass transport, and product selectivity [60]. Aqueous media remain the most widely used due to their low cost, high abundance, high conductivity, and compatibility with a wide range of electrolytes. While sustainable, aqueous systems can promote competing with HER or OER, lowering faradaic efficiency. Organic solvents, in contrast, offer broader electrochemical potential, enhanced solubility for hydrophobic biomass-derived compounds, and stabilization of reactive intermediates. However, these advantages are counterbalanced by reduced ionic conductivity, often requiring the addition of supporting electrolytes or the use of ionic liquids. To combine the strengths of both, biphasic systems are increasingly employed: an aqueous electrolyte phase provides conductivity and electrochemical activity, while an immiscible organic phase extracts or stabilizes desired products in situ [61]. Deep eutectic solvents were first identified as an innovative class of green synthetic solvents, formed by combining two or more components [62]. Such systems suppress overoxidation, improve selectivity, and simplify downstream separation.

In addition to electrode, catalyst, and electrolyte design, operational parameters significantly shape reaction performance. Variations in operational parameters such as pH, temperature, and pressure, along with substrate stability, directly affect the reaction's performance and product distribution. Substrate degradation under harsh electrochemical conditions can reduce yields and produce unwanted by-products, while catalyst degradation decreases active surface area, electron transfer efficiency, and long-term stability [63]. Temperature fluctuations affect reaction kinetics, electrolyte conductivity, and mass transport: higher temperatures generally accelerate reaction rates but may also increase side reactions or cause thermal decomposition of sensitive intermediates. Pressure, though less relevant in standard laboratory cells, can influence gas-evolving reactions and reactant solubility, particularly in volatile species [64,65]. Applied potential and current density must also be carefully optimized. Insufficient potential leads to incomplete conversion, whereas excessively high values can trigger overoxidation, undesired side reactions, or electrolyte breakdown. Likewise, higher current densities can accelerate reaction rates and conversions but may also exacerbate mass transport limitations, generate localized heating, and promote side reactions [66].

3.3.2 Cell configurations and operating modes

Electrochemical biomass conversion is carried out in electrolytic cells, where biomass interacts with electrodes under an applied potential, allowing precise modulation of reaction pathways [67]. A basic electrolytic cell consists of an anode and a cathode separated by an electrolyte that provides ionic conductivity. Depending on the target reaction and products, cells may be configured as single-compartment or divided systems to optimize efficiency, selectivity, and scalability. The simplest design, a single-compartment cell, places both anodic and cathodic reactions in the same electrolyte, simplifying assembly and ensuring easy ion transport, but also causing product crossover and unwanted side reactions [68].

Divided or two-compartment cells address these limitations by separating anodic and cathodic chambers with ion-exchange membranes, glass frits, or porous barriers [69]. These designs prevent product mixing, maintain distinct pH environments, and improve selectivity while reducing cross-contamination. In laboratory research, H-cells with undivided or divided configurations are widely used to study individual anodic and cathodic transformations. At larger scales, flow cells are preferred due to their ability to enable continuous operation, enhance mass transport, distribute current more uniformly, and support automation. A typical three-electrode setup consists of a working electrode (WE), counter electrode, and reference electrode immersed in an electrolyte containing the biomass substrate and supporting electrolyte [70].

Electrochemical biomass conversion can be operated under two main control modes: potentiostatic and galvanostatic [71]. In potentiostatic mode, a fixed potential is applied to the WE relative to a reference electrode, precisely controlling the thermodynamic driving force for redox reactions. This enables selective oxidation or reduction of biomass components while minimizing competing side reactions, making it particularly useful when multiple pathways are possible. In galvanostatic mode, a constant current is maintained, ensuring uniform electron flow between electrodes. This mode is valuable for studying reaction kinetics, optimizing productivity, and scaling processes, as it provides predictable charge passage and reproducible yields. The choice between modes depends on factors such as substrate type, electrode materials, catalyst performance, and electrolyte composition. For example, viscous or heterogeneous biomass slurries may benefit from flow or divided cells to improve mass transport, while reactions requiring high selectivity often favor potentiostatic operation to minimize overoxidation or competing reactions.

To identify optimal operating conditions, electrochemical diagnostic techniques are commonly employed. Cyclic voltammetry and differential pulse voltammetry are used to probe redox behavior, while linear sweep voltammetry and impedance spectroscopy provide insight into electrode kinetics, substrate interactions, and operating voltages or currents. During catalytic studies, chronopotentiometry, chronoamperometry, and chronocoulometry enable real-time monitoring of potential, current, or total charge, offering critical information on reaction kinetics, progression, and efficiency [72,73]. The ultimate objective is to maximize conversion, product yield, and faradaic efficiency; however, achieving all three simultaneously remains challenging due to competing processes, particularly when water serves as the proton source. In summary, the coordinated optimization of electrocatalysts, electrodes, electrolytes, solvent systems, cell design, and operating parameters is essential to achieving efficient, selective, and sustainable electrochemical biomass valorization across both laboratory and industrial scales.

3.4 Electrocatalysts for biomass conversion

Electrode materials play a pivotal role in electrochemical biomass conversion as they facilitate electron transfer, reaction kinetics, selectivity toward targeted products, and overall efficiency. Their properties influence reactant adsorption, stabilization of intermediates, and the transfer of electrons and protons, all essential for effective biomass valorization. Common electrode classes include metals, metal oxides, carbon-based materials, and conductive polymers, each offering distinct advantages discussed below [74,75]. In these systems, the WE serves as the primary catalytic site, making its careful selection for achieving targeted conversions. Proper construction of the WE ensures reproducibility and control, as electro-organic conversion begins with heterogeneous electron transfer at the electrode surface [76]. Electrode materials are thus active participants in electroconversion, and their careful selection, surface engineering, and characterization are essential to controlling pathways, kinetics, and product distributions. The metal type, surface structure, and catalyst modifications collectively determine performance and scalability.

3.4.1 Metal-based electrocatalysts (noble and transition metals)

Metal-based electrocatalysts are broadly classified into noble metals and transition (nonnoble) metals, each offering distinct advantages and limitations depending on the reaction and operating environment. Noble metals such as platinum (Pt), gold (Au), and palladium (Pd) are valued for their high conductivity, strong catalytic activity, and chemical stability [77]. Pt and Pd are frequently employed in alcohol oxidation, hydrogen evolution, and biomass dehydrogenation reactions, while Au is particularly effective for the selective oxidation of polyols and lignin-derived phenolics under mild conditions. Transition metals, including nickel (Ni), copper (Cu), cobalt (Co), and iron (Fe), provide cost-effective alternatives while maintaining reasonable catalytic performance [78]. Ni and Co are especially effective for electrooxidation in alkaline media, such as the oxidation of biomass-derived alcohols and polyols, owing to their stability and favorable adsorption characteristics [79]. Cu-based electrodes are widely used in electroreduction, particularly for the selective hydrogenation of aldehyde groups and the reduction of biomass-derived carbonyl compounds [80]. Fe-based catalysts have also been studied for mediating redox reactions, particularly when incorporated into bimetallic or alloyed systems. Combining metals into bimetallic or alloy catalysts (e.g., Ni–Fe, Ni–Co, Pt–Ir) often yields synergistic effects [81]. These systems can fine-tune adsorption energies for intermediates, improve electron transfer, and reduce overpotentials, ultimately boosting reaction efficiency. While noble metals and their alloys offer superior performance, their cost remains a significant barrier to large-scale applications, reinforcing the importance of developing efficient, durable, and affordable transition-metal-based alternatives.

3.4.2 Metal oxides, hydroxides, and mixed oxides

Metal oxides, hydroxides, and mixed oxides represent a versatile class of electrocatalysts for biomass conversion, valued for their high stability, tunable electronic properties, and ability to promote selective reactions [82,83]. They are especially effective in anodic oxidation, where lignin depolymerization, alcohol oxidation, and conversion of biomass-derived molecules require robust and selective catalysts under electrochemical conditions. Simple oxides such as NiO, Co_3O_4, Fe_2O_3, and MnO_2 demonstrate strong stability in harsh environments and facilitate selective oxidation through surface defects, variable oxidation states, and strong intermediate adsorption. For example, Co_3O_4 and NiO are frequently used for alcohol and polyol oxidation in alkaline media, where their favorable redox behavior yields high activity. Hydroxide-based electrodes, including $Ni(OH)_2$, $Co(OH)_2$, and $Fe(OH)_3$ are also widely used for anodic oxidation in alkaline electrolytes [84]. Their reversible switching between oxidation states provides abundant active sites for electron transfer, improving kinetics. Hydroxides often act as precursors to in situ–formed oxyhydroxides, which are particularly active in selective biomass conversions [85]. Mixed oxides further enhance catalytic performance through synergistic effects [86]. Compared to single-metal oxides, mixed oxides typically exhibit higher conductivity, larger surface area, and improved adsorption properties. The presence of multiple redox-active centers facilitates more efficient electron transfer, lowers overpotentials, and enhances selectivity. For instance, mixed oxides have shown excellent activity for alcohol oxidation and lignin-derived compound conversions while maintaining high stability. Overall, their adjustable surface chemistry, redox flexibility, and compositional diversity enable optimization of selectivity, kinetics, and durability, making them attractive, scalable alternatives to noble metal catalysts for sustainable electrochemical biomass valorization.

3.4.3 Carbon-based electrocatalysts

Carbon-based electrocatalysts have emerged as highly versatile materials due to their excellent electrical conductivity, chemical stability, large surface area, and tunable surface properties. Common examples include graphite, glassy carbon, carbon nanotubes (CNTs), graphene, and heteroatom-doped carbons (e.g., N-, S-, or P-doped) [87]. These materials can function as active catalysts themselves or serve as supports for metal and metal oxide nanoparticles. Their effectiveness lies in the ability to adsorb biomass-derived molecules, stabilize reaction intermediates, and promote rapid electron transfer, thereby improving both kinetics and selectivity. Carbon-based electrocatalysts have shown strong performance in anodic oxidation of alcohols, polyols, and lignin-derived compounds, as well as in cathodic reduction of aldehydes and ketones. Surface functionalization or heteroatom doping further tunes their catalytic activity by modifying electronic structures and active sites, enabling tailored reaction pathways. In addition to high catalytic versatility, carbon-based materials are chemically robust, scalable, and cost-effective, making them attractive for both research and industrial applications [88].

3.4.4 Emerging materials

Emerging electrocatalysts such as single-atom catalysts (SACs), metal–organic frameworks (MOFs), and covalent organic frameworks (COFs) are gaining significant attention due to their unique structural and electronic features [89]. SACs consist of isolated metal atoms anchored on conductive supports, offering maximum atom utilization, uniform active sites, and exceptional selectivity for biomass-derived conversions [90]. MOFs, built from metal nodes and organic linkers, provide highly porous architectures with tunable surface chemistry, enabling efficient substrate adsorption, controlled mass transport, and tailored catalytic activity [91]. COFs, by contrast, are crystalline porous polymers with covalent linkages, which allow precise control over pore size, surface functionality, and electronic properties [92]. These materials can act as independent catalysts or as supports for metal nanoparticles, enhancing electron transfer, stabilizing intermediates, and promoting selective reaction pathways. Collectively, SACs, MOFs, and COFs offer versatile and highly tunable platforms for next-generation electrochemical biomass valorization, with the potential to significantly improve efficiency, product selectivity, and sustainability compared to conventional electrocatalysts.

3.5 Electrochemical conversion techniques

Electrochemical conversion strategies can be broadly categorized into direct electrooxidation, mediated electrooxidation, electroreduction, and hybrid electrochemical–catalytic approaches [93,94]. In direct electrooxidation, biomass-derived substrates undergo oxidation directly at the electrode surface. This method offers precise control but is often constrained by electrode fouling and competing side reactions. Mediated electrooxidation employs soluble redox mediators to transfer electrons between the electrode and substrate, enhancing selectivity, mitigating electrode passivation, and enabling conversions of substrates that are otherwise resistant to direct oxidation. Electroreduction focuses on hydrogenation, deoxygenation, and C–C bond modifications of biomass-derived molecules [44]. However, these reactions frequently compete with the HER, which reduces faradaic efficiency. To overcome such challenges, hybrid systems that couple electrochemical steps with either biocatalysis or traditional thermocatalysis have been developed [95]. These integrated approaches combine the advantages of multiple methodologies, offering improved selectivity, milder operating conditions, and enhanced energy efficiency. A comparative summary of these approaches, highlighting their respective benefits and limitations, is discussed in Table 3.2. Each technique can be fine-tuned by adjusting applied potential or current density, electrolyte composition, and electrode design, underscoring the versatility and sustainability of electrochemical platforms for converting complex lignocellulosic feedstocks into high-value products.

3.6 Integration with renewable energy and circular bioeconomy

Electrochemical valorization of lignocellulosic biomass is naturally aligned with renewable electricity and circular bioeconomy strategies, positioning it as a promising route for sustainable chemicals, biofuel, and biomaterials production while minimizing waste and environmental burdens [96]. Unlike fossil-based thermochemical methods, electrochemical processes are driven by electrons, making their efficiency, scalability, and carbon footprint highly dependent on the energy source. When coupled with renewable energy, these systems not only enhance sustainability but also support the development of climate-neutral biorefineries. In contrast, while linear production systems rely on finite fossil resources and generate significant emissions, circular bioeconomy frameworks prioritize efficient utilization, recycling, and valorization of biological feedstocks. This integration can be considered across two key dimensions: (1) direct coupling with solar, wind, and hydroelectric power, and (2) the use of photoelectrochemical (PEC) and bioelectrochemical (BEC) systems [97,98].

3.6.1 Coupling with solar, wind, and hydroelectric power

The sustainability of electrochemical biomass conversion can be greatly enhanced through integration with renewable power sources such as solar,

TABLE 3.2 Comparative overview of electrochemical conversion strategies for biomass deconstruction, including their operational descriptions, principal advantages, and associated disadvantages.

Electrochemical conversion strategies	Description	Advantages	Disadvantages
Direct electrooxidation	Substrate interacts directly with the electrode surface for charge transfer	• No need for additional mediators • Simple setup • Theoretically applicable to a wide range of substrates • Reaction pathways can be tuned via electrode design • avoids secondary chemicals	• Requires a strong substrate–electrode interaction • Can suffer from poor selectivity and overoxidation • Strong dependence on electrode materials
Mediated electrooxidation	Involves a redox mediator (e.g., metal ions, radicals) that transfers charge between the electrode and substrate	• Enhances reaction rates • Can avoid electrode fouling • High selectivity with suitable mediators • Can target specific functional groups	• Limited choice of mediators • Stability and cost of mediators • Potential contamination of products • Mediator recovery and recycling are needed
Electro-assisted pretreatment	Electrochemical methods (e.g., electro-Fenton, electrolysis) are used to break lignocellulosic recalcitrance before conversion	• Reduces biomass crystallinity and lignin content • Improves enzyme accessibility • Environment-friendly alternative to harsh chemicals	• Energy-intensive • May generate inhibitory byproducts • Challenges in Scale-up
Electrochemical reforming (aqueous phase reforming)	Biomass-derived polyols (e.g., glycerol, sorbitol) are electrochemically reformed to produce H_2, syngas, and light hydrocarbons	• Produces clean hydrogen fuel • Operates at low temperature compared to thermal reforming • Potential integration with renewable electricity	• Low carbon efficiency due to CO_2 release • Catalyst deactivation risk • Competes with direct fuel production

(*continued on next page*)

TABLE 3.2 Comparative overview of electrochemical conversion strategies for biomass deconstruction, including their operational descriptions, principal advantages, and associated disadvantages—cont'd

Electrochemical conversion strategies	Description	Advantages	Disadvantages
Electrocatalytic cleavage of cellulose and hemicellulose	Electrodes or electrocatalysts promote cleavage of β-1,4-glycosidic bonds, producing soluble sugars and oligomers	• Enables direct valorization of lignocellulosic biomass • Can operate under mild conditions • Avoids harsh chemicals used in acid/alkali hydrolysis • compatible with downstream biological upgrading	• Slow kinetics due to polymer crystallinity • Electrode fouling from insoluble residues • Requires optimized catalysts for efficiency • high energy demand for large-scale applications
Coupling electrochemical with biological catalysis (electro-bio catalysis)	Integration of electrochemistry with enzymatic or microbial pathways for selective conversion	• Combines the high specificity of biocatalysts with the tunability of electrochemistry • Can upgrade complex intermediates selectively • Operates at ambient conditions • renewable and eco-friendly	• Enzyme/microbe stability under electrochemical stress • Slow rates compared to chemical catalysis • Scalability challenges
Coupling electrochemical with chemical catalysis (electro-chemo catalysis)	Combines electrochemistry with heterogeneous/homogeneous catalysts for enhanced deconstruction and upgrading	• Expands product scope beyond purely electrochemical routes • Enables cascade reactions • Potential for higher yields and selectivity	• Catalyst cost and deactivation • more complex system design • requires fine-tuned process control

wind, and hydroelectric energy. Because these processes rely on a continuous supply of electricity, direct coupling with renewables provides a pathway toward low-carbon and energy-efficient biomass valorization. Solar energy, delivered via photovoltaic panels or solar-driven PEC cells, offers abundant but intermittent electricity that can directly power electrolysis and enable decentralized biomass conversion units [99]. Wind energy provides another renewable option for large-scale electrochemical operations, though its variability requires buffer systems such as batteries or supercapacitors to maintain stable performance [100]. In contrast, hydroelectric power offers a steady and reliable electricity supply, making it well-suited for continuous industrial-scale electrochemical processes. Hybrid integration combining intermittent solar and wind with the stable baseline of hydropower can improve resilience, reduce operational costs, and enable round-the-clock operation [101].

3.6.2 Photoelectrochemical and bioelectrochemical systems

PEC and BEC systems represent advanced strategies for improving the efficiency and sustainability of electrochemical biomass conversion. PEC systems integrate light-driven energy input with electrochemistry, employing photoactive electrodes to harvest solar energy and drive redox reactions [102]. By converting photons directly into electrochemical potential, PEC processes enable selective oxidation of lignin, cellulose, and hemicellulose under mild conditions, thereby lowering external energy demands. This direct use of solar energy not only enhances energy efficiency but also provides a renewable and environmentally benign route to produce high-value platform chemicals and fuels. BEC systems, including microbial fuel cells (MFCs) [103] and microbial electrolysis cells (MECs), exploit the metabolic activity of microorganisms to mediate electron transfer between biomass and electrodes [104]. Microbial communities catalyze the oxidation of complex organic matter, generating electricity or hydrogen while simultaneously deconstructing biomass into simpler molecules. By combining biological selectivity with electrochemical control, BECs allow the valorization of diverse biomass feedstocks, including lignocellulosic residues and waste streams, under ambient, energy-efficient conditions. Despite their promise, both PEC and BEC approaches face technical hurdles such as electrode fouling, limited current densities, light absorption inefficiencies, and microbial stability issues. Ongoing advances in electrode materials, reactor design, hybrid photo-bioelectrochemical configurations with renewable energy inputs, and integrated biorefinery frameworks are addressing these challenges.

3.7 Products derived from electrochemical conversion of lignocellulosic biomass

Electrochemical conversion of lignocellulosic biomass yields a wide spectrum of products, including platform chemicals, monomers, fuels,

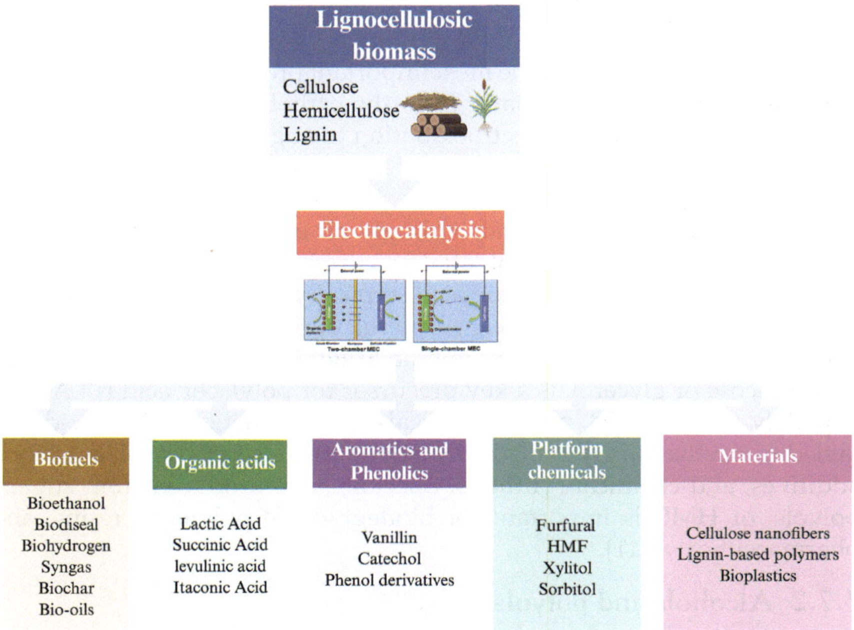

FIGURE 3.1 Overview of value-added bio-based chemicals and fuels derived from lignocellulosic biomass through electrochemical conversion.

and functional materials, as shown in Fig. 3.1. The product distribution depends on both the specific biomass fraction targeted and the electrochemical approach used. Oxidative processes predominantly depolymerize lignin into phenolic and aromatic compounds [105], while reductive processes cleave carbohydrate linkages in cellulose and hemicellulose, generating sugar monomers or platform chemicals with reduced formation of toxic by-products [106]. The major product categories include: Organic acids (e.g., formic, acetic, levulinic), Alcohols and polyols (e.g., ethanol, butanol, glycerol, sorbitol), Furan derivatives (e.g., HMF, furfural, furan dicarboxylic acid), Lignin-derived aromatics (e.g., phenols, aldehydes, ketones), Hydrogen and syngas, Carbon-based functional materials (e.g., biochar-derived electrodes, carbon nanostructures) [107,108]. Together, these products provide sustainable alternatives to fossil-derived chemicals and strengthen the economic and environmental viability of integrated biorefineries operating within a circular bioeconomy framework. Among them, HMF, furfural, and glycerol are particularly prominent, having been extensively studied for their high market value, chemical versatility, and broad industrial applications.

3.7.1 Organic acids

Organic acids are among the most important products of electrocatalytic biomass conversion, particularly from the carbohydrate fractions of cellulose and hemicellulose. Electrooxidation of sugars, sugar alcohols, and polyols can selectively yield a variety of acids, including gluconic, glucaric, xylaric, lactic, glyceric, glycolic, oxalic, formic, acetic, and succinic acids [109]. Their production depends strongly on electrode composition and surface properties, applied potential or current density, electrolyte type, pH, and solvent system. Formic acid, typically produced from biomass-derived polyols or aldehydes (e.g., glycerol, furfural), is valued as a hydrogen carrier, preservative, and antimicrobial agent [110]. Lactic acid, derived from glucose or glycerol, is a key precursor for polylactic acid (PLA) and is widely used in food, pharmaceuticals, and cosmetics. Acetic acid, obtained from ethanol, glycerol, or HMF derivatives, serves in solvents, food additives, and chemical synthesis. Succinic acid, produced from sugars, polyols, or HMF, is important for biodegradable polymers, resins, and pharmaceuticals [111].

3.7.2 Alcohols and polyols

Alcohols and polyols such as ethanol, ethylene glycol, glycerol, sorbitol, and xylitol are versatile platform molecules with broad applications in fuels, solvents, polymers, and fine chemicals. Electrochemical oxidation and reduction provide selective conversions: Ethanol can be anodically oxidized to acetaldehyde or acetic acid, or cathodically reduced to ethylene or ethylamine derivatives [112]. Glycerol, a major biodiesel by-product, undergoes selective oxidation to dihydroxyacetone, glyceraldehyde, glycolaldehyde, glyceric acid, tartronic acid, or formic acid, while electroreduction yields 1,2-propanediol and hydrogen, showcasing its role in power-to-X applications [113]. Ethylene glycol, obtained from cellulose or glycerol, can be oxidized to glycolic acid, oxalic acid, or glyoxal for biodegradable polymers and specific chemicals. Sorbitol and xylitol, with multiple hydroxyl groups, undergo reactions such as hydrogenolysis, dehydration, isomerization, and oxidation, producing acrolein, hydroxyacetone, carboxylic acids, or cyclic carbonates for polymers and green solvents [114]. Their hydrophilic nature, high boiling points, and strong hydrogen-bonding networks influence processing conditions, solubility, and crystallization behavior. Industrial relevance is underscored by the large-scale production of >4 million tons of glycerol, 2.4 million tons of sorbitol, and ~200,000 tons of xylitol annually [115].

3.7.3 Furan derivatives

Furan derivatives, particularly HMF and furfural, are key platform chemicals from the carbohydrate fractions of lignocellulose [116]. HMF is obtained via dehydration of hexoses (e.g., glucose, fructose), while furfural originates from pentoses (e.g., xylose in hemicellulose). HMF oxidation produces 2,5-furandicarboxylic acid (FDCA), a monomer for polyethylene furanoate (PEF), a renewable alternative to PET [117]. Partial oxidation yields intermediates such as 5-formyl-2-furancarboxylic acid, useful for fine chemicals. HMF reduction produces biofuel candidates such as 2,5-dimethylfuran (DMF) and 2-methylfuran, and other fine chemicals [118]. Furfural oxidation yields furoic acid, a precursor for resins, polymers, and pharmaceuticals. Furfural reduction generates furfuryl alcohol, widely used in thermosetting resins and adhesives [119]. Their multifunctionality, aldehyde, hydroxyl, and carboxyl groups make furanics particularly suited to tunable electrochemical valorization, supporting applications in polymers, fuels, solvents, and pharmaceuticals.

3.7.4 Aromatic compounds from lignin depolymerization

Electrochemical valorization provides a sustainable route for producing aromatic compounds from lignin depolymerization [120]. Oxidative cleavage of β–O–4 linkages yields valuable aromatics such as vanillin, syringaldehyde, guaiacol, ferulic acid, and p-hydroxybenzoic acid, useful for flavorings and pharmaceuticals [121]. Reductive strategies convert carbonyl or ether groups into hydrogenated aromatics or phenolic alcohols, useful for polymers, resins, and fine chemicals [122].

3.7.5 Hydrogen and syngas

Electrochemical conversion of lignocellulosic biomass also yields hydrogen (H_2) and syngas ($CO+H_2$), essential as fuels and industrial feedstocks [13,123]. Biomass intermediates can act as electron donors, reducing the overpotential for hydrogen evolution compared with conventional water electrolysis. Hydrogen serves as a clean fuel for fuel cells, as well as a feedstock for ammonia, methanol, and reduction processes. Syngas is generated through partial oxidation or reforming of biomass intermediates. Its composition and H_2/CO ratio can be tuned via reaction conditions, electrode properties, and electrolyte composition. Syngas provides a flexible platform for downstream conversion into fuels and chemicals, such as via Fischer–Tropsch synthesis or methanol production, facilitating integration with bio-based energy systems [124].

3.7.6 Carbon-based functional materials

Electrochemical biomass conversion can also produce carbon-based functional materials, including biochar, activated carbon, carbon nanostructures, and graphitic carbons [125]. These materials exhibit high surface area, tunable porosity, abundant surface functionalities, and excellent conductivity, making them valuable in catalysis, adsorption, energy storage, environmental remediation, and sensing. Anodic/cathodic processes introduce oxygen-containing functional groups (e.g., hydroxyl, carbonyl, carboxyl), enhancing adsorption capacity, wettability, and catalytic activity. Activated carbons are widely applied in CO_2 capture, water purification, and energy storage devices (e.g., batteries, supercapacitors). Carbon nanostructures graphene-like sheets, CNTs, and nanosheets, exhibit superior conductivity and mechanical stability, with applications in electrocatalysis, energy storage, and electronics [126]. Carbon materials also act as catalyst supports, improving the dispersion and durability of metal or metal oxide catalysts in biomass upgrading.

3.8 Challenges and future perspectives

The electrochemical valorization of lignocellulosic biomass presents significant opportunities for sustainable bio-based chemical production. However, several critical challenges must be addressed to enable large-scale, efficient, economically viable, and scalable processes [127]. A primary limitation arises from the intrinsic structural complexity and compositional diversity of biomass, which comprises cellulose, hemicellulose, lignin, and minor components such as ash, extractives, and minerals. This structural diversity, along with impurities, can decrease reproducibility, deactivate electrodes, and lower product yields. Moreover, variations in feedstock composition across sources or seasons further complicate process control [128]. Catalyst and electrode stability remain significant obstacles, as many metals, metal oxides, and carbon-based electrodes are prone to fouling, leaching, or structural degradation during long-term use, especially when exposed to lignin-rich and other complex substrates [129]. Most studies are limited to laboratory-scale setups, and efforts are needed to scale reactors while maintaining high efficiency, selectivity, and energy performance. Achieving high selectivity toward target platform chemicals is also challenging due to the presence of multiple reactive functional groups in biomass that can lead to competing pathways, overoxidation, undesired side reactions, and hydrogen evolution, all of which complicate downstream separation and purification [43]. Additionally, mass and electron transfer limitations hinder performance. Poor solubility of biomass fractions and high viscosity of polyols can reduce conversion efficiency, while scaling up to higher current densities introduces further transport barriers and unstable reaction behavior [130]. Finally, economic and energy considerations, including the costs of catalysts, electrodes, electrolytes, product separation, and reliance on renewable electricity, also remain critical constraints for large-scale applications [131].

Despite these challenges, electrochemical biomass conversion offers numerous future opportunities. Advances in catalyst development and stability are central, with strategies such as heteroatom doping, vacancy engineering, SACs, and nanostructuring being explored to enhance activity, selectivity, and long-term durability. A wide range of materials, including alloys, oxides, (oxy)hydroxides, chalcogenides, and pnictides, has been studied for electrooxidation and electroreduction of lignocellulosic platform molecules for specific target chemicals [132]. Vacancy engineering has shown particular promise by tuning electronic structures, reducing charge transfer resistance, and enabling precise structure–function relationships [77]. Reactor and electrode innovations, including high-surface-area materials, flow-cell designs, biphasic or membrane-separated systems, and continuous-flow operation, can improve mass transport, reduce fouling, and enable in situ product separation. Hybrid electrochemical–biological–photochemical systems expand the diversity of achievable products under milder, more selective conditions [133]. Paired electrolysis, which couples anodic biomass oxidation with value-added cathodic reactions, offers enhanced atom economy and energy efficiency [134]. Digital tools such as process modeling, operando characterization, and machine learning can optimize reaction conditions, elucidate mechanistic pathways, and facilitate scale-up [135,136]. At the systems level, integrating electrochemical biorefineries with renewable energy sources, supported by appropriate policy frameworks and infrastructure, can reduce carbon intensity and improve economic feasibility. Furthermore, the development of sustainable downstream processing methods and comprehensive life cycle and techno-economic analyzes is crucial for evaluating environmental and economic performance and guiding industrial implementation. Addressing these challenges will be pivotal in realizing the full potential of electrochemical strategies for converting lignocellulosic biomass into biofuels and high-value chemicals.

3.9 Conclusion

The rising global energy demand, combined with environmental concerns and depletion of fossil resources, has increased interest in efficient strategies for biomass valorization. Within this context, electrochemical conversion offers a sustainable and versatile platform for lignocellulosic biomass processing, enabling selective production of fuels, organic acids, and platform chemicals. Unlike conventional approaches, the electrochemical method provides precise control over reaction conditions, selectivity, and product distribution, often under mild and environmentally benign operating environments. By directly coupling biomass valorization with renewable electricity, these methods establish a pathway toward low-carbon chemical manufacturing while reducing dependence on harsh reagents and energy-intensive processes. Recent advances in electrocatalyst design, electrode materials, and reaction engineering have significantly improved efficiency, selectivity, and process stability, underscoring the promise of electrooxidation and electroreduction as central strategies for biomass conversion. Nevertheless, several challenges remain, including catalyst degradation, competing side reactions, by-product formation, limitations in product separation, and the scalability of laboratory findings to industrial implementation. Addressing these issues will require continued progress in integrating electrochemical systems with complementary catalytic and biocatalytic processes, developing durable and cost-effective materials, and optimizing operating conditions to enhance faradaic efficiency. With sustained innovation across materials, process design, and system integration, electrochemical biomass conversion has the potential to play a pivotal role in advancing low-carbon circular bioeconomy frameworks, contributing to both sustainable energy systems and environmentally responsible chemical production.

References

[1] K.K. Jaiswal, C.R. Chowdhury, D. Yadav, R. Verma, S. Dutta, K.S. Jaiswal, et al., Renewable and sustainable clean energy development and impact on social, economic, and environmental health, Energy Nexus. 7 (2022) 100118, doi:10.1016/j.nexus.2022.100118.

[2] M. Mujtaba, L. Fernandes Fraceto, M. Fazeli, S. Mukherjee, S.M. Savassa, G. Araujo de Medeiros, et al., Lignocellulosic biomass from agricultural waste to the circular economy: a review with focus on biofuels, biocomposites and bioplastics, J. Clean. Prod. 402 (2023) 136815, doi:10.1016/j.jclepro.2023.136815.

[3] P. Chowdhury, N.A. Mahi, R. Yeassin, N.-U.-R. Chowdhury, O. Farrok, Biomass to biofuel: impacts and mitigation of environmental, health, and socioeconomic challenges, Energy. Convers. Manag. X. 25 (2025) 100889, doi:10.1016/j.ecmx.2025.100889.

[4] A.P. Ingle, S. Saxena, M.P. Moharil, J.D. Rivaldi, L. Ramos, A.K. Chandel, Production of biomaterials and biochemicals from lignocellulosic biomass through sustainable approaches: current scenario and future perspectives, Biotechnol. Sustain. Mater. 2 (2025) 3, doi:10.1186/s44316-025-00025-2.

[5] J. Cai, Y. He, X. Yu, S.W. Banks, Y. Yang, X. Zhang, et al., Review of physicochemical properties and analytical characterization of lignocellulosic biomass, Renew Sustain. Energy Rev. 76 (2017) 309–322, doi:10.1016/j.rser.2017.03.072.

[6] H. Seddiqi, E. Oliaei, H. Honarkar, J. Jin, L.C. Geonzon, R.G. Bacabac, et al., Cellulose and its derivatives: towards biomedical applications, Cellulose. 28 (2021) 1893–1931, doi:10.1007/s10570-020-03674-w.

[7] R. Mathura S, C. Landázuri A, F. Mathura, A.G.A. Sosa, M. Orejuela-Escobar L, Hemicelluloses from bioresidues and their applications in the food industry—towards an advanced bioeconomy and a sustainable global value chain of chemicals and materials, Sustain. Food Technol. 2 (2024) 1183–1205, doi:10.1039/D4FB00035H.

[8] D.S. Bajwa, G. Pourhashem, A.H. Ullah, S.G. Bajwa, A concise review of current lignin production, applications, products and their environmental impact, Ind. Crops Prod. 139 (2019) 111526, doi:10.1016/j.indcrop.2019.111526.

[9] F.H. Isikgor, C.R. Becer, Lignocellulosic biomass: a sustainable platform for the production of bio-based chemicals and polymers, Polym. Chem. 6 (2015) 4497–4559, doi:10.1039/C5PY00263J.

[10] J.J. Bozell, G.R. Petersen, Technology development for the production of biobased products from biorefinery carbohydrates—the US Department of Energy's "Top 10" revisited, Green Chem. 12 (2010) 539–554, doi:10.1039/B922014C.

[11] M. Rojas, R. Manrique, U. Hornung, A. Funke, C.A. Mullen, F. Chejne, et al., Advances and challenges on hydrothermal processes for biomass conversion: feedstock flexibility, products, and modeling approaches, Biomass Bioenergy. 194 (2025) 107621, doi:10.1016/j.biombioe.2025.107621.

[12] Z. Wu, K. Peng, Y. Zhang, M. Wang, C. Yong, L. Chen, et al., Lignocellulose dissociation with biological pretreatment towards the biochemical platform: a review, Mater. Today. Bio. 16 (2022) 100445, doi:10.1016/j.mtbio.2022.100445.

[13] Y. Ramli, V. Chaerusani, Z. Yang, R. Yang, J. Zhang, A. Abudula, et al., Electrochemical conversion of biomass derivatives to value-added chemicals: a review, Green. Carbon. 3 (2025) 113–129, doi:10.1016/j.greenca.2024.10.004.

[14] P. Zhu, M. Shi, Z. Shen, X. Liao, Y. Chen, Electrocatalytic conversion of biomass-derived furan compounds: mechanisms, catalysts and perspectives, Chem. Sci. 15 (2024) 4723–4756, doi:10.1039/D4SC00546E.

[15] X. Liu, Y. Wang, H. Duan, Recent progress in electrocatalytic conversion of lignin: from monomers, dimers, to raw lignin, Precis. Chem. 2 (2024) 428–446, doi:10.1021/prechem.4c00024.

[16] B. Segers, P. Nimmegeers, M. Spiller, G. Tofani, E. Jasiukaitytė-Grojzdek, E. Dace, et al. Lignocellulosic biomass valorisation: a review of feedstocks, processes and potential value chains and their implications for the decision-making process 2024. doi: 10.1039/D4SU00342J.

[17] D.B. Sulis, N. Lavoine, H. Sederoff, X. Jiang, B.M. Marques, K. Lan, et al., Advances in lignocellulosic feedstocks for bioenergy and bioproducts, Nat. Commun. 16 (2025) 1244, doi:10.1038/s41467-025-56472-y.

[18] S. Chaudhari, H. Pawar, K. Pawar, P. More, Recent advances in the measurement of thermochemical, electrochemical and photochemical efficiency of CO_2 conversion to value-added products–its challenges and future directions, Meas Energy. 6 (2025) 100049, doi:10.1016/j.meaene.2025.100049.

[19] H. Durak, Comparative analysis of electrochemical and thermochemical hydrogenation of biomass-derived phenolics for sustainable biofuel and chemical production, Processes. 13 (2025) 1581, doi:10.3390/pr13051581.

[20] S. Poornima, S. Manikandan, R. Prakash, S.R. Deena, R. Subbaiya, N. Karmegam, et al., Biofuel and biochemical production through biomass transformation using advanced thermochemical and biochemical processes—a review, Fuel. 372 (2024) 132204, doi:10.1016/j.fuel.2024.132204.

[21] J. Chen, X. Ma, M. Liang, Z. Guo, Y. Cai, C. Zhu, et al., Physical–chemical–biological pretreatment for biomass degradation and industrial applications: a review, Waste. 2 (2024) 451–473, doi:10.3390/waste2040024.

[22] D.B. Sant'Ana Júnior, M. Kelbert, P.H. Hermes de Araújo, C.J. de Andrade, Physical pretreatments of lignocellulosic biomass for fermentable sugar production, Sustain Chem. 6 (2025) 13, doi:10.3390/suschem6020013.

[23] A. Fernandes, L. Cruz-Lopes, B. Esteves, D.V. Evtuguin, Microwaves and ultrasound as emerging techniques for lignocellulosic materials, Materials. 16 (2023) 7351, doi:10.3390/ma16237351.

[24] A.I. Gavrila, A. Vartolomei, I. Calinescu, M. Vinatoru, O.C. Parvulescu, G. Psenovschi, et al., Ultrasound-assisted alkaline pretreatment of biomass to enhance the extraction yield of valuable chemicals, Agronomy. 14 (2024) 903, doi:10.3390/agronomy14050903.

[25] C. Arce, L. Kratky, Mechanical pretreatment of lignocellulosic biomass toward enzymatic/fermentative valorization, iScience. 25 (2022) 104610, doi:10.1016/j.isci.2022.104610.

[26] N.C. Joshi, S. Sinha, P. Bhatnagar, Y. Nath, B. Negi, V. Kumar, et al., A concise review on waste biomass valorization through thermochemical conversion, Curr. Res. Microb. Sci. 6 (2024) 100237, doi:10.1016/j.crmicr.2024.100237.

[27] Y. Wang, J.J. Wu, Thermochemical conversion of biomass: potential future prospects, Renew. Sustain. Energy Rev. 187 (2023) 113754, doi:10.1016/j.rser.2023.113754.

[28] L. Traven, Sustainable energy generation from municipal solid waste: A brief overview of existing technologies, Case Stud Chem Environ Eng (2023), doi:10.1016/j.cscee.2023.100491.

[29] M. Pahnila, A. Koskela, P. Sulasalmi, T. Fabritius, A Review of pyrolysis technologies and the effect of process parameters on biocarbon properties, Energies. 16 (2023) 6936, doi:10.3390/en16196936.

[30] M. Ivanovski, A. Petrovič, D. Goričanec, D. Urbancl, M. Simonič, Exploring the properties of the torrefaction process and its prospective in treating lignocellulosic material, Energies 16 (2023) 6521, doi:10.3390/en16186521.

[31] A. Mathanker, S. Das, D. Pudasainee, M. Khan, A. Kumar, R. Gupta, A review of hydrothermal liquefaction of biomass for biofuels production with a special focus on the effect of process parameters, co-solvents, and extraction solvents, Energies 14 (2021) 4916, doi:10.3390/en14164916.

[32] R. Muzyka, H. Mumtaz, S. Sobek, S. Werle, J. Adamek, D. Semitekolos, et al., Solvolysis and oxidative liquefaction of the end-of-life composite wastes as an element of the circular economy assumptions, J. Clean. Prod 478 (2024) 143916, doi:10.1016/j.jclepro.2024.143916.

[33] S. Mucci, A. Mitsos, D. Bongartz, Combustion versus gasification in power- and biomass-to-X processes: an exergetic analysis, ACS Omega 9 (2024) 48213–48231, doi:10.1021/acsomega.4c05549.

[34] H. Wang, M. Tucker, Y. Ji, Recent development in chemical depolymerization of lignin: a review, J. Appl. Chem. 2013 (2013) 838645, doi:10.1155/2013/838645.

[35] R.C. Kuhad, A. Rapoport, V. Kumar, D. Singh, V. Kumar, S.K. Tiwari, et al., Biological pretreatment of lignocellulosic biomass: an environment-benign and sustainable approach for conversion of solid waste into value-added products, Crit. Rev. Environ. Sci. Technol. 54 (2024) 771–796, doi:10.1080/10643389.2023.2277670.

[36] S. Nanda, F. Pattnaik, B.R. Patra, K. Kang, A.K. Dalai, A review of liquid and gaseous biofuels from advanced microbial fermentation processes, Fermentation 9 (2023) 813, doi:10.3390/fermentation9090813.

[37] C. González, Y. Wu, A. Zuleta-Correa, G. Jaramillo, J. Vasco-Correa, Biomass to value-added products using microbial consortia with white-rot fungi, Bioresour. Technol. Rep. 16 (2021) 100831, doi:10.1016/j.biteb.2021.100831.

[38] A. Gutierrez-Blanco, C. Mejuto, Understanding biomass valorization through electrocatalysis: transformation of glycerol and furan derivatives, J. Phys. Chem. Lett. 16 (2025) 2785–2792, doi:10.1021/acs.jpclett.4c03553.

[39] M.P.J.M. van der Ham, J. Creus, J.H. Bitter, M.T.M. Koper, P.P. Pescarmona, Electrochemical and non-electrochemical pathways in the electrocatalytic oxidation of monosaccharides and related sugar alcohols into valuable products, Chem. Rev. 124 (2024) 11915–11961, doi:10.1021/acs.chemrev.4c00261.

[40] D.R. Kauffman, J. Thakkar, R. Siva, C. Matranga, P.R. Ohodnicki, C. Zeng, et al., Efficient electrochemical CO_2 conversion powered by renewable energy, ACS Appl. Mater. Interfaces 7 (2015) 15626–15632, doi:10.1021/acsami.5b04393.

[41] P. Prabhu, Y. Wan, J.-M. Lee, Electrochemical Conversion of biomass derived products into high-value chemicals, Matter. 3 (2020) 1162–1177, doi:10.1016/j.matt.2020.09.002.

[42] M. Yoo, D. Choi, M. Shin, S. Jung, J. Lee, H.J. Kim, Selective electrocatalytic oxidation of glycerol to high-value C1–C3 products: from chemistry to scalability, Chem. Eng. J. 518 (2025) 164743, doi:10.1016/j.cej.2025.164743.

[43] H. Luo, J. Barrio, N. Sunny, A. Li, L. Steier, N. Shah, et al., Progress and perspectives in photo- and electrochemical-oxidation of biomass for sustainable chemicals and hydrogen production, Adv. Energy Mater. 11 (2021) 2101180, doi:10.1002/aenm.202101180.

[44] J. Chen, S. Wu, X. Pan, X. Zhou, X. Zhang, Electrochemical reduction hydrogenation, hydrogenolysis and dimerization of bio-derived aldehydes: a review, Renew Sustain. Energy Rev. 207 (2025) 114900, doi:10.1016/j.rser.2024.114900.

[45] G. Zhao, W. Sun, Electrochemical hydrogen evolution reaction, Photo- and Electro-Catalytic Processes: Water Splitting, N2 Fixing, CO_2 Reduction, John Wiley & Sons, Ltd, 2022, pp. 87–122, doi:10.1002/9783527830084.ch3.

[46] M. Zheng, J. Zhang, P. Wang, H. Jin, Y. Zheng, S.-Z. Qiao, Recent advances in electrocatalytic hydrogenation reactions on copper-based catalysts, Adv. Mater. 36 (2024) 2307913, doi:10.1002/adma.202307913.

[47] A. MariaJoseph, S. Nangan, D. Verma, L. Gnanasekaran, S. Rajendran, T. Natesan, et al., Rational coupling of selective electrochemical oxidation and reduction reactions for in-situ value-added chemical generation, Fuel 367 (2024) 131408, doi:10.1016/j.fuel.2024.131408.

[48] M. Zhang, Z. Zheng, X. Zhang, Z. Jiang, X. Yong, K. Li, et al., Pairing electrocatalytic reduction and oxidation of biomass-derived 5-hydroxymethylfurfural into highly value-added chemicals, JACS Au. 5 (2025) 937–947, doi:10.1021/jacsau.4c01135.

[49] J. Jin, L. Sun, Z. Wang, S. Li, L. Shen, H. Yin, Enhanced electrocatalytic performance for selective glycerol oxidation to formic acid at a multiphase AuCu-Ag/AgBr interface, Catalysts. 15 (2025) 831, doi:10.3390/catal15090831.

[50] S. Cheng, H. Zhong, F. Jin, A mini review of electrocatalytic upgrading of carbohydrate biomass—system, path, and optimization, Energy Sci. Eng. 11 (2023) 2944–2965, doi:10.1002/ese3.1487.

[51] D. Gao, D. Ouyang, X. Zhao, Controllable oxidative depolymerization of lignin to produce aromatic aldehydes and generate electricity under mild conditions with direct biomass fuel cells as flexible reactors, Chem. Eng. J. 479 (2024) 147874, doi:10.1016/j.cej.2023.147874.

[52] A. Hermawan, V.N. Alviani, Wibisono, Z.W. Seh, Fundamentals, rational catalyst design, and remaining challenges in electrochemical NOx reduction reaction, iScience 26 (2023) 107410, doi:10.1016/j.isci.2023.107410.

[53] M.-K. Zhang, J. Cai, Y.-X. Chen, On the electrode charge at the metal/solution interface with specific adsorption, Curr. Opin. Electrochem. 36 (2022) 101161, doi:10.1016/j.coelec.2022.101161.

[54] J. Zhang, S. Xia, Y. Wang, J. Wu, Y. Wu, Recent advances in dynamic reconstruction of electrocatalysts for carbon dioxide reduction, iScience. 27 (2024) 110005, doi:10.1016/j.isci.2024.110005.

[55] G.K. Gebremariam, K. Siraj, I.A. Pašti, Tailoring electrocatalytic pathways: a comparative review of the electrolyte's effects on five key energy conversion reactions, Catalysts. 15 (2025) 835, doi:10.3390/catal15090835.

[56] U.R. Dotel, M.O. Sydnes, H. Urkedal, T. Hemmingsen, The effects of acidic, alkaline, and neutral anolytes on electrochemical seawater deoxygenation, Appl. Sci. 8 (2018) 2280, doi:10.3390/app8112280.

[57] G. Marcandalli, A. Goyal, M.T.M. Koper, Electrolyte effects on the faradaic efficiency of CO_2 reduction to CO on a gold electrode, ACS Catal. 11 (2021) 4936–4945, doi:10.1021/acscatal.1c00272.

[58] J.C. Bui, C. Kim, A.J. King, O. Romiluyi, A. Kusoglu, A.Z. Weber, et al., Engineering catalyst–electrolyte microenvironments to optimize the activity and selectivity for the

electrochemical reduction of CO_2 on Cu and Ag, Acc. Chem. Res. 55 (2022) 484–494, doi:10.1021/acs.accounts.1c00650.

[59] B. Markunas, T. Yim, J. Snyder, pH-mediated solution-phase proton transfer drives enhanced electrochemical hydrogenation of phenol in alkaline electrolyte, ACS Catal. 14 (2024) 16936–16946, doi:10.1021/acscatal.4c04874.

[60] M. König, J. Vaes, E. Klemm, D. Pant, Solvents and supporting electrolytes in the electrocatalytic reduction of CO_2, iScience 19 (2019) 135–160, doi:10.1016/j.isci.2019.07.014.

[61] B. Gao, C. Sun, T. Yang, Q. Wen, S. You, Q. Yang, et al., Biphasic solvent systems enabled lignocellulosic biomass fractionation: a pathway towards comprehensive biomass utilization, Ind. Crops Prod. 202 (2023) 117036, doi:10.1016/j.indcrop.2023.117036.

[62] Y. Bao, Y. Wang, C. Yan, Z. Xue, Deep eutectic solvents for fractionation and valorization of lignocellulose, Green Chem. Eng. 6 (2025) 21–35, doi:10.1016/j.gce.2024.04.001.

[63] V. Purohit, A. Datar, Overview of topics in electrocatalysis for sustainability: reactions, electrocatalysts, degradation, and mitigation, Processes 13 (2025) 2659, doi:10.3390/pr13082659.

[64] S. Zong, A. Chen, M. Wiśniewski, L. Macheli, L.L. Jewell, D. Hildebrandt, et al., Effect of temperature and pressure on electrochemical CO_2 reduction: a mini review, Carbon Capture Sci. Technol. 8 (2023) 100133, doi:10.1016/j.ccst.2023.100133.

[65] X. Tian, G. Xu, S. Chen, Q. Li, J. Duan, Progress and prospects of high-pressure CO_2 electrocatalytic reduction, J. Mater. Chem. A 13 (2025) 22295–22323, doi:10.1039/D5TA03027G.

[66] Y. Yao, J. Lyu, X. Li, C. Chen, F. Verpoort, J. Wang, et al., A review of efficient electrocatalysts for the oxygen evolution reaction at large current density, DeCarbon 5 (2024) 100062, doi:10.1016/j.decarb.2024.100062.

[67] S. Liang, N. Altaf, L. Huang, Y. Gao, Q. Wang, Electrolytic cell design for electrochemical CO_2 reduction, J. CO_2 Util. 35 (2020) 90–105, doi:10.1016/j.jcou.2019.09.007.

[68] A. Fathima, I.M.S.K. Ilankoon, Y. Zhang, M.N. Chong, Scaling up of dual-chamber microbial electrochemical systems—an appraisal using systems design approach, Sci. Total Environ. 912 (2024) 169186, doi:10.1016/j.scitotenv.2023.169186.

[69] M.F.A. Magalhães, G.B. Simoso, E.S. de Borba, T.J. Brocksom, G.M. Martins, K.T.de Oliveira, Electrosynthesis reactions with divided cells: unlocking potentials in organic synthesis, Chem. Synth. 5 (2025) N/A-N/A, doi:10.20517/cs.2024.177.

[70] T.R. Garrick, J. Gao, X. Yang, B.J. Koch, Modeling electrochemical transport within a three-electrode system, J. Electrochem. Soc. 168 (2021) 010530, doi:10.1149/1945-7111/abdafb.

[71] T. Yang, Y. Ding, C. Li, N. Yin, X. Liu, P. Li, Potentiostatic and galvanostatic two-step electrodeposition of semiconductor Cu_2O films and its photovoltaic application, J. Alloys Compd. 727 (2017) 14–19, doi:10.1016/j.jallcom.2017.08.123.

[72] M. Rafiee, D.J. Abrams, L. Cardinale, Z. Goss, A. Romero-Arenas, S.S. Stahl, Cyclic voltammetry and chronoamperometry: mechanistic tools for organic electrosynthesis, Chem. Soc. Rev. 53 (2024) 566–585, doi:10.1039/d2cs00706a.

[73] N. Raeisi-Kheirabadi, A. Nezamzadeh-Ejhieh, H. Aghaei, Cyclic and linear sweep voltammetric studies of a modified carbon paste electrode with nickel oxide nanoparticles toward tamoxifen: effects of surface modification on electrode response kinetics, ACS Omega 7 (2022) 31413–31423, doi:10.1021/acsomega.2c03441.

[74] T.-W. Chen, S.-M. Chen, G. Anushya, R. Kannan, G. Al-Sehemi A, S. Alargarsamy, et al., Development of different kinds of electrocatalyst for the electrochemical reduction of carbon dioxide reactions: an overview, Molecules. 28 (2023) 7016, doi:10.3390/molecules28207016.

[75] Y. Xu, S. Yu, H.M. Johnson, Y. Wu, X. Liu, B. Fang, et al., Recent progress in electrode materials for micro-supercapacitors, iScience. 27 (2024) 108786, doi:10.1016/j.isci.2024.108786.

[76] S.B. Beil, D. Pollok, S.R. Waldvogel, Reproducibility in electroorganic synthesis—myths and misunderstandings, Angew. Chem. Int. Ed. 60 (2021) 14750–14759, doi:10.1002/anie.202014544.
[77] D. Bhalothia, A. Beniwal, H. Gurjar, K. Shekhawat, A. Bagaria, T.-Y. Chen. Surface and interface engineering of noble metal heterostructures for superior ORR performance: unlocking ultralow loading and maximum catalyst utilization. Small. e06018. doi:10.1002/smll.202506018.
[78] R. Ge, J. Li, H. Duan, Recent advances in non-noble electrocatalysts for oxidative valorization of biomass derivatives, Sci. China. Mater. 65 (2022) 3273–3301, doi:10.1007/s40843-022-2076-y.
[79] A. Vazhayil, L. Vazhayal, J. Thomas, C.S. Ashok, N. Thomas, A comprehensive review on the recent developments in transition metal-based electrocatalysts for oxygen evolution reaction, Appl. Surf. Sci. Adv. 6 (2021) 100184, doi:10.1016/j.apsadv.2021.100184.
[80] D. Kim, C.S. Kley, Y. Li, P. Yang, Copper nanoparticle ensembles for selective electroreduction of CO_2 to C2–C3 products, Proc. Natl. Acad. Sci. USA. 114 (2017) 10560–10565, doi:10.1073/pnas.1711493114.
[81] C. Especel, G. Lafaye, F. Epron, Bimetallic catalysts for sustainable chemistry: surface redox reactions for tuning the catalytic surface composition, ChemCatChem 15 (2023) e202201478, doi:10.1002/cctc.202201478.
[82] G. Lee, S.E. Jun, Y. Kim, I.-H. Park, H.W. Jang, S.H. Park, et al., Multicomponent metal oxide- and metal hydroxide-based electrocatalysts for alkaline water splitting, Materials. 16 (2023) 3280, doi:10.3390/ma16083280.
[83] A.S. Jamadar, R. Sutar, S. Patil, R. Khandekar, J.B. Yadav, Progress in metal oxide-based electrocatalysts for sustainable water splitting, Mater. Rep. Energy. 4 (2024) 100283, doi:10.1016/j.matre.2024.100283.
[84] L. Chen, X. Dong, Y. Wang, Y. Xia, Separating hydrogen and oxygen evolution in alkaline water electrolysis using nickel hydroxide, Nat. Commun. 7 (2016) 11741, doi:10.1038/ncomms11741.
[85] A.K. Mohammed, J.N. Hausmann, S. Gaber, P.P. Sánchez, F. Gándara, K. Laun, et al., In-situ transformation of a conjugated nickel-organic framework into active nickel oxyhydroxide for electrocatalytic 5-hydroxymethylfurfural oxidation, J. Colloid Interface Sci. 694 (2025) 137630, doi:10.1016/j.jcis.2025.137630.
[86] M. Nozari-Asbemarz, M. Amiri, H. Imanzadeh, A. Bezaatpour, S. Nouhi, P. Hosseini, et al., Mixed metal oxides as efficient electrocatalysts for water oxidation, Int. J. Hydrog. Energy. 47 (2022) 5250–5259, doi:10.1016/j.ijhydene.2021.11.145.
[87] S.M. Mousavi, M.Y. Kalashgrani, M.-K. Tsai, W.-H. Chiang, Carbon-based materials derived from green and sustainable chemistry: current perspectives for electrocatalysis and energy applications, Renew. Sustain. Energy Rev. 222 (2025) 116001, doi:10.1016/j.rser.2025.116001.
[88] W. Li, Y. Xu, G. Wang, T. Xu, K. Wang, S. Zhai, et al., Sustainable carbon-based catalyst materials derived from lignocellulosic biomass for energy storage and conversion: atomic modulation and properties improvement, Carbon Energy. 7 (2025) e708, doi:10.1002/cey2.708.
[89] S. Vallem, M. Gopalan Sibi, R. Patil, V. Goyal, A.G. Babu, E. Lohith, et al., Covalent organic and metal organic frameworks based single atom catalysts for valorisation of CO_2 to value added chemicals, Energy Environ. Sci. (2025), doi:10.1039/D5EE02702K.
[90] J. Li, C. Chen, L. Xu, Y. Zhang, W. Wei, E. Zhao, et al., Challenges and perspectives of single-atom-based catalysts for electrochemical reactions, JACS Au. 3 (2023) 736–755, doi:10.1021/jacsau.3c00001.
[91] D. Li, A. Yadav, H. Zhou, K. Roy, P. Thanasekaran, C. Lee, Advances and applications of metal-organic frameworks (MOFs) in emerging technologies: a comprehensive review, Glob. Chall. 8 (2024) 2300244, doi:10.1002/gch2.202300244.

[92] D. Zamel, A.U. Khan, Covalent organic frameworks: a green approach to environmental challenges, Mater. Today Sustain. 30 (2025) 101096, doi:10.1016/j.mtsust.2025.101096.

[93] Y. Qi, H. Guo, J. Li, L. Ma, Y. Xu, H. Liu, et al., Recent advances in the electrocatalytic oxidative upgrading of lignocellulosic biomass, ChemPhysMater. 3 (2024) 157–186, doi:10.1016/j.chphma.2024.02.001.

[94] Y. Sun, J. Miao, X. Fan, K. Zhang, T. Zhang, Recent progress in electrochemical conversion from biomass derivatives into high-value-added chemicals, Small Struct. 5 (2024) 2300576, doi:10.1002/sstr.202300576.

[95] X. Tan, J. Nielsen, The integration of bio-catalysis and electrocatalysis to produce fuels and chemicals from carbon dioxide, Chem. Soc. Rev. 51 (2022) 4763–4785, doi:10.1039/D2CS00309K.

[96] T. Angelin Swetha, K. Mohanrasu, M. Sudhakar, R. Raja, K. Ponnuchamy, G. Muthusamy, et al., A comprehensive review on techniques used in conversion of biomass into bioeconomy, Sustain. Energy Technol. Assess. 53 (2022) 102682, doi:10.1016/j.seta.2022.102682.

[97] J. Ye, A. Hu, G. Ren, M. Chen, S. Zhou, Z. He, Biophotoelectrochemistry for renewable energy and environmental applications, iScience 24 (2021) 102828, doi:10.1016/j.isci.2021.102828.

[98] S.G. Nnabuife, A.K. Hamzat, J. Whidborne, B. Kuang, K.W. Jenkins, Integration of renewable energy sources in tandem with electrolysis: a technology review for green hydrogen production, Int. J. Hydrog. Energy. 107 (2025) 218–240, doi:10.1016/j.ijhydene.2024.06.342.

[99] R. Boddula, Y.-Y. Lee, S. Masimukku, G.-P. Chang-Chien, R. Pothu, R.K. Srivastava, et al., Sustainable hydrogen production: solar-powered biomass conversion explored through (Photo)electrochemical advancements, Process Saf. Environ. Prot. 186 (2024) 1149–1168, doi:10.1016/j.psep.2024.04.068.

[100] S.O. Ganiyu, C.A. Martínez-Huitle, The use of renewable energies driving electrochemical technologies for environmental applications, Curr. Opin. Electrochem. 22 (2020) 211–220, doi:10.1016/j.coelec.2020.07.007.

[101] J.A. Ferraz de Andrade Santos, P. de Jong, C. Alves da Costa, E.A. Torres, Combining wind and solar energy sources: potential for hybrid power generation in Brazil, Util. Policy. 67 (2020) 101084, doi:10.1016/j.jup.2020.101084.

[102] Z. Xi, M. Liu, Advancing photoelectrochemical systems for sustainable energy and chemical production: challenges and opportunities, Npj Mater. Sustain. 3 (2025) 18, doi:10.1038/s44296-025-00061-z.

[103] P. Jalili, A. Ala, P. Nazari, B. Jalili, D.D. Ganji, A comprehensive review of microbial fuel cells considering materials, methods, structures, and microorganisms, Heliyon 10 (2024) e25439, doi:10.1016/j.heliyon.2024.e25439.

[104] O.L. Adepitan, O.O. Alabi, O.J. Gbadeyan, N. Deenadayalu. Bioelectrochemical systems for waste treatment and valorization: a review of applications and benefits. Biofuels Bioprod. Biorefining. doi: 10.1002/bbb.70035.

[105] M.dK. Islam, J. Ratthiwal, S. Kongparakul, C. Samart, Advances in oxidative fractionation and depolymerization for sustainable lignin valorization, Mol. Catal. 586 (2025) 115436, doi:10.1016/j.mcat.2025.115436.

[106] L.T. Mika, E. Cséfalvay, Á. Németh, Catalytic conversion of carbohydrates to initial platform chemicals: chemistry and sustainability, Chem. Rev. 118 (2018) 505–613, doi:10.1021/acs.chemrev.7b00395.

[107] M. Li, T. Wang, X. Chen, X. Ma, Conversion study from lignocellulosic biomass and electric energy to H2 and chemicals, Int. J. Hydrog. Energy. 48 (2023) 21004–21017, doi:10.1016/j.ijhydene.2022.09.191.

[108] F.H. Isikgor, C.R. Becer, Lignocellulosic biomass: a sustainable platform for the production of bio-based chemicals and polymers, Polym. Chem. 6 (2015) 4497–4559, doi:10.1039/C5PY00263J.

[109] W. Yan, Q. Guan, F. Jin, Catalytic conversion of cellulosic biomass to harvest high-valued organic acids, iScience. 26 (2023) 107933, doi:10.1016/j.isci.2023.107933.

[110] I. Chauhan, P.M. Vijay, R. Ranjan, K.K. Patra, C.S. Gopinath, Electrocatalytic and selective oxidation of glycerol to formate on 2D 3d-metal phosphate nanosheets and carbon-negative hydrogen generation, ACS Mater. Au. 4 (2024) 500–511, doi:10.1021/acsmaterialsau.4c00024.

[111] Y. Zhong, J. Gu, C. Shang, J. Deng, Y. Liu, Z. Cui, et al., Sustainable succinic acid production from lignocellulosic hydrolysates by engineered strains of *Yarrowia lipolytica* at low pH, Bioresour. Technol. 408 (2024) 131166, doi:10.1016/j.biortech.2024.131166.

[112] H.D. Manamperi, M.A. Jackson, E.C. Wegener, K.E. Vermillion, Selective oxidation of ethanol on a nickel foam electrode followed by aldol condensation with furfural: an electrochemical approach for the utilization of biomass-derived molecules, ACS Electrochem. 1 (2025) 369–377, doi:10.1021/acselectrochem.4c00094.

[113] H. Chi, Z. Liang, J. Lin, D. Han, S. Zhang, X. Ma, Electrosynthesis of chemicals from biomass glycerol, JACS Au. 5 (2025) 2977–2991, doi:10.1021/jacsau.5c00644.

[114] C. Xu, E. Paone, D. Rodríguez-Padrón, R. Luque, F. Mauriello, Reductive catalytic routes towards sustainable production of hydrogen, fuels and chemicals from biomass derived polyols, Renew Sustain. Energy Rev. 127 (2020) 109852, doi:10.1016/j.rser.2020.109852.

[115] J.C. Serrano-Ruiz, Selective deoxygenation of biomass polyols into diols, Molecules. 30 (2025) 3559, doi:10.3390/molecules30173559.

[116] P.P. Upare, R.E. Clarence, H. Shin, B.G. Park, An Overview on production of lignocellulose-derived platform chemicals such as 5-hydroxymethyl furfural, furfural, protocatechuic acid, Processes. 11 (2023) 2912, doi:10.3390/pr11102912.

[117] G. Dargó, D. Kis, A. Ráduly, V. Farkas, J. Kupai, Furandicarboxylic Acid (FDCA): electrosynthesis and its facile recovery from polyethylene furanoate (PEF) via depolymerization, ChemSusChem. 18 (2025) e202401190, doi:10.1002/cssc.202401190.

[118] Y. Gao, L. Ge, H. Xu, K. Davey, Y. Zheng, S.-Z. Qiao, Electrocatalytic refinery of biomass-based 5-hydroxymethylfurfural to fine chemicals, ACS Catal. 13 (2023) 11204–11231, doi:10.1021/acscatal.3c02272.

[119] Y. Zhang, Q. Fu, X. Zhao, Current advances in furfural reduction to furfuryl alcohol via thermal- and electrocatalysis, Chem. Eng. J. 521 (2025) 166588, doi:10.1016/j.cej.2025.166588.

[120] T. Peng, W. Zhang, B. Liang, G. Lian, Y. Zhang, W. Zhao, Electrocatalytic valorization of lignocellulose-derived aromatics at industrial-scale current densities, Nat. Commun. 14 (2023) 7229, doi:10.1038/s41467-023-43136-y.

[121] O.Y. Abdelaziz, I. Clemmensen, S. Meier, C.A.E. Costa, A.E. Rodrigues, C.P. Hulteberg, et al., On the oxidative valorization of lignin to high-value chemicals: a critical review of opportunities and challenges, ChemSusChem. 15 (2022) e202201232, doi:10.1002/cssc.202201232.

[122] N. Karanwal, S. Kim, Y. Liyanage, D.K. Lee, J. Kim, Highly efficient electro-reductive conversion of lignin into aromatics and cyclohexenes, Appl. Catal. B Environ. Energy. 381 (2026) 125851, doi:10.1016/j.apcatb.2025.125851.

[123] C. Dolle, N. Neha, C. Coutanceau, Electrochemical hydrogen production from biomass, Curr. Opin. Electrochem. 31 (2022) 100841, doi:10.1016/j.coelec.2021.100841.

[124] N.L. Slatter, B. Vichanpol, J. Natakaranakul, K. Wattanavichien, P. Suchamalawong, K. Hashimoto, et al., Syngas production for Fischer–Tropsch synthesis from rubber wood pellets and eucalyptus wood chips in a pilot horizontal gasifier with CaO as a tar removal catalyst, ACS Omega. 7 (2022) 44951–44961, doi:10.1021/acsomega.2c05178.

[125] M.S. Soffian, F.Z. Abdul Halim, F. Aziz, A. Rahman M, M.A. Mohamed Amin, D.N. Awang Chee, Carbon-based material derived from biomass waste for wastewater treatment, Environ. Adv. 9 (2022) 100259, doi:10.1016/j.envadv.2022.100259.

[126] T.-W. Chen, S.-M. Chen, G. Anushya, R. Kannan, P. Veerakumar, A.G. Al-Sehemi, et al., Electrochemical energy storage applications of functionalized carbon-based nanomaterials: an overview, Int. J. Electrochem. Sci. 19 (2024) 100548, doi:10.1016/j.ijoes.2024.100548.

[127] M.S. Singhvi, D.V. Gokhale, Lignocellulosic biomass: hurdles and challenges in its valorization, Appl. Microbiol. Biotechnol. 103 (2019) 9305–9320, doi:10.1007/s00253-019-10212-7.

[128] M.S. Roni, Y. Lin, D.S. Hartley, D.N. Thompson, A.N. Hoover, R.M. Emerson, Importance of incorporating spatial and temporal variability of biomass yield and quality in bioenergy supply chain, Sci. Rep. 13 (2023) 6813, doi:10.1038/s41598-023-28671-4.

[129] A. Marquez R, T. Bender J, M. Aleman A, E. Kalokowski, T.V. Le, L. Williamson C, et al., Insights into catalyst degradation during alkaline water electrolysis under variable operation, Energy Environ. Sci. 18 (2025) 7170–7187, doi:10.1039/D5EE02194D.

[130] P. Paulsen Thoresen, H. Lange, U. Rova, P. Christakopoulos, L. Matsakas, Role and importance of solvents for the fractionation of lignocellulosic biomass, Bioresour. Technol. 369 (2023) 128447, doi:10.1016/j.biortech.2022.128447.

[131] S. Ergun, A. Dik, R. Boukhanouf, S. Omer, Large-scale renewable energy integration: tackling technical obstacles and exploring energy storage innovations, Sustainability. 17 (2025) 1311, doi:10.3390/su17031311.

[132] Y. Xie, F. Luo, Z. Yang, Strategies for the enhancements in catalytic performance and stability of anodic electrocatalyst in PEM water splitting, Energy Rev. 3 (2024) 100103, doi:10.1016/j.enrev.2024.100103.

[133] J. Bai, C. Huang, Y. Liu, X. Zheng, J. Liu, L. Zhou, et al., Integrating biocatalysis with continuous flow: current status, challenges, and future perspectives, J. Adv. Res. (2025), doi:10.1016/j.jare.2025.07.010.

[134] Z. Zhang, J. Li, S. Yang, T. E, C. Peng, Advances in paired electrolysis for furfural conversion: from design principle, mechanisms to perspectives, Green Chem. 27 (2025) 6967–6988, doi:10.1039/D5GC01306B.

[135] Z. Xie, Y. Liu, L. He, J. Chen, X. Wu, M. Li, et al. In situ/operando characterization techniques for reaction interface in electrocatalytic CO2 reduction. Small. 2502083. doi: 10.1002/smll.202502083.

[136] L.G. de Araujo, L. Vilcocq, P. Fongarland, Y. Schuurman, Recent developments in the use of machine learning in catalysis: a broad perspective with applications in kinetics, Chem. Eng. J. 508 (2025) 160872, doi:10.1016/j.cej.2025.160872.

CHAPTER

4

Electrochemical platforms for biomass valorization and green energy production

Ganesan Kausalya Sasikumar[a], *Jagadeeswari Rangaraman*[b], *Keerthi Kumar Kulanthaigounder Jaganathan*[c], *Manoj Eswaran*[c], and *Ganesamoorthy Jeevitha*[d]

[a]Centre of Research and Development, KPR Institute of Engineering and Technology, Coimbatore, Tamil Nadu, India, [b]Department of Chemistry, KPR Institute of Engineering and Technology, Coimbatore, Tamil Nadu, India, [c]Department of Electronics and Communication Engineering, KPR Institute of Engineering and Technology, Coimbatore, Tamil Nadu, India, [d]Department of Molecular Physics, Saveetha School of Engineering, Saveetha Institute of Medical and Technical Sciences (SIMATS), Chennai, Tamil Nadu, India

4.1 Introduction to biomass

4.1.1 Definition and overview

Biomass is renewable organic matter that comes as a product of plants and animals. It can be used either by burning it to generate heat or it can be used to make liquid and gaseous fuel. Biomass can be used as a heating source, electricity, and transportation fuel. It is an important fuel in many nations, especially in developing nations, to be used in cooking and heating.

Electrochemical Biomass Conversion: Sustainable Processes and Applications for Bioenergy, Biofuels, and Bioproducts
DOI: https://doi.org/10.1016/B978-0-443-26757-4.00020-9

4.1.2 Classification of biomass: Biomass feedstocks

4.1.2.1 *Agricultural residue*

Includes plant materials that were not consumed during the harvest, such as stems, husks, and leaves. Being a second-generation feedstock, it does not have competition with food production. They can be used as biofuels, animal feed, compost, bioplastics, and biochar.

4.1.2.2 *Forestry residue*

It contains treetops, bark, wood chips, and leaves, which contain lots of lignocellulosic matter, and are used as a source of bioenergy and value-added products.

4.1.2.3 *Food waste*

Food waste is produced over the food supply chain and is highly organic. New technologies transform it into biofuels, enzymes, bioactive compounds, and fertilizers.

4.1.2.4 *Municipal wastewater*

Includes biodegradable substances of urban waste. It helps in biogas generation, recovery of nutrients, growing of algae, and treatment of wastewater.

4.1.2.5 *Industrial organic waste*

Food, agricultural, paper, textile, and brewing industries are biodegradable by-products. It is re-processing energy and useful chemicals.

4.1.2.6 *Animal manure*

Livestock feces and bedding are utilized in biogas, fertilizer and soil conditioning.

4.1.2.7 *Theoretical and practical implications of biomass*

Biomass is environmentally friendly, carbon-neutral, and a source of waste valorization. The problems are land usage, conflicting with food, technological constraints, and supportive policies.

4.2 Biomass composition and conversion pathways

The biomass composition and especially the ash components have a great variation because of the difference in the moisture content, the ash yield, and the type of genetic inorganic material present in the biomass. The factors influencing this variation include the origin of the biomass (i.e.,

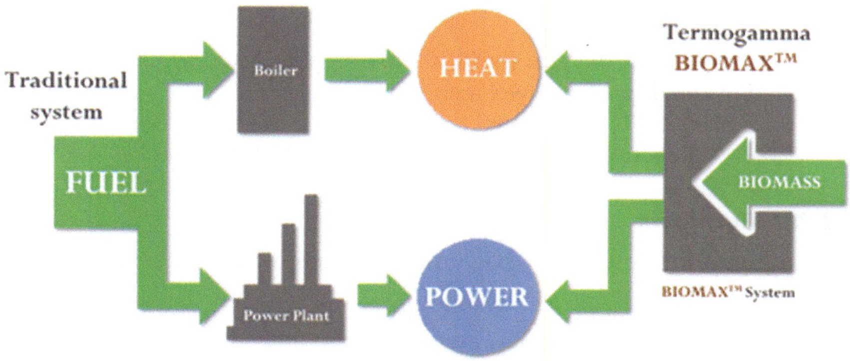

FIGURE 4.1 Biomass cogeneration scheme. *From Dcbert95 (2017).* Biomass Cogeneration Schema, *CC BY-SA 4.0.*

plant, animal or a mixture of both), and the origin. The most commonly occurring elements in biomass, by their abundances, are: major constituents: carbon (C), oxygen (O), hydrogen (H), nitrogen (N) minor constitutions: calcium (Ca), potassium (K), silicon (Si), magnesium (Mg), aluminum (Al), sulfur (S), iron (Fe), phosphorus (P), chlorine (Cl), sodium (Na), manganese (Mn), and titanium (Ti). Concentration and occurrence of these elements depend on the source and origin of the biomass, such as plant, animal, or anthropogenic materials [1].

In the case of conversion pathways of biomass, an emphasis is laid on how biomass can be converted into energy or other products of value. Note: the various avenues allow biomass to be converted into electricity, heat, biofuels, and chemicals to make the energy system more sustainable (Fig. 4.1).

4.2.1 Chemical composition: Cellulose, hemicellulose, and lignin

Biomass mainly consists of cellulose (rigid, crystalline glucose polymer), hemicellulose (easily degradable branched polysaccharide), and lignin (aromatic polymer offering structural integrity and microbial resistance). Their proportions determine biomass digestibility and suitability for conversion [2].

4.2.2 Thermochemical conversion process

High-temperature methods transform biomass into gases, liquids, and solids.

4.2.2.1 *Combustion*

Direct burning in oxygen (~900°C) yields heat, CO_2, water vapor, and ash. Used in heating and power generation.

Reaction: $C_6H_{10}O_5 + O_2 \rightarrow CO_2 + H_2O + \text{Heat}$

4.2.2.2 *Pyrolysis*

Heating in the absence of oxygen produces bio-oil, syngas, and char.

- *Slow*: Solid-heavy.
- *Fast*: Liquid-rich (400°C–750°C).
- *Flash*: Gas-rich (800°C–1000°C).
  Reaction: Biomass → Bio-oil + Char + Syngas

4.2.2.3 *Gasification*

Converts biomass to syngas ($CO + H_2$) via partial oxidation at 700°C to 1000 °C.

Example: $C + H_2O \rightarrow CO + H_2$

4.2.2.4 *Liquefaction*

Uses water and heat (250°C–400°C, high pressure) to convert biomass into bio-oil, enhanced by catalysts.

4.2.3 Biochemical conversion processes

Biochemical conversion is the biological process where biomass is transformed into products such as biofuels ethanol, methane (CH_4), and biochemicals. These conversions are typically done by microbes bacteria, yeast, and fungi, etc., which break down the organic materials of the biomass like carbohydrates, lipids, and proteins into simpler forms. The two major biochemical conversion processes include anaerobic digestion (AD) and fermentation.

4.2.3.1 *Fermentation*

Converts sugars into ethanol, lactic acid, or acetic acid.

Example: $C_6H_{12}O_6 \rightarrow 2C_2H_5OH + 2CO_2$

4.2.3.2 *Anaerobic digestion*

Microbial breakdown in four stages: hydrolysis, acidogenesis, acetogenesis, and methanogenesis, produces biogas and digestate.

4.2.3.3 *Enzymatic hydrolysis*

Enzymes degrade cellulose/hemicellulose into fermentable sugars. Key for bioethanol and biochemical production.

4.2.4 Hybrid conversion technologies

Hybrid conversion technologies integrate various conversion processes, sometimes both biological and thermochemical methods, to optimize the conversion of biomass into energy or value-added products. Hybrid technologies have tried to take advantage of the strengths of one of the technologies and compensate for its shortcomings, increasing efficiency and broadening the selection of feedstock types that may be taken advantage of.

Applications for hybrid conversion technologies

- Integrated biofuel production
- Waste-to-energy systems
- Sustainable chemical production
- Decentralized energy solutions
- Biorefinery applications
- Carbon and nutrient recycling
- Hydrogen production
- Improved biomass utilization

4.2.4.1 *Solar thermal pyrolysis*

Solar thermal pyrolysis applies the large amounts of solar energy to heat biomass at high temperatures in the absence of oxygen to produce biochar, bio-oil, and syngas through thermal decomposition of organic substances. This technology uses solar power rather than fossil fuels to facilitate the pyrolysis reaction and is therefore a greener and sustainable way of converting biomass.

4.2.4.2 *Upgrading of pyrolysis oil*

The upgrade of the pyrolysis oil (also known as bio-oil) is the process of enhancing the quality of the oil that was formed during the pyrolysis of the biomass to make it more useful as a fuel or as a feedstock to make a chemical. Bio-oil is usually high in oxygen, acidity, and moisture, rendering it unstable and inappropriate to be used in direct form. The processes of upgrading are developed to minimize such off-putting properties so that it becomes stable, energy-dense, and burnable.

4.2.4.3 *Combined hydrothermal carbonization and anaerobic digestion system*

Hydrothermal carbonization (HTC) is a wet technology where the biomass is reacted with hot (180°C–260°C), pressurized (2–6 MPa) water and is subjected to chemical reactions such as hydrolysis, condensation, and decarboxylation. The result of this process is the hydrochar, different from biochar, and gaseous and liquid by-products [3–5] Fig. 4.2.

Methods and materials

- *Lab-scale experiment:* Sawdust (a typical byproduct of wood) was put in a mixture with water (1:5 ratio), nitrogen was added, and put under pressure, then heated at 220°C in a mini reactor for 40 minutes. The solid hydrochar was dried and analyzed.
- *HTC-AD system:* HTC was replicated and combined with Rankine and Brayton cycles, and AD. To drive the Brayton cycle, wet hydrochar was dried and fed into the Rankine cycle, and the by-product liquid was fed into AD to generate CH_4 and hydrogen. The gas turbine was used to provide exhaust heat to HTC as well as through the Rankine cycle.

Such an arrangement allows effective production of biofuels and recycling of energy.

4.2.5 Emerging and innovative conversion techniques

New and emerging conversion methods on biomass are aimed at improving the effectiveness and sustainability of biomass exploitation as a source of energy and products. Such methods examine novel technologies or enhance the current ones to transform biomass into biofuels, bio-based chemicals, and other valuable products with a reduced environmental impact in most cases [6].

A list of newer and innovative conversion methods of biomass is as follows:

- *Hydrothermal liquefaction:* The feed is wet biomass (such as algae) that is converted to the bio-crude oil in a process that mimics the formation of fossil fuels at high temperature and pressure.
- *Algae-based biofuel production:* Algae Biofuel Production: Stemming from algae rich in oil, algae is used to produce biofuels such as bioethanol and biodiesel, as well as has the advantage of carbon dioxide (CO_2) capture.
- *Fast pyrolysis:* This is a procedure that involves heating biomass at high rates in the absence of oxygen in order to produce bio-oil. Bio-oil, which is produced, can be then refined to biofuels or used as a chemical feedstock.

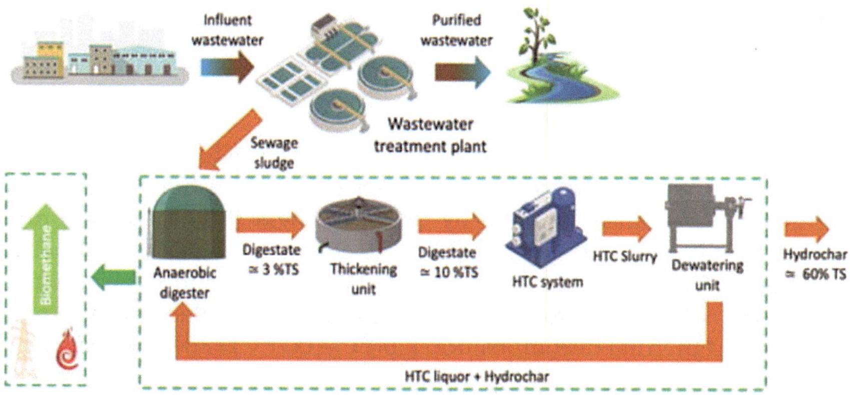

FIGURE 4.2 Combined HTC + AD system flowchart illustrates a sequential HTC process converting biomass into hydrochar and HTC liquor, which are then fed into an AD system to produce biogas and digestate. *From Ferrentino R, Merzari F, Fiori L, Andreottola G. (2020). Coupling hydrothermal carbonization with anaerobic digestion for sewage sludge treatment. Energies. 13(23), 6262, CC-BY 4.0.*

- *Biomass gasification with carbon capture and storage:* Biomass is transformed into power or fuel, and the carbon emissions get captured to minimize the environmental impact.
- *Enzyme-based biochemical conversion:* It decomposes cellulose and hemicellulose in biomass by using enzymes to ferment them to yield fermentable sugars to use in bioethanol or biochemicals.
- *Solar-assisted biomass conversion:* Uses solar energy in enhancing biomass conversion processes, such as pyrolysis, to enhance the efficiency and sustainability of energy.

4.2.5.1 *Nanomaterials for dark fermentation*

Dark fermentation is an anaerobic reaction in which microorganisms ferment organic matter in the dark to produce biohydrogen (bioH_2) as well as alcohols, CO_2, and organic acids. Glucose is a typical one, which is converted to pyruvate, then to acetyl- CoA with the help of such enzymes as PFL and PFOR, which are involved in the production of bioH_2. PFL pathway produces as much as two moles of bioH_2 for each glucose, whereas PFOR is able to produce four moles. Increased bioH_2 content retards production, resulting in byproducts such as ethanol. Depending on the pathway, acetone, butanol, butyric acid, and acetone are formed under strictly anaerobic conditions, and acetic and butyric acids are under different conditions. Lignin makes the conversion of lignocellulosic biomass slow. The butyric pathway generates lower CO_2 but could be inhibited by the acid accumulation, and the acetic pathway is associated with high hydrogen yield but cannot overcome the redox balance because of the CO_2 production.

4.2.5.2 *Biomass-based catalyst*

Waste products such as shells, bones, seed cake, ash, biochar, and activated carbon are used to produce greener catalysts in the production of biodiesels. Nanocatalysts made by using biomass (less than 100 nm) is formed by different methods of thermal, chemical, and physical activation. The CaO is extracted by means of waste shells and bones containing $CaCO_3$, where the process of its extraction takes place in multistage washing and thermal decomposition (300°C–1000°C) is performed, which influences the properties of this substance.

4.2.5.2.1 Types of biomass catalysts

- *Lignocellulosic materials:* Biomass derived from plants consists of lignin, cellulose, and hemicellulose, which can be modified to optimize the conversion process for biofuels, including hydrogen generation.
- *Biochar:* Biochar, a carbon-dense substance derived from pyrolysis, possesses a porous architecture that enhances its role as a catalyst in the transformation of biomass.
- *Enzyme-based catalysts:* Hydrogenases and cellulases are enzymes that decompose organic material to generate hydrogen, with enhancements boosting their efficiency.
- *Heterogeneous catalysts:* Biomass-derived solid-phase catalysts are advantageous due to their ease of separation and reusability, making them highly suitable for the production of biodiesel and biofuels [7].

4.3 Electrochemical platforms for biomass conversion

Electrochemical platforms for biomass conversion, or electrochemical conversion of organic materials from plants and animals into valuable products like biofuels, chemicals, or electricity. Such platforms tap into electrochemical reactions in an electrochemical cell: oxidation and reduction reactions can break down those complex molecules like cellulose and lignin into relatively simpler forms or even produce energy directly. This approach provides advantages such as milder reaction conditions, greater efficiency, and the possibility of integration with renewable sources of energy, so it is a promising technology for sustainable biomass utilization (Fig. 4.3) [8].

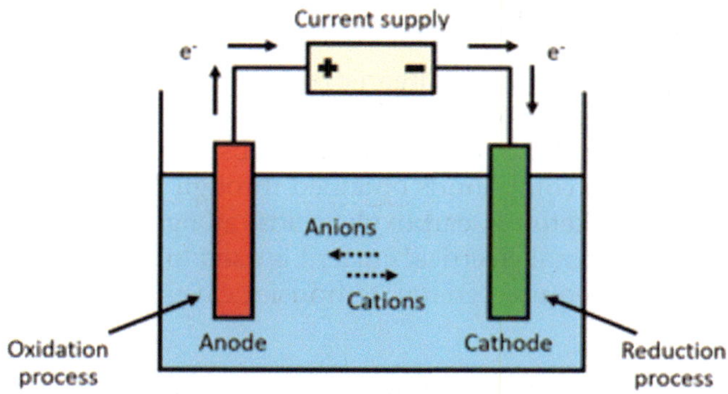

FIGURE 4.3 Electro-oxidation apparatus. *From: Emanuele Oddo (2024). Electro-oxidation apparatus, CC BY-SA 4.0.*

4.3.1 Electrocatalytic biomass conversion

4.3.1.1 Electrooxidation

Oxidation at the anode, where organic substrates lose electrons. Applied in wastewater treatment, fuel cells, and chemical production. Efficiency depends on electrode material and conditions.

Electrooxidation processes are widely used for:

- *Water and wastewater treatment:* Oxidizing organic pollutants.
- *Energy generation:* In fuel cells like direct alcohol fuel cells, where alcohols are oxidized to produce electricity.
- *Chemical production:* Converting raw materials (e.g., biomass) into valuable chemicals.

The performance and selectivity of electrooxidation are influenced by the applied electrode type, voltage, and the reaction conditions [9].

4.3.1.2 Electroreduction

Electroreduction is a reaction which involves the use of electrical energy to reduce chemical species on the cathode. It is a part and parcel of energy storage, conversion of CO_2, use of biomass, and the production of chemicals. Its applications include the process of reducing CO_2 to useful compounds, the transformation of biomass to biofuels, and the conversion of nitrogen to ammonia. They are also used in fuel cells, batteries, and in hydrogen storage systems. Nonetheless, it faces the obstacles as the continuity of catalyst efficiency and scalability. Thus, optimization development is significant due to its large-scale use [10].

4.3.2 Electrochemical organic oxidation

Electrochemical oxidation of organic compounds is the process in which the electrochemical reaction oxidizes organic compounds. In the process, an electrochemical cell that has an organic substrate loses electrons off its anode. The different compounds obtained through this oxidation process include aldehydes, ketones, carboxylic acids, along with other useful organic products. External electrical current is used to facilitate the reaction, which delivers the required energy to transfer electrons.

4.3.2.1 Mediated electro-oxidation

Electrochemical oxidation Mediated electro-oxidation is an electrochemical oxidation method that uses an electron mediator (redox mediator) to oxidize an organic substrate. The redox mediator in this mechanism undergoes a reversible electron transfer at the electrode, and also at the same time, transfers electrons to or off the organic molecule undergoing oxidation. The mediator has an impact on overcoming competition, like poor conductivity of the organic substrate or low current density due to high overpotentials at the electrode surface [11].

4.3.2.2 Direct electro-oxidation

Direct electro-oxidation is also a method of electrochemical reaction where the direct oxidation of the organic substrate that is present on the electrode surface is carried out directly, without the use of a mediator. External electrical current can be used to accelerate the process of oxidation, whereby the organic compound at the anode of the electrochemical cell is deprived of electrons. Electron transfer between the electrode and substrate is direct, leading to the generation of oxidized products (aldehydes, ketones, carboxylic acid, and other organic products) [12].

4.3.3 Selective oxidation of carbon–hydrogen bond

Selective oxidation of carbon-hydrogen (C-H) bonds focuses on oxidizing a certain type of C-H bond in organic molecules, and not on the other bonds and functional groups. This process is a critical challenge of synthetic chemistry because of the solid and stable character of C-H bonds. It allows the transformation of the simple hydrocarbons to more useful chemical molecules, including alcohol, aldehyde, ketone, and carboxylic acid, and other functionalized molecules. The selective oxidation of the C-H bonds can be very useful in organic synthesis, petrochemical processing, and environmental applications [13].

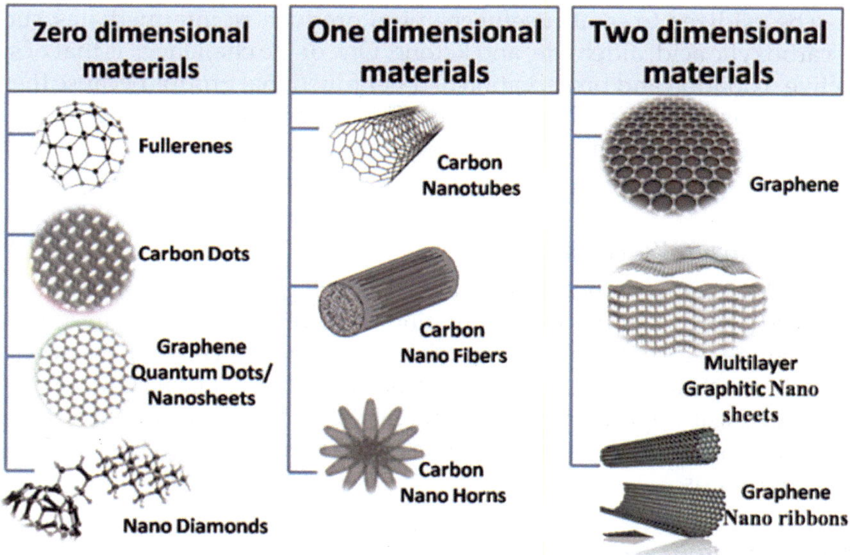

FIGURE 4.4 Classification of carbon nanomaterials by dimensionality: 0D fullerenes, 1D nanotubes, 2D graphene, and 3D bulk carbon. *From Khairunnisa Amreen (2024). Emerging Trends of Miniaturized Carbon Nanomaterial Based Energy Storage Microdevices, CC-BY 4.0.*

4.3.3.1 *Key aspects*

- *Bond activation:* The activation of C-H bonds through oxidation is done in the presence of specific reagents, catalysts, or electrochemical methods.
- *Regioselectivity:* The presence of target bonds in complex molecules is significant to achieve efficiency.
- *Catalysts:* Oxidation includes transition metal catalysts (palladium, platinum, and copper), enzymes (cytochrome P450), and nonmetallic catalysts (molecular oxygen).
- *Tolerance:* Close to functional groups are not affected since the reaction is done in mild conditions.

This crucial approach aids in developing selective oxidation procedures in industries and in laboratories.

4.3.4 Selective oxidation of oxygenated compounds

Selective oxidation of oxygenated compounds involves the selective oxidation of both benzene and alcohols. Selective oxidation of oxygenated compounds plays a significant role in synthetic chemistry and in the industry. Chemicals such as alcohols, aldehydes, ketones, phenols, and ethers may be oxidized to create useful chemical products or intermediates, such as carboxylic acid, aldehyde, and ketone. One of the challenges is that of selective oxidation and preservation of other functional groups because these compounds have complex reactivity. This is commonly applied in pharmaceuticals, agrochemicals, petrochemicals, and environmental chemistry [14].

4.3.4.1 *Alcohol's oxidation*

Alcohol oxidation belongs to the basic changes in the field of organic chemistry, and it is the transformation of alcohols (R-OH) into aldehydes, ketones, carboxylic acid, or other compounds. This is necessary in the manufacture of fine chemicals, pharmaceuticals, agrochemicals, and petrochemical intermediates. Various methods are used in the oxidation process, reaction selectivity and efficiency depend on reagents, catalysts, and conditions.

4.3.4.2 *Furanic oxidation*

The furan compounds are oxidized through furanic oxidation, made up of the oxygen ring consisting of five members (C_4H_4O). This is important in the manufacture of furan-based chemicals and the transformation of natural furan compounds. It is integral to biofuels, pharmaceuticals, and agricultural chemicals, and assists in modifying lignin to be used in biorefineries. The nature of furanic oxidation and the chemicals employed also determine the results.

4.3.5 Selective electrocatalytic carbon-carbon bond cleavage

Selective electrocatalytic carbon-carbon (C-C) bond cleavage includes breaking of C-C bonds in organic products by using an electric current with a catalyst. This method has been of interest in organic chemistry, materials science, and green chemistry as a way to achieve effective reactions at mild conditions. Electrocatalytic systems provide alternatives to traditional methods that are less hazardous in terms of chemical use and energy consumptions and they allow the use of renewable energy [15].

4.3.5.1 *Key aspects*

- *Cleavage mechanism:* The electron transfer triggers breaking of the bonds, and this is influenced by variables such as voltage, structure of the molecules, and the characteristics of the catalyst.
- *Electrocatalysts:* Metal catalysts (Pt, Pd, and Ni), carbon (graphene and nanotubes), and enzyme-like catalysts are employed.

- *Selectivity issues:* Design of catalyst and optimization of reaction conditions: this is necessary to selectively target chemical bonds and exclude side reactions.
- *Uses:* Organic synthesis, biofuel conversion, waste conversion, pharmaceutical production, chemicals, and sustainable energy.
- *Benefits:* Mild conditions usability, sustainability, and high selectivity will be used against environmental impact and improvement of efficiency.

It provides a high-potential direction in the development of green chemistry and the establishment of renewable energy technologies.

4.3.6 Role of electrocatalysts in biomass conversion

Electrocatalysis has become a game-changer in terms of biomass conversion in that it offers a sustainable approach to convert biomass into useful chemicals, biofuels, and energy. Electrochemical processes of breaking down biomass through the use of electrocatalysts comprise complex macromolecules such as cellulose, lignin, and hemicellulose, which are broken down into simpler products. Metal, carbon-based, and metal-free catalysts are used to selectively react in such reactions as bond breaking, hydrogenation, and oxidation under mild conditions. Electrocatalysts aid in overcoming biomass conversion issues, such as the resistance of lignin and the selectivity of products. The inclusion of electrocatalysis has advantages, including the minimization of chemical consumption, decreased energy usage, and valorization of waste. With the development of research, electrocatalysis will have an important role in the production of sustainable energy and chemicals and will facilitate a cyclical economy and decrease the environmental impact of industry [16].

4.3.6.1 Metal-based catalysts

Metal-based catalysts are integral to transforming biomass into valuable chemicals, fuels, and products. These catalysts are used in electrocatalytic processes to facilitate reactions like hydrogenation, oxidation, and C-C bond cleavage, essential for decomposing biomass structures, including cellulose, lignin, and hemicellulose. Their electronic properties, stability, and high surface area make them effective in promoting biomass conversion under mild electrochemical conditions.

4.3.6.2 Nanostructured catalysts

Nanostructured catalysts Fig. 4.4 [1] and [2] have become central in biomass conversion, due to their features like large surface area, enhanced reactivity, and customizable properties. Made from metals, metal oxides, or carbon-based substances, these catalysts effectively transform biomass

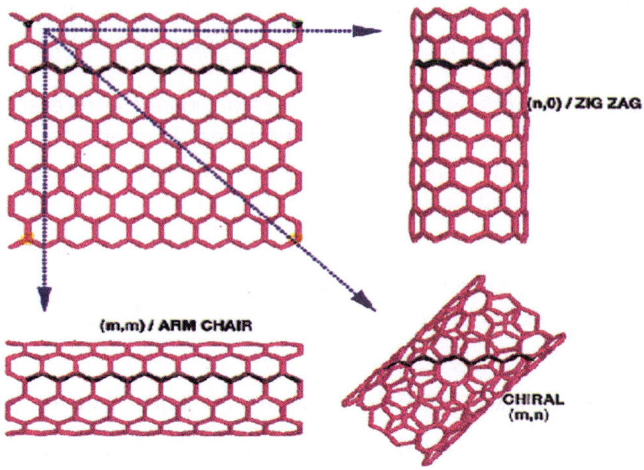

FIGURE 4.5 Structures of carbon nanotubes showing the rolling of graphene sheets into cylindrical forms. *From Iijima, Sumio. "Helical microtubules of graphitic carbon." nature 354, no. 6348 (1991): 56-58. Public Domain CC-BY 4.0.*

feedstocks into valuable chemicals, biofuels, and energy. Their nanoscale size, high dispersion, and capacity to create highly active sites make them especially well-suited for selective biomass transformations under gentle conditions.

4.4 Applications of biomass-derived electrochemical platform

Biomass-based electrochemical platforms have significant and promising research opportunities within sustainable energy and chemical production. Electrochemical processes transform the complex biomass feedstocks into useful chemicals, biofuels, and energy by leveraging the renewability of biomass. Eco-friendly, energy-efficient alternative to the traditionally applied chemical processes relies upon the use of biomass-based electrochemical systems since those usually depend on feedstocks of nonrenewable sources and conditions of aggressive severity.

4.4.1 Biofuels production

Biofuels, derived from renewable biomass resources, are vital for reducing dependence on fossil fuels and mitigating environmental impacts. The production of biofuels from biomass involves various electrochemical and biochemical processes. Key biofuels produced from biomass include hydrogen, bioethanol, and biodiesel.

4.4.1.1 *Hydrogen from biomass electrolysis*

Electrolysis of biomass compounds is converted to hydrogen through an electrochemical reaction with renewable electricity, which is a clean alternative to hydrogen produced by fossil fuels

4.4.1.2 *Bioethanol*

The material is manufactured through fermentation or electrochemical reduction of furfural and is an additive used as fuel or biofuel [17].

4.4.1.3 *Biodiesel*

Biomass oils are reacted with alcohol by transesterification, and electrochemical methods enhance the catalysis and decrease energy consumption [18].

4.4.2 Production of platform chemicals

Renewable biomass possesses platform chemicals that are key in the shift toward a bio-based economy. These chemicals are used as intermediates to produce biodegradable plastics, biofuels, solvents, pharmaceuticals, and food additives. Bio-based alternatives can minimize fossil fuel consumption and impact on the environment. In this section, two major categories are considered: furans and derivatives (5-hydroxymethylfurfural [HMF] and furandicarboxylic acid [FDCA]) and organic acids (succinic, lactic, and acetic acids), which are used as building blocks in industries.

4.4.2.1 *Furans (hydroxymethylfurfural and furandicarboxylic acid)*

Sugar dehydration forms HMF, which undergoes oxidation to form 2, 5-FDCA that forms bio-based plastics such as polyethylene furanoate. These are used in packaging, biofuels, solvents, and resins.

4.4.2.2 *Organic acids*

- *Succinic acid*: Fermented from glucose; used in plastics and pharmaceuticals.
- *Lactic acid*: Produced by Lactobacillus; key for polylactic acid bioplastics.
- *Acetic acid*: Fermented by Acetobacter; used in vinegar, solvents, and acetates [19].

4.4.3 Electrochemical oxidation of biomass-derived molecules to high-value chemicals

The oxidation of biomass molecules by electrochemical methods presents a sustainable solution to the production of high-value chemicals by way of renewable electricity. This is a way of converting biomass compounds into useful platform chemicals by catalysis. The major processes are the HMF to 2,5-FDCA and the oxidation of glycerol to value products [20–23].

4.4.3.1 *5-hydroxymethylfurfural to furandicarboxylic acid*

Electrochemical oxidation converts HMF to FDCA under gentle, energy-saving conditions for eco-friendly plastics.

4.4.3.2 *Glycerol oxidation*

Converts the byproduct of biodiesel, glycerol, to useful products including dihydroxyacetone, formic acid, and tartronic acid.

4.4.4 Applications in the environment

The combination of bioelectrochemical systems (BES) and CO_2 capture technologies is necessary to deal with environmental problems. These methods not only cut down on pollution, but also limit the effects of climate change and also produce certain useful by-products. This part discusses the treatment of the wastewater through BES and the capture and utilization of the CO_2.

4.4.4.1 *Wastewater treatment*

The bio-electrochemical systems eliminate pollutants and produce electricity or chemicals.

4.4.4.2 *Carbon dioxide capture and utilization*

Biomass platforms facilitate the conversion of captured CO_2 into fuels and chemicals, thereby contributing to carbon-neutral cycles.

4.4.5 Hydrogen evolution reaction from biomass

The hydrogen evolution reaction (HER) is a reaction that produces hydrogen gas (H_2) through the use of metal or nonmetal catalysts on water. Plant-based substances have become ecologically friendly HER catalysts that provide a sustainable alternative of obtaining clean hydrogen fuel using renewable resources [24].

4.4.5.1 *Hydrogen evolution reaction mechanism*

Involves water reduction at electrodes using biomass-derived catalysts.

4.4.5.2 *Biomass as hydrogen evolution reaction source*

Biomass intermediates aid H_2 generation with lower overpotentials.

4.4.5.3 *Benefits*

Green hydrogen, reduced emissions, integration with renewable energy.

4.4.5.4 *Challenges*

Catalyst durability, cost, and reaction efficiency.

4.4.6 Oxygen evolution reaction in biomass systems

Oxygen Evolution Reaction (OER) uses water to generate oxygen gas (O_2), which forms the basis in the field of electrolysis and fuel cells. In biomass systems, OER allows the separation of water and conversion of energy, but usually restricts these processes [25].

4.4.6.1 *Oxygen evolution reaction mechanism*

Oxidation of water to oxygen using biomass-derived electrodes.

4.4.6.2 *Biomass catalysts*

Biochar, metal oxides, and doped carbon materials enhance activity.

4.4.6.3 *Benefits*

Supports water splitting, eco-friendly material use, and low-cost catalysis.

4.4.7 Reduction of biomass intermediates to chemicals

The conversion of biomass intermediates to chemicals is vital in the transformation of renewable materials to useful chemicals and fuels, developing sustainable alternatives to the petrochemical-based production. Plant materials and other organic wastes are biomass and contain complex molecules that can be reduced to form simple chemicals through reduction reactions [26].

4.4.7.1 *Biomass intermediates*

Biomass can be converted to intermediates via hydrolysis, pyrolysis, and fermentation. These mediate substances, such as sugars, alcohols, organic acids, lignin compounds, and fatty acids, facilitate chemical production.

4.4.7.2 *Reduction reactions*

Reduction reactions refer to the process where electrons or hydrogen are added to biomass intermediates to form a highly desirable chemical.

4.5 Environmental and economic impact

The transformation of biomass into chemicals and biofuels has a positive impact on the sustainability of the environment and the economy. It assists in the decrease of greenhouse gas emissions and also contributes to energy security and employment. Nonetheless, these advantages are possible to be achieved through sustainable production processes, innovative technologies, and management of cost and infrastructure issues [27].

4.5.1 Life cycle assessment of biomass electrochemical systems

The life cycle assessment (LCA) is an evaluation of the environmental impact of biomass systems, beginning with the purchase of raw materials to their utilization of the final product. Despite the fact that the biomass electrochemical processes can usually lead to less emissions compared to the fossil-based ones, they still need to be evaluated [28,29]:

- *Feedstock collection:* Transport and preprocessing impact.
- *Conversion efficiency:* Energy and catalyst usage.
- *Product utilization:* Emissions from biofuel or chemical use.
- *Waste management:* Disposal or reuse of byproducts like ash or digestate.

An optimized biomass system minimizes the emissions of GHG, reliance on fossil oil, and environmental degradation [30].

4.5.2 Policy and regulatory framework

Supportive policies are crucial for scaling biomass technologies. Key aspects include:

- *Incentives/Subsidies:* For renewable energy projects using biomass.
- *Standards and certifications:* Ensure sustainable sourcing and emissions compliance.
- *Carbon credits/Trading:* Encourage low-carbon innovations.
- *Waste management regulations:* Promote biomass valorization of organic waste.

Comprehensive regulation rules and collaboration between the government and the business sector can contribute to incorporating biomass in the national energy plans and the circular economy [31].

4.6 Conclusion

Plant biomass is converted to valuable chemicals and biofuels, which is a solution to the global energy and environmental challenge. This is achieved by thermal methods, fermentation by microbes, and electrochemical methods. New electrochemical processes have become more efficient. Nevertheless, it is necessary to assess the environmental or economic effects using the LCA. Also, it requires strong policies to ease the adoption of technology and enhance a sustainable circular economy.

Abbreviations

AD	Anaerobic digestion
Al	Aluminum
BES	Bioelectrochemical systems
C	Carbon
C-C	Carbon–carbon bond
C-H	Carbon–hydrogen bond
Ca	Calcium
$CaCO_3$	Calcium carbonate
CaO	Calcium oxide
CH_4	Methane
Cl	Chlorine
CO_2	Carbon dioxide
FDCA	Furandicarboxylic acid
Fe	Iron
GHG	Greenhouse gases
H	Hydrogen
HER	Hydrogen evolution reaction
HMF	5-Hydroxymethylfurfural
H2	Hydrogen gas
HTC	Hydrothermal carbonization
K	Potassium
LCA	Life cycle assessment
Mg	Magnesium
Mn	Manganese
MPa	Megapascal
N	Nitrogen
Na	Sodium
Ni	Nickel
O	Oxygen
OER	Oxygen evolution reaction
P	Phosphorus
Pd	Palladium
Pt	Platinum
S	Sulfur
Si	Silicon
Ti	Titanium

References

[1] S. Vassilev, C. Vassileva, Water-soluble fractions of biomass and biomass ash and their significance for biofuel application, Energy Fuels 33 (4) (2019) 2763–2777, doi:10.1021/acs.energyfuels.9b00081.
[2] M. Stöcker, Perspectives for thermochemical conversions of lignocellulosic biomass, Small 20 (39) (2024) e2302495, doi:10.1002/smll.202302495.
[3] M. Heidari, A. Dutta, B. Acharya, S. Mahmud, A review of the current knowledge and challenges of hydrothermal carbonization for biomass conversion, J. Energy Inst. 92 (2018) 1779–1799.
[4] H.S. Kambo, A. Dutta, A comparative review of biochar and hydrochar in terms of production, physico-chemical properties and applications, Renew Sustain. Energy Rev. 45 (2015) 359–378.
[5] R. Ferrentino, F. Merzari, L. Fiori, G. Andreottola, *Coupling hydrothermal carbonization with anaerobic digestion for sewage sludge treatment*, Energies 13 (23) (2020) 6262.
[6] O.L. Rominiyi, B.A. Adaramola, O.M. Ikumapayi, O.T. Oginni, S.A Akinola, Potential utilization of sawdust in energy, manufacturing and agricultural industry: waste to wealth, World J Eng. Technol. 5 (2017) 526–539.
[7] A. Antelava, S. Damilos, S. Hafeez, G. Manos, S.M. Al-Salem, B.K. Sharma, et al., Plastic solid waste (PSW) in the context of life cycle assessment (LCA) and sustainable management, Environ. Manage. 64 (2) (2019) 230–244, doi:10.1007/s00267-019-01178-3.
[8] J. Zhang, X. Zhang, The thermochemical conversion of biomass into biofuels, in: D Verma, E Fortunati, S Jain, X Zhang (Eds.), Woodhead Publishing Series in Composites Science and Engineering, Biomass, Biopolymer-Based Materials, and Bioenergy, Woodhead Publishing, 2019, pp. 327–368, doi:10.1016/B978-0-08-102426-3.00015-1.
[9] O.K. Lee, D.H. Seong, C.G. Lee, E.Y. Lee, Sustainable production of liquid biofuels from renewable microalgae biomass, J. Ind. Eng. Chem. 29 (2015) 24–31, doi:10.1016/j.jiec.2015.04.016.
[10] Q. Chen, X. Tan, Y. Liu, S. Liu, M. Li, Y. Gu, et al., Biomass-derived porous graphitic carbon materials for energy and environmental applications, J. Mater. Chem. A 8 (2020) 5773–5811.
[11] A. Cork, M. Winter, The electrochemical behaviour of amine-functionalized boron-based molecular sorbents, Electrochim. Acta 45 (24) (2000) 3991–3997.
[12] A.M. Bond, P.G. Symons, S. Fletcher, P.G. Symons, The relationship between the electrochemistry and the crystallography of microcrystals. the case of TCNQ (7,7,8,8-tetracyanoquinodimethane), Analyst 123 (1998) 1891–1904.
[13] Y. Li, M. Bettge, J. Bareño, S.E. Trask, D.P. Abraham, Exploring electrochemistry and interface characteristics of lithium-ion cells with Li1.2Ni0.15Mn0.55Co0.1O2 positive and Li4Ti5O12 negative electrodes, J. Electrochem. Soc. 162 (2015) A7049.
[14] X.H. Chadderdon, D.J. Chadderdon, T. Pfennig, B.H. Shanks, W. Li, Paired electrocatalytic hydrogenation and oxidation of 5-(hydroxymethyl)furfural for efficient production of biomass-derived monomers, Green Chem. 21 (2019) 6210–6219.
[15] X. Li, Z. Xu, J.R. Cort, W.J. Qian, B. Yang, Lipid production from non-sugar compounds in pretreated lignocellulose hydrolysates by *Rhodococcus jostii* RHA1, Biomass Bioenergy 145 (2021) 105970, doi:10.1016/j.biombioe.2021.105970.
[16] T. Zhang, K. Shen, Y. Zhang, W. Tang, W. Dai, W. Zhang, Cathode and plastic corecycling from spent batteries via paired electrolysis, ACS Sustain. Chem. Eng. 13 (43) (2025) 18633–18641, doi:10.1021/acssuschemeng.5c06970.
[17] S.A. Akhade, N. Singh, O.Y. Gutiérrez, J. Lopez-Ruiz, H. Wang, J.D Holladay, et al., Electrocatalytic hydrogenation of biomass-derived organics: a review, Chem. Rev. 120 (20) (2020) 11370–11419, doi:10.1021/acs.chemrev.0c00158.

[18] G. Knothe, Biodiesel and renewable diesel: a comparison, Prog. Energy Combust. Sci. 36 (3) (2010) 364–373, doi:10.1016/j.pecs.2009.11.004.
[19] S. Ramesh, K. Vijayan, Lactic acid production from renewable resources, Green Chem. 20 (2018) 367–380.
[20] M. Wei, L. Xu, Electrochemical oxidation of 5-HMF to FDCA: a green and sustainable approach, J. Power Sources 414 (2019) 103–109.
[21] S. Liu, W. Zhang, Electrochemical oxidation of 5-HMF to FDCA: mechanisms, catalysts, and strategies, J. Mater. Chem. A 8 (24) (2020) 12081–12101.
[22] J. Qiao, Y. Liu, Electrochemical oxidation of glycerol: challenges and opportunities for biofuel production, J. Electroanal. Chem. 818 (2018) 41–47.
[23] C.H. Choi, H.S. Lee, Electrochemical reduction of CO_2 to valuable chemicals, Nat. Sustain. 3 (2) (2020) 109–116.
[24] X. Sun, L. Zhang, Hydrogen evolution from biomass-derived catalysts for sustainable energy production, Renew Energy 154 (2020) 23–32.
[25] Z. Wang, L. Zhang, Biomass-derived catalysts for the oxygen evolution reaction: prospects and challenges, Renew Sustain. Energy Rev. 132 (2020) 110031.
[26] M.E. Himmel, D.K. Johnson, W.S. Adney, E.C. Rydholm, Biomass recalcitrance: deconstructing the plant cell wall for bioenergy, Annu. Rev. Energy Environ. 32 (2007) 145–173.
[27] R. Santos, J.M. Monteiro, Life cycle assessment of bioenergy production from biomass: a review of recent advances, Renew Sustain. Energy Rev. 77 (2017) 20–31.
[28] A.A. Koutinas, S. Papanikolaou, Policy incentives for bioenergy: a global overview, Energy Policy 138 (2020) 111200.
[29] S. Yang, X. Jin, B. Zhu, D. Yang, X. Wan, Y. Dai, Electrocatalytic oxidation of HMF to FDCA over multivalent ruthenium in neutral electrolyte, Molecules 30 (8) (2025) 1780, doi:10.3390/molecules30081780.
[30] Y. Qi, H. Guo, J. Li, L. Ma, Y. Xu, H. Liu, Recent advances in the electrocatalytic oxidative upgrading of lignocellulosic biomass, ChemPhysMater 3 (2) (2024) 157–186, doi:10.1016/j.chphma.2024.02.001.
[31] Y. Zhang, Y. Shen, Electrochemical hydrogenation of levulinic acid, furfural and 5-hydroxymethylfurfural, Appl. Catal. B Environ. 343 (2024) 123576, doi:10.1016/j.apcatb.2023.123576.

Further readings

K. Amreen, Emerging trends of miniaturized carbon nanomaterial-based energy storage microdevices., Univers. J. Carbon Res (2024) 124–136.
S. Iijima, Helical microtubules of graphitic carbon, nature 354 (6348) (1991) 56–58.

CHAPTER

5

Modeling and simulation in biomass electrochemistry

Megha Pramod[a], Taniya Rose Abraham[a], Neelesh Ashok[b], Deepthi R Murali[c], and Sabu Thomas[a]

[a]School of Energy Materials, Mahatma Gandhi University, Kottayam, Kerala, India, [b]Amrita School of Artificial Intelligence, Amrita Vishwa Vidyapeetham, Coimbatore, Tamil Nadu, India, [c]Post Graduate and Research Department of Chemistry, Sree Narayana College for Women, Kollam, Kerala, India

5.1 Introduction

Increasing demands of energy is resulting in the use of renewable energy sources like wind, solar, and nuclear power. However, due to technical and economic limitations, some of these newer energy technologies still face challenges in widespread adoption. Biomass is a natural energy resources in the earth. Ancient humans were widely dependent on biomass. After the energy crisis occurs at 1970, worldwide research has been promoted for a clean energy resource. Biomass is a natural energy resource formed from living organisms. Nowadays, biomass energy can be obtained by pyrolysis, gasification, combustion, and biochemical fermentation. Biomass was formed by the process of photosynthesis, whereby CO_2 and water are converted to sugars with the energy of the sun. Biomass is derived from wood, crops, industrial, and municipal waste [1].

Biochemical conversion and thermochemical conversion are the two main ways of converting biomass energy into biofuels. Biochemical conversions can be done by the conversion of biomass into gaseous fuels by fermentation or anaerobic digestion. Thermochemical conversion can be done by combustion, gasification, and pyrolysis. Gasification is one of

Electrochemical Biomass Conversion: Sustainable Processes and Applications for Bioenergy, Biofuels, and Bioproducts
DOI: https://doi.org/10.1016/B978-0-443-26757-4.00017-9

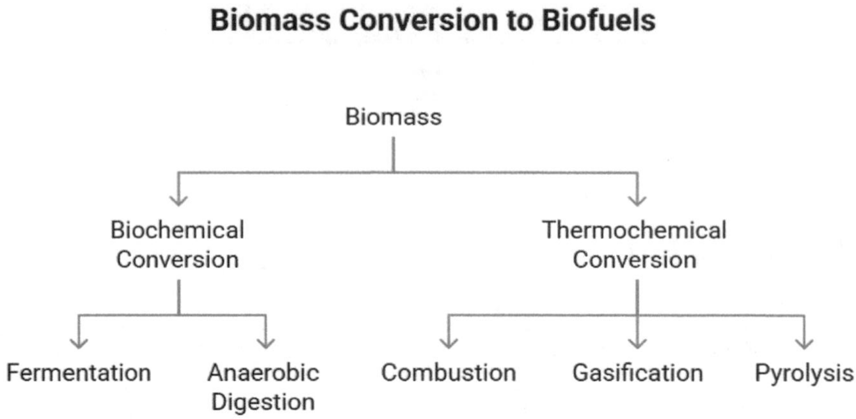

FIGURE 5.1 Biochemical conversion to biofuels.

the major conversion methods, it has the advantage of using low-value feedstocks, producing syngas, offers higher efficiencies, and allows the efficient removal of contaminants like sulfur and nitrogen species from syngas in lower emissions. There are some challenges in gasification, such as inefficiency due to the high temperature required, high capital costs, and technical risks for large-scale commercialization, and high cost of product gas, such as alkali compounds and other contaminants. Main steps in the gasification process are (1) upstream processing prepares the biomass for gasification, (2) gasification is the main process and takes place at a higher temperature (600°C–1000°C) in the presence of a suitable oxidizing agent. The biomass is then converted into gases like carbon monoxide, hydrogen, methane, ash, tar, and minor contaminants. Commonly used reactions for the gasification process are hydrogasification, water gas shift reaction, steam reforming, water gas reaction, and Boudouard reaction (Fig. 5.1) [2].

An example depicts here that biomass-derived molecules, such as glycerol and 5-hydroxymethylfurfural oxidation. Biomass oxidation is integrated with photochemistry, allows for solar-driven reactions, which reduces the energy consumption. This is a photoelectrochemistry device, biomass-derived compounds are selectively oxidized at the photoanode, while hydrogen or other reduction products are generated at the cathode. Glycerol is a byproduct from biodiesel industry. Replacing the thermodynamically demanding Oxygen Evolution Reaction with the oxidation of biomass-derived waste products requires less potential. Biomass oxidation products generally have much higher economic value than oxygen [3]. Studies show that the same glycerol solution in both anolyte and catholyte compartments and optimizing bridge geometry significantly minimizes undesired mixing and reduces ohmic voltage loss. The OPHERA Project, funded by the European Innovation Council, aims to address the challenges through two approaches. The utilization of lead-free halide perovskite nanocrystals is a very highly effective material for photoanode and photocathode. Replacement of the Oxygen Evalution Reaction (OER) with a glycerol oxidation reaction at the anode. This research contributes to the development of more economical and environmental friendly photoelectrochemistry systems for sustainable chemical and fuel production [3].

Another example for modeling carbon electrodes for hybrid supercapacitors using COMSOL Multiphysics 6.2 can be done by Finite Element Method. It is a method used in computational fluid mechanics (CFD) for simulation. This research shows a model for charge storage behavior by the combination of both electric double-layer capacitance and pseudocapacitive mechanisms through cyclic voltammetry (CV). This research aims to provide an understanding of electrode behavior by the optimization of biomass-derived materials for sustainable energy storage systems. CV is a method for electrochemical analysis that shows a variation in a triangular waveform pattern. The CV is mainly used for the studies of ionic motion and reaction rates at the interface, and also for the studies of batteries, supercapacitors, and fuel cells. Simulations of CV can help to increase the storage capacity of these electrochemical devices. The electrodes used in this kind of study can be derived from biomass (carbon) .

Modeling and simulation of bio-electrochemical systems (BES) has gained attention among researchers. Early study shows bioelectrochemical systems mainly focus on biofilm growth, substrate diffusion, and nutrient distribution in biofilm. Nowadays, the modeling of BES can extend its studies to improve the electrode potential and product yield [4]. In this study, a novel 2-thermodynamic model has been developed to understand the relationship between external voltage and microbial yield in BES, and this model integrates Gibbs free energy calculations to predict electron and proton transfers during microbial growth. Modeling and simulation in biomass electrochemistry play a vital role in optimizing BES for efficient energy production and resource utilization. The integration of thermodynamic models, sensitivity analyses, and predictive simulations provides valuable insights into enhancing the performance of biomass conversion processes. These advancements contribute to the development of sustainable energy solutions and the effective utilization of biomass resources [4].

Installing a pilot-plant operations needs to research about modeling approaches, different kinds of simulation approaches, understanding the governing equations, and conducting a validation test to ensure the accuracy of the model. This chapter shows the biomass gasification processes and different modeling approaches, discussing the possibilities of developing biomass gasification application, methods to combine the understanding of physical and statistical models, understanding the governing equations, and the validation of models with experimental data. This chapter presents the combination of physical modeling and statistical modification.

5.2 Theoretical framework

5.2.1 Governing equation

Biomass conversion can be discretized by two models that are continuum model and the discrete models. The continuum model is a classical approach to the numerical model of biomass, where the macroscopic properties are considered in the continues function with respect to time and space. Discrete models are based on the percolation theory, where the medium is represented by 2D or 3D networks. For doing a computational model, the pore size and structure of the biomass are very important factors. The above two models are the recent developments in this field. Modeling biomass electrochemistry involves solving systems of different equations that affect the model. Electrochemical reactions are one of the important factors involved in the reactions. This kind of reaction involves electron transfer and is typically described by the current density (i). Butler–Volmer equation is used for governing the electron transfer.

$$i = i_0\left[\exp\left(\frac{\alpha_a nF\eta}{RT}\right) - \exp\left(-\frac{\alpha_c nF\eta}{RT}\right)\right]. \tag{5.1}$$

Where i and i_0 are the current density and exchange current density, respectively, η is the overpotential, α_a, α_c are the anodic and cathodic transfer coefficients, n is the number of electrons transferred, F is the Faraday constant (96,485 C/mol), R is the universal gas constant, and T is the temperature (K) [5]. Butler–Volmer equation can be reduced to the Tafel equation due to high overpotential. It can be expressed as:

$$i = i_0 \exp\left(\frac{\eta}{b}\right) \tag{5.2}$$

Where b is the Tafel slope [6]. This equation is mainly used for the modeling of biomass electrooxidation to characterize kinetic performance and catalytic efficiency.

Movement of electrons causes the conduction of current. It can be seen as mass transport. Mass transfer modeling can be done by Eq. (5.3).

$$\frac{\partial c_i}{\partial t} = -\nabla \cdot \mathrm{N}_i + R_i \tag{5.3}$$

Where N_i is the flux of i, R_i is the production or consumption of i. Flux can be modeled by the Nernst–Planck equation.

$$N_i = -D_i \nabla c_i - z_i u_i F c_i \nabla \varphi + c_i \mathrm{u} \tag{5.4}$$

z_i is the charge number, u_i is the mobility of the species, c_i is the concentration, D_i is the diffusivity, and u is the fluid bulk velocity [6].

Electrolyte current density can be governed by Eq. (5.5).

$$i_{\text{electrolyte}} = F \sum z_i N_i \tag{5.5}$$

F is the Faradays constant.

The selection of the governing equation used to determine the potential field is by different approaches in each case. Three important models are The Nernst–Planck–Poisson equation (NPP) model, the Nernst–Planck model with the electroneutrality approximation, and the Laplace equations. The NPP equation is used to govern the mass and charge transport of species. This model is a combination of both NPP equations for determining the mass conservation and charge density. NPP is very difficult to compute, so the most easily method is assuming that the electrolyte is electroneutral. The Poisson equation can be expressed as in Eq. (5.6) and electroneutrality can be expressed as in Eq. (5.7).

$$\nabla^2 \Phi = -\frac{\mathrm{F}}{\epsilon} \sum_i z_i c_i \tag{5.6}$$

$$\sum_i z_i c_i = 0 \tag{5.7}$$

Current density in the electrolyte can be expressed by Ohm's Law in Eq. (5.8).

$$i = \kappa \nabla \Phi \tag{5.8}$$

where κ is electrical conductivity [6].

Mass transfer in electrochemistry can be done by three mechanisms, they are diffusion, convection, and migration. Diffusion is the movement of species from high to low concentration driven by a concentration gradient, and this process is governed by Fick's law. Fick's first law states that the flux of particles is proportional to the negative gradient of concentration.

$$J = -D \frac{\partial C}{\partial x} \tag{5.9}$$

where J is the diffusive flux, and D is the diffusion coefficient.

Convection is defined as the movement of species due to bulk fluid flow and is expressed as the convection-diffusion equation.

$$\frac{\partial C}{\partial t} + \vec{u} \cdot \nabla C = D \nabla^2 C + R \tag{5.10}$$

where $D\nabla^2 C$ is the diffusive transport term, C is the concentration of species, and $\vec{u}\cdot\nabla C$ is the convective transport term.

The mass transfer coefficient can be expressed as the rate of mass transfer to the concentration difference (k_m).

$$J = k_m(C_{\text{bulk}} - C_{\text{surface}}) \tag{5.11}$$

Where J is the mass flux, k_m is the mass transfer coefficient, C_{bulk} is the concentration of the species in the bulk fluid, and $C_{surface}$ is the concentration of the species at the electrode or surface.

5.3 Modeling approaches

5.3.1 Thermodynamic models in biomass gasification

Thermodynamic models are used to understand how materials behave when they reach a balanced state (temperature, pressure, and composition). There are two types of thermodynamics model, such as the stoichiometric model and the nonstoichiometric model. Stochiometric models are based on mass balance and equilibrium constant, where nonstochiometric models are based on nonequilibrium constant. Nonstoichiometric models are based on minimizing Gibbs free energy [7]. This kind of model is used to describe the fuel and process parameters. Mass and Energy balance results in a system of nonlinear algebraic equations are found by applying Newton–Raphson methods. The thermodynamic model is time invariant and independent of kinetic factors. This model is simple and easy to implement, flexible to use, provide idea about energy recovery, unit operation, process optimization, and economic analysis [8]. Apart from the pros, there are some disadvantages in this method, they work under many assumptions. When compared to a real case, equilibrium may not be reached in some cases at low temperatures.

The nonstoichiometric method is also known as the quasi-equilibrium approach, which is used to calculate the equilibrium through a virtual model. A model named the quasi-equilibrium approach is used to model the nonstoichiometric approach. This model can be trained to minimize the error (Fig. 5.2) [7,8].

5.3.2 Kinetic models in biomass gasification

The kinetic model is used to calculate gas yield, product composition, temperature inside the module, and gasifier performance. The kinetic model is simple and accurate with compare with thermodynamic models. In a gasification process, reaction kinetics give the information about mass and energy balances. The rate at which char undergoes gasification is

Characteristic	Stoichiometric Model	Non-Stoichiometric Model
Basis	Mass balance and equilibrium constant	Minimizing Gibbs free energy
Equilibrium	Equilibrium-based	Quasi-equilibrium
Time Dependence	Time-invariant	Time-invariant
Kinetic Factors	Independent	Independent

FIGURE 5.2 Difference between stoichiometric and nonstoichiometric model.

typically defined either by the external surface area of the biomass char or relative to the volume of the reactor. This is because, during gasification, the biomass particle loses mass, which can appear as either a decrease in particle size while maintaining the same density or a reduction in density while the size stays constant. When the process involves only char, the reaction rate is generally expressed based on the reactor's volume. To analyze this behavior, models such as the shrinking core model, shrinking particle model, or the volumetric reaction rate model are commonly used [8]. Kinetic models can be classified into these models due to the hydrodynamics 0D, 1D, 2D, and 3D.

Jayah et al. [9] developed a two-zone kinetic model for a downdraft gasifier, which simulates how biomass converts into gas. The first model is the flaming pyrolysis model use basic concepts of chemical equilibrium, mass and energy balance principle, whereas the second model is the gasification zone, which includes the physical and chemical process, flow equations, transport phenomena, and conservation laws. This model also predicts gas composition and temperature. The kinetics of pyrolysis can be described by two models, and the equations are modeled by the Runge–Kutta technique. Researchers develop the sub-topics between each step and obtain the energy efficiency and power output [10].

The first step in kinetic modeling is understand the gasification zones such as drying, pyrolysis, oxidation, and reduction. The second step is identifying key reactions, each reaction is associated with a rate expression, often based on Arrhenius kinetics. The third step is formulated governing equations such as mass conservation, energy conservation, momentum, and transport phenomena, these equations are typically solved in 1D, 2D, or 3D complexity. The fourth step is to determine kinetic parameters. The

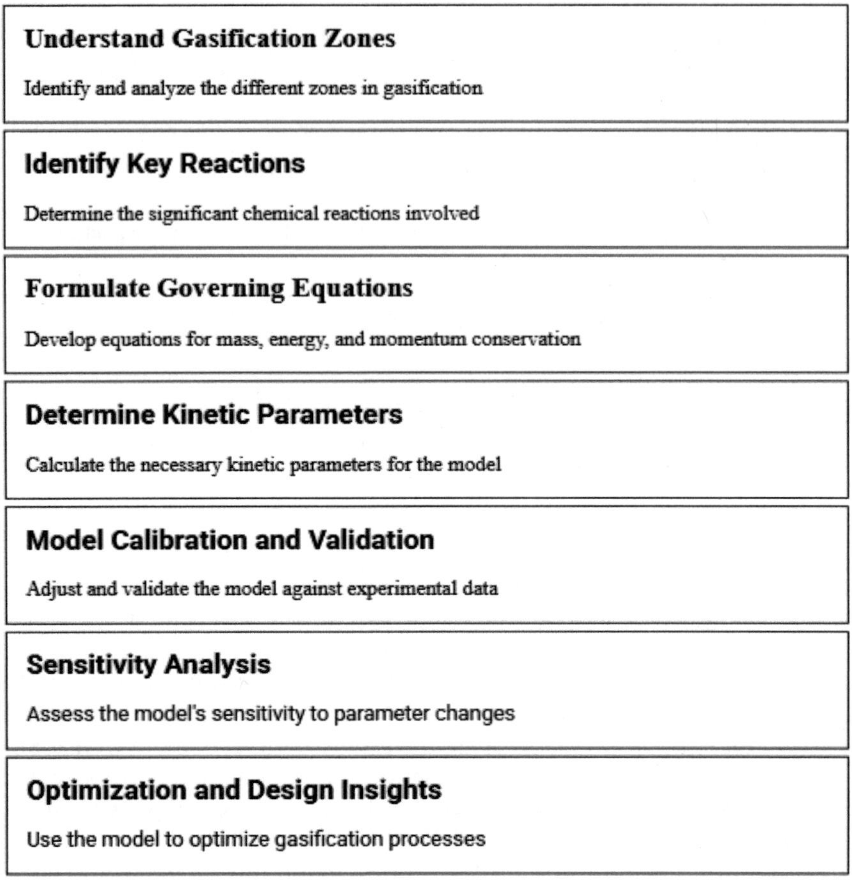

FIGURE 5.3 Steps for kinetic modeling in biomass gasification.

fifth step is model calibration and validation. The seventh step is sensitivity analysis, and the last step is optimization and design insights (Fig. 5.3) [11].

5.3.3 Computational fluid dynamic modeling

CFD is a simulation tool to simulate fluid flow, heat, mass transfer, and chemical reactions. CFD serves as a powerful tool for investigating mechanisms. It helps in obtaining accurate operational parameters for reactors, which are difficult to fully explain or reduce through experience or direct observation. Compared to experimental testing, CFD modeling is more cost-saving, timely, and theoretically easier to scale up. CFD is based on methods like finite difference or finite element, requiring extensive grid division and iterative solutions of node equations, which consume a lot of computing resources compared to kinetic modeling. For example, a CFD-Discrete Element Method (DEM) simulation for a 100 g/h fluidised bed reactor consumed over 320,000 CPU machine hours for the reaction process. This high cost limits its availability to most researchers. CFD simulation in thermochemical biomass includes combustion, pyrolysis, and gasification. It is used to simulate gasifiers in biomass thermochemical modeling. From the discrete element method, particle interactions and hydrodynamics can be calculated. Models like the coarse-grain model or random pore model (RPM) are used to reflect heterogenous gasification reaction more easily. A CFD model is also possible to couple with other models to obtain the result accurately, such as a model called Flamelet-Generated Manifold [12–14].

In biomass gasification, fluidised bed gasifier and fixed-bed gasifier are the most commonly used reactor which use CFD model to optimize. RPM suggested that the reaction rate depends on the surface area inside the pores [15]. This surface area changes due to two opposing effects: pore expansion, which increases the surface area, and pore merging. Total specific surface area can be expressed as in Eq. (5.12).

$$s = s_0(1 - x)\sqrt{1 - \phi \ln(1 - x)} \tag{5.12}$$

where S is the specific surface area, S_O is the true specific surface area of unreacted solid particles per unit volume. x is the reaction conversion rate, which is expressed as in Eq. (5.13).

$$x = 1 - \exp\left[-t_f t(1 + \phi t_f t/4)\right] \tag{5.13}$$

Where t_f is the time factor.

13-reaction studies and 16-reaction studies are used to simulate the partial oxidation zone in a two-stage gasifier [16–18]. On the other hand, detailed kinetic mechanisms often involve hundreds of chemical species and thousands of elementary reactions, making them too complex for direct use with CFD models. As a result, such mechanisms typically require simplification before being incorporated into CFD simulations. However, detailed kinetic mechanisms can still be used independently for optimizing reaction parameters. While CFD models are crucial for analyzing and predicting biomass thermochemical conversion processes, challenges remain—especially in accurately modeling the interactions between multiple phases, such as gases, volatiles, and solid components like biomass and biochar. For example, during pyrolysis, volatiles generated inside biomass particles must pass through the solid matrix and escape, a process governed by complex physical and chemical phenomena [17,18]. In this technique, the most difficult problem is the multiphase interaction between gases, organic volatiles, and solid particles. The coupling interaction and heterogenous reaction mechanism is a challenge in CFD [19].

5.3.4 Artificial neural network modeling

Considering the biomass thermochemical reaction, artificial neural network (ANN) modeling is very easy because it does not consider the detailed pyrolysis or gasification mechanism [20]. It is an AI based methods and is used as a black box model. The major advantage of this method is that it can get output without a comprehensive overview of the input. The term black box refers to the modeling approach where the internal workings of a system are unknown [21]. It is possible to model the kinetic parameters of biomass pyrolysis of cellulose, hemicellulose, and lignin. The ANN consists of neurons and is used to recognize patterns and different functions. It has three layers, which are the input layer, the hidden layer, and the output layer. It is possible to minimize the error by continuous iteration [22]. Steps to train the ANN model are collecting the database, analyzing and preprocessing the data, training the ANN, testing the trained network, and the use of the trained ANN for simulation and prediction [23]. For the biomass modeling, the Levenberg–Marquardt (LM) was chosen because it combines the gradient descent and Gauss–Newton methods. LM is a second-order method and is effective for nonlinear problems. It can be represented as Eq. (5.14).

$$w_{t+1} = w_t - \left(J^T J + \mu I\right)^{-1} J^T e \tag{5.14}$$

Where w is the weight vector, J is the Jacobian matrix, μ is the combination coefficient, I is the identity matrix, and e is the error value [24].

The application of biomass higher heating value prediction is done by a nonlinear model. Support vector machine, Kohonen, Hopfield, radial basis function, Adaptive Neuro, Fuzzy Interface systems, and feedforward back propagation are the different types of ANN. The learning and testing of datasets are based on trial and error method. The number of nodes and hidden neurons can be determined by the rule of thumb [25].

In conclusion, further research is necessary in order to develop universal correlations that can accurately predict the properties of biomass from varied sources. Advances in key areas of model development, such as data partitioning, network architecture design, and modeling techniques for limited data sets (Fig. 5.4) [25].

5.4 Case studies in biomass gasification

Studies shown that carbon electrodes from biomass are used for the supercapacitors, and the CV is simulated through COMSOL Multiphysics. This helps in the modeling of ionic transport, redox reactions, and potential variation. The factors mainly affecting the simulations are secondary current distribution (determines potential distribution), transport of diluted species (determines the ionic transport via diffusion and migration

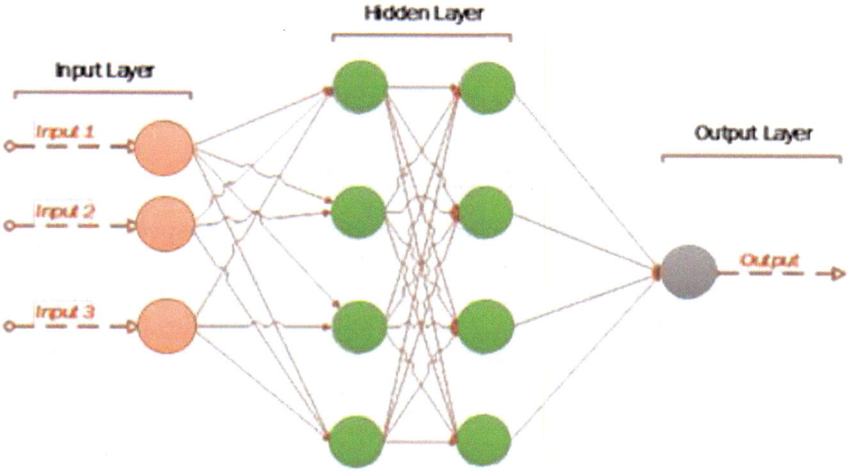

FIGURE 5.4 Schematic showing a multilayer ANN.

of electrolyte), and the electrochemistry aspect (Butler–Volmer equation). The results from studies provide the ion adsorption and improve the surface accessibility for ion adsorption. This offers a better understanding of electrochemical reactions using biomass-derived carbon materials [,26].

Biomass reforming is possible to produce chemicals and fuels, which produces carbon-neutral energy production. Biomass oxidation, which can be combined by photoelectrochemistry, will reduce the energy consumption. Simulation of the model will offer cost efficient and energy consumption model. Photovoltaics integrated with electrolysis require separate investments in electrolyser and conversion devices. This research uses COMSOL for modeling, and the CFD and Euler–Euler Methods are used for the simulation. Euler–Euler method is used for studying the properties of liquid and gas bubble mixtures. This study helps us to understand the feasibility of membrane configurations, material costs, and operational stability [3,27].

Another study shows that electrode materials derived from biomass can be used in supercapacitors or Li-ion batteries. In this study, silica is incorporated with carbon materials, and the simulation is done by density functional theory simulation and molecular dynamics simulation (MD). Density of biochar obtained from the experiment computed with MD simulations. This work provides carbon-based electrode materials for energy storage devices [28].

Researches in high temperature solid oxide fuel cell (SOFC) stacks on biomass syngas is also developed. In this study, combined heat and power is investigated. SOFC converts chemical energy to electrical energy, so it is an electrochemical device. Aspen Plus is used to model the SOFC. The methodology in that work is developing a programming language and link it to Aspen Plus. This model provides an issue with realistic operation, the SOFC stack needs to be operated in a low steam-to-carbon ratio. When integrated with biomass, it has a lack of efficiency [29].

5.5 Conclusion

Mathematical modeling is an important tool in a real aspect to optimize the design and operation, which offers less cost and time-investing. A well-developed mathematical model can quickly determine the optimal feedstock by simulating outputs based on its properties. This type of model is used describe the chemical and physical processes inside the gasifier. Commonly used modeling approaches in biomass gasification studies include thermodynamic modeling, kinetic modeling, CFD, and ANN modeling.

Thermodynamic modeling has proven effective for predicting the performance of downdraft gasifiers, as both pyrolysis and gasification products pass through the high-temperature oxidation zone, allowing equilibrium to be established rapidly. These types of models have some limitations, which can be recovered by integrating empirical correlations from experimental data.

Kinetic modeling combines reaction kinetics and hydrodynamics. Reaction kinetics uses the equation of bed hydrodynamics with mass and energy balances to predict yields of gas, tar, and char under specific conditions, while reactor hydrodynamics accounts for physical mixing.

CFD integrates the strengths of various models to study gasifier design and behavior. Achieving CFD simulations requires chemical reaction mechanisms combined with advanced numerical methods for multiphase flows.

ANN modeling has emerged as an alternative modeling approach for studying biomass gasification, and numerical simulations without the knowledge of a complex mechanism in the model.

This chapter discusses the recent developments in biomass electrochemistry modeling, classifying them based on factors such as gasifier type, feedstock, modeling approach, and evaluated parameters, offering researchers and decision-makers guidance in selecting suitable modeling strategies or gasifier designs for specific applications.

References

[1] S. Zhao, Y. Luo, Multiscale modeling of lignocellulosic biomass thermochemical conversion technology: an overview on the state-of-the-art, Energy Fuels 34 (10) (2020) 11867–11886.

[2] A. Kumar, D.D. Jones, M.A. Hanna, Thermochemical biomass gasification: a review of the current status of the technology, Energies 2 (3) (2009) 556–581.

[3] A.F.P. Torres, H. Kong, S.S. Krishnamurti, F. Liang, S. Gimenéz, R. van de Krol, M. Favaro, Multiphysics modeling of photoelectrochemical devices for simultaneous solar-driven biomass reforming and hydrogen production, Energy Fuels 39 (26) (2025) 12716.

[4] V. Ahmadi, N. Aryal, Bioenergetic modeling of the relationship between voltage and electroactive microbial biomass yield for bioelectrochemical carbon dioxide reduction to methane, Fermentation 11 (1) (2025) 40.

[5] K.M. Fischer, D.J. Batstone, M.C. van Loosdrecht, C. Picioreanu, A mathematical model for electrochemically active filamentous sulfide-oxidising bacteria, Bioelectrochemistry 102 (2015) 10–20.

[6] C.V. Moraes, R.G. Kelly, A comparison of FEM results from the use of different governing equations in a galvanic cell part I: in the presence of a supporting electrolyte, Electrochim. Acta 469 (2023) 143146.

[7] V. Marcantonio, L. Di Paola, M. De Falco, M. Capocelli, Modeling of biomass gasification: from thermodynamics to process simulations, Energies 16 (20) (2023) 7042.

[8] D. Baruah, D.C. Baruah, Modeling of biomass gasification: a review, Renew. Sustain. Energy Rev. 39 (2014) 806–815.

[9] T.H. Jayah, L. Aye, R.J. Fuller, D.F. Stewart, Computer simulation of a downdraft wood gasifier for tea drying, Biomass Bioenergy 25 (4) (2003) 459–469.

[10] B.V. Babu, A.S. Chaurasia, Modeling, simulation and estimation of optimum parameters in pyrolysis of biomass, Energy Convers. Manage. 44 (13) (2003) 2135–2158.

[11] P. Sharma, B. Gupta, M. Pandey, K.S. Bisen, P. Baredar, Downdraft biomass gasification: a review on concepts, designs analysis, modelling and recent advances, Mater. Today: Proc. 46 (2021) 5333–5341.

[12] M. Zhang, Z. Geng, Y. Yu, Density functional theory (DFT) study on the pyrolysis of cellulose: the pyran ring breaking mechanism, Comput. Theor. Chem. 1067 (2015) 13–23.

[13] L. Abella, K. Yamamoto, K. Fukuda, S. Nanbu, N. Oikawa, K. Morita, T. Matsumoto, Ab initio calculations for the reaction paths of levoglucosan: an intermediate of cellulose. https://www.semanticscholar.org/paper/Ab-Initio-Calculations-for-the-Reaction-Paths-of-An-Abella-Yamamoto/04694c7ca3820210fef23cbb050657cfe38546db.

[14] S. Zhao, Y. Luo, Multiscale modeling of lignocellulosic biomass thermochemical conversion technology: an overview on the state-of-the-art, Energy Fuels 34 (10) (2020) 11867–11886.

[15] S.K. Bhatia, D.D. Perlmutter, A random pore model for fluid-solid reactions: isothermal, kinetic control, AlChE J. 26 (3) (1980) 379–386.

[16] X. Yu, P.H. Blanco, Y. Makkawi, A.V. Bridgwater, CFD and experimental studies on a circulating fluidised bed reactor for biomass gasification, Chem. Eng. Process.-Process Intensif. 130 (2018) 284–295.

[17] L. Gerun, M. Paraschiv, R. Vijeu, J. Bellettre, M. Tazerout, B. Gøbel, U. Henriksen, Numerical investigation of the partial oxidation in a two-stage downdraft gasifier, Fuel 87 (7) (2008) 1383–1393.

[18] Y. Su, Y. Luo, Y. Chen, W. Wu, Y. Zhang, Experimental and numerical investigation of tar destruction under partial oxidation environment, Fuel Process. Technol. 92 (8) (2011) 513–552.

[19] D. Xu, L. Yang, K. Ding, Y. Zhang, W. Gao, Y. Huang, et al., Mini-review on char catalysts for tar reforming during biomass gasification: the importance of char structure, Energy Fuels 34 (2) (2019) 1219–1229.

[20] G. Özbay, E.S. Kökten, Modeling of bio-oil production by pyrolysis of woody biomass: artificial neural network approach, Politeknik Dergisi 23 (4) (2020) 1255–1264.

[21] C. Li, D. Guo, Y. Dang, D. Sun, P. Li, Application of artificial intelligence-based methods in bioelectrochemical systems: recent progress and future perspectives, J. Environ. Manage. 344 (2023) 118502.

[22] X. Chen, H. Zhang, Y. Song, R. Xiao, Prediction of product distribution and bio-oil heating value of biomass fast pyrolysis, Chem. Eng. Process.-Process Intensif. 130 (2018) 36–42.

[23] T. Parthiban, R. Ravi, N. Kalaiselvi, Exploration of artificial neural network [ANN] to predict the electrochemical characteristics of lithium-ion cells, Electrochim. Acta 53 (4) (2007) 1877–1882.

[24] Ş.T. Güner, M.J. Diamantopoulou, K.P. Poudel, A. Çömez, R. Özçelik, Employing artificial neural network for effective biomass prediction: an alternative approach, Comput. Electron. Agric. 192 (2022) 106596.

[25] O. Obafemi, A. Stephen, O. Ajayi, M. Nkosinathi, A survey of artificial neural network-based prediction models for thermal properties of biomass, Procedia Manuf. 33 (2019) 184–191.

[26] H. Kim, B.N. Popov, A mathematical model of oxide/carbon composite electrode for supercapacitors, J. Electrochem. Soc. 150 (9) (2003) A1153.

[27] V. da Fonseca Dias, J. Dias da Silva, Mathematical modelling of the solar-driven steam reforming of methanol for a solar thermochemical micro-fluidized bed reformer: thermal performance and thermochemical conversion, J. Braz. Soc. Mech. Sci. Eng. 42 (9) (2020) 447.

[28] J. Muniz, N.D. Espinosa-Torres, A. Guillen-Lopez, A. Longoria, A.K. Cuentas-Gallegos, M. Robles, Insights into the design of carbon electrodes coming from lignocellulosic components pyrolysis with potential application in energy storage devices: a combined in silico and experimental study, J. Anal. Appl. Pyrolysis 139 (2019) 131–144.

[29] W. Doherty, A. Reynolds, D. Kennedy, Modelling and simulation of a biomass gasification-solid oxide fuel cell combined heat and power plant using Aspen Plus, Energy 35 (12) (2009) 4545–4555.

CHAPTER

6

Electrochemical fermentation of biomass

Andal Gopal Venugopal[a], *Suba Kannaiyan*[b], *Priya Bijesh*[a], *Selvaraj Vaithilingam*[a,c], *Vetha Gopal*[a], *and Jayanthi Gopal*[a]

[a]Department of Chemistry, KCG College of Technology, Chennai, Tamil Nadu, India, [b]Department of Chemistry, Sri Sairam Engineering College, Chennai, Tamil Nadu, India, [c]Department of Chemistry, University College of Engineering, Villupuram, Tamil Nadu, India

6.1 Introduction

Fermentation is a fascinating metabolic process that allows organisms to convert sugars, primarily glucose, into energy in the absence of oxygen. Fermentation is a vital biological process that not only allows various organisms to survive in anaerobic conditions but also plays a significant role in food production and biotechnology. Its versatility and efficiency make it an essential topic in both microbiology and food science. But the conventional fermentation processes, while widely used for food and beverage production, face several limitations that can affect efficiency, scalability, time-consumption, complexity, cost, and product quality [1,2]. Electrochemical fermentation (EF) has emerged as an innovative method to address the limitations of conventional fermentation processes. By integrating electrochemistry with traditional fermentation, EF offers several advantages that enhance efficiency and product yield [3]. In order to reduce the usage of fossil fuels and the use of biomass as feedstock for chemical industries [4].

Biomass, derived from renewable organic materials of plant and animal origin, offers a promising avenue for reducing carbon dioxide emissions

Electrochemical Biomass Conversion: Sustainable Processes and Applications for Bioenergy, Biofuels, and Bioproducts
DOI: https://doi.org/10.1016/B978-0-443-26757-4.00018-0

[5]. The production of chemicals from biomass, also known as bio-based chemicals, is critical in creating a sustainable and environmentally friendly future, particularly as the world strives to reduce dependence on fossil fuels [6]. Microbial fermentation is indeed used to produce a wide variety of chemicals in bio-based industries [7].

Microbial fermentation is a biotechnological process that converts sugars derived from lignocellulosic biomass into various biofuels and value-added biochemicals. This process utilizes microorganisms such as fungi, yeast, and bacteria to break down complex carbohydrates into simpler sugars, which are then fermented into desired products. Microbial fermentation, while promising for biofuel production, faces several challenges across various stages of the process. These challenges include scalability, cost, product quality, and technological gaps [8, 9].

EF presents several advantages over conventional fermentation methods for biomass conversion into biofuels and biochemicals. The key benefits include

Higher carbon efficiency: EF improves carbon utilization by enhancing microbial metabolism through controlled electron transfer, leading to increased yields of target products compared to traditional fermentation processes [10].

Improved selectivity: The application of an electrical field can steer metabolic pathways, allowing for the production of specific biochemicals with greater selectivity and reduced formation of unwanted by-products [11].

Regulation of redox balance: EF allows for better control of redox conditions, which can stabilize microbial growth and enhance product formation. This is particularly beneficial in handling complex substrates, such as waste materials [12].

The present chapter deals with redox restriction, the principle of EF, various types of electro-fermentation systems, application, value-added fuels and chemical synthesis, challenges and opportunities, and future perspectives.

6.2 Redox restriction in conventional fermentation

The redox reaction involves the electron transfer process, in which the oxidizing one (Ox) attracts electrons and then becomes the reducing one (Red). The "redox restriction" in conventional fermentation refers to the inherent requirement of microbes to preserve intracellular redox equilibrium in a closed system devoid of an external electron acceptor such as oxygen [13]. The redox reaction requires a specific reduction-oxidation potential to avoid sluggish fermentation. The oxidation-reduction potential (ORP)

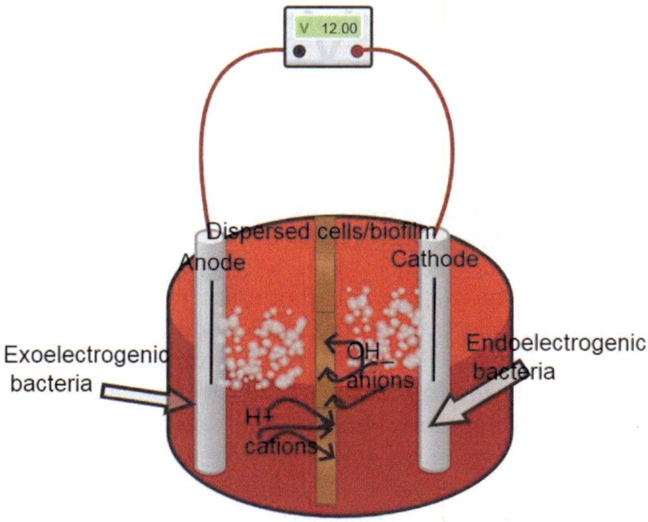

FIGURE 6.1 The principle of electromagnetic fermentation.

can be visualized by recording a graph (time-series) during fermentation to monitor microbial activity. The high oxidation-reduction potential will always support aerobic bacteria during fermentation, whereas the low oxidation-reduction potential supports anaerobic bacteria. When the microbial fermentation is active, the ORP shows a bathtub structure. When the fermentation is restricted or sluggish, there is a decrease or drop in ORP. To overcome the redox restriction in conventional fermentation, a hybrid process combining microbial fermentation with electrochemical technology is applied. The electric potential will favor the artificial electrons to meet the demand of the bioreactor [14].

6.3 Principle of electrochemical fermentation

EF is a frontier research in which the electrodes act as artificial electron donors to control the microbial metabolism. By using the electrodes, the researchers control the metabolic pathway of microorganisms to achieve the required product, which is a notable advantage over conventional fermentation [15]. The principle of EF is the combined process of microbial fermentation and electrochemical systems to provide a controlled redox imbalance, higher product yield, and increased biomass [16]. The principle of electromagnetic fermentation is represented in Fig. 6.1. Electrofermentation is used for electricity generation, treatment of waste effluents in wastewater, bioremediation, and generation of high-value products [17].

6.4 Electro fermentation types and mechanisms

Electro fermentation types are categorized based on their electrode functions, electron transfer mechanisms, microorganism culture, reactor configuration, and applications.

Based on the electrode function, such as electron donor or acceptor, the electro fermentation is classified as cathode electro fermentation and anodic electro fermentation [18].

- Cathode electro fermentation: In cathode electro fermentation (CEF), the electrode supplies electrons to the microorganisms to drive the redox process. This enhances the NaDH/NaD+ ratio, increasing the yield of reducible products such as butanol, caproate, or methane.
- Anodic electro fermentation: In the anodic electro fermentation (AEF) process, the microorganism transfers excess electrons to the anode, which enhances the anaerobic production of chemicals. Microbes such as Actinobacillus succinogenes, Enterobacter aerogenes, and Corynebacterium glutamicum are used to synthesize succinate, ethanol, lysine, etc. [19].

Based on the electron transfer mechanism between the electrode and the microbes the electro fermentation can be classified into two types:-

- Direct electron transfer: The microbes transfer the electrons through the outer cell wall directly to the electrode, where they are adhered. This is also called direct extracellular electron transfer.
- Shuttle-mediated electron transfer: When the direct electron transfer is insufficient, the redox-active compounds act as mediators to transfer the electrons between microbes and the electrode [20].

Electro fermentation is classified into two types based on its microorganism culture:-

- Pure culture: This method utilizes a single isolated species of microorganism to control the metabolic pathway and yields a highly selective and pure product.
- Mixed culture: This method utilizes mixed cultures obtained from waste material as substrates to control the metabolic pathway to yield higher production [21].

Electro fermentation is classified into three types based on the design of the reactor:-

- Single chamber reactor: In this process, both electrodes are in the same chamber, and the setup is low-cost, but the products formed at both electrodes will interact with each other.

- Double chamber reactor: In this process, a proton exchange membrane is used as a separator between the electrodes, and its setup is costlier, and it has greater control over the metabolic pathway and yields a pure product.
- Hybrid tractor: The reactor chamber is designed based on the applications into tubular, stacked, and baffled for enhanced efficiency in various industrial processes [22].

6.4.1 Electro fermentation based on its applications

- Microbial fuel cells (MFC)
- Microbial electrolysis cells (MEC)
- Microbial remediation cells (MRC)
- Microbial desalination cells (MDC)
- Microbial electrosynthesis (MES)

6.5 Application of electro fermentation in various industrial processes

6.5.1 Chemical and biochemical production

Electro fermentation is used to produce value-added chemicals such as biofuels, biochemicals, and biopolymers. The application of electro fermentation is represented in the Fig. 6.2. Biofuels meet the energy demands of today's society by reducing greenhouse gases. Biofuels are an alternative to fossil fuels, which include biodiesel, bioethanol, and biomethanol.

6.5.2 Bioethanol

Bioethanol production from biomass is a sustainable way of reducing the depletion of energy and carbon

Footprint. Recently, bioethanol has been used worldwide, and the raw materials used to produce bioethanol are starch, sugar, and biomass [23]. Microbial electrolysis cell technology is used for the production of bioethanol at low cost from lignocellulose bio oil [24].

6.5.3 Biohydrogen

In recent years, microbial electrolysis cells have been developed as a novel technique for the synthesis of hydrogen from organic waste [25]. The advantages of this process are that they used organic waste as an electron source, it overcomes the conventional fermentation, and the yield was high compared to other H_2 production [26, 27]. Substrates that are used for the generation of hydrogen are listed in Table 6.1.

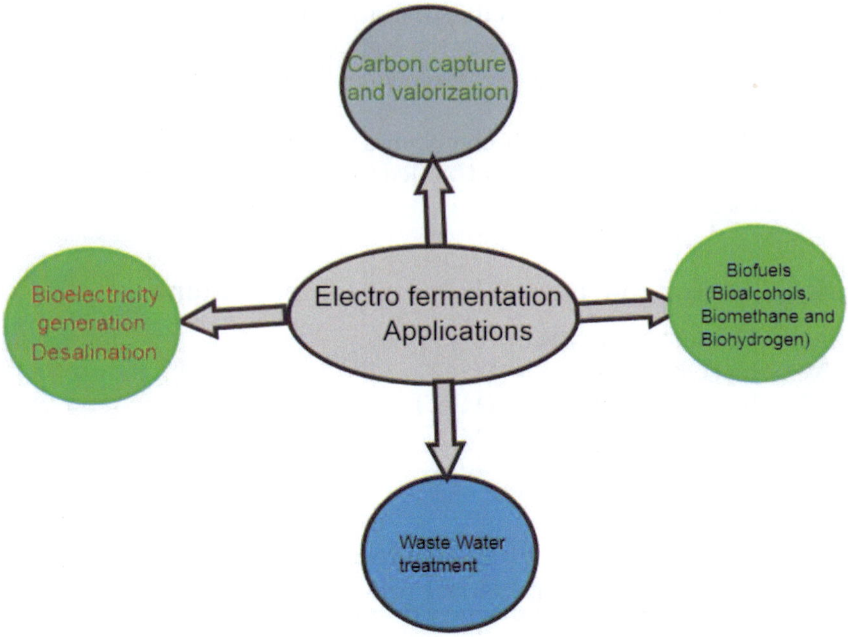

FIGURE 6.2 Application of electro fermentation.

TABLE 6.1 List of substrates used for hydrogen production.

S.No	Substrate Used for Hydrogen Generation	Reference
1	Acetate	[27]
2	Glucose, sucrose	[27]
3	Xylose	[27]
4	Formate	[27]
5	Xanthate	[27]
6	Wastewaters (domestic, municipal, brewery, distillery, pharmaceutical, landfill)	[27,28]
7	Ethanol effluent, slurry	[27]

To overcome the high operational costs associated with electro-fermentation for hydrogen production, researchers are exploring various cost-effective and efficient alternative electrode materials to replace traditional, expensive ones. Alternative anode materials, which were used for electro fermentation for hydrogen production, are recycled carbon fiber, stainless steel, and modified carbon-based materials [29]. Recently, carbon-based materials have been used as an anode material due to their low cost, high surface area, high chemical and mechanical stability, biocompatibility, and electron transfer kinetics. Commonly explored materials by researchers are carbon felt, rod, fiber, cloth, mesh, paper, activated carbon cloth, glassy carbon, brushes, reticulated vitreous carbon, and graphite. Researcher Wang et al. [30] have observed that doped carbon cloth exhibits a high current efficiency of 2777.7 mW/m^2 due to the nitrogen doping. Similarly, graphite-doped materials are used as anode materials, which increases the bacterial contact by reducing the electrode potential [31]. Carbon nanoparticles shows highest current density of 7135 ± 110 mA/m^2, which is cost-effective [32].

6.6 Desalination

Microbial desalination cells are an important type of electro-fertilization for desalinating seawater and generating electricity. This technology addresses two important global issues waste water treatment and freshwater scarcity. This technology also provides clean water for arid and underserved areas. The factors that affect the microbial desalination cells are temperature, pH, substrate concentration, and electrode materials. Recently, research was carried out to generating electricity from dairy wastewater by using ZnO, carbon electrodes, and potassium permanganate. The researchers observed that increasing $KMNO_4$ enhances the voltage [33]. Temperature, pH, ionic strength, and salinity are the most important factors in the growth of bacteria, which affect the performance of microbial desalination cells. Material and construction of the anode, cathode, and membrane are also influential and must be considered in designing the microbial desalination cells due to their cost and performance [34].

6.7 Challenges and opportunities

Despite all the developments in electro-fermentation, the scientific community still faces many challenges and hurdles in terms of electricity generation and water pollutant treatment. Many types of electro-fermentation have been developed, but the practical implementation has not been achieved. The main reason for it is economic feasibility. Future research will focus on the electro-fermentation scalability for the production of chemicals and value-added products.

6.7.1 Concluding remarks

This chapter summarizes the electro fermentation principle, types, redox restriction, and applications. We discussed biofuels, biohydrogen generation, recent research, and their importance. We discussed types of materials that serve as anodes in MFCs, such as carbon-based materials. MDC is an advanced technology of electro-fermentation, which performs desalination simultaneously with wastewater treatment and power generation.

References

[1] N. Abbaspour, Fermentation's pivotal role in shaping the future of plant-based foods: an integrative review of fermentation processes and their impact on sensory and health benefits, Appl. Food Res. 4 (2) (2024) 100468, doi:10.1016/j.afres.2024.100468.

[2] S.M. Mirsalami, M. Mirsalami, Advances in genetically engineered microorganisms: transforming food production through precision fermentation and synthetic biology, Future Foods 11 (2025) 100601, doi:10.1016/j.fufo.2025.100601.

[3] A. Schievano, T.P. Sciarria, K. Vanbroekhoven, H. De Wever, S. Puig, S.J. Andersen, et al., Electro-fermentation–merging electrochemistry with fermentation in industrial applications, Trends Biotechnol. 34 (11) (2016) 866–878, doi:10.1016/j.tibtech.2016.04.007.

[4] D.T. Troiano, T. Hofmann, S. Brethauer, M.H-P. Studer, Toward optimal use of biomass as carbon source for chemical bioproduction, Curr. Opin. Biotechnol. 81 (2023) 102942, doi:10.1016/j.copbio.2023.102942.

[5] F. Ali, A. Dawood, A. Hussain, et al., Fueling the future: biomass applications for green and sustainable energy, Discov. Sustain. 5 (2024) 156, doi:10.1007/s43621-024-00309-z.

[6] P. Chowdhury, N.A Mahi, R. Yeassin, N.-U.r-R. Chowdhury, O. Farrok, Biomass to biofuel: impacts and mitigation of environmental, health, and socioeconomic challenges, Energy Convers. Manag.: X 25 (2025) 100889, doi:10.1016/j.ecmx.2025.100889.

[7] P. Mena, B. Singh, Microbial fermentation in food and beverage industries: innovations, challenges, and opportunities, Foods 14 (1) (2025) 114, doi:10.3390/foods14010114.

[8] L. Liu, F. Wang, L. Wang, X. Jiang, H. Zhang, Addressing challenges in microbial manufacturing: systematic microbial biotechnology, Innovation (Camb) 6 (6) (2025) 100871, doi:10.1016/j.xinn.2025.100871.

[9] A. Pandey, C. Larroche, S.C. Ricke, C.-G. Dussap, E. Gnansounou, Microbial engineering in biofuel production—a global outlook, advances, and roadmap, Biofuels: Alternative Feedstocks and Conversion Processes for the Production of Liquid and Gaseous Biofuels, Elsevier, 2024, pp. 547–593, doi:10.1007/978-3-031-52167-6_22.

[10] S. Goswami, H. Rana, S. Chatterjee, Anjana, A. Sharma, Electro-fermentation for Circular Carbon Economy, Springer, 2025, doi:10.1007/978-981-96-6358-3_10.

[11] C. Liu, Y. Yin, C. Chen, X. Zhang, J. Zhou, Q. Zhang, et al., Advances in electricity-steering organic waste bio-valorization for medium chain carboxylic acids production, Energies 16 (2023) 2571, doi:10.3390/en16062571.

[12] N. Wang, M. Gao, S. Liu, W. Zhu, Y. Zhang, X. Wang, et al., Electrochemical promotion of organic waste fermentation: research advances and prospects, Environ. Res. 244 (2024) 117422, doi:10.1016/j.envres.2023.117422.

[13] E.B. Hansen, Redox reactions in food fermentations, Curr. Opin. Food Sci. 19 (2018) 98–103, doi:10.1016/j.cofs.2018.03.004.

[14] K. Chandrasekhar, A.N. Kumar, G. Kumar, D.-H. Kim, Y.-C. Song, S.-H. Kim, Electro-fermentation for biofuels and biochemicals production: current status and future directions, Bioresour. Technol. 323 (2021) 124598, doi:10.1016/j.biortech.2020.124598.

[15 G. Kumar, R.G. Saratale, A. Kadier, P. Sivagurunathan, G. Zhen, S.-H. Kim, et al., A review on bio-electrochemical systems (BESs) for the syngas and value added biochemicals production, Chemosphere 177 (2017) 84–92, doi:10.1016/j.chemosphere.2017.02.135.
[16] S. Nayak, S. Sevda, Basic principles of electro-fermentation process, in: S. Sevda (Ed.), Electro-Fermentation for Synthesis of Chemicals, Biochemicals and Biofuels, Springer, Singapore, 2025, doi:10.1007/978-981-96-6358-3_1.
[17] D.D. Bhagchandanii, R.P. Babu, J.M. Sonawane, N. Khanna, S. Pandit, D.A. Jadhav, et al., A comprehensive understanding of electro-fermentation, Fermentation 6 (2020) 92, doi:10.3390/fermentation6030092.
[18] B. Virdis, R.D. Hoelzle, A. Marchetti, S.T. Boto, M.A. Rosenbaum, R. Blasco-Gómez, et al., Electro-fermentation: sustainable bioproductions steered by electricity, Biotechnol. Adv. 59 (2022) 107950, doi:10.1016/j.biotechadv.2022.107950.
[19] I. Vassilev, N.J.H. Averesch, P. Ledezma, M. Kokko, Anodic electro-fermentation: empowering anaerobic production processes via anodic respiration, Biotechnol. Adv. 48 (2021) 107728, doi:10.1016/j.biotechadv.2021.107728.
[20] Z. Gong, H. Yu, J. Zhang, F. Li, H. Song, Microbial electro-fermentation for synthesis of chemicals and biofuels driven by bi-directional extracellular electron transfer, Synth. Syst. Biotechnol. 5 (4) (2020) 304–313, doi:10.1016/j.synbio.2020.08.004.
[21] P. Kumar, K. Chandrasekhar, A. Kumari, E. Sathiyamoorthi, B.S. Kim, Electro-fermentation in aid of bioenergy and biopolymers, Energies 11 (2018) 343, doi:10.3390/en11020343.
[22] P. Pilania, K. Bhushan, U.G. Phutela, Electro-fermentation for biofuel and biochemical production, Fermentation 11 (2025) 219, doi:10.3390/fermentation11040219.
[23] M. Romans-Casas, E. Perona-Vico, P. Dessì, Boosting ethanol production rates from carbon dioxide in MES cells under optimal solventogenic conditions, Sci. Total Environ. 856 (2023) 159124, doi:10.1016/j.scitotenv.2022.159124.
[24] A.V. Pandit, R. Mahadevan, In silico characterization of microbial electrosynthesis for metabolic engineering of biochemicals, Microb. Cell Fact 10 (2011) 76, doi:10.1186/1475-2859-10-76.
[25] R.A. Rozendal, H.V.M. Hamelers, G.J.W. Euverink, S.J. Metz, C.J.N. Buisman, Principle and perspectives of hydrogen production through biocatalyzed electrolysis, Int. J. Hydrogen Energy. 31 (12) (2006) 1632–1640.
[26] A. Kadier, Y. Simayi, P. Abdeshahian, N.F. Azman, K. Chandrasekhar, M.S. Kalil, Comprehensive review of microbial electrolysis cells (MEC) reactor designs and configurations for sustainable hydrogen gas production, Alexandria Eng. J. 55 (1) (2016) 427–443.
[27] A. Kadier, Y. Simayi, M. Sahaid Kalil, P. Abdeshahian, A. Abdul Hamid, A review of the substrates used in microbial electrolysis cells (MECs) for producing sustainable and clean hydrogen gas, Renew. Energy 71 (2014) 466–472, doi:10.1016/j.renene.2014.05.052.
[28] H. Liu, S. Grot, B.E. Logan, Electrochemically assisted microbial production of hydrogen from acetate, EnvironSci Technol 39 (2005) 4317–4320.
[29] A.A. Yaqoob, M.NM. Ibrahim, M. Rafatullah, Y.S Chua, A. Ahmad, K. Umar, Recent advances in anodes for microbial fuel cells: an overview, Materials (Basel) 13 (9) (2020) 2078, doi:10.3390/ma13092078.
[30] X.-D. Wang, Y.-F. Xu, H.-S. Rao, W.-J. Xu, H.-Y. Chen, W.-X. Zhang, et al., Novel porous molybdenum tungsten phosphide hybrid nanosheets on carbon cloth for efficient hydrogen evolution, Energy Environ. Sci. 9 (2016) 1468–1475, doi:10.1039/C5EE03801D.
[31] N.G. Yasri, G. Nakhla, Electrochemical behavior of anode-respiring bacteria on doped carbon electrodes, ACS Appl. Mater. Interfaces. 8 (2016) 35150–35162, doi:10.1021/acsami.6b09907.

[32] S. Singh, P.K Bairagi, N. Verma, Candle soot-derived carbon nanoparticles: an inexpensive and efficient electrode for microbial fuel cells, Electrochim. Acta 264 (2018) 119–127, doi:10.1016/j.electacta.2018.01.110.

[33] A.S. Shenoy, B. Biju, S. Athulya, T.N. Harshananda, Energy recovery and water desalination using microbial fuel cell, Int. J. Creat. Res. Thoughts 13 (3) (2025) 22–30.

[34] H. Luo, P. Xu, T.M Roane, P.E Jenkins, Z. Ren, Microbial desalination cells for improved performance in waste water treatment, electricity production, and desalination, Bioresour Technol 105 (2012) 60–66.

CHAPTER

7

Electrocatalytic oxidation of biomass

Surbhi Dhadda[a], Nikita Choudhary[a], Sonali Sharma[a], Sajay K. Meena[a], Himanshu Sharma[b], Harsh Pandey[c], Mahipal Singh Sankhla[d], Garima Awasthi[d], and Kumud Kant Awasthi[d]

[a]Vivekananda Global University, Jaipur, Rajasthan, India, [b]Department of Chemistry, Science College, Udaipur, Rajasthan, India, [c]Department of Biotechnology and Chemical Engineering, Manipal University, Jaipur, Rajasthan, India, [d]School of Basic & Applied Sciences, K. R. Mangalam University, Gurugram, Haryana, India

7.1 Introduction

7.1.1 Overview of biomass valorization: Significance in sustainable energy and chemicals production

Biomass valorization is a cornerstone of the sustainable energy and chemicals sector, offering a renewable alternative to fossil-based resources [1]. Electrocatalytic oxidation has emerged as a promising route for upgrading biomass-derived feedstocks, enabling selective conversion into value-added chemicals and clean energy carriers under mild, electricity-driven conditions. This method not only minimizes carbon emissions but also integrates seamlessly with renewable energy systems, facilitating decentralized and environmentally benign production [2]. Ongoing advances in catalyst design, reaction engineering, and process integration are rapidly enhancing the efficiency and scalability of this approach, reinforcing its role in the transition toward a circular bioeconomy.

Electrochemical Biomass Conversion: Sustainable Processes and Applications for Bioenergy, Biofuels, and Bioproducts
DOI: https://doi.org/10.1016/B978-0-443-26757-4.00008-8

TABLE 7.1 Examples of electrocatalytic biomass conversions.

S. No.	Biomass feedstock	Electrocatalytic reaction pathway	Main products obtained	Industrial/potential applications
1.	**Glycerol** (byproduct of biodiesel)	Partial oxidation of polyol groups	Glyceric acid, tartronic acid, di-hydroxyacetone	Food additives, pharmaceuticals, fine chemicals
2.	**Sorbitol** (from glucose hydrogenation)	Selective oxidation of hydroxyl groups	Gluconic acid, sorbose	Sweeteners, polymer precursors
3.	**Ethylene glycol** (from syngas / biomass)	Oxidation of terminal alcohol groups	Glycolic acid, oxalic acid	Cosmetics, biodegradable polymers
4.	**5-HMF** (5-hydroxymethylfurfural)	Anodic oxidation at furan ring and aldehyde site	2,5-furandicarboxylic acid (FDCA)	Bioplastics (PEF as PET alternative)
5.	**Xylitol** (from hemicellulose hydrolysis)	Controlled oxidation of alcohol groups	Xylonic acid, arabinonic acid	Biodegradable polymers, chelating agents
6.	**Lignin-derived aromatics** (e.g., vanillin, syringol)	Electro-oxidation of phenolic/aromatic groups	Vanillic acid, syringaldehyde	Flavors, resins, pharmaceuticals
7.	**Glucose**	Selective oxidation at aldehyde group	Gluconic acid, glucaric acid	Food industry, corrosion inhibitors, polymers

Biomass valorization represents a pivotal strategy in advancing a circular and low-carbon economy, enabling the conversion of abundant, renewable organic resources into high-value fuels, chemicals, and energy carriers. Unlike fossil-based feedstocks, biomass is inherently carbon-neutral, offering a sustainable pathway to mitigate greenhouse gas emissions while meeting global energy demands [3]. Among emerging conversion technologies, electrocatalytic oxidation has gained particular attention due to its ability to selectively transform biomass-derived molecules under mild conditions, driven by renewable electricity. This approach not only reduces the energy intensity of traditional thermochemical processes but also cogenerates value-added chemicals alongside hydrogen or other clean energy vectors. By integrating tailored electrocatalysts with optimized reaction environments, electrocatalytic biomass upgrading can achieve high efficiency, selectivity, and scalability. Consequently, it holds significant potential to bridge the gap between green energy production and sustainable chemical manufacturing, positioning itself as a key enabler of the bio-based economy (Table 7.1).

7.2 Electrocatalytic oxidation

Electrocatalytic oxidation is a process in which an electrochemical reaction is accelerated by a catalyst, enabling the oxidation of reactant molecules at an electrode surface with improved efficiency and lower

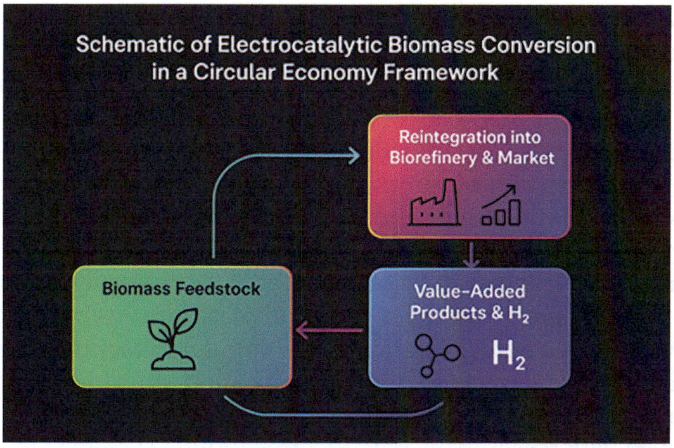

FIGURE 7.1 Schematic diagram of electrocatalytic biomass conversion in a circular economy framework.

energy input. In this process, the catalyst facilitates electron transfer between the reactant and the electrode, reducing the activation energy and enhancing reaction selectivity (Fig. 7.1).

7.2.1 Fundamental principles of electrocatalytic oxidation

Electrocatalytic oxidation is a process in which electrical energy is used to drive oxidation reactions at the interface between an electrode and an electrolyte. This is achieved through the application of a controlled electrical potential, which facilitates the transfer of electrons from chemical species to the electrode surface, effectively oxidizing the target compound. At the heart of this process is an electrocatalyst—a material, often a transition metal or metal oxide, that enhances the reaction kinetics without being consumed in the process. These catalysts are typically deposited on conductive substrates, allowing for efficient electron flow and selective activation of chemical bonds. Unlike conventional oxidation methods that rely on thermal energy or biological agents, electrocatalysis relies on electron transfer mechanisms, making it possible to conduct reactions at ambient or moderately elevated temperatures. The applied potential can be finely tuned, giving the operator precise control over reaction pathways, product formation, and selectivity.

- *Advantages over thermochemical oxidation:* Thermochemical oxidation, commonly used in industrial settings, involves the use of heat often in combination with oxidants like air, oxygen, or steam—to initiate and sustain oxidation reactions. While widely practiced, this method presents several limitations, especially from environmental and operational standpoints.

- *Lower energy requirements:* Electrocatalytic processes can operate effectively at room temperature or slightly elevated temperatures, eliminating the need for the high thermal inputs characteristic of thermochemical systems.
- *Enhanced selectivity:* By adjusting the electrode potential, specific oxidation pathways can be promoted while suppressing unwanted side reactions, enabling cleaner and more targeted transformations.
- *Reduced environmental impact:* Because the process does not rely on combustion or high-temperature operations, it significantly reduces greenhouse gas emissions and the generation of hazardous by-products.
- *Improved process control:* The ability to regulate reaction rates through electrical inputs allows for rapid start-up and shutdown, contributing to operational flexibility and safety.
- *Integration with renewable energy:* Electrocatalytic systems are inherently compatible with intermittent energy sources such as solar or wind, making them well-suited for future sustainable chemical manufacturing.
- *Advantages over biological oxidation:* Biological oxidation typically employs microorganisms or isolated enzymes to catalyze oxidation reactions, often in the context of waste treatment, biodegradation, or biofuel production. While biologically based processes are generally low-energy and environmentally benign, they suffer from notable limitations in terms of reaction speed, substrate specificity, and operational robustness.

7.2.2 Role in integrated biorefineries and circular economy strategies

Integrated biorefineries aim to maximize biomass utilization similar to petroleum refineries. Electrocatalytic oxidation is a key enabler of sustainable biorefineries, allowing the selective conversion of biomass-derived feedstocks into value-added chemicals while cogenerating green hydrogen. Operating under mild conditions and powered by renewable electricity, it reduces energy demand and avoids the use of stoichiometric oxidants. Integrated into biorefinery platforms, this approach maximizes feedstock utilization by transforming lignocellulosic residues and platform molecules (e.g., glycerol, 5-HMF) into fuels, fine chemicals, and polymer precursors, thereby minimizing waste. From a circular economy perspective, electrocatalytic oxidation supports closed-loop resource flows, reduces dependence on fossil resources, and aligns industrial production with carbon-neutrality goals—making it an essential strategy for next-generation sustainable chemical manufacturing [4–6].

Upgrading intermediates: Conversion of glycerol, sorbitol, and HMF into fine chemicals.

Distributed modularity: Compact electrocatalytic units enable decentralized processing.

Renewable integration: Direct coupling with solar, wind, or hydro power.

Economic synergy: Hydrogen generation enhances profitability.

Electrocatalytic oxidation supports circular economy principles [7] through:

- *Waste-to-value:* Valorization of crude glycerol from biodiesel or lignin-derived aromatics.
- *Carbon neutrality:* Minimal CO_2 emissions when powered by renewable electricity.
- *Closed material loops:* Substituting fossil-based chemicals with biomass-derived equivalents.
- *Hybrid processes:* Synergies with enzymatic and thermochemical routes.

7.3 Fundamentals of electrocatalytic oxidation

Electrocatalytic oxidation of biomass-derived molecules represents a sustainable alternative to the conventional oxygen evolution reaction (OER) in electrolyzers [8]. By replacing OER with biomass oxidation at the anode, overall cell voltages are reduced, and value-added products are generated alongside hydrogen. Understanding the underlying reaction pathways and thermodynamic driving forces is critical for catalyst development and process optimization (Fig. 7.2).

7.3.1 Electrochemical reaction pathways and thermodynamics

Biomass-derived molecules contain oxygenated functionalities such as hydroxyl (–OH), aldehyde (–CHO), and carboxyl (–COOH) groups, enabling multiple oxidation routes. These transformations are typically multielectron, multiproton processes that depend on the electrode potential, catalyst composition, and electrolyte environment [9].

7.3.2 Alcohol oxidation

Primary alcohols:

$$R\text{–}CH_2OH \rightarrow R\text{–}CHO + 2H^+ + 2e^- \text{ (aldehyde)}$$

$$R\text{–}CHO + H_2O \rightarrow R\text{–}COOH + 2H^+ + 2e^- \text{ (carboxylic acid)}$$

Secondary alcohols:

$$R\text{–}CHOH\text{–}R' \rightarrow R\text{–}C(=O)\text{–}R' + 2H^+ + 2e^- \text{ (ketone)}$$

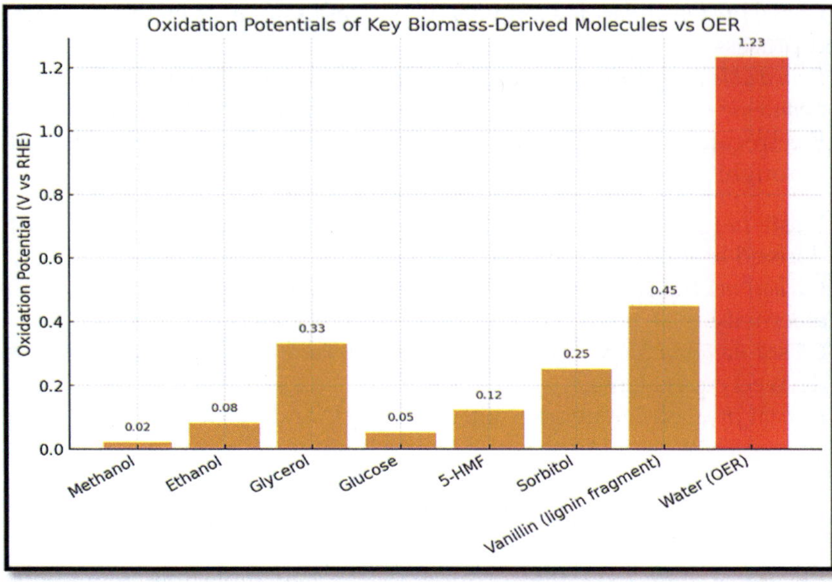

FIGURE 7.2 Figure comparative bar chart showing the oxidation potentials of key biomass-derived molecules versus the oxygen evolution reaction (OER).

This pathway is highly relevant for biomass-derived polyols like glycerol and sorbitol.

7.3.3 Polyols

Polyhydric alcohols undergo selective or deep oxidation depending on potential:

- *Glycerol ($C_3H_8O_3$):* Pathways lead to glyceraldehyde, dihydroxyacetone (DHA), glyceric acid, tartronic acid, glycolic acid, or oxalic acid.
- *Sorbitol ($C_6H_{14}O_6$):* Oxidizes to gluconic acid or sorbose, with applications in sweeteners and polymer precursors. Selectivity is strongly influenced by catalyst choice: Pt favors C–C bond cleavage, Au promotes selective partial oxidation, and Ni-based catalysts are effective in alkaline media.

7.3.4 Furanic compounds

- *5-Hydroxymethylfurfural (HMF):* Sequential oxidation of hydroxymethyl (–CH_2OH) and aldehyde (–CHO) groups produces 2,5-furandicarboxylic acid (FDCA), a precursor for bioplastics. Side reactions include ring-opening to levulinic acid and polymerization to humins.

TABLE 7.2 Standard electrode potentials E° (V vs RHE) of electrocatalytic oxidation.

Reaction	E° (V vs. RHE)
Methanol → Formaldehyde	~0.02
Ethanol → Acetaldehyde	~0.08
Glycerol → Glyceraldehyde	~0.33
Glucose → Gluconic acid	~0.05
HMF → FDCA	~0.10–0.15
Water oxidation (OER)	1.23

7.3.5 Aromatic compounds (from lignin)

Phenolic compounds undergo oxidative transformations such as:
Vanillin → Vanillic acid
Syringol → Syringaldehyde → Syringic acid
This valorizes lignin, one of the most underutilized biomass fractions.

7.3.6 Sugars

Glucose: oxidizes at the aldehyde group to gluconic acid and further to glucaric acid.

Xylose: converts into xylonic acid with potential in biodegradable polymers.

Thermodynamics of electrocatalytic oxidation: The thermodynamic feasibility of electrocatalytic oxidation is reflected in its standard electrode potentials (E° vs Reversible Hydrogen Electrode (RHE)) (Table 7.2).

Oxidation potentials of key biomass-derived molecules (e.g., alcohols, polyols, lignin fragments): Biomass oxidation requires much lower potentials than OER.

- Most substrates (methanol, ethanol, glucose, HMF, sorbitol, glycerol, vanillin) oxidize below 0.5 V versus RHE, while OER requires 1.23 V.
- This means lower energy input when biomass replaces OER in electrolysis.

Variation across substrates

Lowest potentials: Methanol (~0.02 V), ethanol (~0.08 V), glucose (~0.05 V)—very easy to oxidize.

Moderate potentials: HMF (~0.12 V), sorbitol (~0.25 V), glycerol (~0.33 V).

Highest among biomass substrates: Vanillin (~0.45 V), due to its aromatic/phenolic structure (more stable, harder to oxidize).

Thermodynamic advantages: By replacing OER with biomass oxidation, cell voltage can drop by 0.8 to 1.0 V, saving energy and producing value-added chemicals instead of just O_2.

Proton-coupled electron transfer (PCET) mechanisms

Electrocatalytic oxidation of biomass is governed by PCET processes, where electron transfer and proton transfer occur in a coupled fashion. Since most biomass molecules contain hydroxyl, carbonyl, and phenolic groups, their oxidation requires simultaneous removal of electrons (e−) and protons (H^+) [10].

Stepwise vs. concerted pathways

Stepwise PCET: Proton transfer precedes or follows electron transfer → can result in unstable intermediates (e.g., alkoxide, carbocation).

Concerted PCET: Proton and electron transfer occur simultaneously → typically more efficient and avoids high-energy intermediates.

Example: glycerol oxidation

On Au or Pt catalysts: $-CH_2OH \rightarrow -CHO$ involves removal of $2e^- + 2H^+$.

PCET pathway determines whether glyceraldehyde, glyceric acid, or complete oxidation products form.

Role of catalysts

Metals like Pt, Pd, and Au facilitate C–H bond cleavage through adsorption and PCET. Transition metal oxides/hydroxides ($Ni(OH)_2$, CoOOH) operate via lattice oxygen participation, where proton release is coupled to surface redox cycling (Ni^{2+}/Ni^{3+}).

Implication: PCET efficiency strongly dictates the rate, selectivity, and overpotential of biomass oxidation.

Influence of electrolyte pH and ionic strength

Electrolyte pH controls both the proton availability and the thermodynamic potentials of oxidation reactions (since standard potentials shift with pH according to the Nernst equation) [11].

Acidic media (pH < 7): Proton-rich environment → favorable for proton release steps. Noble metals (Pt, Pd) remain stable, but selectivity is often poor due to strong adsorption of intermediates (CO, formate).

Neutral media (pH ~7): More sustainable but less conductive. Some biomass molecules (e.g., 5-HMF to FDCA) show higher selectivity in near-neutral conditions.

Alkaline media (pH > 7): Hydroxide ions participate in PCET, stabilizing reaction intermediates. Strongly lowers overpotentials for alcohol and polyol oxidation. Ni, Co, and Cu oxides/hydroxides are highly active in alkaline electrolytes. Many biomass oxidation studies report superior performance in alkaline KOH or NaOH solutions.

Example: Glycerol oxidation potential drops from ~0.6 V (acidic) to ~0.3 V (alkaline). Au and Ni-based catalysts show up to 10 times higher activity in alkaline media compared to acidic.

Ionic strength: Higher ionic strength improves conductivity and reduces ohmic drop, enhancing reaction rates. However, too high concentrations may alter the solubility of biomass substrates. Electrolyte Anions can adsorb onto catalyst surfaces, blocking or promoting active sites (e.g., Cl– poisoning of Pt, carbonate buffering in alkaline solutions). Buffering Capacity is Essential for reactions producing/consuming protons; phosphate buffers are commonly used in neutral media for glucose/HMF oxidation.

7.4 Electrocatalyst materials

Electrocatalyst selection is a critical factor governing the activity, selectivity, and stability of biomass electro-oxidation. Biomass-derived substrates are structurally diverse (alcohols, polyols, sugars, lignin-derived aromatics), and their conversion involves multiple PCET steps. Therefore, catalysts must efficiently activate C–H, O–H, and sometimes C–C bonds while suppressing undesired side reactions (e.g., complete oxidation to CO_2). Broadly, catalysts fall into two categories: noble metals and transition metal-based systems [12] (Table 7.3).

- Metal-based catalysts: noble metals (Pt, Pd, and Au), transition metals (Ni, Co, and Cu)
  - *Noble metal catalysts (Pt, Pd, Au):* Noble metals are the most widely studied due to their high intrinsic activity and well-understood surface chemistry.
  - *Platinum (Pt):* Exhibits strong activity for C–H bond cleavage in alcohols and polyols. Common in acidic media (PEM fuel cells), but prone to CO poisoning from strongly adsorbed intermediates (e.g., formate, CO). Alloying with Ru (Pt–Ru) or Sn (Pt–Sn) mitigates poisoning and enhances tolerance.
  - *Palladium (Pd):* Highly effective in alkaline media for glycerol, glucose, and ethanol oxidation. Less susceptible to CO poisoning compared to Pt. Pd-based nanostructures (Pd–Ni and Pd–Cu) show improved activity and durability.
  - *Gold (Au):* Exceptional selectivity in mild/alkaline conditions for partial oxidation of biomass (e.g., glycerol → glyceric acid, HMF → FDCA). Limited activity in acidic electrolytes. Au nanoparticles and supported Au catalysts demonstrate high selectivity to aldehydes/acids without overoxidation.
  - *Transition metal catalysts (Ni, Co, and Cu):* Transition metals provide a low-cost alternative and often exhibit excellent activity in alkaline electrolytes due to their redox-active oxide/hydroxide layers.
  - *Nickel (Ni):* One of the most studied nonnoble metals. In alkaline media, Ni is converted to NiOOH, which serves as the active species

TABLE 7.3 Electrooxidation of biomass substrates: catalysts, potentials, and conditions.

Substrate	Anodic half-reaction	E° (V vs RHE)	Typical catalysts	Conditions
Methanol	$CH_3OH \rightarrow HCHO + 2H^+ + 2e^-$	0.02	Pt, Pt–Ru alloys	Acidic/neutral; PEM fuel cells
Ethanol	$C_2H_5OH \rightarrow CH_3CHO + 2H^+ + 2e^-$	0.08	Pt, Pd, Au-based alloys	Acidic (Pt) or alkaline (Pd, Au) media
Glycerol	$C_3H_8O_3 \rightarrow$ Glyceraldehyde $+ 2H^+ + 2e^-$	0.33	Au, Pd, Pt, Ni-based catalysts	Strongly alkaline (KOH) enhances activity
Glucose	$C_6H_{12}O_6 + H_2O \rightarrow$ Gluconic acid $+ 2H^+ + 2e^-$	0.05	Au, Pt, $Ni(OH)_2$	Alkaline electrolytes preferred
5-HMF	HMF $\rightarrow$ FDCA $+ 4H^+ + 4e^-$	0.12	Au, Pd, NiFe layered oxides	Alkaline or neutral media
Sorbitol	$C_6H_{14}O_6 \rightarrow$ Gluconic acid $+ 2H^+ + 2e^-$	0.25	Au, Ni, Pd	Alkaline favored for higher selectivity
Vanillin (lignin fragment)	$C_8H_8O_3 \rightarrow$ Vanillic acid $+ 2H^+ + 2e^-$	0.45	Ni, Co, Cu oxides; Au–Pd alloys	Alkaline/neutral media
Water (reference)	$2H_2O \rightarrow O_2 + 4H^+ + 4e^-$	1.23	IrO_2, RuO_2, NiFe oxides	Acidic/alkaline

for PCET-driven oxidation. Highly active for glycerol, glucose, sorbitol, and lignin model compounds. NiFe layered double hydroxides (LDHs) exhibit synergistic effects, boosting both activity and stability.
  - *Cobalt (Co):* Forms $CoOOH/CoO_2$ active layers under anodic polarization. Effective for oxidation of alcohols and polyols, though generally less active than Ni. Often combined with Ni or Fe to improve conductivity and durability.
  - *Copper (Cu):* Active in alkaline electrolytes, especially for oxidation of carbohydrates and polyols. Tends to undergo surface restructuring, forming Cu_2O/CuO species that participate in oxidation. Cu-based bimetallics (e.g., Cu–Ni, Cu–Pd) improve stability and selectivity.

- *Metal oxides and hydroxides (NiOOH, CoOOH, and MnOx):* Metal oxides and hydroxides act as redox-active catalytic layers in alkaline media, where they undergo in-situ electrochemical transformations that generate highly active oxygenated species. Unlike metallic surfaces, these catalysts operate via a lattice-oxygen–mediated mechanism or surface redox cycling, which facilitates PCET during biomass oxidation.
  - *Nickel oxyhydroxide (NiOOH) formation and active phase:* Ni-based catalysts oxidize under anodic potentials to form NiOOH (β- or γ-phase), which is the true active species for alcohol/polyol oxidation. The redox couple $Ni^{2+} \leftrightarrow Ni^{3+}$ drives PCET and stabilizes reaction intermediates.
  - *Activity:* Highly active for polyols (glycerol and sorbitol) and sugars (glucose). Selectivity can be tuned by alloying (NiFe, NiCo, and NiCu).
  - *Advantages:* Abundant, inexpensive, and shows superior activity in alkaline KOH solutions. NiFe LDHs show a synergistic effect, where Fe incorporation enhances conductivity and lowers overpotential.
  - *Challenges:* Limited stability under prolonged operation due to structural reconstruction.
  - *Cobalt oxyhydroxide (CoOOH) formation and active phase:* Co-based catalysts undergo transformation to $CoOOH/CoO_2$ at anodic potentials. The Co^{3+}/Co^{4+} redox couple plays a key role in oxidation steps.
  - *Activity:* Effective for **alcohol and phenolic compound oxidation**, including lignin model compounds. Often paired with Ni or Fe to improve activity and electronic conductivity.
  - *Advantages:* High intrinsic activity toward C–H and O–H activation. Better selectivity than pure Ni in some cases (e.g., aromatic alcohols).
  - *Challenges:* Stability lower than Ni-based systems; can suffer from dissolution or deactivation in prolonged cycling.
  - *Manganese oxides (MnO_x) structures:* Exist in multiple oxidation states (Mn^{2+}, Mn^{3+}, and Mn^{4+}), giving rise to polymorphs such as MnO_2,

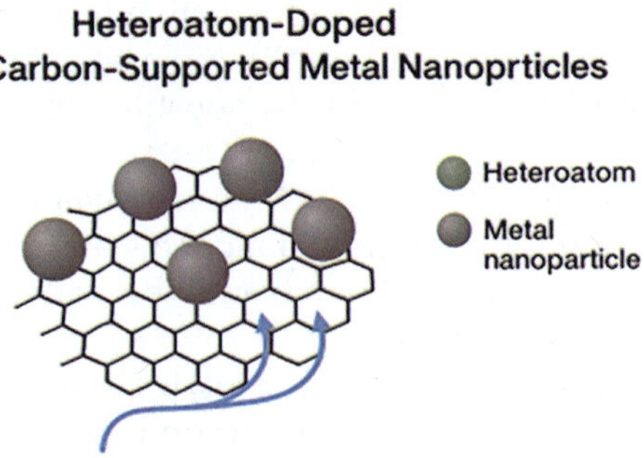

FIGURE 7.3 Schematic representation of heteroatom-doped carbon-supported metal nanoparticles for electrocatalysis.

  Mn_3O_4, and Mn_2O_3. Exhibit strong redox flexibility, enabling diverse catalytic pathways.
  - *Activity:* Active for biomass-derived aromatic molecules and polyols. MnO_2 nanostructures (α-, β-, γ-phases) show promising oxidation of 5-HMF to FDCA.
  - *Advantages:* Abundant, environmentally benign, and low cost. Often synergistic when combined with Ni or Co (NiMn and CoMn oxides).
  - *Challenges:* Lower conductivity than Ni/Co oxides. Requires conductive supports (carbon, graphene, and CNTs) for optimal performance.
- *Heteroatom-doped carbon-supported metal nanoparticles:* Heteroatom-doped carbon-supported metal nanoparticles are composite catalytic materials in which metal nanoparticles are dispersed on a carbon matrix that has been doped with heteroatoms such as nitrogen, sulfur, phosphorus, or boron. The heteroatom doping modifies the electronic structure and surface chemistry of the carbon support, creating anchoring sites for the nanoparticles, improving their dispersion, and preventing agglomeration. This synergistic architecture enhances charge transfer, optimizes adsorption energies of reactants and intermediates, and increases catalytic activity, selectivity, and stability. Such materials are widely employed in electrocatalysis, including biomass oxidation reactions, oxygen reduction, and hydrogen evolution, due to their high conductivity, tunable surface properties, and ability to integrate earth-abundant metal species (Fig. 7.3).
- *Hybrid and composite electrocatalysts:* Hybrid and composite electrocatalysts integrate two or more active components—such as metals, oxides, and conductive carbons—into a single architecture to exploit synergistic effects. These systems enhance electron transfer, improve structural stability, and optimize reaction pathways, leading to higher activity, selectivity, and durability. They are particularly suited for biomass electrooxidation, where multistep reactions benefit from complementary active sites and tunable electronic properties.

7.5 Reaction selectivity and product distribution

Reaction selectivity and product distribution describe the ability of an electrocatalytic system to favor the formation of a desired product over competing pathways, thereby controlling the relative proportions of all reaction products. High selectivity is crucial for maximizing process efficiency, minimizing downstream separation costs, and achieving economically viable biomass valorization [13].

- *Controlling partial oxidation versus complete mineralization:* A key challenge in electrocatalytic biomass valorization is steering the reaction toward partial oxidation products—such as acids, aldehydes, or ketones—rather than complete mineralization to CO_2. Achieving this control requires precise tuning of catalyst composition, applied potential, and reaction environment to selectively break specific bonds while preserving valuable carbon frameworks.
- *Factors affecting selectivity (potential, temperature, and catalyst surface structure):* Reaction selectivity in electrocatalytic biomass oxidation is governed by a combination of electrochemical and materials factors. The applied potential dictates which reaction pathways are thermodynamically and kinetically accessible, with higher potentials often promoting overoxidation and undesired mineralization to CO_2. Temperature influences reaction rates, adsorption–desorption equilibria, and intermediate stability, thereby modulating product distribution. Catalyst surface structure—including crystallographic facets, defect density, and heteroatom doping—plays a pivotal role in stabilizing specific intermediates and lowering activation barriers for target products. Together, these parameters must be carefully tuned to achieve high selectivity toward value-added chemicals while minimizing energy losses and side reactions.
- *Target products:* Electrocatalytic biomass oxidation can be selectively steered toward target products such as organic acids, aldehydes, ketones, furans, and other value-added intermediates, which serve as versatile building blocks for polymers, fuels, and fine chemicals.

7.6 Advanced electrode architectures

Advanced electrode architectures are crucial for enhancing the efficiency of electrocatalytic biomass oxidation. Three-dimensional, hierarchically porous electrodes maximize electrochemically active surface area, improve mass transport, and facilitate bubble release, leading to higher current densities [14,15]. Nanostructured coatings and heterostructure interfaces can stabilize active phases, promote selective intermediate adsorption, and reduce overpotentials. Conductive scaffolds such as metal foams, carbon cloth, and 3D-printed current collectors offer mechanical robustness and enable scalable, flow-cell-compatible designs. Together, these innovations address kinetic limitations and bring laboratory performance closer to industrial relevance.

- *Nanostructured electrodes for enhanced surface area and mass transport:* Nanostructured electrodes offer dramatically increased electrochemically active surface area, enabling a higher density of accessible active sites and improved catalytic turnover. Their hierarchical porosity facilitates rapid reactant diffusion and product desorption, minimizing concentration gradients and mass transport limitations. As a result, nanostructured architectures deliver superior current densities, lower overpotentials, and improved selectivity compared to conventional planar electrodes, making them highly promising for scalable biomass electro-oxidation.
- *Binder-free electrodes and 3D printed structures:* Binder-free electrodes eliminate the need for polymeric binders that can block active sites, introduce resistive interfaces, or degrade under electrochemical conditions. Directly grown or electrodeposited catalytic layers on conductive substrates ensure intimate electrical contact, improved mechanical stability, and enhanced mass transport. Meanwhile, 3D-printed electrodes allow precise control over geometry, porosity, and flow channels, enabling tailored current distribution and scalable, reactor-compatible designs. Together, these approaches advance high-performance, durable electrode systems for industrial biomass electro-oxidation.
- *Flow-cell and membrane electrode assemblies (MEA) for continuous production:* Flow-cell reactors and MEAs are pivotal for translating electrocatalytic biomass oxidation from batch studies to continuous, industrially relevant operation. Flow-cells enhance mass transport by maintaining a constant supply of reactants and rapid removal of products, enabling higher current densities and space–time yields. MEAs integrate catalyst layers, gas diffusion media, and ion-conducting membranes into a compact architecture that minimizes ohmic losses and allows zero-gap operation. These systems are compatible with renewable energy inputs and scalable to pilot-scale production, representing a critical step toward commercial deployment.

7.7 Process optimization

Flow-cell reactors and MEAs are pivotal for translating electrocatalytic biomass oxidation from batch studies to continuous, industrially relevant operation [16]. Flow-cells enhance mass transport by maintaining a constant supply of reactants and rapid removal of products, enabling higher current densities and space–time yields. MEAs integrate catalyst layers, gas diffusion media, and ion-conducting membranes into a compact architecture that minimizes ohmic losses and allows zero-gap operation. These systems are compatible with renewable energy inputs and scalable to pilot-scale production, representing a critical step toward commercial deployment.

- *Electrolyte composition, temperature, and applied potential control:* The efficiency and selectivity of electrochemical reactions are governed by electrolyte composition, temperature, and applied potential. The electrolyte determines ionic conductivity, pH, and reaction pathways; temperature influences reaction rates, product distribution, and system stability; and applied potential directs reaction selectivity and electrode behavior [17]. Together, precise control of these factors is essential for optimal electrocatalytic performance.
- *Coupling oxidation with hydrogen evolution (electrochemical coproduction of fuels and chemicals):* Coupling oxidation with hydrogen evolution refers to the simultaneous electrochemical generation of value-added chemicals at the anode and hydrogen gas at the cathode. By pairing selective oxidation of organic substrates (e.g., alcohols, biomass-derived compounds) with the hydrogen evolution reaction (HER), this approach enhances energy efficiency, reduces overall cell voltage, and enables the coproduction of fuels and chemicals in a sustainable and economically attractive manner.
- *Energy efficiency and technoeconomic (TEA) considerations:* Energy efficiency and TEA feasibility are critical in electrochemical processes. Maximizing energy efficiency requires minimizing overpotentials and optimizing conditions such as applied potential, temperature, and electrolyte composition to convert electrical energy into chemical energy effectively. Simultaneously, TEA considerations—including capital and operational costs, scalability, electrode durability, and product value—determine whether a process is industrially viable. Together, these factors ensure that electrochemical technologies are both sustainable and economically practical.

7.8 Case studies

Electrocatalytic oxidation of biomass-derived platform molecules such as glycerol, HMF, and lignin-derived phenolics demonstrates the potential of this approach to produce high-value chemicals like lactic acid, FDCA, and vanillin under mild, sustainable conditions (Table 7.4).

- *Glycerol oxidation to lactic acid, DHA:* Glycerol, a major by-product of biodiesel production, serves as a model polyol for electrocatalytic

TABLE 7.4 Case studies with performance metrics.

Biomass substrate/Reaction	Catalyst/System	Key metrics (Yield, FE, current density, etc.)	Notes and challenges
HMF → FDCA	Mn-doped NiS nanosheets on 3D graphite felt ($Mn_{0.2}NiS/GF$) [21]	Selectivity ≈ 98.3%, Yield ≈ 97.6%, Faradaic efficiency ≈ 94.2% at 500 $mA{\bullet}cm^{-2}$ current density; production rate ~4.56 $g{\bullet}h^{-1}$ in H-cell; upscaled continuous-flow operation gave ~44.32 $g{\bullet}h^{-1}$.	Very good industrial-level current density; scaling works. Stability over 10 cycles.
HMF → FDCA	Co-doped Ni_3S_2 on Ni foam ($Co_{0.4}NiS@NF$) [22]	Onset potential ~0.9 V vs RHE; current density ~497 $mA{\bullet}cm^{-2}$ at 1.45 V vs RHE; FDCA yield rate ~330.4 $\mu mol{\bullet}cm^{-2}{\bullet}h^{-1}$; H_2 co-generation ~1000 $\mu mol{\bullet}cm^{-2}{\bullet}h^{-1}$.	Large current densities, dual product generation, and challenges in maintaining high FE under continuous operation.
HMF → FDCA	Co_3O_4–NiO@C bimetallic oxide heterojunctions [23]	HMF conversion ≈ 99.1%, FDCA yield ≈ 99.1%, Faradaic efficiency ≈ 98.9%.	Very high-performance metrics; demonstrates the value of heterojunction design.
Glycerol oxidation (to glycerate/lactate/formate, etc.)	Pd/Ni foam (Pd/NF) for Glycerol Oxidation Reaction (GOR) [24]	Geometric current density ~235 $mA{\bullet}cm^{-2}$ at ~0.92 V vs RHE in 3 M KOH + 1 M glycerol; Pt-mass normalized current ~3.71 $A{\bullet}mg_{)}Pt_{)}^{-1}$; >92% cumulative selectivity toward C_3 products (lactic acid major) at 0.65 V over 5 hours.	Good catalyst activity, but observed drop in current over long time (to ~35% of initial) likely due to poisoning, etc.
Glycerol → Formate	Cu-doped NiCo alloy on Ni foam (Cu-NiCo/NF) [25]	At 1.23 V vs RHE in alkaline medium with glycerol, a faradaic efficiency for formate ≈ 93.8%; good stability over 144 h.	Very promising for value-added chemical coupled with hydrogen-related energy conversion.
Lignin → Vanillin (Aromatics)	Electro-depolymerization of Kraft lignin in a high-temperature electrolysis cell [26]	Vanillin obtained with ~67% efficiency under optimized electrolytic conditions at 160°C in an undivided cell.	Good yield/selectivity, but high temperature and special cell needed; separation and impurity control remain issues.

valorization. Its electro-oxidation can selectively yield value-added chemicals such as lactic acid and DHA under mild conditions [18]. Noble-metal catalysts like Pt, Pd, and Au supported on carbon or heteroatom-doped frameworks exhibit high selectivity toward DHA at controlled potentials, whereas bimetallic systems (e.g., Pt–Bi, Pd–Cu) enhance lactic acid production. These transformations provide a dual benefit—glycerol utilization and simultaneous hydrogen generation at the cathode—offering an attractive route to integrated power-to-chemicals processes.

- *HMF oxidation to FDCA:* Electrocatalytic oxidation of HMF to FDCA represents one of the most extensively studied biomass-to-bioplastic pathways. FDCA is a key monomer for polyethylene furanoate (PEF), a renewable alternative to PET. Transition-metal-based catalysts such as Ni, Co, and Cu oxides, as well as NiFe LDHs, have achieved near-quantitative Faradaic efficiencies (>95%) and high selectivity under alkaline conditions. Recent advances include operando-characterized NiFeOOH active phases and flow-cell reactor configurations, which demonstrate industrially relevant current densities (>100 mA cm^{-2}) and stability over extended operation [19].
- *Lignin-derived phenolic oxidation to aromatic aldehydes:* Lignin, the most abundant renewable aromatic polymer, can be selectively depolymerized and electrochemically upgraded to aromatic aldehydes such as vanillin and syringaldehyde [20]. Electrocatalytic oxidation offers a greener alternative to harsh oxidative routes by enabling selective C–C and C–O bond cleavage at ambient conditions. Catalysts based on RuO_2, MnO_x, or heteroatom-doped carbons show promising selectivity for aldehyde production. Integrating lignin valorization with cathodic hydrogen evolution further improves process economics and supports circular biorefinery models.

7.9 Challenges and future perspectives

Despite notable advances, electrocatalytic biomass oxidation continues to confront several critical challenges. First, catalyst design must reconcile high catalytic activity, selectivity toward desired oxidation products, and long-term stability all while relying on earth-abundant, low-cost materials. For example, electrode materials often suffer from corrosion, metal leaching, or loss of active sites under operational conditions. Second, issues of scale-up and reactor engineering are nontrivial: lab-scale systems often exhibit ideal mass transfer, mixing, and uniform current distribution, but these properties degrade in larger reactors, leading to losses in energy efficiency and diminished product yields. Third, feedstock heterogeneity differences in biomass composition, impurity content, moisture content, and pretreatment history—complicates reaction kinetics, catalyst fouling, and reproducibility. Fourth, rigorous TEA and life-cycle assessments (LCA) are still sparse and often fail to capture full upstream and downstream costs (pretreatment, separation, electrolyte recycling, etc.), making the true sustainability and cost-competitiveness of proposed systems uncertain.

- *Catalyst stability and deactivation pathways:* Catalyst stability is a critical parameter determining the long-term performance of electrocatalytic systems. During biomass electrooxidation, catalysts are exposed to harsh electrochemical environments that can lead to nanoparticle agglomeration, leaching of active metals, surface oxidation or reduction, carbon support corrosion, and poisoning by reaction intermediates or impurities. These deactivation pathways reduce active-site availability, alter electronic structures, and ultimately diminish catalytic activity and selectivity. Understanding the kinetics and mechanisms of these degradation processes through in-situ/operando spectroscopy and electrochemical analysis is essential for designing robust, durable catalysts capable of maintaining performance over extended operational cycles.
- *Scale-up challenges and integration into industrial processes:* Translating electrocatalytic biomass oxidation from the laboratory to industrial scale poses significant challenges. Mass and heat transfer limitations become pronounced in larger systems, often leading to uneven current distribution, concentration gradients, and reduced selectivity. Electrode architecture and reactor design must be optimized to handle continuous operation, minimize energy losses, and ensure efficient product separation. Additionally, the variability of biomass feedstock complicates process control, requiring adaptive strategies to maintain consistent performance. For successful industrial integration, electrocatalytic systems must be compatible with existing biorefinery infrastructure, operate under cost-competitive conditions, and be powered by renewable electricity to ensure carbon-neutrality. Process intensification through continuous-flow reactors and digital process control, combined with TEA and LCA, will be critical to achieving commercial viability.
- *Future research directions: in-situ characterization, computational design, machine learning-driven catalyst discovery:* Future efforts should focus on **in-situ and operando characterization techniques** to monitor reaction intermediates and active-site evolution in real time, enabling mechanistic insight. **Computational modelling and density functional theory** will play a central role in predicting structure–activity relationships, while **machine learning–driven catalyst discovery** can accelerate the identification of optimal compositions and morphologies. Together, these approaches will guide the rational design of next-generation electrocatalysts for efficient biomass valorization.

Acknowledgment

We would also like to thank our colleagues for their cooperation, constructive discussions, and assistance. Finally, I acknowledge all the authors and researchers whose work has inspired and guided this study.

References

[1] D. Ravelli, C. Samori, Biomass Valorisation, Wiley-VCH, Weinheim, Germany, 2021.
[2] M. Mennani, A.A. Benhamou, M.E. Achaby, A. Moubarik, Z. Kassab, Recent trends in biomass valorization for energy, in: S. Kandasamy, M.P. Shah, K. Subbiah, N. Manickam (Eds.), Microbial Niche Nexus Sustaining Environmental Biological Wastewater and Water-Energy-Environment Nexus, Springer Nature Switzerland, 2025, pp. 453–488.
[3] M.Y. Alazaiza, T.M Alzghoul, M.B Ramu, D.E Nassani, Catalysis in biofuel production and biomass valorization: trends, challenges, and innovations through a bibliometric analysis, Catalysts 15 (3) (2025) 227.
[4] S. Li, L. Ye, W. Cen, D. Sun, Electrocatalytic biomass upgrading coupled with hydrogen evolution and CO_2 reduction, Nanoscale 17 (11) (2025) 6308–6328.
[5] X. Zhang, F. Kong, X. Liang, X. Wang, M. Ji, M. Wang, The combination of Ni nanoparticles and single Ni atoms realizes co-electrolysis of CO_2 and biomass-derived sugar and polyols, Chem. Eng. J. 498 (2024) 155565.
[6] Y. Pan, H. Zhang, B. Zhang, F. Gong, J. Feng, H. Huang, S. Vanka, R. Fan, Q. Cao, M. Shen, Z. Li, Renewable formate from sunlight, biomass and carbon dioxide in a photoelectrochemical cell, Nat. Commun. 14 (1) (2023) 1013.
[7] Z. Chen, S. Yun, L. Wu, J. Zhang, X. Shi, W. Wei, Y. Liu, R. Zheng, N. Han, B.J Ni, Waste-derived catalysts for water electrolysis: circular economy-driven sustainable green hydrogen energy, Nano-Micro Lett. 15 (1) (2023) 4.
[8] C.F Zinola Sánchez, M.E Martins, E. Pastor Tejera, N. Pimenta Neves, Electrocatalysis: fundamentals and applications, Int. J. Electrochem. 2012 (2012) 1–2.
[9] F.W Lucas, R.G Grim, S.A Tacey, C.A Downes, J. Hasse, A.M Roman, C.A Farberow, J.A Schaidle, A. Holewinski, Electrochemical routes for the valorization of biomass-derived feedstocks: from chemistry to application, ACS Energy Lett. 6 (4) (2021) 1205–1270.
[10] C. Costentin, Electrochemical approach to the mechanistic study of proton-coupled electron transfer, Chem. Rev. 108 (7) (2008) 2145–2179.
[11] E. Aranda-García, L. Morales-Barrera, G. Pineda-Camacho, E. Cristiani-Urbina, Effect of pH, ionic strength, and background electrolytes on Cr (VI) and total chromium removal by acorn shell of Quercus crassipes Humb. & Bonpl., Environ. Monit. Assess. 186 (10) (2014) 6207–6221.
[12] V.T. Nguyen, K. Cho, Y. Choi, B. Hwang, Y.K. Park, H. Nam, D. Lee, Biomass-derived materials for energy storage and electrocatalysis: recent advances and future perspectives, Biochar 6 (1) (2024) 96.
[13] L. Wang, N. Li, Y. Lu, R. Zhang, Z. Sun, S. Niu, Y. Luo, Product distribution from pyrolysis of large biomass particle: effects of intraparticle secondary reactions, Fuel 325 (2022) 124851.
[14] J. Jiang, Y. Li, J. Liu, X. Huang, C. Yuan, X.W. Lou, Recent advances in metal oxide-based electrode architecture design for electrochemical energy storage, Adv. Mater. 24 (38) (2012) 5166–5180.
[15] X. Zhang, Z. Ju, Y. Zhu, K.J Takeuchi, E.S. Takeuchi, A.C. Marschilok, G. Yu, Multiscale understanding and architecture design of high energy/power lithium-ion battery electrodes, Adv. Energy Mater. 11 (2) (2021) 2000808.

[16] M. Yang, Z. Yuan, R. Peng, S. Wang, Y. Zou, Recent progress on electrocatalytic valorization of biomass-derived organics, Energy Environ. Mater. 5 (4) (2022) 1117–1138.
[17] L. Li, S. Lu, L. Fang, Y. Wei, S. Yang, Review on biomass-derived carbon-based materials for electrocatalytic hydrogen production: state of the art and outlook, Energy Fuels 37 (23) (2023) 18485–18501.
[18] S. Lux, M. Siebenhofer, Synthesis of lactic acid from dihydroxyacetone: use of alkaline-earth metal hydroxides, Catal. Sci. Technol. 3 (5) (2013) 1380–1385.
[19] M. Sajid, X. Zhao, D. Liu, Production of 2, 5-furandicarboxylic acid (FDCA) from 5-hydroxymethylfurfural (HMF): recent progress focusing on the chemical-catalytic routes, Green Chem. 20 (24) (2018) 5427–5453.
[20] L. Martínková, M. Grulich, M. Pátek, B. Křístková, M. Winkler, Bio-based valorization of lignin-derived phenolic compounds: a review, Biomolecules 13 (5) (2023) 717.
[21] S. Li, S. Wang, Y. Wang, J. He, K. Li, Y. Xu, M. Wang, S. Zhao, X. Li, X. Zhong, J. Wang, Doped Mn enhanced NiS electrooxidation performance of HMF into FDCA at industrial-level current density, Adv. Funct. Mater. 33 (24) (2023) 2214488.
[22] L.Q. Lee. Lignocellulosic biomass reform for co-producing green chemicals and green hydrogen Nanyang Technological University, Singapore.
[23] Z. Cheng, J. Hu, W. Zhou, W. Deng, M. Ma, Y. Tan, Carbon layer confined Co–Ni bimetallic oxide heterojunctions for high-efficiency electrosynthesis of 2, 5-furandicarboxylic acid, J. Mater. Chem. A 12 (22) (2024) 13400–13408.
[24] Y.X. Leiu, K.M. Lim, Z.J. Chiah, E.S. Mah, W.J. Ong, Plastic-to-Treasure: innovative advances in photo/electro-catalytic upcycling technologies for commodity chemicals and fuels, EcoEnergy 3 (2) (2025) 217–253.
[25] C. Li, H. Li, B. Zhang, H. Li, Y. Wang, X. Wang, P. Das, Y. Li, X. Wu, Y. Li, Y. Cui, Efficient electrocatalytic oxidation of glycerol to formate coupled with nitrate reduction over Cu-doped NiCo alloy supported on nickel foam, Angew. Chem. Int. Ed. 63 (46) (2024) e202411542.
[26] Y.P. Wijaya, K.J. Smith, C.S. Kim, E.L. Gyenge, Electrocatalytic hydrogenation and depolymerization pathways for lignin valorization: toward mild synthesis of chemicals and fuels from biomass, Green Chem. 22 (21) (2020) 7233–7264.

CHAPTER

8

Electrocatalytic hydrogenation of biomass

Omvir Singh[a,b], *Shubhi Gupta*[c], *Bipul Sarkar*[b], *and Prasenjit Mondal*[d]

[a]Department of Sciences and Humanities, Rajiv Gandhi Institute of Petroleum Technology, Jais, Amethi, Uttar Pradesh, India, [b]Catalytic Depolymerization Area, CSIR-Indian Institute of Petroleum, Dehradun, Uttarakhand, India, [c]Department of Fuel, Minerals and Metallurgical Engineering, Indian Institute of Technology (Indian School of Mines), Dhanbad, Jharkhand, India, [d]Department of Chemical Engineering, Indian Institute of Technology Roorkee, Roorkee, Uttarakhand, India

8.1 Hydrogenation reaction

Hydrogenation is an addition reaction of a hydrogen molecule (H_2) to an unsaturated reactive functionality. It is one of the several essential procedures employed in industries for the synthesis and processing of various fine chemicals, such as food additives, pharmaceuticals, perfumes, and dyes. The hydrogenation reaction usually involves the dissociation of hydrogen gas in the presence of a suitable catalyst to reduce compounds with multiple bonds. The hydrogenation processes have three reaction components: an unsaturated substrate, H_2, and a suitable catalyst capable of dissociating the gaseous H_2 molecule. The largest class of hydrogenation catalysts is metal catalysts, which typically consist of highly scattered fine metal particles on oxide supports. Many hydrogenation reactions require a high temperature for completion, and even a catalyst to operate at a higher temperature. H_2 is dissociated on the active metal that may "spillover" from the metal onto the catalyst support. Herein, this chapter summarizes the electrocatalytic hydrogenation (ECH) processes [1].

Electrochemical Biomass Conversion: Sustainable Processes and Applications for Bioenergy, Biofuels, and Bioproducts
DOI: https://doi.org/10.1016/B978-0-443-26757-4.00006-4

Electrochemical hydrogenation (ECH) offers an essential understanding of electron transfer pathways and the production of selected compounds. Electrochemical technology has recently garnered significant attention. The electrochemical method refers to a process induced or accompanied by the flow of electric current, involving the transfer of electrons between two substances, namely the electrode and the electrolyte [2]. The electrochemical process has extensive applications in diverse domains, including fuel cells, chemical synthesis, photocatalysis, energy storage, and gas production. The benefits of electrochemical technology encompass gentle working conditions, environmental sustainability, and cost-effectiveness [3]. The task of enhancing ECH of biomass must concurrently address the following characteristics: (1) the electro-interaction dynamics of an organic molecule; (2) the impact of electron transfer between the electrode surface and biomass molecules; (3) the application of external potential for a chemical reaction; (4) the regulatory influence of specific substrates (such as surface area, particle size, etc.) active in aqueous and organic compounds; less rigorous requirements for catalysts and electrocatalysts, which can be deposited on the working electrode, result in enhanced reactivity and catalytic efficiency; the chemical redox species are supplied internally [4]. Consequently, electrochemical technology is a viable approach for enhancing the application of renewable electrical energy in chemical processes, making it realistic for industrial use. This method seeks to mitigate energy-related climate issues by synthesizing biomass feedstock by electrocatalytic reduction using renewable energy sources, such as lignocellulosic biomass. ECH of biomass is a method for producing valuable compounds through reductive processing [5].

8.2 Valorization of lignocellulosic biomass via electrochemical hydrogenation conversion

8.2.1 Fundamental and mechanism of electrochemical hydrogenation

ECH has been investigated for more than a century. The electrochemical synthesis of organic molecules offers distinct advantages over thermocatalytic hydrogenation (TCH) techniques. The direct reduction of organic compounds, without adsorption onto the electrocatalyst surface, is hypothesized to need a greater cathodic potential than the surface hydrogenation pathway and occurs solely on electrodes exhibiting high overpotentials [6]. The simultaneous occurrence of hydrogenolysis and/or hydrodeoxygenation (HDO) of organic compounds in the ECH process can effectively reduce the oxygen-to-carbon ratio, presenting an additional advantage of the ECH process.

8.2.1.1 Prospects of electrochemical hydrogenation

Key advantages of the ECH process compared to traditional TCH processes encompass (1) mild reaction conditions (reduced temperature and pressure, eliminating the necessity for energy-intensive operations), (2) continuous hydrogen generation (derived from water splitting, negating the requirement for external H_2 gas supply), (3) straightforward operating conditions (manageable current/voltage input), and (4) a cleaner process with facile product separation (enhanced carbon recovery and improved H/C ratios). Furthermore, the energy required for the ECH process can be amalgamated with several renewable electric energy sources, including solar, wind, and hydroelectric electricity. In this context, the enzymatic conversion of lignocellulosic biomass advances a sustainable method that can be linked with renewable energy, enhancing the total efficiency of the biorefinery process [7].

8.2.1.2 Mechanistic pathways of electrochemical hydrogenation

The ECH was derived from the electrochemical hydrogen evolution reaction (HER), functioning as a cathodic reaction through a two-electron transfer mechanism. The reaction involving the HER can occur in various pH electrolytes, depending on either the Volmer–Heyrovsky or Volmer–Tafel mechanisms [8]. The Volmer reaction involves the interaction of protons with the catalyst surface by electrochemical adsorption of hydrogen atoms [9]. The Heyrovsky reaction involves electrochemical desorption of hydrogen atoms through the interaction of a proton and an electron. The Tafel reaction is a chemical process involving hydrogen desorption to generate H_2. The Volmer–Heyrovsky–Tafel reaction mechanism elucidates the HER and ECH, which are significantly influenced by the electrolyte employed in the process [10].

Understanding the significance of the Volmer–Heyrovsky–Tafel reaction in relation to the reaction parameters is crucial for gaining insights into the progression of the reaction. The relationships between current density and surface coverage with respect to overpotential will be assessed for the Volmer–Heyrovsky and Volmer–Tafel mechanisms. The Tafel slopes offer insights into the assessment of rate-determining stages, which typically presume maximal coverage of the adsorbed species, whereas the Volmer equation highlights electron transfer kinetics [11]. Chemisorbed hydrogen and adsorbed organic reactants are pivotal in the ECH process, influencing the faradaic efficiency (FE). The applied reduction potential in the reaction system often influences the rate of ECH and may dictate hydrogen (H_2) adsorption, whereas the capacity for H_2 adsorption inversely correlates with negative overpotential [12]. Overpotential is an activity parameter at a specified current density. The potential beyond the thermodynamic redox potential (the voltage measured for a redox reaction in relation to hydrogen at standard state) is referred to as overpotential, which accelerates reaction speeds. Furthermore, the elevated overpotential is contingent upon the electrocatalyst, which is associated with its form and surface area. Consequently, selecting suitable electrocatalysts and electrolytes results in enhanced reaction rates and conversions in ECH [13]. Most of the energy conversion reactions are redox in nature, involving the transfer of both electrons and protons during the process. Likewise, ECH also involves electrocatalytic reduction of organic compounds by electronation followed by protonation reactions. It also involves simultaneous hydrogenolysis and HDO to lower oxygen content that can be an added advantage of ECH [14]. Two different mechanisms are proposed in the literature to illustrate electron transfer during ECH: (1) direct electronation-protonation (EP), and (2) the ECH pathway or mechanism. The former case involves direct interaction between the electro-reduced organic substrate and the proton (H^+) generated from the electrolyte solution (Eqs. 1–4). In this, the organic substrate is firstly reduced to an intermediate, which is then followed by its simultaneous protonation-electronation with H^+ generated from the bulk electrolyte solution to form corresponding hydrogenated species [15]. In contrast to direct EP, the ECH pathway involves chemisorbed hydrogen on the electrocatalyst surface to interact with the adsorbed organic substrate. It can proceed either through the Eley–Rideal mechanism (E–R) or the Langmuir–Hinshelwood (L–H) mechanism. Herein, the organic substrate gets absorbed on the catalyst active site, the hydrogen atom produced through water dissociation gets adsorbed on another metal active site on the catalyst surface via Volmer reaction (Eqs. 8.5–8.7) [16]. Both the adsorbed organic substrate and chemisorbed hydrogen then react with each other to form reduced species, which then desorb to give the hydrogenated organic compound and free catalyst sites for another set of reactions (Eqs. 8.8–8.11). The occurrence of ECH either through the E–R and L–H route depends on the catalyst properties and operating conditions such as potential, pH, temperature, and electrolyte concentration [17]. Even one ECH reaction can involve both routes. Direct EP can occur via either inner sphere electron transfer or outer sphere electron transfer, whereas ECH involves an inner sphere electron transfer mechanism. Direct EP is constrained by a larger hydrogen overpotential and low hydrogen coverage. Furthermore, excessive chemisorption of hydrogen compared to organic substrate would lead to HER through the Heyrovsky reaction, thereby reducing process conversion (Eqs. 8.12–8.13). Whereas, a strong bond between the targeted product and the catalyst site might lead to excessive hydrogenation reaction, thereby reducing the selectivity of the desired or targeted product.

The electron transfer during ECH has been described using the following steps:

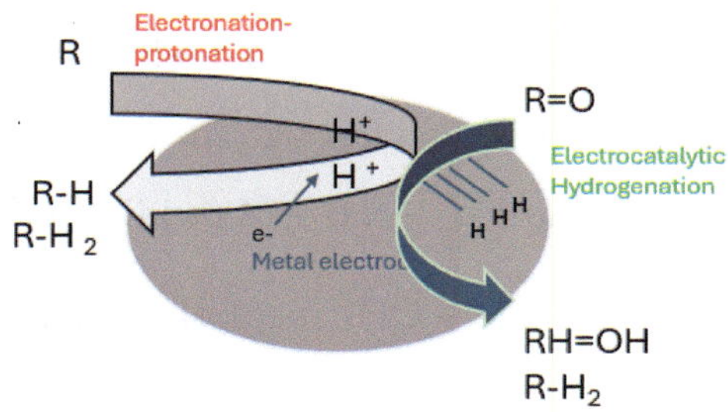

FIGURE 8.1 Electron transfer mechanism in electrochemical hydrogenation reactions.

Direct electronation-protonation (EP):

$$S + e^- \leftrightarrow S^- \tag{8.1}$$

$$S^- + H^+ \leftrightarrow SH \tag{8.2}$$

$$SH + e^- \leftrightarrow SH^- \tag{8.3}$$

$$SH^- + H^+ \leftrightarrow SH_2 \tag{8.4}$$

Here, S represents the organic substrate or reactant.

Electrocatalytic hydrogenation (ECH):

Electron transfer mechanism in ECH reactions, which was shown in Fig. 8.1.

The following equations represent a simplified and plausible ECH reaction mechanism involving the elementary steps:

Chemisorbed hydrogen generation on the catalyst metal active site (A) by Volmer reaction:

$$\mathrm{H^+ + e^- + A \rightarrow HA\ (acidic\ media)} \tag{8.5}$$

$$\mathrm{H_2O + A + e^- \rightarrow HA + OH^-(neutral\ or\ basic\ media)} \tag{8.6}$$

Adsorption of unsaturated organic substrate on another catalyst active site (B):

$$\mathrm{(S = O) + B \leftrightarrow (S = O)B} \tag{8.7}$$

Reaction between chemisorbed hydrogen and adsorbed organic substrate:

$$\mathrm{(S = O)B + 2HA \leftrightarrow (SH - OH)B + 2A} \tag{8.8}$$

$$\mathrm{(SH - OH)B + 2HA \leftrightarrow (S - H_2)B + 2A + H_2O} \tag{8.9}$$

Desorption of the target product

$$\mathrm{(SH - OH)B \leftrightarrow (SH - OH) + B} \tag{8.10}$$

$$(S - H_2)B \leftrightarrow (S - H_2) + B \quad (8.11)$$

Chemisorbed hydrogen desorption through the Heyrovsky reaction:

$$HA + (H^+)aq + e^- \rightarrow H_2 + M \text{ (acidic media)} \quad (8.12)$$

$$HA + H_2O + e^- \rightarrow H_2 + M + OH^- \text{ (neutral or basic media)} \quad (8.13)$$

8.3 The roles of electrocatalyst and electrolyte in electrochemical hydrogenation

Both the electrocatalyst and electrolyte play a crucial role in the ECH process. Electrocatalysts offer a large surface area to carry out the reaction between adsorbed organic substrate and chemisorbed hydrogen species, whereas electrolytes provide electrical conductivity for hydrogen radical evolution through the water splitting reaction. The synergism involved between the electrocatalyst and electrolyte impacts the Faraday efficiency of the process. To increase ECH process efficiency, high hydrogen coverage and organic substrate coverage on the catalyst are essential. The optimal surface coverage would favor ECH over HER. Different types of electrocatalysts have shown significant activity toward ECH, such as metal (e.g., Pt, Pd, Raney Ni, Cu) based carbon or metal oxide catalysts, metal organic frameworks, bimetallic electrocatalysts (Pt-Co, Pd-Rh, Pd-Pt, Ru-Ir), metal phosphides and sulfides, metals doped conductive polymeric membrane, etc [18].

The hydrogen radical generation reactions can be carried out at varying electrolyte pH. Depending upon the nature of the electrolyte (for instance, acidic, neutral, or alkaline), there are different mechanisms for the adsorbed hydrogen radical formation, as shown in Eqs. 8.5–8.6. Acidic electrolytes involve the generation of adsorbed hydrogen radicals through proton electroreduction owing to the abundance of protons in acidic media, whereas neutral or alkaline media involve water reduction to simultaneously generate hydroxide radicals with the hydrogen radical. Compared to acidic media, hydrogen evolution kinetics is more sluggish and complicated in alkaline or neutral media because of the involvement of additional water dissociation reaction and hydroxide radical generation that may act as a catalyst poison by occupying surface active sites [19]. Therefore, acid-tolerant electrocatalysts are more effective for the ECH process. Noble metal catalysts show very high activity in the ECH process. However, these catalysts show poor performance in the alkaline or neutral media, demanding high overpotential. Distinct electrocatalyst and electrolyte, as well as the synergism involved between them, impact the efficiency of the ECH process.

The solubility of organic feedstock is another parameter significantly influencing the ECH rate. Higher dissolution of organic feedstock is desirable for a higher reaction rate [20]. Biomass constituents like cellulose and hemicellulose dissolve effectively in most of the organic solvents and acidic solutions. However, lignin is hard to disperse in these solvents owing to its complex structure, consequently resulting in lower activity. Alkaline solvents show comparatively better dissolubility and thus ECH rate for lignin compared to other solvents. Besides this, ionic liquids also showed high effectiveness as electrolytes due to their high dissolution, high proton conductivity, and high stability [21]. However, high cost and environmental toxicity provide limitations on their usage. Deep eutectic solvents (DES) are the quaternary ammonium salts along with a metal salt or a hydrogen bond donor. Being green solvents, low cost, low toxicity, and biodegradable, DES are also a good choice for ECH of lignocellulosic biomass [22].

8.4 Electrocatalytic hydrogenation of cellulose, hemicellulose, and lignin derivatives

Lignocellulosic biomass, consisting primarily of cellulose, hemicellulose, and lignin, is a highly abundant and renewable source of feedstock that can be converted into biofuels and valuable chemicals. ECH is emerging as a powerful method to convert these biomass derivatives into useful products, taking advantage of renewable electricity and hydrogen under mild reaction conditions. Currently, most of the published studies on ECH are based on model compounds such as phenol, furfural, guaiacol, syringol, etc [21]. Very few studies have attempted the ECH of bio-oil and bio-oil aqueous fractions. Cellulose is the most abundant polysaccharide on earth and is composed of glucose monomers linked by $\beta(1\rightarrow4)$ glycosidic bonds. Its direct conversion into bio-based chemicals remains a challenge due to its crystalline structure, which makes it difficult to solubilize and break down. However, after hydrolysis to glucose, the ECH of cellulose-derived glucose offers a viable pathway to produce sorbitol, mannitol, and other sugar alcohols used in various industries, including pharmaceuticals and food additives [21]. Different electrocatalysts, including Pd, RaNi, Pb, Ag, Au, Cu, Ni, Zn, PtRu/C, and ZnFe/CNT, showed significant activity for ECH of glucose and fructose [23]. 5-hydroxymethylfurfural (HMF) is obtained through hydrolysis of glucose or fructose through acid catalysts and can be hydrogenated to give a variety of value-added chemicals and fuels. Zhao et al. [19] reported the development of a Cu-Fe bimetallic electrocatalyst, which significantly enhanced the hydrogenation of 5-HMF, to produce 2,5-dihydroxymethylfurfural (DHMF) with 90.1% yields while maintaining a selectivity of 93.3% under mild conditions, as shown in Table 8.1.

The Fe component in the catalyst regulated the electronic structure of copper sites, which increased the hydrogen activation and HMF adsorption and consequently hindered the dimerization reactions. Hemicellulose, a heterogeneous polymer composed of pentose sugars such as xylose, arabinose, and mannose, is more easily hydrolyzed than cellulose due to its amorphous structure. Xylose, in particular, can be converted into xylitol, a sugar alcohol widely used as a low-calorie sweetener in food products. Kasad et al.[28] showed that Ru/C catalyzed the hydrogenation of furfural to tetrahydrofurfuryl alcohol (THFA) with high yields under mild electrochemical conditions (less than 50°C temperature and 1 atm pressure) [28]. It involved the formyl group reduction and saturation of the heteroaromatic ring using hydrogen equivalent generated through in-situ water splitting reaction. Although Ru is effective, its cost and availability have led to efforts in developing cheaper and more abundant transition metal catalysts. Lignin, the most recalcitrant component of lignocellulosic biomass, is a complex aromatic polymer formed from phenolic monomers. Its depolymerization and hydrogenation are challenging due to the diversity of chemical bonds present in the polymer, including ether linkages and carbon–carbon bonds. However, the valorization of lignin is of great interest, as it offers a rich source of aromatic chemicals that can be hydrogenated into phenols, benzyl alcohols, and cyclohexanols, which have applications in fuel production, resins, and specialty chemicals. The ECH of lignin-derived compounds is typically performed using noble metal catalysts such as platinum (Pt), ruthenium (Ru), and palladium (Pd) [17] as shown in Table 8.2.

8.5 Comparison of two technologies

Electrochemical reduction of organic compounds has several advantages over the thermochemical reduction process, which was shown in Table 8.3. The former is less energy-intensive than the latter due to the requirement of mild operational conditions in terms of lower temperature and pressure. Thermochemical reduction necessitates an external hydrogen supply to carry out hydrogenation reactions, whereas electrochemical reduction involves in-situ hydrogen generation through the water splitting process, thereby avoiding the need of external hydrogen supply. This also avoids problems related to hydrogen transportation and storage. Electrochemical reduction is more controllable and cleaner than the thermochemical process. Also, it offers more efficient product separation with improved hydrogen to carbon ratio and greater carbon recovery. Furthermore, the electrochemical process can be made more sustainable and cleaner by

TABLE 8.1 Summary of the literature on various catalysts for electrocatalytic hydrogenation(ECH) of 5-hydroxymethyl furfural.

Substrate	Electrocatalyst /Electrode	Membrane/Cell type	Electrolyte/ Solvent	Conditions	Products	Ref.
5-hydroxymethyl furfural	Pt	Undivided cell type (3 electrode system)	0.2 M sulfate buffer (pH=2.0)	0.02 M HMF, E = −1.2 V, j=57.4 mA cm^{-2}	2,5-dihydroxymethylfuran (DHMF), S = 100%, FE=0%	[24]
	Au			0.02 M HMF, E = −1.2 V, j=50.1 mA cm^{-2}	2,5-hexanedione (S=6.92%), hydroxyl-2,5-hexanedione (S=6.91%), 2,5-di (hydroxymethyl)furan (S=42.8%), FE=2.39%	
	Cu			0.02 M HMF, E = −1.2 V, j=39.6 mA cm^{-2}	2,5-hexanedione (S=8.13%), hydroxyl-2,5-hexanedione (S=12.37%), 2,5-di (hydroxymethyl)furan (S=39.5%), FE=3.75%	
	Zn			0.02 M HMF, E = −1.2 V j=14.4 mA cm^{-2}	2,5-hexanedione (S=81.6%), hydroxyl-2,5-hexanedione (S=8.98%), 2,5-di (hydroxymethyl)furan (S=3.19%), FE=72.4%	

(*continued on next page*)

TABLE 8.1 Summary of the literature on various catalysts for electrocatalytic hydrogenation(ECH) of 5-hydroxymethyl furfural—cont'd

Substrate	Electrocatalyst /Electrode	Membrane/Cell type	Electrolyte/ Solvent	Conditions	Products	Ref.
5-hydroxymethyl furfural	Aggd (prepared by glvanic deposition)	Undivided cell type (3 electrode system), Pt counter electrode, and an Ag/AgCl (4 M KCl) reference electrode	0.5 M borate buffer solution (pH=9.2)	0.02 M HMF, j=6.9 $mAcm^{-2}$	2,5-dihydroxymethylfuran (DHMF), S = 99% Production rate=0.127 $mmol/hcm^2$, FE=99%	[25]
	Agsp			0.02 M HMF, j=2.53 $mAcm^{-2}$	2,5-dihydroxymethylfuran (DHMF), S = 84% Production rate (mmol/h cm2) =0.034, FE=72%	
	Cu			0.02 M HMF, j=1.32 $mAcm^{-2}$	2,5-dihydroxymethylfuran (DHMF), S = 39% Production rate (mmol/h cm^2) =0.0047, FE=19%	
5-hydroxymethyl furfural	Cu	Nafion-117	0.1 M H_2SO_4 in a 1: 1 mixture of water and ethanol.	RT, j=10 $mAcm^{-2}$	2,5-dimethylfuran (DMF), S = 35.6%, 2,5-di(hydroxymethyl)furan (S=33.8%), 5-methylfurfuryl alcohol (S=11.1%) and 5 methylfuran-2-carbaldehyde (S= 0.5%)	[26]
	Pd/VN/CF VN-3D vanadium nitride		0.2 M $HClO_4$	i =20 mA, 10 mM HMF, 55 min.	Conversion rate of 1.24 mmol h-1 2,5-di(hydroxymethyl)furan (S≥88%), 5-methylfurfural (MF), 5-methylfurfural alcohol (MFA), and 2,5-dimethylfuran (DMF)	
5-hydroxymethyl furfural	CF- Carbon fiber	Experiments carried out on a membrane electrode assembly electrolyser (MEA)	0.2 M $HClO_4$		HMF conversion ≥ 90%, FE ≥ 90%	[27]
	Ru/VN		0.2 M $HClO_4$	i =20 mA, 10 mM HMF, 55 min	HMF conversion = 30%,	
	Pt/VN		0.2 M $HClO_4$	i =20 mA, 10 mM HMF, 55 min.	HMF conversion = 24%,	

FE, *faradaic efficiency;* RT, *room temperature;* S, *selectivity of the product;* X, *conversion of the substrate;* Y, *yield of the product.*

TABLE 8.2 Various catalysts for electrocatalytic hydrogenation of biomass model compounds.

Substrate	WE/CE/R E	Catalyst	Electrolyte	Conditions	Products	Refernces
Phenol (2 mL)	Reticulated vitreous carbon (RVC) / Pt wire / Ag/AgCl	Rh/C, Pt/C, Pd/C / nafion	Anolyte: 0.1 M of H_2SO_4; Catholyte: Acetic acid, H_3PO_4, H_2SO_4	Current: −40, −60 mA; time: 2 h; temp.: 5 80°C; stirring: 500 rpm	Cyclohexano l, cyclohexano ne; conv. rate: 26%–63%	[29]
Benzene	Rh–Pt / not mention / non	Rh–Pt / Nafion, Du pont	Water	Surface area: 7.07 cm^2; atmosphere: H_2 with atmospheric pressure	Cyclohexane (16%)	[30]
Guaiacol	Ru|ACC / Pt wire / non	3-CE-NH_3 / Nafion-117	Catholyte: 0.2 M HCl; anolyte: 0.2 M phosphate buffer	100 mA, 80°C, 2 h	2 methoxycycl ohexanol (47%), cyclohexanol (53%); conv. rate: 75%	[1,31]
Guaiacol	PtRhAu-C felt / Pt / RHE reference	PtRhAu deposited on carbon felt / Nafion-117	0.2 M $HClO_4$	Current density: 60 mA/cm2, time: 1 h; stirring: 1000 rpm; surface area: 1 cm2	conversion rate; 2 methoxy-4-propylcyclohexanol (68% of selectivity)	[32]
Syringol	Ru|ACC / Pt wire / non	1.5-CE NH_3 / nafion-117	Catholyte: 0.2 M HCl; anolyte: 0.2 M phosphate buffer	Current: 100 mA; temp.: 80°C; time: 2 h	2 methoxycycl ohexanol (36%), cyclohexanol (35%), guaiacol (15%) conv. rate: 58%	[31]

TABLE 8.3 The main features of conventional thermochemical hydrogenation and electrochemical hydrogenation processes.

Thermal Catalytic Hydrogenation	Electrocatalytic Hydrogenation
High temperature (200°C–400°C)	Mild temperature (RT-80°C)
High pressure (1–10 MPa)	Ambient pressure
Short reaction time	Long reaction time
Side reaction: coke formation	Side reaction: H_2 evolution reaction
Thermal energy consumption	Electric energy consumption
Catalyst deactivation at high temperatures	Acid and alkali corrosive-resistant catalyst
Hydrogen gas as a hydrogen donor	Hydrogen atom protonation from the electrolyte

integrating the process's power requirement with renewable energy, for instance, solar, wind, hydropower, etc. Thus electrochemical reduction provides a cleaner and more sustainable pathway for the production of valuable chemicals than the thermochemical process.

8.6 Challenges and future scheme

A promising strategy for biomass utilization focuses on maximizing the energy density of the desired products while minimizing carbon losses during processing. Among various emerging methods, the electrochemical conversion of biomass stands out as a sustainable alternative aligned with global energy and environmental policies. Traditional thermal catalytic hydrogenation methods, though effective, require high energy inputs and operate under harsh conditions. In contrast, ECH offers a low-energy pathway by utilizing renewable electricity, allowing reactions to proceed under mild or ambient conditions. This makes it highly attractive for transforming biomass-derived molecules into fuels and valuable chemicals with reduced environmental impact. Electrochemical treatment of cellulose and hemicellulose at room temperature can prevent undesirable polymerization of intermediates, ensuring higher selectivity toward alcohols such as sorbitol and xylitol through the selective reduction of carbonyl groups. Furthermore, ECH allows the simultaneous oxidation and reduction of biomass-derived compounds like furfural, 5-HMF, and levulinic acid in separate electrolytic chambers, enhancing product diversity and overall yield. Despite these benefits, several practical challenges remain. The complexity of electrochemical cell design, product separation, and analysis still limits large-scale adoption and integration with existing biorefinery systems.

In the case of lignin, integrating ECH with thermochemical processes such as fast pyrolysis presents a particularly promising route. Lignin, being a highly aromatic and recalcitrant polymer, often exhibits low conversion efficiency and a tendency to form soot during pyrolysis. However, pretreatment via hydrogenation can modify its structure, increasing the hydrogen-to-carbon ratio and improving olefin yields. Previous studies show that thermal hydrogenation increased the olefin yield from 5.3 wt.% to about 9.9 wt.%. ECH of lignin, offering better control and higher hydrogen incorporation (an H/C ratio increase of 0.04), is projected to further boost olefin yield to approximately 15 wt.%. Such advancements could significantly enhance the valorization of lignin, a typically underutilized byproduct of biomass refining into high-value hydrocarbon feedstocks. Nonetheless, several barriers hinder the widespread application of ECH for biomass conversion. The high cost and scarcity of noble metal catalysts limit scalability, making the development of transition-metal-based or carbon-supported catalysts crucial. However, these cheaper alternatives often suffer from reduced selectivity and stability. Research into alloy formation, heteroatom doping, and catalyst surface modification may overcome these drawbacks. Additionally, optimizing operational parameters such as electrolyte composition, pH, applied potential, and temperature is essential to improve reaction efficiency and product selectivity. The intrinsic complexity of biomass feedstocks, which contain varying proportions of cellulose, hemicellulose, and lignin, adds another layer of difficulty in achieving consistent product distributions.

Future research should aim to design multifunctional catalysts capable of performing hydrogenation alongside other key transformations, such as dehydration and oxidation, to enable direct conversion of complex biomass mixtures into value-added chemicals and fuels. Integrating ECH with renewable energy sources could also help develop decentralized, low-carbon biorefineries. Moreover, addressing challenges related to reactor design, process intensification, and product separation will be crucial for the industrial implementation of ECH. Overall, while ECH of biomass remains an evolving technology, it holds immense potential to revolutionize sustainable fuel and chemical production by bridging renewable electricity with biomass valorization.

Declaration of competing interest

The authors declare that they have no known competing financial interests or personal relationships that could have appeared to influence the work reported in this paper.

Acknowledgments

O.S acknowledges the financial support of the Rajiv Gandhi Institute of Petroleum Technology. BS acknowledges Gratitude is extended to CSIR-IIP for facilities to carry out collaboration with RGIPT under the joint research collaboration program.

References

[1] L. Zhang, Y. Ren, W. Liu, A. Wang, T. Zhang, Single-atom catalyst: a rising star for green synthesis of fine chemicals, Natl. Sci. Rev. 5 (2018) 653–672.

[2] M.A. Alkhadra, X. Su, M.E. Suss, H. Tian, E.N. Guyes, A.N. Shocron, et al., Electrochemical Methods for Water Purification, Ion Separations, and Energy Conversion, Chem. Rev. 122 (2022) 13547–13635.

[3] A.M. Sadeq, R.Z. Homod, A.K. Hussein, H. Togun, A. Mahmoodi, H.F. Isleem, et al., Hydrogen energy systems: Technologies, trends, and future prospects, Sci. Total Environ. 939 (2024) 173622.

[4] W. Xu, C. Yu, J. Chen, Z. Liu, Electrochemical hydrogenation of biomass-based furfural in aqueous media by Cu catalyst supported on N-doped hierarchically porous carbon, Appl. Catal. B 305 (2022) 121062.

[5] R. Xia, S. Overa, F. Jiao, Emerging Electrochemical Processes to Decarbonize the Chemical Industry, JACS Au 2 (2022) 1054–1070.

[6] G. Han, G. Li, Y. Sun, Electrocatalytic Hydrogenation Using Palladium Membrane Reactors, JACS Au 4 (2024) 328–343.

[7] A. Ugwu, F. Donat, A. Zaabout, C. Müller, K. Albertsen, S. Cloete, et al., Hydrogen production by water splitting using gas switching technology. *Powder Technology*, Powder Technol. 370 (2020) 48–63.

[8] N. Dubouis, A. Grimaud, The hydrogen evolution reaction: from material to interfacial descriptors, Chem. Sci. 10 (2019) 9165–9181.

[9] S. Anantharaj, Hydrogen evolution reaction on Pt and Ru in alkali with volmer-step promotors and electronic structure modulators, Curr. Opin. Electrochem. 33 (2022) 100961.

[10] A. Kazemi, F. Manteghi, Z. Tehrani, Metal Electrocatalysts for Hydrogen Production in Water Splitting, ACS Omega 9 (2024) 7310–7335.

[11] R.R. Raja Sulaiman, A. Hanan, W.Y. Wong, R. Mohamad Yunus, K. Shyuan Loh, R. Walvekar, et al., Structurally Modified MXenes-Based Catalysts for Application in Hydrogen Evolution Reaction: A Review, Catalysts 12 (2022) 1576.

[12] D.R. Aireddy, K. Ding, Heterolytic Dissociation of H2 in Heterogeneous Catalysis, ACS Catal. 12 (2022) 4707–4723.

[13] A. Raveendran, M. Chandran, R. Dhanusuraman, A comprehensive review on the electrochemical parameters and recent material development of electrochemical water splitting electrocatalysts, RSC Adv. 13 (2023) 3843–3876.

[14] J.A. Lopez-Ruiz, E. Andrews, S.A. Akhade, M.-S. Lee, K. Koh, U. Sanyal, et al., Understanding the Role of Metal and Molecular Structure on the Electrocatalytic Hydrogenation of Oxygenated Organic Compounds, ACS Catal. 9 (2019) 9964–9972.

[15] J. Lessard *Encyclopedia of Applied Electrochemistry* Eds. G. Kreysa, K.-I. Ota, R.F. Savinell, Springer New York, New York, 2014, pp. 443–448.

[16] L. Zhou, Y. Li, Y. Lu, S. Wang, Y. Zou, pH-Induced selective electrocatalytic hydrogenation of furfural on Cu electrodes, Chin. J. Catal. 43 (2022) 3142–3153.

[17] [a] C.H. Lam, W. Deng, L. Lang, X. Jin, X. Hu, Y. Wang, Minireview on Bio-Oil Upgrading via Electrocatalytic Hydrogenation: Connecting Biofuel Production with Renewable Power, Energy Fuels 34 (2020) 7915–7928; [b] Y.P. Wijaya, K.J. Smith, C.S. Kim, E.L. Gyenge, Electrocatalytic hydrogenation and depolymerization pathways for lignin

valorization: toward mild synthesis of chemicals and fuels from biomass, Green Chem. 22 (2020) 7233–7264.

[18] [a] S.A. Akhade, N. Singh, O.Y. Gutiérrez, J. Lopez-Ruiz, H. Wang, J.D. Holladay, et al., Electrocatalytic Hydrogenation of Biomass-Derived Organics: A Review, Chem. Rev. 120 (2020) 11370–11419; [b] E. Andrews, J.A. Lopez-Ruiz, J.D. Egbert, K. Koh, U. Sanyal, M. Song, et al., Performance of Base and Noble Metals for Electrocatalytic Hydrogenation of Bio-Oil-Derived Oxygenated Compounds, ACS Sustain. Chem. Eng. 8 (2020) 4407–4418; [c] R.J. Dixit, K. Bhattacharyya, V.K. Ramani, S. Basu, Electrocatalytic hydrogenation of furfural using non-noble-metal electrocatalysts in alkaline medium, Green Chem. 23 (2021) 4201–4212.

[19] Z. Zhao, R. Yu, S. Wang, L. Guo, L. Zhang, M. Chen, J. Chen, et al., Introducing Fe into Cu-based catalyst to boost the electrocatalytic hydrogenation of 5-hydroxymethylfurfural, ChemSusChem. (2024) e202401278.

[20] M.A. Shah, W. Farooq, T. Shahnaz, M. Akilarasan, bioenergy and value-added chemicals derived through electrocatalytic upgradation of biomass: a critical review, BioEnergy Res. 17 (4) (2024) 2029–2049.

[21] [a] Y. Xiao, C. Shen, Z. Xiong, Y. Ding, L. Liu, W. Zhang, et al., Comprehensive Study Addressing the Challenge of Efficient Electrocatalytic Biomass Upgrading of 5-(Hydroxymethyl)Furfural (HMF) with a CH(3) NH(2) Ionic Liquid on Metal-Embedded Mo(2) B(2) MBene Nanosheets, Small 19 (2023) e2302271; [b] L. Zhang, T.U. Rao, J. Wang, D. Ren, S. Sirisommboonchai, C. Choi, et al., Fuel Process. Technol. 226 (2022) 107097.

[22] [a] J. Vander Steen, F. Boissou, M. Luhmer, C. Buess-Herman, S. Baranton, et al., Furfural electroreduction in choline-glycerol deep eutectic solvent, J. Electroanal. Chem. 933 (2023) 117269; [b] R. Wang, X. Ma, D. Ding, B. Huang, Z. Zhu, T. Su, et al., Inorg. Chem. Front. 10 (2023) 5420–5429; [c] M.S. Álvarez, M.A. Longo, A. Rodríguez, F.J. Deive, The role of deep eutectic solvents in catalysis. A vision on their contribution to homogeneous, heterogeneous and electrocatalytic processes, J. Ind. Eng. Chem. 132 (2024) 36–49; [d] Y. Shen, H. Zhou, X. He, F. Shen, Z. Xu, B. Yang, L. Kong, et al., Deep eutectic solvent-mediated hydrothermal treatment for biomass conversion: a review, Green Chem. 26 (2024) 8123–8144.

[23] [a] Y. Kwon, M.T.M. Koper, Electrocatalytic Hydrogenation and Deoxygenation of Glucose on Solid Metal Electrodes, ChemSusChem 6 (2013) 455–462; [b] J.P. Oña, R.-M. Latonen, N. Kumar, M. Peurla, I. Angervo, H. Grénman, Electrochim. Acta 426 (2022) 140754; [c] J. Lee, S. Jung, Y.T. Kim, H.J. Kim, K.H. Kim, Renew. Sustain. Energy Rev. 181 (2023) 113337.

[24] J.J. Roylance, K.S. Choi, Electrochemical reductive biomass conversion: direct conversion of 5-hydroxymethylfurfural (HMF) to 2,5-hexanedione (HD) via reductive ring-opening, Green Chem. 18 (2016) 2956–2960.

[25] J.J. Roylance, T.W. Kim, K.S. Choi, Efficient and Selective Electrochemical and Photoelectrochemical Reduction of 5-Hydroxymethylfurfural to 2,5-Bis(hydroxymethyl)furan using Water as the Hydrogen Source, ACS Catal. 6 (2016) 1840–1847.

[26] P. Nilges, U. Schröder, Electrochemistry for biofuel generation: production of furans by electrocatalytic hydrogenation of furfurals, Energy Environ. Sci. 6 (2013) 2925–2931.

[27] S. Li, X. Sun, Z. Yao, X. Zhong, Y. Cao, Y. Liang, Z. Wei, et al., Interfacial Interaction-Mediated Regulation of Metal Oxidation States for Enhanced CO2 Reduction, Adv. Funct. Mater. 29 (2019) 1904780.

[28] M.R. Kasad, J.E. Jackson, C.M. Saffron, Electrocatalytic hydrogenation of the formyl group and heteroaromatic ring in furfural on activated carbon cloth-supported ruthenium, RSC Sustain. (2024).

[29] Y. Song, O.Y. Gutiérrez, J. Herranz, J.A. Lercher, Aqueous phase electrocatalysis and thermal catalysis for the hydrogenation of phenol at mild conditions, Appl. Catal. B 182 (2016) 236–246.

[30] N. Itoh, W.C. Xu, S. Hara, K. Sakaki, Electrochemical coupling of benzene hydrogenation and water electrolysis, Catal. Today 56 (2000) 307–314.
[31] Z. Li, M. Garedew, C.H. Lam, J.E. Jackson, D.J. Miller, et al., Electrocatalytic Upgrading of Phenolic Compounds Observed after Lignin Pyrolysis, Green Chem. 14 (2012) 2540–2549.
[32] C. Ruan, R. Akutsu, K. Yang, N.M. Zayan, J. Dou, J. Liu, et al., Hydrogenation of bio-oil-derived oxygenates at ambient conditions via a two-step redox cycle, Cell Rep. Phys. Sci. 4 (2023) 101506.

CHAPTER

9

Noble metals based electrocatalysts for biomass conversion

Naveen Thakur[a], *Ramasamy Ganesamoorthy*[b], *Muthu Senthil Pandian*[c], *Rajender Boddula*[d], *Kuljit Kaur*[e], *Karchiyappan Thirugnanasambandham*[f], *and Ashwani Tapwal*[a]

[a] ICFRE-Himalayan Forest Research Institute, Shimla, Himachal Pradesh, India, [b] Department of Chemistry, Vinayaka Mission's Kirupananda Variyar Arts and Science College, Vinayaka Mission's Research Foundation (Deemed to be University), Salem, Tamil Nadu, India, [c] Department of Physics, SSN Research Centre, Sri Sivasubramaniya Nadar College of Engineering, Kalavakkam, Chennai, Tamil Nadu, India, [d] School of Sciences, Woxsen University, Hyderabad, Telangana, India, [e] School of Natural Sciences, GNA University, Sri Hargobindgarh, Phagwara, Punjab, India, [f] Nammazhvar Organic Farming Research Centre, Tamil Nadu Agricultural University, P N Pudur, Coimbatore, Tamil Nadu, India

9.1 Introduction

The shift toward sustainable and carbon-neutral technologies has sparked growing interest in using renewable resources to produce fuels and valuable chemicals [1]. Among these resources, biomass stands out

Electrochemical Biomass Conversion: Sustainable Processes and Applications for Bioenergy, Biofuels, and Bioproducts
DOI: https://doi.org/10.1016/B978-0-443-26757-4.00002-7

as the most plentiful renewable carbon source, offering a promising alternative to fossil fuels. It can be converted into essential building-block chemicals like glycerol, glucose, and 5-hydroxymethylfurfural (5-HMF), helping drive the growth of a circular bioeconomy [2–3]. Traditional methods for processing biomass—such as pyrolysis, hydrothermal techniques, and enzymatic catalysis—often come with drawbacks like high energy demands, long reaction times, or the need for costly biological agents. In contrast, electrocatalytic conversion is gaining traction as a more sustainable option. It can run under mild conditions, uses electricity from renewable sources, and offers precise control over which products are formed [4–5]. This makes it particularly suitable for smaller, decentralized production setups powered by solar or wind energy. Noble metals such as platinum (Pt), palladium (Pd), gold (Au), ruthenium (Ru), and iridium (Ir) have shown excellent performance in electrocatalytic applications, especially in converting molecules derived from biomass [6–8]. These metals are highly effective due to their unique electronic properties, strong ability to bind molecules, and resistance to corrosion. They play a key role in enabling complex chemical reactions in water-based environments. Noble metal nanoparticles, in particular, are central to many clean energy applications, including fuel cells (which generate electricity through hydrogen and oxygen reactions), water splitting (for hydrogen production), CO_2 reduction (turning carbon dioxide into fuels or chemicals), and nitrogen reduction (for ammonia synthesis) [9–10]. The same catalytic strengths are being used to advance the electrocatalytic conversion of biomass-derived compounds, making these materials essential to the future of green energy and chemical manufacturing.

However, the broad use of noble metals is limited by their scarcity, high price, and vulnerability to performance decline over time [11–12]. These challenges have driven research into improving catalyst design, such as creating bimetallic systems, core–shell structures, single-atom catalysts (SACs), and incorporating conductive materials to reduce noble metal usage while maintaining high performance [13–14]. This chapter explores how noble metal-based electrocatalysts function in biomass conversion, focusing on how they help turn renewable feedstocks into useful chemicals and fuels under environmentally friendly conditions. It also reviews recent developments, ongoing challenges such as cost and long-term stability, and potential strategies to create efficient, scalable solutions for bio-based industries. (Fig. 9.1).

9.2 Key principles of electrocatalytic biomass conversion

Electrocatalytic biomass conversion is a method that turns renewable, plant-based materials into useful fuels and chemicals through

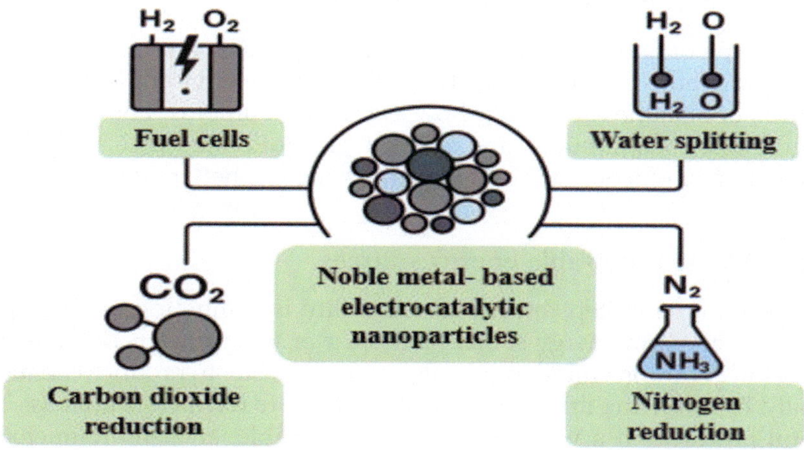

FIGURE 9.1 Schematic illustration of primary electrocatalytic functions of noble metal nanoparticles in energy conversion.

electrochemical reactions. This approach combines clean energy, eco-friendly chemistry, and efficient catalysts to provide a greener option compared to conventional methods that rely on heat and fossil fuels [15–16]. The key principles behind this approach are explained below.

9.2.1 Selectivity and efficiency of electrocatalysts

Designing effective electrocatalysts requires a careful balance between selectivity and efficiency. Biomass-derived compounds often contain multiple functional groups and are rich in oxygen, which makes it challenging to control the reaction pathway. For instance, when converting 5-HMF into 2, 5-furandicarboxylic acid (FDCA), using noble metals like platinum or ruthenium, the catalyst surface must be precisely engineered to promote oxidation while preserving the carbon structure.

9.2.2 Reaction thermodynamics and kinetics

Thermodynamics and kinetics play a crucial role in how electrochemical reactions unfold. These reactions take place where the electrode meets the electrolyte, and they are driven by an external voltage that causes either oxidation or reduction. Take glycerol oxidation on nickel-based electrodes, for example, while it's thermodynamically favorable, its reaction rate can be slow. Improving the surface properties of the catalyst is key to speeding things up and boosting product yield.

9.2.3 Importance of electrolyte and pH

The type of electrolyte used and its pH level play a key role in determining the outcome of the reaction. Whether the environment is acidic, neutral, or alkaline can change how the reaction unfolds. For example, in glucose electrooxidation, using an alkaline solution tends to support the production of gluconic acid and generally leads to better efficiency.

9.2.4 Using renewable energy sources

One important step toward sustainability is combining biomass processing with clean energy sources like solar or wind power. This not only helps lower carbon emissions but also makes it easier to produce chemicals locally. By following this approach, it's possible to turn waste biomass into useful products in a way that's efficient, scalable, and environmentally friendly—contributing to a more circular and sustainable economy.

9.3 Applications of noble metals

Noble metal-based electrocatalytic NPs have attracted considerable interest in recent years due to their remarkable catalytic activity, durability, and surface reactivity in a wide range of electrochemical processes [17]. Metals NPs form, functioning as efficient electrocatalysts in key applications, including fuel cells, water splitting, and carbon dioxide conversion. Table 9.1 presents a summary of the primary uses, types of nanoparticles, and representative electrochemical reactions catalyzed by noble metal-based systems, particularly in the context of biomass conversion and other sustainable technologies. For example, Pt nanoparticles supported on carbon (Pt/C) are extensively used in proton exchange membrane fuel cells, where they facilitate both hydrogen oxidation and oxygen reduction due to their exceptional electrocatalytic properties. In water electrolysis, Pt serves as a highly effective catalyst for the hydrogen evolution reaction (HER), while IrO_2 and RuO_2 nanoparticles are recognized for their efficiency in catalyzing the oxygen evolution reaction (OER) [18–19]. Additionally, Au and Ag nanoparticles contribute significantly to the electrochemical reduction of CO_2 into CO and various hydrocarbons, presenting a viable approach to sustainable carbon recycling [20–21]. In nitrogen reduction reactions (NRR), Ru nanoparticles demonstrate promising activity for the ambient conversion of atmospheric nitrogen (N_2) into ammonia (NH_3), offering a greener alternative to the conventional, energy-intensive Haber–Bosch process [22–23]. Although noble metal nanoparticles are costly, their catalytic efficiency can be further enhanced through alloying strategies or by dispersing them on conductive supports, thus improving their viability for practical applications in clean energy and environmental technologies

TABLE 9.1 Summary of noble metal nanoparticles used in energy electrocatalysis.

Noble metal	NP type	Key applications	Example reactions
Platinum (Pt)	Pt NPs, Pt/C, Pt-alloys	Fuel cells, HER, ORR, sensors	$H_2 \rightarrow 2H^+ + 2e^-$ (HOR); $O_2 + 4H^+ + 4e^- \rightarrow 2H_2O$ (ORR)
Palladium (Pd)	Pd NPs, Pd-Au, Pd-Cu	CO_2 reduction, methanol oxidation, hydrogen sensing	$CO_2 + 2H^+ + 2e^- \rightarrow HCOOH$
Gold (Au)	Au NPs, Au@Pd, Au nanocages	CO_2 reduction, electrochemical sensors, Cancer therapy	$CO_2 + 2e^- \rightarrow CO + O^{2-}$
Silver (Ag)	Ag NPs, Ag/C, Ag-Au alloys	CO_2 reduction, ORR, antimicrobial electrocatalysis	$CO_2 + e^- \rightarrow CO$; $O_2 + 2H^+ + 2e^- \rightarrow H_2O_2$
Iridium (Ir)	IrO_2 NPs, Ir-alloys	Oxygen evolution reaction (OER), acidic water splitting	$2H_2O \rightarrow O_2 + 4H^+ + 4e^-$
Ruthenium (Ru)	RuO_2 NPs, Ru-based alloys	OER, NRR, supercapacitors	$N_2 + 6H^+ + 6e^- \rightarrow 2NH_3$
Rhodium (Rh)	Rh NPs, Rh-Pt	Hydrogen oxidation, ethanol oxidation	$CH_3CH_2OH \rightarrow CH_3COOH + 4H^+ + 4e^-$
Osmium (Os)	Os NPs (less common)	High-stability sensors, specialty catalysis	H_2O_2 sensing is less explored in mainstream applications

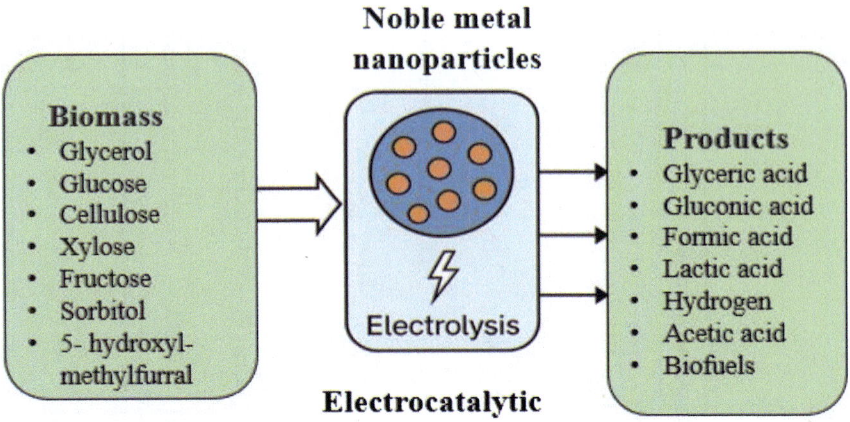

FIGURE 9.2 Electrocatalytic conversion of biomass compounds using noble metal NPs.

9.4 Biomass conversion by noble metals

Noble metal-based nanoparticles are increasingly recognized for their effectiveness in promoting sustainable biomass conversion. These catalysts, typically composed of Pt, Pd, Au, Ru, and Rh, demonstrate exceptional activity in electrochemical transformations. As illustrated in Fig. 9.2, biomass derived from plant-based compounds such as glycerol, glucose, and 5-HMF can be converted into valuable chemicals and biofuels through electrochemical processes [24]. These transformations are often conducted under mild conditions using renewable electricity, contributing to both energy efficiency and environmental sustainability. For example, Pt and Pd nanoparticles facilitate the electrooxidation of glycerol to produce compounds like glyceric acid and formic acid, which hold significance in fuel cell applications. Similarly, glucose can undergo oxidation in the presence of Au nanoparticles, yielding gluconic acid, a widely used product in the food and pharmaceutical sectors. Additionally, Ru and Rh nanoparticles are effective in reducing levulinic acid to γ-valerolactone (GVL), a key intermediate for biofuel production. Another prominent example involves the transformation of HMF into 2,5FDCA using Pt or Ir catalysts. FDCA is a bio-based alternative to traditional petrochemical-derived plastics. Beyond individual reactions, noble metal nanoparticles offer the advantage of fine-tuning reaction selectivity, enabling targeted product formation while minimizing undesired by-products. Their high surface area and adaptable surface chemistry further improve catalytic efficiency. As interest grows in decentralized and sustainable chemical manufacturing, the use of noble metal-based electrocatalysts presents a compelling approach for converting low-value biomass into high-value, eco-friendly fuels and materials.

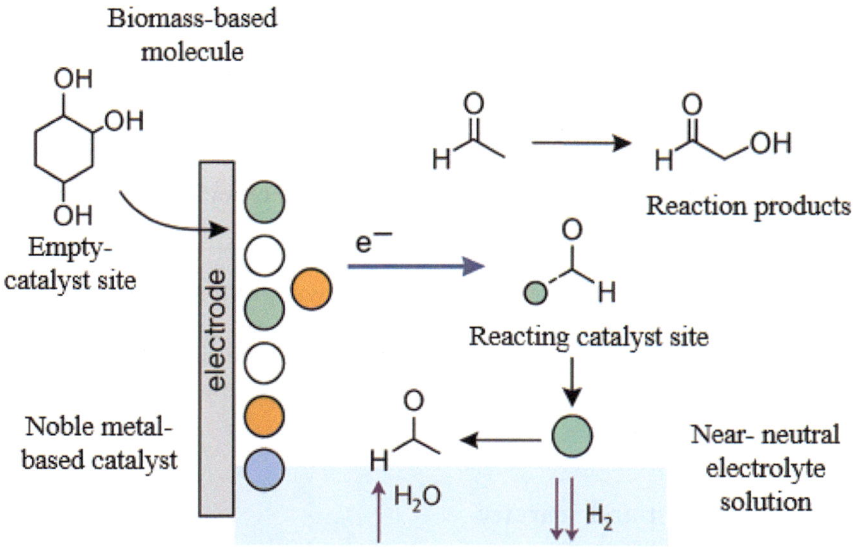

FIGURE9.3 Electrochemical pathways for biomass conversion facilitated by noble metal-based electrocatalysts.

9.5 Sustainable biomass conversion of noble metals

Noble metal nanoparticles are proving to be highly effective in converting plant-based biomass into valuable products. These metals serve as catalysts in electrochemical reactions that transform compounds like glycerol, glucose, and 5-HMF into biofuels and useful chemicals [25]. This transformation happens under mild conditions using electricity from renewable sources, making the process both energy-efficient and environmentally responsible. During the reaction, molecules from biomass attach to the surface of these metal catalysts. Through electron transfer, they are then converted into more valuable chemical substances, often in near-neutral electrolytes. One added benefit of this method is the simultaneous generation of hydrogen gas, which adds to its sustainability (Fig. 9.3). For examples highlight how different metals contribute to various transformations. Pt and Pd nanoparticles are known to drive the oxidation of glycerol, producing compounds like glyceric acid and formic acid, both of which are important for fuel cell applications [26]. Gold nanoparticles can help convert glucose into gluconic acid, a substance widely used in food and pharmaceuticals. Ruthenium and rhodium nanoparticles are effective in turning levulinic acid into γ-GVL, a promising biofuel [27]. Additionally, HMF, a commonly studied biomass molecule, can be transformed into 2,5-FDCA using Pt, Au, or Ru catalysts—offering a sustainable alternative to conventional plastics [28]. So, this method provides a cleaner, more selective way to convert low-cost biomass into high-value products while reducing environmental impact.

9.6 Current challenges in emerging electrocatalysts for biomass conversion

Although noble metal catalysts have shown promising results, their widespread application faces several key limitations. These include high costs associated with their limited availability and the risk of deactivation over extended periods of use, both of which compromise their durability and long-term efficiency.

9.6.1 High cost and scarcity

A persistent and critical limitation in the application of noble metals such as platinum (Pt), palladium (Pd), and iridium (Ir) is their high cost and restricted global availability. These elements are geologically scarce and predominantly extracted from a limited number of regions, resulting in concentrated supply chains that are highly susceptible to geopolitical disruptions and market volatility. Consequently, such economic and logistical constraints significantly hinder their scalability and long-term sustainability in industrial applications [29].

9.6.2 Catalysts deactivation and poisoning

During the electrooxidation of biomass, catalyst performance often declines due to the accumulation of strongly adsorbed species such as carbon monoxide (CO) or polymeric byproducts on the active sites. These surface-bound intermediates hinder the availability of active catalytic sites, thereby diminishing overall activity and posing significant challenges for their implementation in large-scale industrial processes.

9.6.3 Limited stability in harsh conditions

Although certain noble metals demonstrate strong catalytic activity, they often suffer from limited durability when exposed to highly acidic or alkaline environments necessary for specific biomass conversion processes. These conditions can lead to metal dissolution or agglomeration under electrochemical stress, ultimately compromising both the reproducibility and long-term stability of their performance.

9.6.4 Selectivity challenges in biomass electro conversion

Biomass-derived compounds typically exhibit significant structural complexity and contain multiple functional groups. This complexity presents major challenges in steering reactions toward the formation of a specific, high-value product, as numerous competing pathways may occur simultaneously. As a result, achieving high selectivity demands precise control over both catalyst design and reaction conditions [30].

9.7 Future strategies in biomass conversion

Future strategies in electrocatalysis for biomass conversion focus on enhancing efficiency, sustainability, and cost-effectiveness.

- *Alloying with earth-abundant metals*: Forming bimetallic or trimetallic systems (e.g., Pt-Ni, Pd-Cu) reduces noble metal content while maintaining or even enhancing catalytic activity.
- *Single-atom catalysts*: By isolating metal atoms on suitable supports, SACs maximize atom efficiency and allow for tunable electronic environments that improve selectivity and activity [31].
- *Advanced supports and architectures*: Using conductive supports like carbon materials, metal oxides, or metal-organic frameworks aids in preventing nanoparticle agglomeration, promoting uniform dispersion, and stabilizing active catalytic sites, which collectively enhance catalytic efficiency and durability.
- *Integration with renewable energy*: Coupling electrocatalysis with solar, wind, or bioelectricity sources further enhances the sustainability of biomass conversion. (Fig. 9.4).

9.8 Conclusion

Noble metal-based electrocatalysts play a key role in the sustainable conversion of biomass into valuable chemicals and fuels. These catalysts offer high efficiency, good stability, and excellent selectivity under mild reaction conditions. They are effective in converting compounds into useful products for clean energy and chemical industries. Although their high cost and limited supply are major challenges, recent improvements such as bimetallic systems, core–shell structures, and single-atom catalysts are helping to reduce the use of noble metals and enhance performance. By combining electrocatalysis with renewable electricity sources, these materials support the development of eco-friendly and decentralized chemical production. Overall, noble metal-based electrocatalysts provide a strong platform for green technologies and better use of biomass in the future.

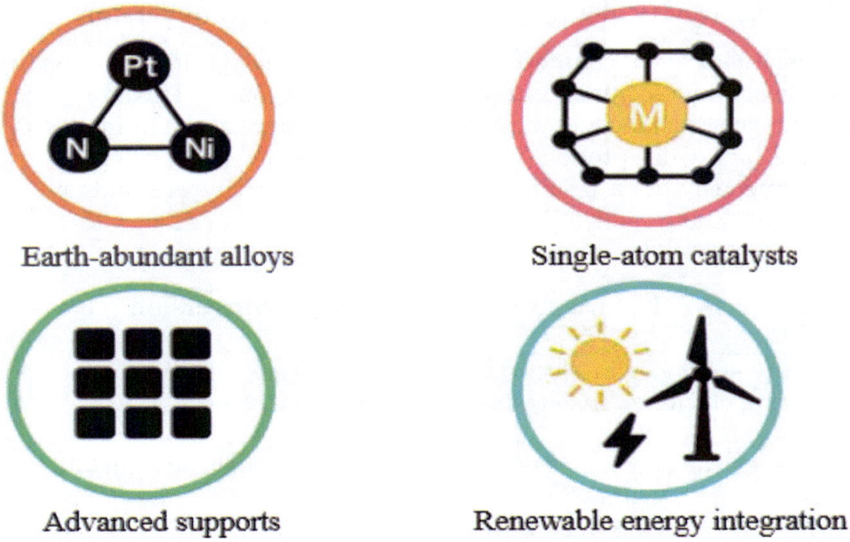

FIGURE 9.4 Future directions for noble metal-based electrocatalysts in biomass conversion.

References

[1] M.A. Shah, W. Farooq, T. Shahnaz, M. Akilarasan, Bioenergy and value-added chemicals derived through electrocatalytic upgradation of biomass: a critical review, BioEnergy Res 17 (4) (2024) 2029–2049.
[2] L. Cherwoo, I. Gupta, G. Flora, R. Verma, M. Kapil, S.K. Arya, et al., Biofuels an alternative to traditional fossil fuels: a comprehensive review, Sustain. Energy Technol. Assess. 60 (2023) 103503.
[3] S. Dutta, Sustainable synthesis of drop-in chemicals from biomass via chemical catalysis: scopes, challenges, and the way forward, Energy Fuels 37 (4) (2023) 2648–2666.
[4] B. Zhang, B.K. Biswal, J. Zhang, R. Balasubramanian, Hydrothermal treatment of biomass feedstocks for sustainable production of chemicals, fuels, and materials: progress and perspectives, Chem Rev 123 (11) (2023) 7193–7294.
[5] W.M. Lewandowski, M. Ryms, W. Kosakowski, Thermal biomass conversion: a review, Processes 8 (5) (2020) 516.
[6] P. Zhou, J. Zhang, Electrochemical transformation of biomass-derived oxygenates, Sci China Chem 66 (4) (2023) 1011–1031.
[7] A.A. Turkin, E.V. Makshina, BF. Sels, Catalytic hydroconversion of 5-HMF to value-added chemicals: insights into the role of catalyst properties and feedstock purity, ChemSusChem 15 (13) (2022) e202200412.
[8] S.M. Ali, Biomass derived carbon/platinum nanoparticles as electrocatalyst for the hydrogen evolution, Sci Rep 15 (1) (2025) 4602.
[9] T. Noor, L. Yaqoob, N. Iqbal, Recent advances in electrocatalysis of oxygen evolution reaction using noble-metal, transition-metal, and carbon-based materials, ChemElectroChem 8 (3) (2021) 447–483.
[10] H. Kim, T.Y. Yoo, M.S. Bootharaju, J.H. Kim, D.Y. Chung, T. Hyeon, Noble metal-based multimetallic nanoparticles for electrocatalytic applications, Adv Sci 9 (1) (2022) 2104054.

[11] F. Zhang, Y. Zhu, Q. Lin, L. Zhang, X. Zhang, H. Wang, Noble-metal single-atoms in thermocatalysis, electrocatalysis, and photocatalysis, Energy Environ Sci 14 (5) (2021) 2954–3009.

[12] H. Lu, J. Tournet, K. Dastafkan, Y. Liu, Y.H. Ng, S.K. Karuturi, et al., Noble-metal-free multicomponent nanointegration for sustainable energy conversion, Chem Rev 121 (17) (2021) 10271–10366.

[13] W. Li, Z. Guo, J. Yang, Y. Li, X. Sun, H. He, J. Zhang, Advanced strategies for stabilizing single-atom catalysts for energy storage and conversion, Electrochem Energy Rev 5 (3) (2022) 9.

[14] C. Wang, C. An, C. Qin, H. Gomaa, Q. Deng, S. Wu, N. Hu, Noble metal-based catalysts with core-shell structure for oxygen reduction reaction: progress and prospective, Nanomaterials 12 (14) (2022) 2480.

[15] M. Yang, Z. Yuan, R. Peng, S. Wang, Y. Zou, Recent progress on electrocatalytic valorization of biomass-derived organics, Energy Environ. Mater. 5 (4) (2022) 1117–1138.

[16] L. Liu, D. Pant, Integrative electrochemical and biological catalysis for the mild and efficient utilization of renewable electricity and carbon resources, Sustain. Energy Fuels. 8 (3) (2024) 460–480.

[17] Y. Sharma, P. Kapoor, Noble metal nanostructures for various applications, Mater. Today Proc. (2023) 1–10 In press, doi:10.1016/j.matpr.2023.03.628.

[18] X. Tan, S. Shahgaldi, X. Li, The effect of non-spherical platinum nanoparticle sizes on the performance and durability of proton exchange membrane fuel cells, Adv. Appl. Energy 4 (2021) 100071.

[19] T.D. Le, D.S. Kim, T.V. Tran, B. Urupalli, G.S. Shin, G.J. Oh, Y.T. Yu, Electronic structure engineering of Pt–Ni alloy NPs by coupling of gold single atoms on N-doped carbon for highly efficient oxygen reduction reaction and hydrogen evolution reaction, Small 20 (34) (2024) 2311971.

[20] D. Bhalothia, D.W. Lee, G.P. Jhao, H.Y. Liu, Y. Jia, S. Dai, et al., Reaction pathways for the highly selective and durable electrochemical CO_2 to CO conversion on ZnO supported Ag nanoparticles in KCl electrolyte, Appl Surf Sci 608 (2023) 155224.

[21] Y.J. Kim, J.Y. Maeng, S.Y. Hwang, C.K. Rhee, Y. Sohn, Electrochemical reduction of CO2 and CO using interface-engineered Au/Ti electrodes for long-chain hydrocarbon production, Appl Catal B 338 (2023) 123017.

[22] H. Fang, D. Liu, Y. Luo, Y. Zhou, S. Liang, X. Wang, et al., Challenges and opportunities of Ru-based catalysts toward the synthesis and utilization of ammonia, ACS Catal 12 (7) (2022) 3938–3954.

[23] R. Akter, S.S. Shah, M.A. Ehsan, M.N. Shaikh, M.H. Zahir, M.A. Aziz, AS. Ahammad, Transition-metal-based catalysts for electrochemical synthesis of ammonia by nitrogen reduction reaction: advancing the green ammonia economy, Chem Asian J 19 (16) (2024) e202300797.

[24] X. Li, M. Xu, Y. Su, X. Zheng, Y. Qin, F. Zhang, et al., Metal-based materials for electrocatalytic biomass conversion, Functional Materials for Electrocatalytic Energy Conversion, 2025, pp. 509–537.

[25] M.U. Reddy, J. Pachiyappan, S. Sivamani, Plant-biomass-based nanomaterials for sustainable energy, Biomass Conversion through Nanomaterials, Elsevier, 2025, pp. 1–25.

[26] T. Li, DA. Harrington, An overview of glycerol electrooxidation mechanisms on Pt, Pd and Au, ChemSusChem 14 (6) (2021) 1472–1495.

[27] N. Schlegel, G.K. Wiberg, M. Arenz, On the electrooxidation of glucose on gold: towards an electrochemical glucaric acid production as value-added chemical, Electrochim Acta 410 (2022) 140023.

[28] S.A. Sorokina, S.P. Mikhailov, N.V. Kuchkina, A.V. Bykov, A.L. Vasiliev, M.G. Ezernitskaya, et al., Ru@hyperbranched polymer for hydrogenation of levulinic acid to gamma-valerolactone: the role of the catalyst support, Int J Mol Sci 23 (2) (2022) 799.

[29] C. Zhou, W. Deng, X. Wan, Q. Zhang, Y. Yang, Y. Wang, Functionalized carbon nanotubes for biomass conversion: base-free aerobic oxidation of 5-hydroxymethylfurfural to 2,5-furandicarboxylic acid over Pt/CNT catalyst, ChemCatChem 7 (18) (2015) 2853–2863.
[30] S. Hameed, L. Lin, A. Wang, W. Luo, Recent developments in metal-based catalysts for the catalytic aerobic oxidation of 5-hydroxymethyl-furfural to 2,5-furandicarboxylic acid, Catalysts 10 (1) (2020) 120.
[31] S. Pandey, M. Mottoul, V. Orsat, J.F. Morin, M.J. Dumont, Base-free synthesis of renewable furan-2,5-dicarboxylic acid (FDCA) over an earth-abundant magnetic catalyst, J Environ Chem Eng 12 (3) (2024) 112763.

CHAPTER

10

Nonnoble metal-based electrocatalysts for biomass conversion: Sustainable and cost-effective approaches

Vijayakrishnan Jeyavani[a], *Ramyakrishna Pothu*[b,c]

[a]Department of Physics, Rathinam College of Arts and Science, Coimbatore, Tamil Nadu, India, [b]Centre for Research and Development Cell, Sharda University, Greater Noida, Uttar Pradesh, India, [c]School of Physics and Electronics, College of Chemistry and Chemical Engineering, Hunan University, Changsha, Hunan, China

10.1 Introduction

In recent days, converting biomass into fuels and chemicals with added value provides a sustainable way to produce materials and energy. The nonnoble metals-based electrocatalysts are showing promising substitutes for precious metals due to their efficiency, cost-effectiveness, and availability. With an emphasis on their design, synthesis, reaction processes, and use in electrochemical biomass valorization, this chapter examines current developments in nonnoble metal-based electrocatalysts. The transition metal catalysts, such as Fe, Co, Ni, Cu, and Mn, transition metal oxides, transition sulfides, carbides, and doped carbon compounds are specifically highlighted. Further, the future scope of the research, challenges, and perspectives are also highlighted. In contrast to fossil fuels, which are limited and carbon-intensive, biomass is a renewable, carbon-rich feedstock that is regenerated quickly through photosynthesis. It includes a variety of

Electrochemical Biomass Conversion: Sustainable Processes and Applications for Bioenergy, Biofuels, and Bioproducts
DOI: https://doi.org/10.1016/B978-0-443-26757-4.00003-9

resources, including municipal organic waste, forestry waste, energy crops, and agricultural wastes. Its closed carbon loop, which allows for carbon-neutral or even carbon-negative processes when combined with carbon capture technologies, is its inherent benefit. The CO_2 released during biomass consumption was first absorbed from the atmosphere during biomass growth. Because of this, biomass becomes a sustainable substitute carbon source for materials, chemicals, and fuels, lowering dependency on carbon generated from fossil fuels and reducing greenhouse gas emissions [1–3].

10.2 Fundamentals of electrochemical biomass conversion

Biomass conversion unlocks various platform molecules that serve as foundational building blocks to produce value-added chemicals and biofuels, bridging raw biomass feedstock and final products. The key examples include 5-Hydroxymethylfurfural (HMF), which is formed via acid-catalyzed dehydration of hexose sugars. HMF is a versatile intermediate convertible into derivatives like 2,5-furandicarboxylic acid (FDCA), a monomer for bio-based plastics, and biofuels such as 2,5-dimethylfuran (DMF) with high energy density [4–6]. Methods for enhancing HMF yields include novel heterogeneous catalysis and integration of biochemical and thermochemical routes [4, 7]. Second, furfural, generated mainly from pentoses by acid-catalyzed dehydration, acts as a precursor to resins, solvents, and biofuels. Advances in catalytic processes and integrated production methods seek to enhance yields and sustainability, improving furfural competitiveness relative to petroleum-based chemicals [4, 7, 8]. Glycerol is a biodiesel byproduct that can be catalytically upgraded into chemicals such as propanediols, acrolein, and epichlorohydrin, enabling value-added utilization and reducing waste streams [4, 9]. Traditional bioethanol production through sugar fermentation provides both a biofuel and a chemical feedstock convertible into ethylene and acetaldehyde. Engineered microbial strains demonstrate the potential for consolidated bioprocessing directly from lignocellulosic biomass, enhancing efficiency and sustainability [2, 3]. These platform molecules establish biomass conversion as a versatile alternative feedstock strategy, allowing the synthesis of fuels and chemicals that compete with petroleum-derived analogs by integrating catalytic upgrading and biotechnological innovations [4, 5, 9].

10.3 Role in circular economy and decarbonization

Biomass conversion is foundational to a circular bioeconomy, emphasizing the valorization of renewable biomass and organic waste streams to minimize resource depletion and environmental impacts. The circular model encourages reuse, recycling, and upgrading of biomass-derived products within sustainable supply chains, reducing dependency on virgin fossil materials [1, 10]. From a decarbonization perspective, biomass contributes by providing carbon-neutral fuels and chemicals that can displace fossil-based carbon sources. The net-zero greenhouse gas emissions potential arises because biomass captures atmospheric CO_2 during growth, balancing emissions on utilization. Furthermore, coupling biomass conversion with carbon capture and storage or utilization enhances the potential for carbon-negative processes, an essential component to meet stringent climate targets [1–3]. In addition, shifting chemical industry feedstocks from fossil hydrocarbons to biomass reduces the carbon intensity of production processes. Catalytic routes that efficiently upgrade biomass-derived carbohydrates into hydrocarbons, aromatics, and biofuels with targeted molecular structures support the replacement of petroleum in diverse applications, ranging from transportation fuels to industrial chemicals, thereby enabling systemic decarbonization [2, 9]. By providing renewable carbon feedstocks, generating versatile platform molecules, and embedding circular economy principles, biomass conversion represents a keystone technology for sustainable energy and chemical sectors. Its integration with advanced catalytic and biotechnological methods propels the transition from fossil dependency toward low-carbon, renewable, and circular industrial ecosystems vital for climate change mitigation and sustainable development [1–5].

10.4 Electrocatalytic conversion of biomass using noble metals

Electrocatalysis has garnered substantial interest as a clean and efficient approach for driving chemical transformations, particularly in energy conversion and biomass upgrading processes. Its advantages over conventional thermocatalysis are principally linked to milder operational conditions and superior control over reaction pathways. In thermocatalysis, reactions commonly require elevated temperatures (often above 300°C) and pressures to overcome activation barriers inherent to biomass-derived substrates or small molecules such as CO_2 and N_2 [1, 2]. By contrast, electrocatalysis operates predominantly at ambient temperature and pressure, leveraging electrical energy to activate reactants through controlled electron transfer steps. This results in reduced energy inputs, enhanced safety, and simplified reactor design, which are critically beneficial for integrating into distributed or modular production systems [3].

Additionally, the electrochemical driving force offers unparalleled tunability in reaction parameters such as applied potential, current density, electrolyte composition, and pH that can be dynamically adjusted to selectively target desired reaction intermediates or products. This tunability enables higher selectivity compared to thermally driven routes, lowering unwanted side reactions and thus improving energy efficiency and product purity. For example, in the electrochemical conversion of biomass-derived platform molecules like 5-HMF and furfural, fine control over the potential can steer the pathway toward hydrogenation, oxidation, or coupling products [3, 4]. Such selectivity and adaptability are challenging to achieve using conventional catalysis, where reaction conditions are often fixed and less specific to the catalyst surface state. Despite the exceptional catalytic activity and well-understood behavior of noble metals such as platinum (Pt), palladium (Pd), and ruthenium (Ru), their widespread use in electrocatalysis is hindered by critical drawbacks. First, their scarcity and high market cost limit large-scale deployment, particularly in applications aiming for sustainable and economically viable energy and chemical production pathways [5]. The finite supply of these metals also raises long-term supply security concerns, which are especially acute for global decarbonization efforts requiring massive catalyst volumes.

Second, noble metal catalysts often suffer from susceptibility to poisoning by reaction intermediates or contaminants. For instance, CO and sulfur-containing species can strongly adsorb on Pt-group metals, blocking active sites and leading to rapid performance decay, necessitating costly regeneration strategies or catalyst replacement [5]. Noble metals are benchmark catalysts for electrochemical biomass conversion due to their high catalytic activities and selectivities. However, their widespread application is hindered by several factors, such as the h**igh cost and scarcity of** Pt, Pd, and Ru, which pose scalability and economic barriers for large-scale biomass electroconversion. Second, **the** catalytic surfaces of noble metals often suffer deactivation by strongly adsorbed intermediates such as CO, sulfur species, or other biomass-derived impurities, leading to rapid loss of activity during prolonged operation. The noble metals can undergo morphological changes such as agglomeration or dissolution under reactive electrochemical environments, compromising long-term durability. These challenges provide the impetus for developing nonnoble metal-based electrocatalysts that are earth-abundant, cost-effective, and can offer competitive catalytic performance with enhanced resistance to poisoning and corrosion. This influences the catalyst durability and operational uptime, increasing the overall cost and complexity. Furthermore, the environmental sustainability of using noble metals is questionable given the energy-intensive mining and refining processes involved, as well as challenges in recycling and recovery pathways at end-of-life [6]. These issues motivate the urgent development of noble-metal-free electrocatalysts with competitive performance.

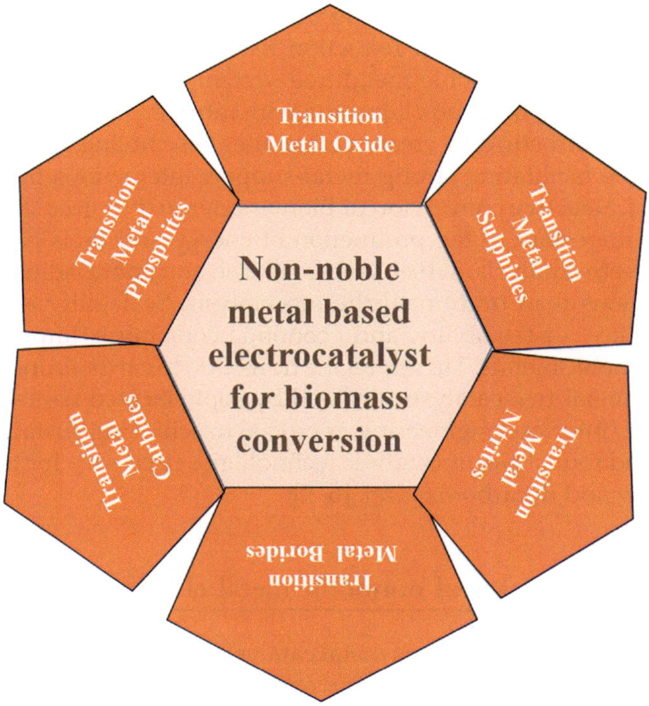

FIGURE 10.1 Nonnoble metal-based electrocatalyst for biomass conversion with transition metal composition.

10.5 Nonnoble metal electrocatalysts for biomass conversion

Recent advances have highlighted alternative materials based on earth-abundant transition metals such as iron (Fe), cobalt (Co), nickel (Ni), copper (Cu), and manganese (Mn) [6, 7]. These transition metal oxides, sulfides, carbides, nitrides, borides, and doped carbon compounds provide a more effective, sustainable, and cost-effective foundation, and when they are in nanomaterial form, heteroatom doping and support engineering, their intrinsic catalytic activities and selectivity can be noticeably enhanced. Fig. 10.1 represents the different nonnoble metal (based on transition metal) compositions used for biomass conversion. Further, metal-free catalysts such as doped graphitic carbons and conducting polymers have shown promising catalytic behavior for key electrocatalytic reactions (e.g., oxygen reduction and hydrogen evolution) with high stability and low cost [7]. Hybrid composites integrating earth-abundant metal centers coordinated within carbonaceous frameworks (e.g., M-N-C structures) have achieved activity approaching or sometimes surpassing that of Pt-group metals for the oxygen reduction reaction (ORR), highlighting the practical feasibility of noble-metal-free systems [5, 6]. For example, the heteroatom doping of carbon supports with nitrogen or sulfur can modify the electronic structure to create active sites resembling those on noble metals, while facilitating strong metal-support interactions that improve stability [7]. Selective conversion of biomass-derived molecules and small molecules important to the production of energy and fuels is made possible by electrocatalysis, which offers milder and more adjustable reaction conditions than traditional thermocatalysis. Scalability is hampered by availability, sustainability, and economic constraints imposed by the reliance on noble metals. Therefore, the transition to earth-abundant metal-based and metal-free catalysts, enabled by sophisticated nanostructuring and surface/interface engineering, is crucial to achieving sustainable, economical, and robust electrocatalytic technologies necessary for the circular bioeconomy and decarbonization. [6, 7].

10.6 Classes of nonnoble metal electrocatalysts

A wide range of nonprecious metal catalysts with appealing activity and stability for biomass electrooxidation and conversion have been produced by advances in material chemistry, particularly in alkaline media, where many biomass oxidation processes are kinetically advantageous.

10.6.1 Ni, Co, and Fe-based transition metal oxides and hydroxides

Nickel- and cobalt-based layered double hydroxides (LDHs) and mixed oxides form one of the most extensively studied classes of nonnoble electrocatalysts. **Ni-based LDHs and NiOOH** are recognized as highly active catalysts for glycerol and HMF electrooxidation, enabling selective C–C bond cleavage and higher oxidation state product formation like FDCA with high Faradaic efficiencies [8]. In particular, the $Ni^{(III)}$ active sites formed during anodic polarization facilitate electron transfer and intermediate stabilization. The **mixed** transition **metal oxides,** such as NiFe or NiCo LDHs, improve electrical conductivity and synergistically tune the electronic environment, enhancing kinetics for biomass-derived substrate oxidation. The Fe or Co incorporation modulates redox potentials and oxygen vacancy concentration, boosting catalytic turnover. These metal oxides also exhibit good stability due to self-passivation by surface oxide/hydroxide formation, mitigating metal leaching under oxidative potentials.

10.6.2 Transition metal phosphides

Metal phosphides have emerged as promising electrocatalysts for dual-function biomass electrooxidation and hydrogen evolution reaction (HER) due to their high electrical conductivity arising from strong metal–phosphorus covalent bonding, facilitating rapid charge transfer. Ni_2P and CoP specifically stabilize key adsorbed intermediates through d-electron density tuning, enhancing rates of oxidation for alcohol substrates like ethanol and glycerol. These phosphide catalysts often show good tolerance against poisoning species such as CO and H_2S compared with noble metals due to modified surface electronic properties [9].

10.6.3 Single-atom catalysts

In general, metal–nitrogen–carbon (M–N–C) combination is known as single-atom catalysts (SACs). Nitrogen-doped carbon matrices containing atomically distributed Fe or Co sites imitate enzyme active sites and offer distinct catalytic conditions. These catalysts optimize the use of active sites and modify electronic structures to facilitate the oxidation of biomass. Fe-N-C catalysts have shown exceptional selectivity and endurance in the moderate oxidation of furfural to furoic acid and HMF to FDCA. By forming anchoring sites, the nitrogen dopants extend the catalyst's lifetime and stop metal agglomeration [10, 11].

10.6.4 Spinel and perovskite oxides

Perovskite oxides and spinels are complementary catalyst platforms for biomass conversion: perovskites offer tunable oxygen nonstoichiometry and the capacity to dissolve active, sinter-resistant metal nanoparticles, while spinels provide structural robustness and multisite redox flexibility [12]. Combining meticulous defect and interface engineering, operando diagnostics that record dynamic active-site evolution, and catalytic testing on realistic feeds to address deactivation pathways and scale-up issues will yield the most promising advancements. The complementary spinels like $NiFe_2O_4$ or $CoFe_2O_4$, $CuFe_2O_4$, and Lanthanum-based perovskite $La_{1-x}Sr_xBO_{3-\delta}$ with B = Ni/Fe/Co) offer flexible oxidation states and lattice oxygen activation. The presence of oxygen vacancies and redox-active sites promotes the adsorption and activation of biomass molecules. Strong metal-support interactions allow stabilization of reaction intermediates and promote electron transfer. Such materials are less costly and structurally robust, suitable for harsh electrochemical regimes, especially for water oxidation coupled with biomass electrooxidation in hybrid systems [13].

10.6.5 Bimetallic and heterostructured catalysts

In bimetallic systems, the combination of two metals leads to modified electronic and geometric environments that improve the activation of C–O and C–C bonds, enabling efficient hydrodeoxygenation, hydrogenation, and reforming of biomass-derived molecules such as glycerol, furfurals, and bio-oils. Synergistic effects are exploited by engineered heterostructures and alloys such as Ni–Cu and $Ni_2P/Co(OH)_2$. By adjusting intermediate adsorption energies, synergistic interfaces encourage more effective biomass oxidation routes. While Ni provides OH^- adsorption sites to optimize kinetics, Cu addition helps break C–C bonds in glycerol. Layered and core–shell nanoscale structures improve mass transport, conductivity, and active surface areas [14]. These catalysts also exhibit strong metal–support interactions that enhance durability and suppress coke formation, making them suitable for converting cellulose, lignin, and platform molecules into valuable fuels and chemicals. Overall, the interplay of dual metals or coupled phases provides superior catalytic performance compared to conventional single-component catalysts, positioning bimetallic and heterostructured materials as promising candidates for sustainable and efficient biomass valorization.

10.6.6 Catalyst supports, functionalization, and structural engineering

The roles of catalyst supports, functionalization, and structural properties are to enhance the efficiency and selectivity of biomass conversion processes. The schematic representaion of various fuctionalization techniques are given in Fig. 10.2. **Carbon-based supports (graphene, carbon nanotubes, and biomass-derived porous carbon)** improve electrical conductivity and prevent agglomeration of nonnoble metal nanoparticles [15]. **Doping of supports with heteroatoms (N, P, S)** enhances electron density and provides anchoring sites for metal species, increasing stability and catalytic efficiency [12]. Nanostructuring into porous, hierarchical, and yolk–shell morphologies improves surface accessibility for large biomass molecules, enhancing reactant diffusion and turnover rates. By integrating these three aspects, catalysts can achieve enhanced activity, resistance to deactivation, and improved tolerance to water and impurities commonly present in biomass feedstocks. Together, support design, functionalization, and engineered structures create robust catalytic systems capable of handling the multistep and heterogeneous nature of biomass conversion pathways, ultimately improving yield, sustainability, and overall process efficiency [16–18].

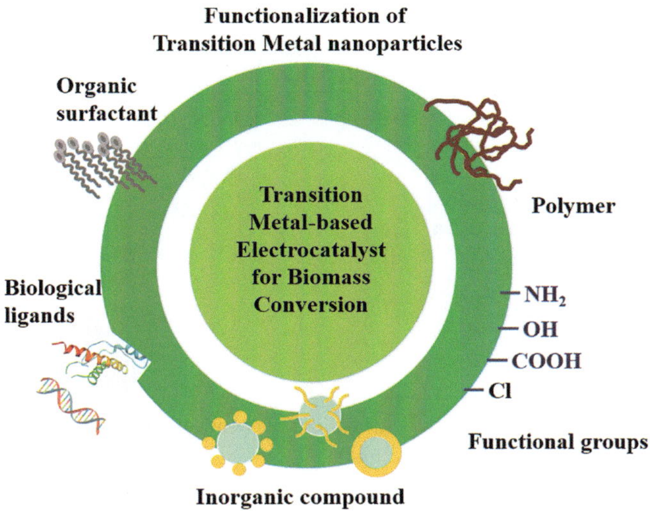

FIGURE 10.2 Various techniques of functionalizing transition metal-based electrocatalysts for biomass conversions.

10.7 Electrochemical biomass conversion reaction pathways using nonnoble metal compositions

10.7.1 Electrooxidation pathways

Electrooxidation in biomass valorization primarily targets the anodic conversion of alcohols, polyols, and aldehydes into valuable chemicals such as aldehydes, acids, or further oxidized carbon products (e.g., CO_2). The electrochemical transfermations of different chemical compounds are given in the Fig. 10.3. This involves multielectron/proton transfer reactions driven at controlled electrode potentials and heavily dependent on catalyst surface chemistry and electrolyte environment [19].

10.7.1.1 Oxidation of alcohols

Biomass-related alcohols (e.g., methanol, ethanol, furfuryl alcohol, glycerol, HMF) typically undergo sequential dehydrogenation and oxidation steps, generically represented as below.

(i) $\text{R-CH}_2\text{OH} \xrightarrow[\textbf{Agent}]{\textbf{Mild Oxidizing}} \text{R-CHO}$ (Primary alcohol to aldehyde)

(ii) $\text{R-CHO} \xrightarrow[\textbf{Agent}]{\textbf{StrongOxidizing}} \text{R-COOH}$ (Aldehyde to carboxylic acid)

(iii) $\text{R-COOH} \xrightarrow[\textbf{Condition}]{\textbf{Decarboxylation}} CO_2$ (Carboxylic acid to Carbon dioxide)

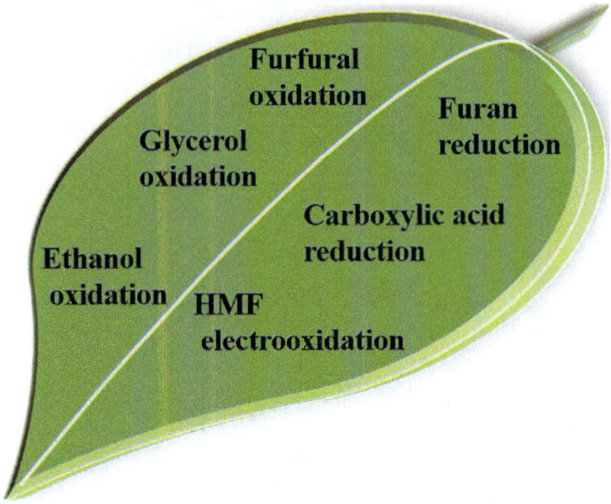

FIGURE 10.3 Electrochemical transformations of chemical species via oxidation or reduction routes.

The key features include:

- **Selection of catalyst:** Noble metals (Pt, Pd, Au) traditionally excel at activity but suffer from CO intermediate poisoning and cost issues. Nonnoble catalysts such as nickel hydroxides ($Ni(OH)_2$/NiOOH), LDHs, and transition metal oxides offer enhanced poisoning tolerance and cost benefits, particularly in alkaline media where hydroxide ions facilitate alcohol deprotonation and intermediate stabilization [20].
- **The effect of pH:** Oxidation kinetics improve greatly in alkaline electrolytes (NaOH, KOH) due to easier removal of protons and formation of surface oxyhydroxide species that act as active sites (e.g., NiOOH). In contrast, neutral media are challenging, but emerging strategies (e.g., Ru SACs on NiO) have shown selective aldehyde production by promoting water dissociation, thereby producing surface OH^* that aids oxidation [21].
- **Pathways and intermediates:** Adsorbed intermediates such as aldehydes (RCHO), carboxylates, and adsorbed CO are critical. Reaction selectivity depends on balancing adsorption strength—too weak adsorption limits activation, while too strong adsorption causes poisoning [21, 22].

10.7.1.2 Oxidation of polyols

The polyols (e.g., glycerol, sorbitol) introduce complexity due to multiple hydroxyl groups, allowing diverse oxidation pathways and C–C bond cleavage. Typical pathways include partial oxidation: conversion of one or more –OH groups to aldehyde or acid functional groups (e.g., glyceraldehyde, glyceric acid). C–C cleavage yielding smaller acids: formate, glycolate, oxalate, or eventual mineralization to CO_2. The selective oxidation hinges on the catalyst's ability to adsorb multiple sites of polyol molecules in defined geometries and promote multielectron processes with controlled intermediate desorption.

10.7.1.3 Oxidation of aldehydes

The aldehydes (e.g., HMF, furfural) inherently exhibit higher electrophilicity on the carbonyl carbon, facilitating oxidation steps:

(I) HMF oxidation of different functional groups:

(i) Hydroxymethyl group oxidation: HMF → 5-hydroxymethyl-2-furancarboxylic acid (HMFCA).
(ii) Aldehyde group oxidation: HMFCA → 5-formyl-2-furancarboxylic acid (FFCA).
(iii) Final oxidation: FFCA → FDCA.

10.8 Electroreduction pathways

The cathodic reduction reactions favor biomass-derived molecules such as furans and carboxylic acids, typically involving proton-coupled multielectron transfers that compete with hydrogen (H_2 fuel) evolution. The electrochemical oxidation and reduction of biomass derivatives using transition metal oxide based catalysts are tabulated in Table 10.1.

10.8.1 Reduction of furans

In general, furans like furfural and HMF can be selectively reduced to products including furfuryl alcohol, 2-methylfuran (2-MF), and DMF. In the primary step, reduction of the aldehyde takes place, where C=O is converted to furfuryl alcohol. Further reduction or hydrogenation of the furan ring may yield methylated derivatives. The catalytic factors are very important in the reduction process. Transition metal-based catalysts show better electrochemical reduction, such as Cu, Ni, and SACs. The simple Cu-based metals show selectivity toward furfuryl alcohol by favoring aldehyde reduction over ring hydrogenation. In other words, bimetallic catalysts (Cu/Ni, Cu/Co) tailor adsorption energies and hydrogenation pathways to access deeper reduction products. SAC catalysts provide single-atom sites that enable selective adsorption, modulating proton and electron transfer pathways to control site-specific reduction [23–25]. The electrolyte environments (acidic or neutral) govern proton availability, influencing the relative rates of reduction versus competing HER. Acidic or neutral electrolytes (e.g., $NaClO_4$) often facilitate furfural reduction but require balancing HER suppression.

TABLE 10.1 Summary table of electrooxidation and electroreduction of biomass derivatives.

Substrate	Reaction	pH	Catalyst	Key Intermediates	Products	Electrode Potential (V vs. RHE)
Alcohols (methanol, HMF)	Electrooxidation	Alkaline (NaOH, KOH)	$Ni(OH)_2$/NiOOH, Pd, Au, Ru single atoms	Aldehydes, Adsorbed CO	Aldehydes, acids, CO_2	1.2 to 1.6
Polyols (glycerol)	Electrooxidation	Alkaline	NiFe-LDH, CoOOH, phosphides	Glyceraldehyde, formate	Partial acids, cleavage products	~1.3 to 1.5
Aldehydes (HMF, furfural)	Electrooxidation	Alkaline/neutral	NiFe-LDH, Co-OOH, Ru_1–NiO, M-N-C	FFCA, HMFCA	FDCA, furoic acid, aldehydes	1.2 to 1.6
Furans (furfural)	Electroreduction	Neutral/acidic	Cu, Cu–Ni, Co–N–C	Adsorbed furan radicals	Furfuryl alcohol, 2-MF	−0.3 to −0.6
Acids (levulinic acid)	Electroreduction	Neutral/acidic	MoS_2, CuSn, CuPd	Hydroxyvaleric acid	γ-Valerolactone (GVL)	−0.4 to −0.7

10.8.2 Reduction of carboxylic acids

The biomass-derived acids, such as levulinic acid, can be electrochemically transformed into valuable chemicals like γ-valerolactone (GVL) via reduction of the ketone C=O group after proton or electron transfer steps, producing hydroxyvaleric acid, which spontaneously lactonizes to GVL. The transition metal-based catalysts, such as CuSn alloys, MoS_2, and CuPd bimetallics, provide active sites for selective reduction, with proton-coupled electron transfer controlling the reaction kinetics and overpotential windows.

10.9 Factors affecting electrochemical biomass conversion using nonnoble metals

The biomass conversion using nonnoble metal-based catalysts is also affected by factors such as electrolyte pH, electrode potential, catalyst substrate interaction, and intermediate or pathway mechanism.

10.9.1 Electrode potential

The controls over thermodynamics and kinetics can be achieved. In general the oxidation potentials are typically between 1.2 and 1.6 V vs. RHE for substrates like HMF on NiFe-LDH catalysts, where intermediate formation and desorption are optimized. The reduction potentials for furfural range approximately from −0.3 to −0.6 V vs. RHE on Cu-based catalysts, balancing selectivity versus competitive HER. The excessive potential results in overoxidation or side-reactions [26].

10.9.2 Electrolyte pH and composition

The pH of the electrolyte solution plays a vital role in electrochemical reactions. Mostly, **alkaline electrolytes** favor alcohol or polyol oxidation by providing abundant OH^- ions, facilitating deprotonation and intermediate stabilization [27]. Whereas **neutral or acidic electrolytes** generally favor reductions by providing protons, which require catalyst designs that suppress HER side reactions. The supporting ions (Cl^-, SO_4^{2-}) may adsorb on catalyst surfaces, influencing activity and poisoning (for example, Cl^- can poison noble metals).

10.9.3 Catalyst–substrate interactions

The adsorption geometry dictates the oxidation pathway, e.g., which C–OH or C=O group is activated. Whereas the moderate adsorption energy favors both activation and facile desorption of products. The surface functional groups (–OH, –COOH) on substrates influence electronic distribution and adsorption modes. Nanostructuring catalysts into core–shell, alloyed, or supported composites can tune adsorption strength and electronic properties, as learned from magnetic core–shell materials and Au-Fe_3O_4@MOF nanocomposites that combine catalytic activity with accessible surface area and recyclability [27].

10.9.4 Reaction intermediates and mechanism pathways

The reaction intermediates (aldehydes, carboxylates, formates, and radicals) can be tracked by in-situ spectroscopies like Fourier Transform-Infrared (FT-IR), Surface-Enhanced Infrared Absorption Spectroscopy (SEIRAS), Raman spectroscopy, and X-ray Photoelectron Spectroscopy (XPS) to elucidate pathways and optimize selectivity. In addition, density functional theory (DFT) offers insight into energy barriers and preferred binding modes, guiding catalyst design to stabilize desired intermediates while avoiding poisoning species such as adsorbed CO [27, 28]. In polyol oxidation, surface-bound glycolate and formate represent key steps toward partial or complete oxidation products. In furan reduction, adsorbed carbenium radicals or ring-bound species precede selective hydrogenation. Core–shell architectures combining conductive cores (Fe_3O_4, Au) with catalytically active shells (TiO_2, MOFs) enable synergistic electronic effects that can influence intermediate binding energies and electron transfer kinetics [6].

10.10 Catalyst design and functionality

The research articles focusing on magnetic core–shell Fe_3O_4-based nanostructures, Au-Fe_3O_4@MOF composites, and related multifunctional catalysts reveal several relevant design principles of **core–shell architectures and metal–oxide interfaces** [29, 30]. **Core–shell structures** offer spatial separation of functionalities, a magnetic core for facile recovery, and an outer catalytically active shell with tunable thickness and porosity controlling substrate access and electronic interactions. In addition **metal–oxide interfaces** in composites (e.g., Fe_3O_4@birnessite) display synergistic catalytic activities superior to individual components, attributed to enhanced charge transfer and surface oxygen vacancy effects that could benefit oxidation half-reactions in biomass electroconversion. The incorporation of **conductive supports** such as carbon nanotubes into oxide-based catalysts improves electron transport and structural robustness, which are essential for multielectron biomass oxidation/reduction reactions [5]. The surface functionalization via **bio-inspired polymers (e.g., polydopamine)** enables stable immobilization of enzymes or catalytic species, suggesting routes to hybrid bio-electrocatalysts for selective substrate transformation [10]. The adjustable optical and electronic properties of core–shell nanoparticles (e.g., Fe_3O_4/Au) suggest possibilities for photoelectrochemical enhancement of oxidation/reduction pathways, though further integration into biomass conversion requires development [6, 8, 11]. These insights into nanocomposite catalyst engineering directly contribute to improved catalyst substrate interactions, optimized reaction kinetics, and enhanced catalyst stability necessary for practical biomass electrochemical valorization.

10.11 Stability and durability of nonnoble metal-based electrocatalysts

Although nonnoble metals are more abundant, their long-term stability remains a central challenge because these metals can dissolve under anodic potentials and undergo restructuring. **The passivation layer formation** may block active sites if uncontrolled. Strategies such as **in-situ formation of stable oxide/hydroxide layers (e.g., NiOOH)**, strong support interactions, and metal–ligand chelation mitigate these effects. 3D architectures on conductive substrates like Ni foam improve mechanical stability against collapse and loss of active surface area during cycling [31].

10.12 Conclusion

It is important to replace noble metals (Pt, Pd, Ru) with nonnoble metal-based catalysts for biomass electroconversion, which addresses issues like cost, scarcity, and poisoning susceptibility. In recent days, transition metal oxides, sufides, hydroxides, phosphides, M–N–C (SACs), and complex oxide structures offer diverse, tunable platforms exhibiting high catalytic activity, selectivity, and operational stability. The rational design combining nanoscale morphology control, conductive and doped supports, and heterostructured interfaces can yield nonnoble electrocatalysts that rival or surpass noble metal benchmarks. By coupling biomass electrooxidation to hydrogen evolution, which can enhance the energy efficiency and sustainability. Challenges such as long-term stability and mechanistic understanding remain active research frontiers, with promising approaches including advanced spectroscopic characterization and green synthesis methods. Simultaneously, knowledge from noble metal core–shell nanocomposites and green functionalization strategies informs design criteria for advanced nonnoble metal-based catalysts, paving the way toward scalable, cost-effective biomass electrochemical conversion technologies. The discussion integrates mechanistic insights, catalyst-substrate interactions, and reaction intermediates, complemented by relevant findings and design principles informed by recent advances in core–shell and nanocomposite catalysts from recent research works. The electrochemical valorization of biomass-derived molecules encompasses intricate anodic oxidation of alcohols, polyols, and aldehydes, alongside cathodic reductions of furans and carboxylic acids. The reaction pathways are substrate-specific and catalyst-specific surface functions, governed critically by electrode potential, pH, electrolyte composition, and catalyst substrate interactions at the nanoscale.

Advanced catalysts leveraging, such as surface atomic and interface engineering, core–shell and nanocomposite architectures offering synergetic electronic effects, and enhanced mass transport. The functionalization from green sources (e.g., polyphenols) for stabilization and selectivity of the biomass conversion, which enables the selective formation of value-added chemicals with high activity and sustainability. The knowledge derived from magnetic core–shell and metal oxide composite nanomaterials underscores the importance of combining catalytic activity, electrical conductivity, and facile catalyst recovery parameters central to practical biomass electroconversion technologies. The optimal control of the electrode environment with electrode potential, pH, catalyst designs, nanostructure, surface chemistry, and understanding of intermediates are collectively indispensable for advancing the efficiency and selectivity of biomass electrooxidation and electroreduction reactions.

References

[1] A. Osman, N. Mehta, A.M. Elgarahy, A. Al-Hinai, A.H. Al-Muhtaseb, D.W. Rooney, Conversion of biomass to biofuels and life cycle assessment: a review, Environ. Chem. Lett. 19 (2021) 4075–4118. https://doi.org/10.1007/s10311-021-01273-0.

[2] D.M. Alonso, J.Q. Bond, J.A. Dumesic, Catalytic conversion of biomass to biofuels, Green Chem. 12 (2010) 1493–1513. https://doi.org/10.1039/C004654J.

[3] D. Chung, M. Cha, A.M. Guss, J. Westpheling, Direct conversion of plant biomass to ethanol by engineered Caldicellulosiruptor bescii, Proc. Natl Acad. Sci. 111 (2014) 8931–8936. https://doi.org/10.1073/pnas.1402210111.

[4] C. Xu, E. Paone, D. Rodriguez-Padron, R. Luque, F. Mauriello, Catalytic routes for preparation and upgrading of furfural and HMF, Chem. Soc. Rev. 49 (2020) 4273–4306. https://doi.org/10.1039/d0cs00041h.

[5] B.R. Caes, R.E. Teixeitra, K.G. Knapp, R.T. Raines, Biomass to furanics, ACS Sustain. Chem. Eng. 3 (2015) 2591–2605. https://doi.org/10.1021/acssuschemeng.5b00473.

[6] O.O. James, S. Maity, L.A. Usman, K.O. Ajanaku, T.O. Siyanbola, S. Sahu, et al., Conversion of carbohydrate biomass feedstocks to biofuels via HMF, Energy Environ. Sci. 3 (2010) 1833–1850. https://doi.org/10.1039/B925869H.

[7] L. Hu, G. Zheo, W. Hao, X. Tang, Y. Sun, L. Lin, et al., Catalytic conversion of biomass-derived carbohydrates into fuels and chemicals via furanic aldehydes, RSC Adv. 2 (2012) 11184–11206. https://doi.org/10.1039/C2RA21811A.

[8] C.M. Cai, T. Zhang, R. Kumar, C.E. Wyman, Integrated furfural production from lignocellulosic biomass, J. Chem. Technol. Biotechnol. 89 (2014) 2–10. https://doi.org/10.1002/jctb.4168.

[9] H. Zhang, Y.-T. Cheng, T.P. Vispute, R. Xiao, G.W. Huber, Catalytic conversion of biomass to olefins and aromatics with ZSM-5, Energy Environ. Sci. 4 (2011) 2297–2307. https://doi.org/10.1039/C1EE01230D.

[10] T. Ennaert, J.V. Aelst, J. Dijkmans, R. Clercq, W. Schutyser, M. Dusselier, et al., Challenges of zeolite chemistry in catalytic biomass conversion, Chem. Soc. Rev. 45 (2016) 584–611. https://doi.org/10.1039/c5cs00859j.

[11] M. Shao, Q. Chang, J.P. Dodelet, R. Chenitz, Recent advances in electrocatalysts for oxygen reduction reaction, Chem. Rev. 116 (2016) 3594–3657. https://doi.org/10.1021/acs.chemrev.5b00462.

[12] C.-X. Zhao, J.-N. Liu, J. Wang, D. Ren, B-Q. Li, Q. Zhang, Noble-metal-free bifunctional ORR/OER electrocatalysts, Chem. Soc. Rev. 50 (2021) 7745–7778. https://doi.org/10.1039/d1cs00135c.

[13] Y. Zhu, C. Guo, Y. Zheng, S.Z. Qiao, Surface and interface engineering of noble-, metal-free electrocatalysts, Acc. Chem. Res. 50 (2017) 915–923. https://doi.org/10.1021/acs.accounts.6b00635.

[14] A.A. Feidenhans'l, Y.N. Regmi, C. Wei, D. Xia, J. Kibsgaard, L.A. King, Precious-metal-free hydrogen evolution catalysts, Chem. Rev. 124 (2024) 5617–5667. https://doi.org/10.1021/acs.chemrev.3c00712.

[15] G. Elmacı, C.E. Frey, P. Kurz, B.Z. Karan, Water oxidation catalysis by birnessite@iron oxide nanocomposites, Inorg. Chem. 54 (2015) 2734–2741. https://doi.org/10.1021/ic502908w.

[16] R. Ge, Y. Wang, Z. Li, M. Xu, S.-M. Xu, H. Zhou, et al., Selective electrooxidation of biomass-derived alcohols in neutral media using Ru SAC on NiO, Angew. Chem. 61 (19) (2022) e202200211. https://doi.org/10.1002/anie.202200211.

[17] Z.J. Xu, Y. Hou, S. Sun, Magnetic core-shell Fe_3O_4/Au nanostructures with plasmonic properties, J. Am. Chem. Soc. 129 (2007) 8698–8699. https://doi.org/10.1021/ja073057v.

[18] F. Ke, L. Wang, J. Zhu, Au-Fe_3O_4@MOF multifunctional core-shell nanocomposites, Nanoscale 7 (2015) 1201–1208. https://doi.org/10.1039/C4NR05421K.

[19] M. Stefan, O. Pana, C. Leostean, D. Bele, M. silipas, E. Gautron, Synthesis and characterization of Fe_3O_4-TiO_2core–shell nanoparticles, J. Appl. Phys. 116 (2014) 114312. https://doi.org/10.1063/1.4896070.

[20] M. Cargnello, M. Grzelczak, R.G. Benito, Z. Syrgiannis, K. Bakhmutsky, V.L. Parola, et al., CNT-driven activity in metal@oxide core–shell catalysts, J. Am. Chem. Soc. 134 (2012) 11760–11766. https://doi.org/10.1021/ja304398b.

[21] M. Martín, P. Salazar, R. Villalonga, S. Campuzano, J.M. Pingarron, J. Gonzalez-Mora, Fe_3O_4@polydopamine magnetic nanoparticles for biosensors, J. Mater. Chem. B 2 (2014) 739–746. https://doi.org/10.1039/C3TB21171A.

[22] V. Salgueiriño-Maceira, M. Correa-Duarte, Magnetic core–shell nanocomposites for biological applications, Adv. Mater. 19 (23) (2007) 4131–4144. https://doi.org/10.1002/adma.200700418.

[23] E. Shevchenko, M.I. Bodnarchuk, M.V. Kovalenko, D.V. Talapin, R.K. Smith, S. Aloni, et al., Gold/iron oxide core–hollow shell nanoparticles, Adv. Mater. 20 (22) (2008) 4323–4329. https://doi.org/10.1002/adma.200702994.

[24] H. Lu, S.-Y. Tang, J. Zhu, X. Huang, H. Forgham, X. Li, et al., Liquid metal core–shell nanoengineering, Adv. Funct. Mater. (2023). https://doi.org/10.1002/adfm.202311300.

[25] J.H. Advani, G. More, R. Srivastava, Spinel-based catalysts for biomass valorisation, Green Chem. (2022). https://doi.org/10.1039/d1gc04592j.

[26] J. Zhu, Z. Li, L. Zhong, P. Xiao, X. Yang, Z. Zhao, et al., Perovskite oxides: preparation and catalytic applications, ACS Catalysis 4 (2014) 2917–2940. https://doi.org/10.1021/cs500606g.

[27] Y. Wang, Z. Yang, L. Liu, Perovskite catalysts for syngas upgrading via intensified steam reforming of biomass tar, ACS Sustain. Chem. Eng. (2024). https://doi.org/10.1021/acssuschemeng.4c04571.

[28] K. Lee, Y. Jing, Y. Wang, N. Yan, Unified view on catalytic conversion of biomass and waste plastics, Nat. Rev. Chem. 6 (2022) 635–652. https://doi.org/10.1038/s41570-022-00411-8.

[29] F. Lin, M. Xu, K.K. Ramasamy, Z. Li, J.L. Klinger, J.A. Schaidle, et al., Catalyst deactivation and mitigation during biomass conversion, ACS Catalysis (2022). https://doi.org/10.1021/acscatal.2c02074.

[30] K. Barta, P. Ford, Catalytic conversion of woody biomass to organic liquids, Acc. Chem. Res. 47 (2014) 1503–1512. https://doi.org/10.1021/ar4002894.

[31] Q. Sun, S. Wang, B. Aguila, X. Meng, S. Ma, F-S. Xiao, Solvation environments in heterogeneous catalysts for biomass conversion, Nat. Commun. (2018) 9. https://doi.org/10.1038/s41467-018-05534-5.

CHAPTER

11

Carbon materials for electrochemical biomass conversion

Joyeeta Das[a], *Vaibhav Sharma*[b], *Monika Kumari Prajapati*[c], *Poonam Kumari*[d], *Varad Nagar*[a], *Garima Awasthi*[e], *Harsh Pandey*[f], *Kumud Kant Awasthi*[e], *and Mahipal Singh Sankhla*[e]

[a]Department of Chemistry, Center of Forensic Science, Banaras Hindu University, Varanasi, Uttar Pradesh, India, [b]Department of Medical Elementology and Toxicology, School of Chemical and Life Sciences, Jamia Hamdard (Deemed to be University), New Delhi, India, [c]Department of Forensic Science, School of Forensic Sciences, Galgotias University, Greater Noida, Uttar Pradesh, India, [d]School of Allied Health Sciences, Jaipur National University, Jaipur, Rajasthan, India, [e]School of Basic & Applied Sciences, K.R. Mangalam University, Gurugram, Haryana, India, [f]Department of Biotechnology and Chemical Engineering, Manipal University, Jaipur, Rajasthan, India

11.1 Introduction

11.1.1 Global context

We are at a precipice where the world is waking up to the fact that how we produce and consume energy is going to have to change [1].

Electrochemical Biomass Conversion: Sustainable Processes and Applications for Bioenergy, Biofuels, and Bioproducts
DOI: https://doi.org/10.1016/B978-0-443-26757-4.00001-5

Fossil fuels are not only finite, but they also produce greenhouse gas emissions that cause climate change [1,2]. Addressing the dual challenge of increasing energy demand and decarbonization requires solutions that are renewable and sustainable [3]. Huge amounts of biomass are produced annually from photosynthesis, and unlike fossil fuels, its use may be close to carbon neutral—any CO_2 emitted upon conversion could potentially be balanced by that taken up during growth [1,3]. Agricultural residues, forestry residue, and food waste are abandoned or burned, which not only is an act of pollution but also a loss of valuable resources [4,5]. The transformation of such streams to fuels, chemicals, or advanced materials is value-added and minimizes environmental loads [4,6]. Crucially, biomass can complement variable renewables—like solar and wind—by storing intermittent energy in a chemical form [2,6].

11.1.1.1 Conversion approaches in perspective

The potential energy and chemical energy in biomass can be extracted in various ways. The three main pathways include thermochemical, biochemical, and electrochemical conversion [1,7]. Thermochemical methods such as combustion, gasification, and pyrolysis are also commonly recommended, since they can handle high volumes of biomass in a short amount of time and lead to the generation of heat, syngas, and bio-oils [7]. The drawback is that they work at high temperatures and many are nonselective, complicating the downstream processing [8]. Under gentle conditions and good selectivity, the Biochemical (enzyme-microorganism-based) process produces significantly low chemical "waste" [4]. However, this pathway tends toward sluggish conversion and relatively high impurity sensitivity; its raw biomass needs expensive pretreatment mostly [5].

Appealing electrochemical transformation protocol, operates under relatively gentle conditions, frequently at room temperature and ambient pressure, where the scientists control reactions toward certain products by changing electrode materials and applied potentials [6]. It is a possible connector of green electricity to sustainable chemical synthesis [8]. Electrochemo conversion is the most attractive of the three conversion routes owing to the mildness of operation, controllability, and its compatibility with renewable energy [2,6]. Together with its proper catalysts, especially carbon-based materials, it provides a way for producing clean fuels and high-value chemicals in an economic and rational fashion [3].

11.1.2 Essential characteristics of carbon materials

At the core of the electrochemical conversion of biomass are the catalysts, which allow the reactions to occur [9]. Carbon—in its many iterations— has proven to be a versatile, sustainable choice of material [2,3]. Due to high electronic conductivity, efficient charge transfer, which

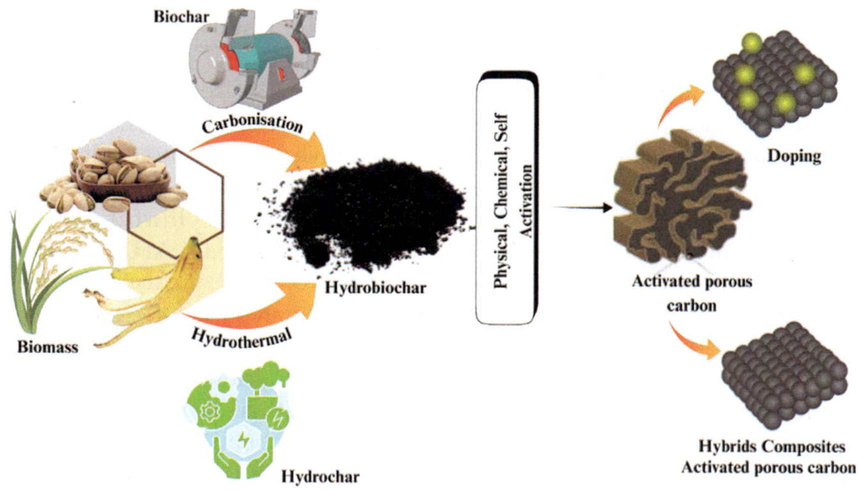

FIGURE 11.1 Schematic representation of carbon production via carbonization and hydrothermal methods of biomass to produce biochar, hydrochar, and hydro-biochar. Future activation and substitutions, including doping and the formation of a hybrid composite, result in highly porous activated carbon materials for their advanced use.

is essential for electrochemical reaction, can be achieved [2]. Apart from the accessibility, the tunability of carbon is interesting, as by manipulating its porosity or by doping heteroatoms such as nitrogen, sulfur, or phosphorus, researchers can significantly tailor its catalytic activity, ion transport, and product selectivity [3,9]. Biomass—from agricultural waste to forestry residues—can be converted into high-performing carbons through relatively simple methods, for example, pyrolysis, hydrothermal carbonization, or chemical activation, shown in Fig. 11.1[10]. This not only provides a low-cost feedstock but also turns waste into value, aligning perfectly with the principles of a circular economy [8]. Unlike carbons derived from fossil resources, biomass-derived carbons (BDC) are inherently sustainable [3]. They recycle atmospheric CO_2 and can often be produced under less energy-intensive conditions, yet still deliver comparable or even superior electrochemical performance [2,6]. Their structural diversity, shaped by the wide variety of natural precursors, enables the design of carbons ranging from hierarchical porous frameworks to graphene-like nanosheets [2,3]. These materials have already shown promise in supercapacitors, batteries, sensors, and catalytic electrodes for key reactions such as oxygen reduction and biomass upgrading [9]. This chapter, therefore, highlights carbon not only as a functional material but also as a sustainable solution in its own right. By bridging renewable electricity with waste valorization, carbon-based materials capture the spirit of the energy transition: efficient, selective, and environmentally responsible.

11.2 Fundamentals of electrochemical biomass conversion

Electrochemical biomass conversion is the use of redox reactions on the surface of electrodes to upcycle renewable feedstocks. Biomass oxidation occurs at the anode during the conversion of molecules such as alcohols and polyols to more oxygenated products (e.g., glycerol to dihydroxyacetone [DHA]/glyceric acid, 5-hydroxymethylfurfural [HMF] to2,5-furandicarboxylic acid [FDCA]) under low potentials. At the cathode, one has a reduction reaction: water is reduced to H_2 or substrates derived from biomass may be electrochemically hydrogenated (using in situ H and e^-) to produce saturated products. As an example, the organics can be reduced to carbonyl or nitro groups to alcohols or amines [11]. These half reactions are commonly combined with one another in practice through biomass electrosynthesis. One can achieve useful chemistry at both anode (replacing the anodic O_2 evolution with organic oxidation) and cathode (combining it with useful cathodic chemistry, that is, H_2 evolution or organic reduction), both of which hold promise in chemical production. Such a paired electrolysis scheme can almost double the total Faradaic charge usage (near 200% joint total apparent Faradaic efficiency [FE]) when compared with the nonpaired systems. That is, both sides of the cell could channel all their electrons to form a product [12].

11.2.1 Core electrochemical pathways

11.2.1.1 Oxidation (anodic)

At the anode, plain organic oxidation breaks through C-H / C-C vegetal or incorporates oxygen into the molecules of the biomass. Classic ones would be the oxidation of primary alcohols to carboxylic acids or carbonyls, or polyols to carboxylic acids or carbonyls. Electrochemical oxidation or hydrogenation has also been found to be a practicable means of performing targeted upgrading using small molecules (methanol, ethanol, glycerol, glucose, furfural, and HMF) as given in Table 11.1. As an example, DHA or glyceric acid may be anodically oxidized to glycerol and furoic acid, or 2,5-furandicarboxylic acid may be anodically oxidized to furfural, usually with high selectivity [13,14].

11.2.1.2 Reduction/hydrogenation (cathodic)

Protons in the electrolyte (or water) gain electrons at the cathode to make H_2 or hydrogenate organics. Hydrogenation by means of electrochemical activity exploits the formation of in situ H-atoms in order to unearth unsaturated bonds. As an example, furfural may be hydrogenated to furfuryl alcohol, nitrobenzene to aniline, or HMF to 2,5-bishydroxymethylfuran in mild conditions. These cathodic hydrogenations are value adding, in that

TABLE 11.1 Electrochemical conversion of biomass-derived feedstocks on carbon-supported catalysts, showing reaction types, conditions, and main products.

Biomass feedstock	Main component(s)	Electrochemical reaction type	Operating conditions	Main products	References
Glycerol	Polyol ($C_3H_8O_3$)	Electrooxidation on carbon-supported Pt/Pt-alloys (Pt/C, Pt–Bi/C)	Alkaline (pH ~13), potentials around −0.35 to +0.9 V vs Ag/AgCl on carbon-supported Pt or Pt–Sb/Pt–Bi; carbon cloth/paper as support.	Dihydroxyacetone (DHA), glycerate/formate (pathway-dependent)	[15]
Furfural	Furan aldehyde ($C_5H_4O_2$)	Electroreduction (ECH) on Cu/C or doped-carbon catalysts	Near-neutral pH ~7; −0.4 V vs RHE, typical for Cu-based cathodes or doped carbon; carbon electrodes used.	Furfuryl alcohol (FA) (high selectivity)	[16]
5-HMF	Furanic platform ($C_6H_6O_3$)	Electro-oxidation on Au/Au-on-carbon	Alkaline electrolyte (e.g., 0.1–1 M KOH); Au single-crystal/carbon-supported Au; potential region where HMF→FDCA occurs before OER.	2,5-Furandicarboxylic acid (FDCA) (polymer monomer)	[17]
5-HMF	Furanic platform ($C_6H_6O_3$)	Electrohydrogenation (ECH) on Cu/Cu-foam-derived carbon supports	Mildly acidic/neutral to alkaline; −0.3 to −0.8 V vs RHE; oxide-derived Cu or Cu on carbon enables high BHMF selectivity.	2,5-Bis(hydroxymethyl)furan (BHMF)	[18]

(*continued on next page*)

TABLE 11.1 Electrochemical conversion of biomass-derived feedstocks on carbon-supported catalysts, showing reaction types, conditions, and main products—cont'd

Biomass feedstock	Main component(s)	Electrochemical reaction type	Operating conditions	Main products	References
Cellulose hydrolysate	Glucose, xylose	Electrooxidation on Au/C, Ni/C	Alkaline media (e.g., 0.1–1 M KOH); ~0.4–0.7 V vs Hg/HgO or RHE; carbon paper/cloth electrodes with Au/Ni.	Gluconic acid → glucaric acid (two-step)	[19]
Lignin model compounds	Phenolics (e.g., phenol)	Electrocatalytic hydrogenation (ECH) on Pd/C, Pt/C	Aqueous media; room T to mild heat; potentials typically negative of HER onset (e.g., −0.2 to −0.7 V vs RHE); carbon supports (Vulcan, carbon cloth).	Cyclohexanol/cyclohexanone (selectivity catalyst-dependent)	[20]
Levulinic acid	Keto acid ($C_5H_8O_3$)	Electroreduction/ECH on carbon-supported metals (e.g., Cu, Ni; nonaqueous often used)	Acidic to nonaqueous proton-donor electrolytes; around −0.5 V vs RHE (or equivalent) depending on medium/electrode; carbon electrodes common.	Γ-Valerolactone (GVL) and related hydrogenated products.	[21]

the C=O or C=C bonds are reduced to -OH/-CH_2, frequently at lower energy than in thermal hydrogenation [22,23].

11.2.1.3 *Paired electrolysis*

Combining the oxidation of biomass at the anode together with a compatible cathodic reaction (e.g., H_2 production or reduction of biomass) leads to maximization of overall efficiency. This two-fold upgrading process implies that both of the electrodes will yield useful products other than O_2, so that the consumption of electrons is probably doubled. As an example, glycerol may be simultaneously oxidized at the anode and H_2 formed at the cathode, utilizing all of the charge into chemicals and adding significantly to energy and electron economy over water splitting [24].

11.2.2 Common biomass-derived feedstocks

11.2.2.1 *Alcohols (C_1–C_5)*

Biomass simple alcohols (bioethanol, isobutanol, etc.) can be oxidized to acids. As an example, branched C_4 alcohol, isobutanol, can be electrocatalytically oxidized in a matter of seconds using nano nickel-based catalysts with FEs >90% in the product, isobutyric acid, under mild conditions. Using paired schemes, such alcohols are also occasionally reduced or dehydrogenated at the cathode [4].

11.2.2.2 *Polyols (sugar alcohols)*

A typical biorefinery output is multihydroxyl compounds such as glycerol, sorbitol, and ethylene glycol. A typical example is glycerol, which is produced as a byproduct of biodiesel: it may undergo either oxidation to glyceraldehyde or DHA, or (through the partial dehydration process) the reduction to acetol. The polyols are also appealing due to the possibility of converting their multiple -OH groups to a range of oxygenates selectively, and the high oxygen content frequently creates products with high selectivity [25,26].

11.2.2.3 *Furans*

Sugarcane-based heterocyclic derivatives such as HMF (cellulose) and furfural (hemicellulose) are optimal platforms. HMF carries as functional groups both aldehyde and alcohol, and can be electro-oxidized to 2,5-furandicarboxylic acid or reduced to 2,5-Bis(hydroxymethyl)furan (BHMF). Furfural may be reduced to furfuryl alcohol, or it may be hydrogenated to tetrahydrofurfuryl alcohol. These furanic substrates hold high oxygen content, hence they are convertible in electroconversion [27].

11.2.2.4 *Lignin fragments*

Lignin depolymerization produces phenolic monomers (e.g., vanillin, guaiacol, and syringol). These aromatic compounds may be electrochemically oxidized (e.g., breaking C-O bonds to acids or to ketones) or be reduced to cyclohexanols. Valorization of such lignin is difficult, and in paired electrolysis schemes, oxidation of lignin is commonly present as the anode half reaction. As a matter of fact, Huddleston et al. [12] unambiguously mention glycerol, HMF, furfural, and lignin as important substrates of paired biomass electrocatalysis.

11.2.3 Performance metrics

11.2.3.1 *Faradaic efficiency*

FE is a fundamental electrochemical metric that quantifies how effectively the input electrons are utilized in generating the desired product. It is defined as the ratio of the number of electrons consumed in forming the target product to the total number of electrons passed through the system. In practice, FE is often expressed as a percentage and is determined by comparing the experimentally measured product yield with the theoretical value expected from Faraday's laws of electrolysis. High FE values indicate that electron transfer is being directed efficiently toward the intended chemical transformation rather than side reactions. For instance, in electrocatalytic biomass conversion or CO_2 reduction, electrons may also drive hydrogen evolution (HER) or other competing pathways, lowering the FE for the target reaction. Hence FE provides a direct measure of the electron economy of the process, linking the electrochemical input to chemical output [12,28].

11.2.3.2 *Selectivity*

Selectivity refers to the preference of a catalytic system for producing one specific product over other possible by-products. While FE measures electron utilization toward a particular product, selectivity accounts for the distribution of products formed from the same feedstock. It is often expressed as the FE of the target product divided by the sum of FE values for all products generated. In catalytic systems dealing with complex biomass feedstocks, multiple products (e.g., alcohols, acids, gases) may be formed simultaneously. High selectivity is essential for practical applications since purification costs and downstream processing depend heavily on product distribution. For example, in the electrochemical upgrading of biomass-derived molecules such as furfural, achieving high selectivity toward alcohols rather than unwanted polymeric by-products enhances the overall process value. Thus selectivity, together with FE, gives insight into how "cleanly" electrons and substrates are converted into a desired product [12].

11.2.3.3 *Turnover frequency*

Turnover frequency (TOF) represents the intrinsic activity of the catalyst, defined as the number of product molecules (or moles) formed per active catalytic site per unit time (commonly s^{-1}). Unlike FE or selectivity, which describe efficiency in electron allocation, TOF reflects how fast the catalyst can generate product once an electron is supplied. Determining TOF requires accurate knowledge of the number of active sites—something that can be challenging in heterogeneous catalysts like carbon-based electrocatalysts, where surface defects, heteroatom dopants, and pore structures act as catalytic centres. Nevertheless, TOF is a crucial metric for comparing different materials on a fair basis, independent of catalyst loading or electrode area. For example, nitrogen-doped porous carbons with high defect density often exhibit higher TOF values than graphitic carbons, due to enhanced electron transfer and active site accessibility. High TOF is desirable as it means greater product output per unit of catalyst, reflecting the catalyst's intrinsic efficiency [28].

11.2.3.4 *Stability (durability)*

Stability—or durability—is a measure of how long a catalyst can maintain its activity and selectivity under operational conditions. It is usually reported as lifetime (hours of continuous operation), number of operational cycles, or the rate of activity loss over time. For biomass conversion electrocatalysts, stability is critical because operating conditions (e.g., alkaline or acidic environments, fluctuating voltages, and continuous exposure to biomass-derived intermediates) can degrade catalyst performance. Carbon-based catalysts are particularly attractive in this context due to their corrosion resistance, tunable surface chemistry, and low cost. However, issues such as pore blockage, carbon oxidation, or leaching of dopants/metal atoms can reduce stability. Studies often use chronoamperometry (CA) or chronopotentiometry (CP) to assess durability, monitoring current retention or voltage shift over long-term operation. A stable catalyst ensures consistent product yields and minimizes replacement costs, making it a key requirement for commercial viability [28].

11.2.3.5 *Energy efficiency*

Energy efficiency relates the chemical energy stored in the products to the total electrical energy input required to generate them. It accounts for both FE (how well electrons are utilized) and the overall cell voltage (how much energy is needed to drive the reaction). A process may have high FE but still suffer from poor energy efficiency if the applied potential is much higher than the thermodynamic requirement, leading to unnecessary energy losses as heat. Improving energy efficiency is therefore crucial for sustainable and economical electrocatalysis. Strategies such as paired electrolysis—where both electrodes simultaneously produce valuable products—can dramatically increase overall energy efficiency by nearly doubling the effective FE. For example, coupling anodic biomass oxidation with cathodic fuel or chemical production maximizes electron utilization and reduces wasteful side reactions. Ultimately, energy efficiency determines the practicality of scaling carbon-based electrocatalysts for biomass conversion, as it directly influences product cost and environmental impact [12,28].

11.3 Types of carbon materials

11.3.1 Activated carbon

Activated carbon (AC) is a widely available porous carbon obtained from heat-treating organic precursors (wood, coal, nutshells, etc.) followed by chemical or steam activation. It features very high porosity and surface area (often >1000 m^2/g) with a dominant microporous network. These features make AC excellent for electrochemical double-layer applications: AC electrodes can store large charges at the carbon/electrolyte interface, yielding supercapacitor capacitances on the order of hundreds of F/g (e.g., ~480 F/g was reported for KOH-activated banana-stem carbon). The material is also very low-cost and sustainable (e.g., coconut or wood-derived AC). However, the ubiquitous micropores in AC can impede ion diffusion, limiting high-rate performance. Also, native AC typically has only moderate electrical conductivity (due to partly disordered carbon), so full graphitization or hybridization is often required for high-power uses. To enable catalysis (beyond capacitive storage), AC is usually doped with heteroatoms (N, B, etc.) or combined with metal nanoparticles. For example, nitrogen-doped AC electrodes from biomass have demonstrated oxygen reduction reaction (ORR) activity approaching that of platinum on carbon (Pt/C) in alkaline media. In summary, AC's advantages are its huge internal surface area, low cost, and environmental friendliness, which translate to very high double-layer capacitance and facile adsorption of reactants. Its limitations include limited inherent conductivity and predominantly microporous architecture, which can slow kinetics. By carefully tuning activation and doping, AC-based electrodes have achieved impressive electrochemical performance (e.g., high specific capacitance and competitive ORR/HER activity) [29,30].

11.3.2 Graphene/graphitic carbons

Graphene (single-layer graphite) and related graphitic carbons form an sp^2-hybridized carbon lattice with exceptional conductivity and strength.

These materials have extremely high intrinsic carrier mobility and a large accessible surface area (graphene's 2D structure). Graphene can be readily modified: incorporation of heteroatoms (N, B, and P, etc.) or defects creates catalytic sites for ORR/OER/HER. Bulk graphite (e.g., graphite fibre or foam) is also highly conductive, mechanically robust, and widely used in electrodes (graphite is the standard lithium-ion [Li-ion] anode material with 372 mAh/g capacity). The main drawback is restacking: graphene sheets tend to aggregate via π–π stacking, reducing their effective surface area and pore volume. Creating three-dimensional graphene networks or foam architectures is often necessary to expose the full surface [29,31]. Pure graphite is also less porous than ACs, which can limit reactant access [30]. Additionally, high-quality graphene production can be expensive, although advances in large-scale methods (e.g., chemical vapor deposition [CVD], exfoliation) are mitigating this. Graphene/graphitic carbons exhibit outstanding electrochemical performance when properly engineered. For instance, heteroatom-doped graphene nanosheets have matched Pt catalysts in ORR: one study reported B, N-codoped graphene onset potential ~0.94 V versus reversible hydrogen electrode (RHE) (half-wave 0.82 V), nearly identical to Pt/C. Graphite and graphene are also superb for charge storage: 3D graphene electrodes and graphite foams are used in supercapacitors and batteries for high capacity and long cycle life [32].

11.3.3 Carbon nanotubes

Carbon nanotubes (CNT) are cylindrical graphene sheets (single- or multiwalled) that combine metal-like conductivity with exceptional mechanical strength. Their one-dimensional morphology provides highly efficient electron pathways, and CNTs can form conductive networks even at low loading. As-synthesized CNTs have few catalytic sites, but heteroatom-doped CNTs (N, P, and S, etc.) become highly active electrocatalysts. CNTs can also be grown directly from biomass-derived char, providing a renewable CNT source. As supports, CNTs improve catalyst dispersion and durability. On the other hand, CNTs tend to bundle together, which lowers accessible surface area; and their external surface is relatively smooth (so surface area per mass is modest compared with ACs) [33]. Manufacturing CNTs (especially single-walled) can be costly and require metal catalysts. Bare CNT materials have limited porosity unless templated, so they often need to be integrated with high-porosity matrices for high capacitive or catalytic performance. Despite these challenges, CNT-based materials show exceptional electrochemical results when functionalized. For example, N- and P-codoped multiwalled CNTs (made by pyrolyzing biomass with melamine) exhibited very low overpotential for hydrogen evolution (167 mV at –10 mA/cm^2 in acid). In ORR, CNTs derived from

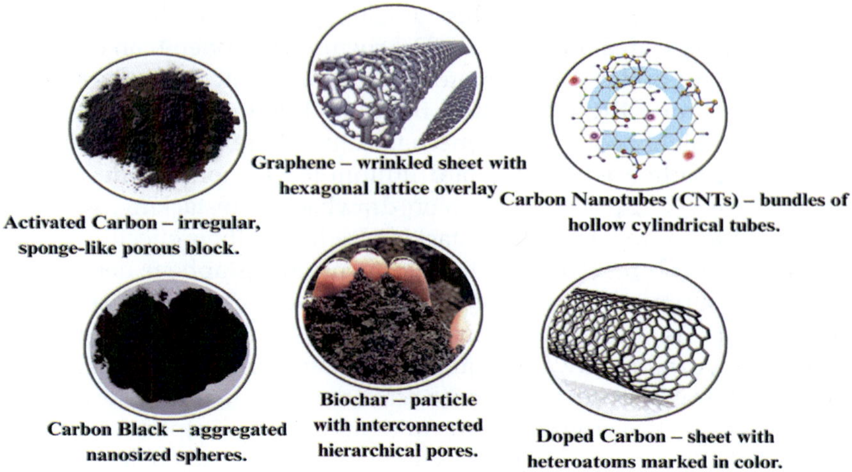

FIGURE 11.2 Morphological and structural variety of carbon-based substances such as activated carbon, biochar, carbon black, graphene, carbon nanotubes (CNTs), and doped carbon. The forms have distinct surface properties, pore structure, and functional characteristics, which are essential in energy storage and catalytic processes.

waste sludge (N-doped) attained an onset potential of ~0.99 V versus RHE (4-electron pathway), comparable to Pt. In summary, CNTs bring ultrahigh conductivity and strength (and tunability via doping), at the cost of aggregation and lower inherent surface area. Optimized CNT electrodes and hybrids routinely reach catalyst-level activities in both fuel-cell and energy-storage reactions [34,35].

11.3.4 Carbon black

Carbon black is a cost-effective, conductive carbon typically produced as an amorphous powder, as shown in Fig. 11.2. It has a moderately high surface area (hundreds of m^2/g) and is widely used as a conductive additive or catalyst support in fuel cells and batteries. The advantages of carbon black include very low cost and ease of large-scale production, plus good electrical conductivity and chemical stability. It often serves as the baseline support for Pt in PEM fuel cell catalysts and is stable under typical operating conditions [19]. However, carbon black's porosity is mostly microporous, which can limit the mass transport of reactants. It also lacks intrinsic catalytic functionality, so it typically requires posttreatment or supports metal nanoparticles to become active. By itself, pure carbon black shows only modest capacitance or ORR activity. Nevertheless, modified carbon blacks have delivered strong performance. For instance, a Vulcan

XC-72 treated with Fe/Zn and organic ligands (to introduce C–O and N–C sites) achieved an ORR half-wave potential of ~0.858 V versus RHE, with high durability. This demonstrates that even simple carbon black, when properly doped, can approach the activity of precious catalysts [29,36].

11.3.5 Biochar-derived carbons

Biochar-derived carbons are made by thermally converting biomass (e.g., wood chips, agricultural waste, biopolymers) in low-oxygen conditions. The resulting biochar is carbon-rich and hierarchical, containing amorphous carbon, graphitic islands, and residual mineral/heteroatom species from the biomass, as shown in Fig. 11.2. Key advantages are renewability and sustainability: biochar recycles waste into value-added materials. They inherently possess a hierarchical porous structure (from decomposition of volatiles) and often retain functionalities (–COOH, –OH, N or S groups) that can serve as catalytic sites. Raw biochar, however, is typically amorphous and has limited conductivity. Its surface area depends on pyrolysis conditions and is usually lower than that of AC, so activation (KOH, steam, etc.) is often needed to develop additional porosity. The chemical composition is variable (depending on feedstock and treatment), which can make performance inconsistent. Activated and doped biochars have shown very promising electrochemical behaviors. For example, nitrogen-doped carbon cubes derived from amino-acid precursors exhibited an ORR onset potential of ~0.92 V versus RHE (only ~60 mV behind Pt/C). Another study produced 3D nitrogen-rich biochar (from plant biomass + urea) with ORR long-term stability and methanol tolerance. In one case, biomass-derived CNTs (from sludge) with graphitic-N reached ORR onset ~0.99 V. These results underscore that biochar carbon—though simple and low-cost—can achieve catalyst-level performance once properly doped and structured. In short, biochars' advantages are their eco-friendly feedstock and innate porous/functional character; their limitations are lower conductivity and variability, which are addressed by posttreatments. When optimized, biochar-carbons combine sustainability with high electrochemical activity [34,37].

11.4 Synthesis and functionalization

11.4.1 Methods

11.4.1.1 Pyrolysis

BDCs are typically generated by thermal or hydrothermal treatment of organic precursors, often followed by activation to develop porosity. In pyrolysis, raw biomass is heated (typically 400°C–1000°C) under an inert atmosphere to produce carbon-rich char. Activating agents such as KOH, zinc chloride ($ZnCl_2$), or phosphoric acid (H_3PO_4) are often mixed with the biomass before or during pyrolysis to etch the carbon matrix and create a high surface area. For example, Zhu et al. activated legume root nodule biomass with $ZnCl_2$ during pyrolysis, yielding a carbon with a record surface area of ~1835 m^2/g. This ultrahigh-SSA carbon exhibited excellent ORR activity (half-wave potentials $E_{1/2}\approx0.868$ V versus RHE in alkaline) and even outperformed Pt/C in alkaline solution [38].

11.4.1.2 *Hydrothermal carbonization*

In hydrothermal carbonization (HTC), biomass is treated in hot pressurized water (typically 180°C–250°C) to form a "hydrochar." HTC tends to preserve lignin-like aromatic structures while partially breaking down cellulose/hemicellulose, often yielding oxygen-rich, polymeric carbon particles [29]. The resultant hydrochar can be further activated to produce porous carbon. For instance, one study subjected various waste biomasses to HTC at 250°C for 12 h, then KOH-activated the hydrochars. The sample carbonized at 250°C (12 h) produced by HTC + KOH showed the highest ORR activity. SEM analysis confirmed that this two-step process (HTC followed by KOH activation) created a well-developed pore network: HTC alone generated some porosity, but KOH activation "opened" additional micropores that HTC could not achieve. In fact, precarbonization via HTC helps stabilize the carbon matrix and promotes micropore formation during subsequent activation [39].

11.4.1.3 *Chemical/physical activation*

Activation is critical in all cases. In physical activation, the char is exposed to oxidizing gases (steam or CO_2) at high temperature (600°C–1000°C), converting part of the carbon to CO/CO_2 and leaving a porous structure. Chemical activation involves impregnating the precursor or char with a chemical agent (e.g., potassium hydroxide [KOH], sodium hydroxide [NaOH], $ZnCl_2$, H_3PO_4) and heating, which yields high porosity at lower temperatures. The two approaches have complementary trade-offs: physical activation increases polar surface groups but requires very high temperatures and gives less control over pore size, whereas chemical activation allows fine control (by chemical ratio and conditions) but uses corrosive reagents. Often, a combined strategy is used: for example, lab-scale carbons may be first HTC-treated or pyrolyzed and then chemically activated. This approach was shown to dramatically increase pore volume and surface area. In summary, pyrolysis (with or without HTC pretreatment) followed by activation is a versatile route to highly porous BDCs. The combination of these steps enables tuning of porosity, surface chemistry, and conductivity, which are key to electrocatalytic performance [40].

11.4.2 Advanced approaches

11.4.2.1 Templating approaches

Template methods are widely used to engineer pore architecture in BDCs. In the hard-template approach, a removable scaffold (e.g., silica, salt, or polymer beads) is mixed with the biomass precursor and carbonized; the template is then etched away to leave a replica of the voids. After silica removal and KOH activation, the carbon had a specific surface area (SSA) of ~1749 m^2/g and combined micro- and mesopores. This sample (called ANPC-3) showed outstanding ORR activity: half-wave potential 0.79 V versus RHE and limiting current ~5.09 mA/cm^2, attributed to the accessible high-surface-area network. Soft templates (e.g., surfactants, block copolymers) can also direct mesopore formation. For instance, Wu et al. (2020) used the surfactant SDS as a soft template and doped phosphorus (via H_3PO_4) to create hollow P-doped carbon spheres (P–CHS). The resulting P–CHS material had ~654 m^2/g area and pore volume 0.36 cm^3/g, and exhibited an ORR onset potential of 0.89 V (RHE) with current density 5.6 mA/cm^2 and near-4e– pathway (n≈3.86). In general, templating (hard, soft, or dual methods) allows precise control over porosity and morphology, providing electrolyte reservoirs and short diffusion paths that enhance electrocatalytic rates [40].

11.4.2.2 Dye-assisted synthesis

Dye-assisted synthesis, or more generally, the use of organic additives as structure-directors, is an emerging approach. This strategy involves adsorbing dye molecules or polymeric organics onto biomass or templates before carbonization. These additives can introduce additional heteroatoms (e.g., nitrogen from azo dyes or phthalocyanines) or serve as lightweight templates. For instance, adding nitrogen-rich compounds (urea, melamine, or aromatic amines) to the precursor can significantly increase N-doping during pyrolysis and yield heteroatom-enriched carbon structures. Such methods remain under active development, with potential to embed functional dyes or organic motifs directly into the carbon matrix, thereby improving the electrocatalytic activity [41,42].

11.4.2.3 Solvent-free pyrolysis

Solvent-free pyrolysis has emerged as a scalable and environmentally friendly route, eliminating the need for dissolving biomass in solvents.

Instead, solid mixtures (e.g., biomass powder mixed with urea or other N-sources) are directly heated. For example, Sathiskumar et al. [43] prepared an N-doped porous carbon (N-PC) by solvent-free pyrolysis of "golden shower" pod biomass with urea. XRD, XPS, and SEM confirmed high N incorporation and sheet-like porosity. The Ni-loaded version (N-PC@Ni) acted as a trifunctional catalyst for ORR/OER/HER: it delivered an ORR half-wave potential of 0.76 V (4e– pathway, $n \approx 3.6$) and excellent OER/HER activity (overpotentials η_{10} = 314 mV for OER and 179 mV for HER). This solvent-free strategy demonstrates potential for sustainable, low-cost synthesis of heteroatom-doped porous carbons [41].

11.4.2.4 *Microwave-assisted approaches*

Microwave-assisted carbonization has attracted increasing attention for rapid, uniform heating. Microwave irradiation can accelerate HTC or pyrolysis by exciting polar biomass components. However, care must be taken: a recent study compared conventional HTC (T-HTC) versus microwave-assisted HTC (MW-HTC) of glucose [43]. The MW-HTC produced carbon more quickly but with a different surface chemistry. In fact, the MW-derived carbon had fewer surface functional groups conducive to ORR, and showed worse ORR performance than the conventionally hydrothermally carbonized sample. Microwave-assisted synthesis and other advanced routes (e.g., flash Joule heating) can finely tailor graphitization, defect density, and electronic properties on very short timescales.

11.4.3 Doping strategies (e.g., N, S, and B) to modulate activity

Incorporation of heteroatoms (especially N, S, B, and P) into the carbon lattice is a powerful way to tune electrocatalytic properties. Heteroatom doping perturbs the electronic structure of carbon, creating active sites with altered charge/spin density that favor reaction intermediates. Nitrogen is the most widely studied dopant. N-doping introduces sites such as pyridinic-N and graphitic-N that bind oxygen species strongly, as shown in Fig. 11.3. For example, enhancing the graphitic and pyridinic N content in biomass carbon was shown to improve O_2 adsorption and ORR kinetics [44]. In practice, N-doped carbons often approach the performance of precious catalysts: He et al. [42] achieved a half-wave potential of 0.884 V (RHE) in 0.1 M KOH with an N-doped coconut shell carbon (using flash Joule heating), effectively matching commercial Pt/C. Similarly, hydrothermally derived carbons rich in surface N/O groups show high ORR activity because the functionalized surface can facilitate O_2 reduction via deprotonation steps. N-doping also benefits HER/OER: in the Ni-loaded N-doped carbon (N-PC@Ni), the pyridinic and graphitic

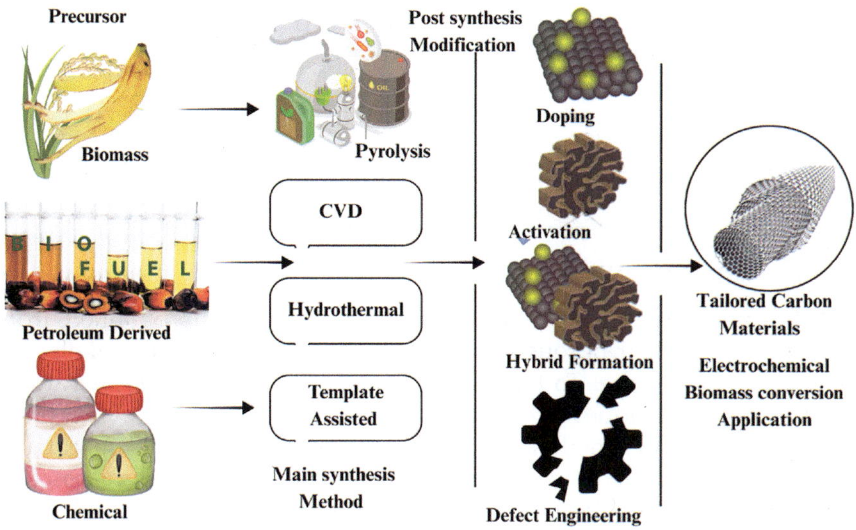

FIGURE 11.3 Synthesis routes for tailored carbon materials from biomass, petroleum-derived, and chemical precursors. Postsynthesis-modifications (doping, activation, hybrid formation, and defect engineering) are performed after a method such as pyrolysis, CVD, hydrothermal, and template-assisted synthesis to improve electrochemical performance.

N sites were credited with lower overpotentials (η_{10}, OER = 314 mV; η_{10}, HER = 179 mV) [27]. Sulfur doping is another common strategy, often used in concert with N. Sulfur has an electronegativity similar to carbon but a larger size, which induces charge delocalization and spin density in the carbon π-system. Codoping with S and N is known to be synergistic: Li et al. [33] synthesized N, S-co-doped mesoporous carbon (from glucose, using SBA-15 templating and thiophene/ammonia as precursors). This N–S–co-doped carbon achieved an ORR onset of ~0.92 V and $E_1/_2 = 0.73$ V (vs RHE) in alkaline media, significantly higher than singly-doped analogues. The extra S atoms increased the positive charge and asymmetry on adjacent carbons, enhancing oxygen binding. Phosphorus doping has a similar role: the P-doped hollow spheres (P–CHS) mentioned above gave an ORR onset of 0.89 V and current 5.6 mA/cm^2, illustrating that heteroatoms with lone pairs (S, P) can tune the ORR active sites. Boron doping (electron-deficient B in the graphitic matrix) is less common in biomass systems but has also been reported. B-sites withdraw electron density, which can catalyze the 2e$^-$ ORR pathway to H_2O_2 or modify adsorption energies. For example, a B-doped biomass carbon achieved very high H_2O_2 selectivity (90%+) in alkaline ORR and an onset ~0.83 V (albeit favoring 2e reduction). In general, any heteroatom doping creates defects and polar sites; graphitic (sp^2) structure with controlled dopant placement often outperforms amorphous carbon. Overall, N, S, P, or B dopants in BDCs significantly improve ORR, and often also lower the overpotential for OER/HER by altering binding strengths of reaction intermediates [33,41].

11.5 Characterization–performance correlation in carbon materials

11.5.1 Experimental approaches

A fundamental question regarding carbon materials obtained from biomass is how their porous structure can be related to their real electrochemical performance. To that end, scientists take advantage of various experimental techniques, which in concert create a full view of morphology, chemistry, and functionality [2]. Microscopy techniques, in particular scanning electron microscopy (SEM) and transmission electron microscopy (TEM), have virtually become the gold standard for initial imaging steps. The SEM offers, in general, a limited perception of surface roughness, particle size, and porosity distribution when compared with TEM, which provides sharper images displaying nanosheet thickness, lattice defects, and pore connectivity. [3]. X-ray diffraction (XRD) is also used for the examination of crystallinity and graphitic domains, as described in Table 11.2. On the other hand, characterization of defects is now routinely performed with Raman spectroscopy: comparing the ratio of the D- and G-bands, it is possible to rapidly fingerprint disorder and graphitization [6]. The texture properties are also equally important. In addition, Brunauer–Emmett–Teller (BET) analysis gives information on the surface area and its pore size distribution, allowing for estimating how much of the carbon structure is accessible for the reactions. Micropores typically enhance the active site density, mesopores favor electrolyte transport and mitigate against diffusion limitations [8]. To aid this, XPS provides information on functional groups and heteroatom dopants (e.g., pyridinic-N or thiophenic-S) that are critical factors to tune the reactivity [9]. Cyclic voltammetry (CV) indicates the catalytic activity and redox behavior, whereas electrochemical impedance spectroscopy (EIS) offers information about the conductivity and charge transfer resistance [3]. The durability and TOF over extended operation are usually tested by galvanostatic charge–discharge (GCD) cycling or polarization curves [5].

11.5.2 Computational approaches

Experiments show us what exists, but theory explains why it works. Computational-based approaches have dominated the last decade in the understanding of carbon catalysts at the atomic scale [2,9]. Density

TABLE 11.2 Techniques and methods employed to correlate the structural parameters with performance, bringing out the key descriptors and their functional role in structure-property correlations.

Technique type	Method	Parameter/descriptor	Role in linking structure to performance (with examples)	References
Experimental	SEM/TEM	Morphology, pore structure, lattice defects	Reveals nanoscale architecture and defect density that control mass transport; e.g., TEM showed interconnected mesopores in hydrothermally carbonized biomass, enhancing ion accessibility.	[2]
	XRD/Raman	Crystallinity, stacking order, ID/IG ratio	Correlates graphitization with conductivity; Raman ID/IG analysis in N-doped carbons showed that moderate defects improve ORR activity.	[6]
	BET/Gas sorption	Surface area, pore-size distribution	Links hierarchical porosity to performance; BET confirmed that micropore–mesopore balance in activated carbons improved HMF electrooxidation selectivity.	[5]
	XPS/UPS	Surface functional groups, heteroatom speciation, work function	Identifies active dopants; XPS revealed pyridinic-N correlates with higher ORR activity, while thiophenic-S improved furfural reduction.	[9]

(*continued on next page*)

TABLE 11.2 Techniques and methods employed to correlate the structural parameters with performance, bringing out the key descriptors and their functional role in structure-property correlations—cont'd

Technique type	Method	Parameter/descriptor	Role in linking structure to performance (with examples)	References
	CV/LSV	Catalytic activity, redox kinetics, and onset potential	Relates electrode design to kinetics; CV of porous carbons derived from cellulose showed lower onset potential for ORR, indicating efficient charge transfer.	[3]
	EIS/GCD	Charge transfer resistance, durability, diffusion resistance	EIS confirmed reduced R_ct in graphitized carbons, while GCD cycling validated their superior long-term durability in biomass electrooxidation.	[5]
Computational	DFT adsorption energy	Binding energies of intermediates	Explains dopant effects; DFT showed N, S codoping lowers the adsorption energy of furfural intermediates, enhancing reaction rates.	[6]
	DOS/Band structure	Electronic distribution near the Fermi level	DOS analysis indicated that N-doping increases states near the Fermi level, improving electron transfer and catalytic efficiency.	[6]

(*continued on next page*)

	Reaction pathway modeling	Free energy diagrams, PDS identification	Pathway modeling demonstrated that C=O bond activation in HMF is the rate-determining step, aligning with CV results.	[5]
	AIMD/MD	Catalyst stability under dynamic conditions	AIMD revealed that heteroatom-doped carbons maintain structural stability under solvated electrolyte conditions, supporting their long-term durability.	[45]
	Machine learning/HTS	Descriptor mapping, performance prediction	High-throughput screening predicted optimal pore sizes and dopant levels for selective ORR pathways, validated by XPS and electrochemical testing.	[46]

Functional Theory (DFT) lets researchers calculate adsorption energies and reaction barriers [6]. Such a mechanism would account for the experimentally observed differences [3]. An analogous method, density of states (DOS) analysis, has been used to bridge doping or defects to electronic structure changes. DOS analysis can indicate the source of catalytic hotspots that are otherwise challenging to observe directly [6]. More sophisticated works employ charge density maps and Bader charge calculations to illustrate electron transfer at the defect site [8]. In this way, ab-initio molecular dynamics (AIMD) methodology is now employed here to go beyond static images and interrogate the behavior of such sites under real conditions. The thermal stability, interactions of electrolyte, and ion migration inside carbon with pore over time by AIMD can be simulated [45]. They complement each other and form a reinforcing loop, accelerating the development of catalyst design by comparing computational prediction with XPS or Raman results [9].

11.5.3 Structure–performance relationships

Based on this understanding, some reliable rules have been proposed for carbon material design. Nanocarbons with large SSA and hierarchical porosity invariably yield superior electrochemical performance, as they achieve maximized available active sites, for which good ion transport properties are maintained [8]. By contrast, moderate density of defects could promote reactivity due to the formation of effective edges and the modifications in local electronic structure [6]. But the network ends up being interrupted by too many defects, reducing overall effectiveness. This implies that there is a value of disorder that optimizes the interplay between conductivity and catalytic activity [6]. Another common method is heteroatom doping. The electron density near the nitrogen tends to rise, leading to a better ORR activity. Sulfur, phosphorus, and boron can warp the carbon lattice in ways that open up new, selective catalytic routes. Individually, the contributions of a pair of introduced heteroatoms often are greater than the sum of their parts [9].

Finally, the degree of graphitization is a key factor in both the conductivity and durability. High-graphitized carbons provide fast transportation capabilities and are also stable in long-term cycling ability [5]. The most suitable carbons are usually those whose graphitic order is combined with engineered defects and doping [8]. Finally, one can say that the superior-performance activated BDCs should be engineered with an enhanced accessible surface, a precisely regulated number of defects, a targeted heteroatom doping, and a moderately graphitic nanostructure. These rules guide the transition from trial-and-error synthesis to predictive and rational catalyst design [9].

11.6 Application in biomass conversion

11.6.1 Electro-oxidation: Alcohols, polyols, and sugars

Electro-oxidation of biomass–derived molecules—such as alcohols, polyols, and monomeric sugars—provides a green route for generating valuable oxidation products like formate, gluconic acid, or FDCA. Li et al. (n.d.) demonstrated a stacked parallel continuous stacked flow reactor (SPCFR) that converted glucose to formate and HMF to FDCA with remarkable single-pass conversion efficiencies (81.8% and 95.8%, respectively) and high selectivity (formate: 76.5%, FDCA: 96.9%)—even at industrially relevant concentrations and scales (>1 kg of product) [47]. However, challenges such as non-Faradaic degradation—especially in alkaline media—can significantly reduce efficiency; the SPCFR addressed this by separator feeding and minimizing residence time. A survey of catalysts for electro-oxidation (e.g., for glycerol, glucose, furfural, HMF) shows transition metals (Ni, Co) as cost-effective alternatives to noble metals, though they operate at higher onset potentials (>1 V vs RHE). Designing BDC-supported Ni or Cobased electrocatalysts could reduce overpotentials and improve adsorption dynamics, thereby optimizing activity and selectivity in biomass electro-oxidation reactions.

11.6.2 Electro-reduction: Furfural, 5-Hydroxymethylfurfural, organic acids

Electro-reduction of biomass platform molecules—such as furfural and HMF—yields important chemicals like furfuryl alcohol, levulinic acid, etc. under controlled potentials. Explicit studies utilizing BDC-supported electrodes for these reductions are scarce, but the broader benefits of BDCs for electrocatalysis are well documented. Nguyen et al.[2] reviewed BDCs for energy storage and catalytic applications, underscoring their hierarchical structures, heteroatom doping, and tailored porosity as beneficial for electrocatalytic performance. Translating this design logic to furfural/HMF electro-reduction could potentially enhance selectivity, lower overpotentials, and enable more sustainable electrode materials. This remains a promising yet underexplored research avenue.

11.6.3 Hydrogenation: Lignin-derived compounds, levulinic acid

Traditional conversion of lignin-derived fragments or levulinic acid often relies on hydrogen gas and metal catalysts, which may require high temperature or pressure. Electrochemical hydrogenation (ECH) presents a greener alternative—producing reactive H-atoms in situ at the cathode

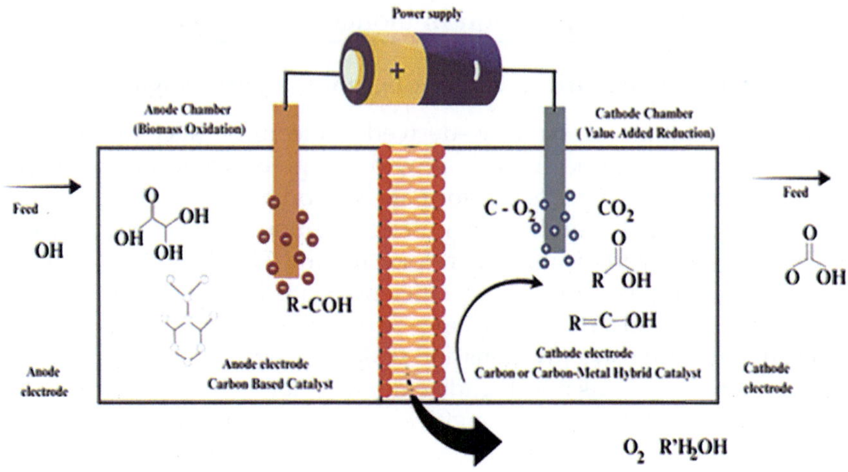

FIGURE 11.4 Conversion of biomass to electrochemical products with carbon or carbon-metal hybrid catalysts as electrodes. It works by oxidation of biomass at the anode and by value-added reduction reaction at the cathode to produce sustainable energy and generate chemicals.

while coupling to anodic reactions. BDCs offer ideal support characteristics for such approaches due to their porous frameworks and conductive networks, which can modify electron transfer and adsorption profiles. Khandekar et al. [3] discussed the design flexibility of BDCs for sustainable energy applications, citing their tunable porosity and rich surface functionality. While their focus was broader (e.g., supercapacitors, general electrocatalysis), the same features could be instrumental in enabling electrohydrogenation of lignin-derived or levulinic acid compounds. This represents a strategic research direction toward valorizing biomass through electrochemical pathways [3].

11.6.4 Paired electrolysis: Coupling biomass oxidation with reduction processes (e.g., hydrogen evolution and carbon dioxide reduction reaction)

Paired electrolysis offers simultaneous valorization of biomass (anodic oxidation) and valuable reduction reactions at the cathode—e.g., hydrogen evolution reaction (HER) or CO_2 reduction reaction (CO_2RR). Coupling reductive and oxidative biomass-driven reactions—especially replacing sluggish OER with organic oxidations—can result in high-value cogeneration with better energy efficiency [48]. Additionally, studies coupling CO_2RR with biomass oxidation demonstrate reduced overall energy demand, expanding the conceptual scope of paired systems as shown in Fig. 11.4.

In such contexts, BDC materials—thanks to their multifunctionality, doping flexibility, and porous structure—are excellent candidates as dual-function electrodes or supports for both oxidative and reductive processes.

11.7 Case studies and performance highlights

11.7.1 High-performance systems

11.7.1.1 Biomass-derived activated carbons

BDCs have become a fundamental component in sustainable electrochemical energy storage materials owing to their renewable characteristics, affordability, and inherent chemical composition abundant in heteroatoms and aromatic structures [49,50]. Agricultural residues, lignocellulosic biomass, and food-processing byproducts provide a natural amalgamation of cellulose, hemicellulose, and lignin, which, upon carbonization, produce high-carbon frameworks with adjustable pore structures [51–53]. A typical case includes carbon made from corn stover through hydrothermal liquefaction and then chemical activation with KOH. This multistep process first breaks down the biomass into hydrochar by hydrolysis and partial carbonization in water that is not too hot (180°C–250°C). Then KOH is added to the hydrochar, and the mixture is heated to 700°C–850°C in an inert atmosphere [49,50]. When activated, redox reactions between KOH and the carbon matrix create micropores by etching carbon atoms and intercalating potassium. This releases CO and CO_2 gases that further open the pore network. The resulting material had specific capacitances of 270–350 F g^{-1} at 0.5 A g^{-1} and kept more than 90% of its performance after 10,000 cycles. This was possible because it had a high micropore volume (~0.8–1.0 cm^3 g^{-1}), an optimized pore size distribution (0.7–3 nm), and nitrogen doping that created pyridinic- and pyrrolic-N sites to improve pseudocapacitance [49,51,52]. Industrial byproducts such as coffee grounds and sugar-processing residues have been turned into ACs with very high SSA values of up to 3300 m^2/g^{-1} and capacitances of about 236 F/g^{-1}, which is almost twice as good as the commercial benchmark YP80F [54–56]. Long-term cycling stability surpassed 10,000 charge–discharge cycles without discernible degradation, indicating structural integrity [55,56].

Beyond these instances, a variety of other biomass precursors, including sawdust, coconut shells, and rice husks, processed by chemical activation ($ZnCl_2$, H_3PO_4) and physical activation (CO_2 or steam), have consistently produced high-performance supercapacitor electrodes, demonstrating that pore architecture tuning in conjunction with precursor selection is essential for optimizing capacitance and energy density [53,55].

11.7.1.2 *Structural engineering for enhanced performance*

High SSA and microporosity are important, but new research shows that hierarchical pore structures (micropores + mesopores + macropores) and controlled morphology can make electrochemical performance even better. To create hollow carbon spheres (HCSs), which are made using hard-template techniques like SiO_2 nanoparticle templating, the silica cores are coated with a carbon precursor (such as sucrose, phenolic resin, or biomass tar), carbonized, and then the template is selectively removed (such as by HF or NaOH etching) [57–59]. Large electrolyte-accessible surfaces, short ion diffusion pathways, and void spaces that protect against structural stress during rapid charge-discharge are all features of these hollow architectures. HCSs benefit from improved wettability and redox-active surface functionalities when doped with nitrogen or sulfur, exhibiting capacitances >300 F g^{-1} and ~98% retention after 10,000 cycles [58,59]. Molten salt-assisted synthesis, another structural innovation, involves carbonising biomass precursors at 700°C–900°C in a eutectic mixture (e.g., sodium–potassium chloride mixture [NaCl–KCl]) [60,61]. Carbons produced by this process have well-developed pore networks, increased interlayer spacings, and high graphitization degrees. Because of their enhanced charge-transfer kinetics, peanut shell-derived carbons, for instance, achieved ~285 F g^{-1} with excellent rate capability and power density retention [61]. These structural engineering techniques demonstrate that, in addition to high SSA, balancing micro- and mesoporosity, managing morphology, and causing partial graphitization are essential for producing electrodes with high capacitance and superior high-rate performance [53,58,59].

11.7.1.3 *Role of heteroatoms*

Heteroatoms drastically change the electronic structure, surface chemistry, and electrolyte interaction characteristics of carbon matrices. Through reversible redox transitions of pyridinic and pyrrolic nitrogen species, nitrogen doping adds electronegative sites that raise electron density close to the Fermi level, improving intrinsic conductivity and introducing faradaic pseudo capacitance [62,63]. Due to sulfur's larger atomic radius, sulfur doping causes local lattice distortions and increases interlayer spacing, which promotes faster ion diffusion and enhances electrolyte accessibility [64,65]. Codoping (e.g., N–S, N–P) synergistically combines these advantages, while phosphorus doping adds electron-rich sites and changes the electronic band structure of the carbon, improving wettability and catalytic activity [66–68].

Because of the coexistence of high conductivity from nitrogen sites and improved ion-accessibility from sulfur-induced interlayer expansion, N, S-co-doped carbons made from sawdust, for instance, demonstrated 310 F g^{-1} with >96% retention after 10,000 cycles, outperforming single-doped analogues [64,67,68]. These studies' XPS and Raman analyses verified that doping improves electrochemical performance and increases electrocatalytic activity for ORR and OER, a feature that is used in multifunctional devices (see Section 11.7.3) [53,64,65].

11.7.2 Machine learning–optimized syntheses of hard carbon

11.7.2.1 Data-driven synthesis optimization

A paradigm shift from empirical trial-and-error to predictive, data-driven materials design is represented by the incorporation of machine learning (ML) into the synthesis optimization of biomass-derived hard carbons [50,66,69]. Iterative experimentation with many variables, including precursor type, pretreatment, carbonization temperature, activation agent concentration, dwell time, and posttreatment, is a time-consuming and resource-intensive process in traditional carbon electrode development [50,69]. The models frequently make use of supervised regression algorithms, including multilayer perceptrons (artificial neural networks), gradient boosting machines, and random forest regressors. The output targets are electrochemical performance metrics like specific capacity, Coulombic efficiency (CE), and capacity retention, while the input features are usually quantitative synthesis parameters (e.g., carbonization temperature in °C, activation ratio in g/g, precursor C/H/O/N content) and resulting material properties (SSA, total pore volume, micropore percentage, and interlayer spacing) [50,69]. For instance, a recent study used Bayesian optimization to direct the synthesis of hard carbons derived from lignin for sodium-ion batteries. It found that the ideal carbonization temperature and activation ratio were around 950°C, which produced reversible capacities of over 350 mAh g^{-1} with >90% CE in the first cycle [50,69]. The most significant predictors of Na^+ insertion performance, according to ML feature importance analysis, were controlled defect densities and expanded interlayer distances (~0.38–0.40 nm). These characteristics increase capacity and efficiency by facilitating the intercalation of desolvated sodium ions into the turbostratic carbon layers without causing an excessive amount of solid-electrolyte interphase to form [50,69].

11.7.2.2 Closed-loop experimental integration

In sophisticated implementations, ML predictions are coupled with closed-loop experimental systems, where each synthesis cycle is followed by a direct feed of in situ and operando characterization data into the predictive model [50,66]. Methods such as N_2 adsorption–desorption for SSA and pore-size distribution, Raman spectroscopy for disorder quantification (I_D/I_G ratio), and X-ray diffraction (XRD) for crystallinity/interlayer spacing offer quick feedback that allows the model parameters to be adjusted in real time. Compared with conventional one-variable-at-a-time optimization, studies show that 40%–60% fewer experimental iterations are needed to reach target performance [69]. Carbons made from cellulose and lignin blends for potassium-ion batteries served as an example. Because of engineered mesopore fractions (20%–35%) that improved K^+ diffusion without sacrificing volumetric energy density, the closed-loop ML framework found a synthesis route that increased rate capability by >35% at 10 A g^{-1} in comparison to baseline carbons [66,69].

11.7.2.3 *Impact on battery performance*

Due to fine-tuned pore size distributions in the subnanometer to mesopore range (~0.5–2 nm), which achieve an ideal balance between ion accessibility and surface area utilization, ML-optimized hard carbons have demonstrated notable rate capability improvements [50,66]. Furthermore, ML-guided defect engineering has produced carbons with high defect densities (~3%–5% edge plane exposure), which serve as active sites for ion adsorption and intercalation but are managed to prevent excessive irreversible capacity losses. After 5000 cycles, ML-designed electrodes in sodium-ion capacitors have maintained over 90% capacity retention, achieving energy densities of up to 65 Wh kg^{-1} at power densities exceeding 10 kW kg^{-1}. These findings highlight how big data, algorithmic prediction, and rapid prototyping can be combined to speedup the development of carbon electrodes for a variety of energy storage chemistries.

11.7.3 Electrocatalytic coupling strategies and real device demonstrations

11.7.3.1 *Multifunctional electrode design*

Carbons derived from biomass, particularly those doped with heteroatoms, have been designed to serve as both electrocatalysts in conversion reactions and electrodes for energy storage [64,67,68,70]. N-doped carbons from wood precursors delivered specific capacitances >250 F g^{-1} in symmetric supercapacitors and showed ORR onset potentials of ~0.92 V vs RHE, which are comparable to commercial Pt/C catalysts [70]. The use of sulfur-phosphorus codoped carbons in rechargeable zinc-air batteries is made possible by their bifunctional ORR/OER activity, low overpotentials (<350 mV for OER at 10 mA cm^{-2}), and high-rate capacitance retention (>90% after 5000 cycles) as given in Table 11.3 [64,67,68].

11.7.3.2 *Coupled storage—conversion devices*

The zinc–air supercapacitor hybrid, which uses the same carbon material as both the ORR/OER catalyst and the supercapacitor electrode, is among the most striking examples of multifunctionality. Energy densities

TABLE 11.3 Selected case studies of carbon architecture to show correlations between fabrication routes, physicochemical properties, and improved functional results.

Material type	Precursor	Synthesis method	Performance	SSA ($m^2 g^{-1}$)	Special features	References
Hard carbon (biomass)	Corn stover	Hydrothermal liquefaction + KOH activation	270–350 F g^{-1} 0.5 A g^{-1}; >90% retention after 10,000 cycles	~2000	High micropore volume, N-doping	[49–52]
Industrial byproduct activated carbon	Coffee/sugar residues	KOH activation (800°C)	~236 F g^{-1} 1 A g^{-1}; Stable >10,000 cycles	3300	Ultrahigh SSA, narrow pore size	[54–56]
Hollow carbon spheres (HCS)	Biomass tar	SiO_2 template + N/S doping	305–320 F g^{-1}; ~98% retention after 10,000 cycles	1500–2200	Hierarchical porosity, pseudocapacitance	[57–59]
Molten salt–derived carbon	Peanut shells	NaCl–KCl molten salt carbonization	~285 F g^{-1}; >95% retention after 5000 cycles	~1600	High graphitization, fast ion transport	[61]
N, S-co-doped carbon	Sawdust	Direct pyrolysis + thiourea treatment	~310 F g^{-1}; >96% retention after 10,000 cycles	1700	Expanded interlayer spacing, high conductivity	[64,67,68]

(*continued on next page*)

TABLE 11.3 Selected case studies of carbon architecture to show correlations between fabrication routes, physicochemical properties, and improved functional results—cont'd

Material type	Precursor	Synthesis method	Performance	SSA ($m^2\ g^{-1}$)	Special features	References
ML-optimized hard carbon	Lignin	Carbonization + KOH (ML-guided)	>350 mAh g^{-1} (Na-ion battery); >90% CE, stable 200 cycles	~1800	Expanded d-spacing (~0.39 nm), controlled defects	[50,69]
N-doped carbon catalyst	Wood	Pyrolysis + NH_3 doping	>250 F g^{-1}; ORR onset ~0.92 V vs RHE	~1500	Dual ORR + storage function	[67,70]
N, S-HCS hybrid	Biomass tar	Template + dual doping	65–80 Wh kg^{-1} (Zn–air SC hybrid); >95% retention (5000 cycles)	2000	Multifunctional, hierarchical porosity	[64,67,68,70]

of 65–80 Wh kg^{-1} have been achieved by N, S-co-doped HCS, which have demonstrated >95% retention over 5000 hybrid cycles [64,68,70]. While dopant-induced active sites offer high catalytic activity for oxygen reactions, hierarchical porosity guarantees quick ion diffusion for capacitive storage.

Exploration of sodium-ion(Na-ion) hybrid systems that couple a carbon negative electrode derived from biomass with a catalytic positive electrode for CO_2 reduction (CO_2RR) [67]. These designs use a shared electrochemical cell where the positive electrode catalyses the conversion of CO_2RR to fuels and the negative electrode manages reversible ion storage.

11.7.3.3 *Practical device prototypes*

Coin-cell tests have been replaced by pouch-cell and module-scale prototypes in a number of studies [67,70]. Biomass carbon electrodes are frequently pressed onto nickel foam or slurry-coated onto copper/aluminium current collectors to assemble these devices. For higher-voltage operation, electrolytes can be anything from organic carbonates to aqueous KOH and sodium sulfate ($NaSO_4$). Practical load profiles, such as variable current cycling, pulse discharge, and temperature-dependent performance testing (−20°C to 60°C), are used to validate durability. From lab-scale materials to scalable, multipurpose energy devices, the results consistently show that carbons derived from biomass can satisfy the mechanical, electrochemical, and catalytic stability requirements for real-world applications [64,67,68,70]. These case studies demonstrate that converting carbons derived from biomass from lab-scale materials to scalable, multipurpose devices is feasible, as shown in Table 11.3.

11.8 Challenges and future perspectives in biomass electroconversion

11.8.1 Catalyst stability and deactivation

Catalyst stability is one of the most critical challenges in biomass electroconversion. Biomass-derived feedstocks, such as lignin, hemicellulose, and carbohydrate derivatives, often contain impurities such as sulfur, nitrogen, and halides that can poison the catalyst surface, leading to deactivation over time. This deactivation can be due to:

- *Surface fouling*: Adsorption of reaction intermediates or by-products such as humins, which block active sites.
- *Corrosion*: Particularly in acidic or alkaline electrolytes, leading to loss of catalytic activity.

- *Particle agglomeration*: Nanostructured catalysts tend to sinter during prolonged reactions, reducing surface area.

Noble metal catalysts such as Pt and Pd, while highly active for alcohol electrooxidation, suffer from poisoning by CO-like intermediates, requiring strategies such as alloying (e.g., Pt–Ru) or dynamic potential cycling to regenerate active surfaces [71].

11.8.2 Selectivity and product distribution

Achieving high selectivity toward desired products remains a major hurdle. Biomass molecules are polyfunctional, meaning they contain multiple reactive groups. This can lead to a complex product mixture and low yields of target chemicals.

- *Overoxidation or overreduction*: Excessive reaction times or high potentials may convert valuable intermediates (e.g., 5-HMF) into undesired by-products (e.g., CO_2 or humins).
- *Competing reactions*: For example, during electrooxidation, the oxygen evolution reaction (OER) can compete with biomass conversion, lowering FE.
- *Solution*: Use of tailored catalysts with specific active sites, and engineering reaction environments (pH, temperature, electrolyte composition).

Selective electroreduction of furfural to furfuryl alcohol can be improved by using Cu-based catalysts with oxide-derived surfaces [72].

11.8.3 Electrode design and mass transport limitations

Efficient biomass electroconversion requires electrodes with high surface area, controlled porosity, and enhanced electrical conductivity to facilitate both electron transfer and mass transport.

- *Mass transport issues*: Biomass-derived molecules are often large and less mobile in aqueous solutions compared with small molecules such as H_2O or carbon dioxide (CO_2).
- *Electrode fouling*: Deposition of polymeric by-products can clog pores and hinder diffusion.
- *Advanced designs*: Use of 3D-structured electrodes, carbon paper, or metal foams to increase contact between reactants and active sites.

Flow-cell configurations with structured catalysts have been shown to enhance furfural electroreduction efficiency by improving reactant accessibility [73].

11.8.4 Integration with renewable energy sources

The sustainability of biomass electroconversion depends heavily on coupling with renewable electricity sources such as solar or wind power.

- *Intermittent power supply*: Renewable sources are variable, which can cause fluctuations in reaction rates and affect catalyst lifetime.
- *Energy storage integration*: Pairing with batteries or supercapacitors can stabilize power input for continuous biomass conversion.
- *Techno-economic challenge*: Large-scale systems require cost-effective catalyst materials and scalable reactor designs.

Solar-driven electrooxidation of glycerol using NiFe layered double hydroxide catalysts has been demonstrated with >85% FE under simulated sunlight [74].

11.9 Conclusion

The chapter critically analyzes the development of BDC material as a multinavigate and sustainable experiment for biomass electrochemical conversion. Through the rational investigation of their synthesis pathways, architectural guidance of substrateral structure, and electrochemical design, one can see that attributes such as hierarchical porosity, controlled graphitization, and selective heteroatom doping determine the catalytic performance and stability. The combination of modern methods of characterization with computer modeling has taken definite correlations between structure and performance to the next level of catalyst design, progressing the field bigger to the picture rather than on the basis of empirical synthesis to rational and predictive skew in the design of catalysts. Such observations highlight the promising nature of BDCs, which can be utilized to effect selective reactions involving electrooxidation, reduction, and hydrogenation reactions, as well as be utilized to engage with paired electrolysis systems designed to optimize energy savings and the valorization of a product. No less important are the recent case studies that indicate that BDCs cannot only compete with the performance of current benchmarks, but also sustain the added benefits of renewability, low cost, and environmental compatibility. The incorporation of ML and data-driven synthesis optimization further signals a paradigm shift toward accelerated material discovery and scalable applications. Although there remain significant stability problems associated with the use of catalysts, selectivity, and scaling-up with integration into continuous renewable energy systems, the research direction shows that these materials have a deep potential in forming the basis of next-generation sustainable energy and chemical formation technologies. In this respect, biomass-made carbons can reflect the requirements of green chemistry and the circular economy, which is why they will become pillars in the globalization of energy.

11.10 Future outlooks

11.10.1 Scalable fabrication and process integration

The development of solvent-free pyrolysis, molten-salt templating, or machine-learning-optimized activation procedures might be transferred to continuous or modular production schemes. Focus is given to low-energy, feedstock-flexible processes that can preserve surface functionality and pore hierarchy when scaled up without increasing cost or carbon footprint.

11.10.2 Oxygen evolution and reduction (integration) with molybdenum, integration of O_2 with silicon

Development of highly efficient O_2 evolution (OER) and reduction (ORR) chemistry to be used in bioelectroconversion systems is still a challenging area. Rational doping of carbon with N, S, P, and composites based on these doped carbons opens avenues to create bifunctional electrodes that are capable of facilitating the oxidative as well as reductive half-reactions. Future work could also consider electrode pairings such as biomass oxidation in combination with O_2 reduction or production, leading to more efficient electron economy and potentially achieving self-sustained electrochemical loops.

11.10.3 Techno-economic feasibility and life-cycle viability

To truly assess sustainability improvements in a system such as that including a platinized carbon or nickel-based catalyst supported on biomass-derived C, life-cycle assessments become of value. The objective is to achieve a competitive cost per kWh or per kg of product, and to provide carbon neutrality throughout the chain.

11.10.4 Integration at the 35 energy system level with renewables

Next-generation systems will need to be directly interfaced with solar, wind, or bioelectric power sources in a way that permits robust operation while accommodating intermittent conditions. Such integrated storage–conversion platforms would allow the electrochemical reactors to be coupled with supercapacitive or battery modules sharing the same carbon materials, and optimize energy efficiency and reliability in green manufacturing contexts.

Ethical declarations

None.

Conflict of interest

The authors declare no known competing financial interests.

Acknowledgments

None.

References

[1] A.I. Osman, N. Mehta, A.M. Elgarahy, A. Al-Hinai, Al-M. AaH, D.W. Rooney, Conversion of biomass to biofuels and life cycle assessment: a review, Environ. Chem. Lett. 19 (6) (2021) 4075–4118.

[2] V.-T. Nguyen, K. Cho, Y. Choi, B. Hwang, Y.-K. Park, H. Nam, et al., Biomass-derived materials for energy storage and electrocatalysis: recent advances and future perspectives, Biochar 6 (1) (2024) 96.

[3] T. Khandaker, T. Islam, A. Nandi, M. Anik, M.S. Hossain, M.K. Hasan, et al., Biomass-derived carbon materials for sustainable energy applications: a comprehensive review, Sustain. Energy. Fuels. 9 (3) (2025) 693–723.

[4] P. Prabhu, Y. Wan, J.-M. Lee, Electrochemical conversion of biomass-derived products into high-value chemicals, Matter 3 (4) (2020) 1162–1177.

[5] M. Zhang, J. Zhang, S. Ran, W. Sun, Z. Zhu, Biomass-Derived sustainable carbon materials in energy conversion and storage applications: status and opportunities. A mini review, Electrochem. Commun. 138 (2022) 107283.

[6] Á. Ramírez, M. Muñoz-Morales, F.J. Fernández-Morales, J. Llanos, Innovative carbon materials from lignocellulosic wastes for electrochemical hydrogen peroxide production: bridging biomass conversion and material properties, J. Environ. Chem. Eng. 12 (3) (2024) 112985.

[7] Y.A. Begum, S. Kumari, S.K. Jain, M.C. Garg, A review on waste biomass-to-energy: integrated thermochemical and biochemical conversion for resource recovery, Environ. Sci.: Adv. 3 (9) (2024) 1197–1216.

[8] K.A. Sanoja-López, C.W. Guamán-Marquines, R. Luque, Advanced processes in biomass/waste valorization: a review, Sustain. Chem. Pharm. 41 (2024) 101704.

[9] K.M. Maliutina, J.E. Omoriyekomwan, C. He, L. Fan, A. Folli, Biomass-derived carbon nanostructures and their applications as electrocatalysts for hydrogen evolution and oxygen reduction/evolution, Front. Environ. Eng. 2 (2023) 1228992.

[10] C. IPdP, P.A.M. Mourão, J.E. Castanheiro, G. PF, S.R. Suero, B.L. Cano, A Review of the Biomass Valorization Hierarchy, Sustainability 17 (1) (2025) 335.

[11] Y. Sun, J. Miao, X. Fan, K. Zhang, T. Zhang, Recent progress in electrochemical conversion from biomass derivatives into high-value-added chemicals, Small Structures 5 (7) (2024) 2300576.

[12] M. Huddleston, Y. Sun, Biomass Valorization via paired electrocatalysis, ChemSusChem 18 (8) (2025) e202402161.

[13] Y. Yang, T. Mu, Electrochemical oxidation of biomass derived 5-hydroxymethylfurfural (HMF): pathway, mechanism, catalysts and coupling reactions, Green Chem. 23 (12) (2021) 4228–4254.
[14] Y. Holade, N. Tuleushova, S. Tingry, K. Servat, T.W. Napporn, H. Guesmi, et al., Recent advances in the electrooxidation of biomass-based organic molecules for energy, chemicals and hydrogen production, Catal. Sci. Technol. 10 (10) (2020) 3071–3112.
[15] C. Pereira, V. Maia, P. Zambiazi, A.E. de Souza R, A. Neto, PtSb/C electrocatalysts for glycerol oxidation in alkaline electrolyte, Results. Chem. 4 (2022) 100375.
[16] S. Liu, Z. Mukadam, S.B. Scott, S.C. Sarma, M.-M. Titirici, K. Chan, et al., Unraveling the reaction mechanisms for furfural electroreduction on copper, Ees Catalysis 1 (4) (2023) 539–551.
[17] Y. Ren, S. Fan, X. Yu, S. Shi, J. Wang, J. Zeng, et al., Advances in the energy-saving electro-oxidation of 5-hydroxymethylfurfural to 2, 5-furandicarboxylic acid, Adv. Sustain. Sys. 9 (4) (2025) 2400959.
[18] P. Hauke, T. Merzdorf, M. Klingenhof, P. Strasser, Hydrogenation versus hydrogenolysis during alkaline electrochemical valorization of 5-hydroxymethylfurfural over oxide-derived Cu-bimetallics, Nat. Commun. 14 (1) (2023) 4708.
[19] N. Schlegel, G.K. Wiberg, M. Arenz, On the electrooxidation of glucose on gold: towards an electrochemical glucaric acid production as value-added chemical, Electrochim. Acta 410 (2022) 140023.
[20] U. Sanyal, J. Lopez-Ruiz, A.B. Padmaperuma, J. Holladay, O.Y. Gutiérrez, Electrocatalytic hydrogenation of oxygenated compounds in aqueous phase, Org. Process Res. Dev. 22 (12) (2018) 1590–1598.
[21] G.V. Burmakina, D.V. Zimonin, V.V. Verpekin, V.V. Sychev, A.I. Rubaylo, A comparative study of electrochemical reduction of levulinic acid on various electrodes in organic solvents, ChemPhysChem 25 (18) (2024) e202300900.
[22] M.T. Bender, X. Yuan, M.K. Goetz, K-S. Choi, Electrochemical hydrogenation, hydrogenolysis, and dehydrogenation for reductive and oxidative biomass upgrading using 5-hydroxymethylfurfural as a model system, ACS Catalysis 12 (19) (2022) 12349–12368.
[23] S.A. Akhade, N. Singh, O.Y. Gutiérrez, J. Lopez-Ruiz, H. Wang, J.D. Holladay, et al., Electrocatalytic hydrogenation of biomass-derived organics: a review, Chem. Rev. 120 (20) (2020) 11370–11419.
[24] Z. Zhang, J. Li, S. Yang, C. Peng, Advances in paired electrolysis for furfural conversion: from design principle, mechanisms to perspectives, Green Chem. 27 (24) (2025) 6967–6988.
[25] F.W. Lucas, R.G. Grim, S.A. Tacey, C.A. Downes, J. Hasse, A.M. Roman, et al., Electrochemical routes for the valorization of biomass-derived feedstocks: from chemistry to application, ACS Energy Letters 6 (4) (2021) 1205–1270.
[26] J. Carneiro, E. Nikolla, Electrochemical conversion of biomass-based oxygenated compounds, Annu. Rev. Chem. Biomol. Eng. 10 (1) (2019) 85–104.
[27] G. Zhang, X. Liu, L. Wang, H. Fu, Recent advances of biomass derived carbon-based materials for efficient electrochemical energy devices, J. Mater. Chem. A 10 (17) (2022) 9277–9307.
[28] P.A. Kempler, A.C. Nielander, Reliable reporting of Faradaic efficiencies for electrocatalysis research, Nat. Commun. 14 (1) (2023) 1158.
[29] Y. Dessie, E. Tilahun, T.H. Wondimu, Functionalized carbon electrocatalysts in energy conversion and storage applications: a review, Heliyon 10 (20) (2024).
[30] K.M. Maliutina, J.E. Omoriyekomwan, C. He, L. Fan, A. Folli, Biomass-derived carbon nanostructures and their applications as electrocatalysts for hydrogen evolution and oxygen reduction/evolution, Front. Environ. Eng. 2 (2023) 2023.

[31] X. Qin, O. Ola, J. Zhao, Z. Yang, S.K. Tiwari, N. Wang, et al., Recent progress in graphene-based electrocatalysts for hydrogen evolution reaction, Nanomaterials 12 (11) (2022) 1806.

[32] H. Huang, M. Yan, C. Yang, H. He, Q. Jiang, L. Yang, et al., Graphene nanoarchitectonics: recent advances in graphene-based electrocatalysts for hydrogen evolution reaction, Adv. Mater. 31 (48) (2019) 1903415.

[33] W. Li, Y. Xu, G. Wang, T. Xu, K. Wang, S. Zhai, et al., Sustainable carbon-based catalyst materials derived from lignocellulosic biomass for energy storage and conversion: atomic modulation and properties improvement, Carbon Energy 7 (5) (2025) e708.

[34] Y. Wang, H. Zhong, W. Yang, Y. Feng, N. Alonso-Vante, Recent advances with biomass-derived carbon-based catalysts for the high-efficiency electrochemical reduction of oxygen to hydrogen peroxide, Adv. Energy. Sustain. Res. 4 (12) (2023) 2300128.

[35] C. Hu, R. Paul, Q. Dai, L. Dai, Carbon-based metal-free electrocatalysts: from oxygen reduction to multifunctional electrocatalysis, Chem. Soc. Rev. 50 (21) (2021) 11785–11843.

[36] C. Ouyang, B. Ni, Z. Sun, J. Zhuang, H. Xiao, X. Wang, Boosting the ORR performance of modified carbon black via C–O bonds, Chem. Sci. 10 (7) (2019) 2118–2123.

[37] N. Bhuvanendran, S. Ravichandran, M. Narayanan, B. Paulraj, S. Kumarasamy, H. Su, et al., Emerging trends in biomass-derived carbon-supported metal nanostructures as efficient electrocatalysts for critical electrochemical reactions in low temperature fuel cell applications, Biorenewable Nanocomposite Materials, Vol 1: Electrocatalysts and Energy Storage, American Chemical Society, Washington, DC, 2022, pp. 225–256.

[38] M. Hao, R. Dun, Y. Su, L. He, F. Ning, X. Zhou, et al., In situ self-doped biomass-derived porous carbon as an excellent oxygen reduction electrocatalyst for fuel cells and metal–air batteries, J. Mater. Chem. A 9 (25) (2021) 14331–14343.

[39] H.O. Panganoron, J.D.A. Pascasio, E.A. Esparcia, J.A.D. del Rosario, J.D. Ocon, Hydrothermally carbonized waste biomass as electrocatalyst support for α-mno2 in oxygen reduction reaction, Catalysts 10 (2) (2020) 177.

[40] S. Das, S. Ghosh, T. Kuila, N.C. Murmu, A. Kundu, Biomass-derived advanced carbon-based electrocatalysts for oxygen reduction reaction, Biomass 2 (3) (2022) 155–177.

[41] A. Girimonte, A. Stefani, C. Mucci, R. Giovanardi, A. Marchetti, M. Innocenti, et al., Electrochemical performance of metal-free carbon-based catalysts from different hydrothermal carbonization treatments for oxygen reduction reaction, Nanomaterials 14 (2) (2024) 173.

[42] J. He, Y. Zhao, Y. Li, Q. Yuan, Y. Wu, K. Wang, et al., Joule heating-driven sp2-C domains modulation in biomass carbon for high-performance bifunctional oxygen electrocatalysis, Nano-Micro Lett. 17 (1) (2025) 221.

[43] C. Sathiskumar, S. Ramakrishnan, M. Vinothkannan, A. Rhan Kim, S. Karthikeyan, D.J. Yoo, Nitrogen-doped porous carbon derived from biomass used as trifunctional electrocatalyst toward oxygen reduction, oxygen evolution and hydrogen evolution reactions, Nanomaterials 10 (1) (2020) 76.

[44] L. Yan, J. Yu, J. Houston, N. Flores, H. Luo, Biomass derived porous nitrogen doped carbon for electrochemical devices, Green Energy. Environ. 2 (2) (2017) 84–99.

[45] D. Cantador-Fernandez, D. Suescum-Morales, D. Esquivel, J.R. Jiménez, J.M. Fernández-Rodriguez, CO_2 adsorption by ethane periodic mesoporous organosilica at low temperatures and high pressure, J. Environ. Chem. Eng. 11 (5) (2023) 110582.

[46] W. Li, Y. Xu, G. Wang, T. Xu, K. Wang, S. Zhai, et al., Sustainable carbon-based catalyst materials derived from lignocellulosic biomass for energy storage and conversion: atomic modulation and properties improvement, Carbon Energy 7 (5) (2025) e708.

[47] H. Zhou, Y. Ren, B. Yao, Z. Li, M. Xu, L. Ma, et al., Scalable electrosynthesis of commodity chemicals from biomass by suppressing non-Faradaic transformations, Nat. Commun. 14 (1) (2023) 5621.

[48] H. Liu, W. Li, Recent advances in paired electrolysis of biomass-derived compounds toward cogeneration of value-added chemicals and fuels, Curr. Opin. Electrochem. 30 (2021) 100795.

[49] K.M. Shell Biomass-derived electrode materials and sustainable processes for supercapacitors. 2021. doi: 10.25772/TBJ1-H495.

[50] S.Y. Owusu, Machine learning prediction of the reversible capacities of a biomass-derived hard carbon anode for sodium-ion batteries, RSC Sustain 3 (7) (2025) 3133–3143.

[51] D. Zhong, R. Lin, D. Liu, X. Cai, Structure optimization of anode parallel flow field for local starvation of proton exchange membrane fuel cell, J. Power Sources 403 (2018) 1–10.

[52] T. Morresi, A. Pedrielli, S. Beccara, R. Gabbrielli, N.M. Pugno, T.S. Structural, electronic and mechanical properties of all-sp2 carbon allotropes with density lower than graphene, Carbon 159 (2020) 512–526.

[53] M. Sevilla, A.B. Fuertes, Sustainable porous carbons with a superior performance for CO_2 capture, Energy Environ. Sci. 4 (5) (2011) 1765–1771.

[54] J. Selinger, The potential of lignocellulosic materials for supercapacitors and hydrogen storage: activated carbon synthesis and cellulose separator development, Finland, 2025.

[55] K. Pandey, H.K. Jeong, Coffee waste-derived porous carbon based flexible supercapacitors, Chem. Phys. Lett. 809 (2022) 140173.

[56] A.R. Urade, R. Joshi, K. Suresh, I. Lahiri, Graphene from Sugar and Sugarcane Extract: Synthesis, Characterization, and Applications, Graphene from Natural Sources, CRC Press, London, 2022, pp. 1–28.

[57] J. Li, R. Li, T. Li, Y. Ma, Advancements in the utilization of nanocarbon sphere composites in supercapacitor, Adv. Compos. Hybrid. Mater. 8 (1) (2025) 103.

[58] W. Wei, X. Cui, W. Chen, D.G. Ivey, Manganese oxide-based materials as electrochemical supercapacitor electrodes, Chem. Soc. Rev. 40 (3) (2011) 1697–1721.

[59] J. Liu, X. Wang, J. Gao, Y. Zhang, Q. Lu, M. Liu, Hollow porous carbon spheres with hierarchical nanoarchitecture for application of the high performance supercapacitors, Electrochim. Acta 211 (2016) 183–192.

[60] F. Zhu, J. Ge, Y. Gao, S. Li, Y. Chen, J. Tu, et al., Molten salt electro-preparation of graphitic carbons, Exploration 3 (1) (2023) 20210186.

[61] M.-F. Wu, C.-H. Hsiao, C.-Y. Lee, N-H. Tai, Flexible supercapacitors prepared using the peanut-shell-based carbon, ACS Omega 5 (24) (2020) 14417–14426.

[62] M. Sevilla, R. Mokaya, Energy storage applications of activated carbons: supercapacitors and hydrogen storage, Energy Environ. Sci. 7 (4) (2014) 1250–1280.

[63] Z.S. Wu, K. Parvez, X. Feng, K. Müllen, Graphene-based in-plane micro-supercapacitors with high power and energy densities, Nat. Commun. 4 (1) (2013) 2487.

[64] E. Taer, A. Apriwandi, R. Taslim, Heteroatom-doped biomass-derived porous carbon for supercapacitor applications: a review, Trends Sci. 22 (8) (2025) 10115.

[65] D. Guo, R. Shibuya, C. Akiba, S. Saji, T. Kondo, J. Nakamura, Active sites of nitrogen-doped carbon materials for oxygen reduction reaction clarified using model catalysts, Science 351 (6271) (2016) 361–365.

[66] F. Yu, S. Li, W. Chen, T. Wu, C. Peng, Biomass-derived materials for electrochemical energy storage and conversion: overview and perspectives, Energy Environ. Mater. 2 (1) (2019) 55–67.

[67] J. Song, L. Guo, Chapter 12 – Carbon nanomaterials for metal–air batteries, in: R. Paul, V. Etacheri, Y. Wang, C.-T. Lin (Eds.), Carbon-Based Nanomaterials for Advanced Thermal and Electrochemical Energy Storage and Conversion, Elsevier, The University of Manchester (United Kingdom), 2019, pp. 311–333.

[68] J. Zhang, L. Dai, Heteroatom-doped graphitic carbon catalysts for efficient electrocatalysis of oxygen reduction reaction, Acs Catalysis 5 (12) (2015) 7244–7253.

[69] T. Qi, X. Zhang, K. Xiong, H. Yang, S. Zhang, H. Chen, Prediction of discharge performance of hard carbon materials for high-performance sodium-ion batteries based on machine learning, J. Mater. Chem. A 13 (23) (2025) 17748–17771.

[70] Z. Gao, Y. Zhang, N. Song, X. Li, Biomass-derived renewable carbon materials for electrochemical energy storage, Mater. Res. Lett. 5 (2) (2017) 69–88.

[71] Y. Wang, S. Zou, W-B. Cai, Recent advances on electro-oxidation of ethanol on Pt- and Pd-based catalysts: from reaction mechanisms to catalytic materials, Catalysts 5 (3) (2015) 1507–1534.

[72] J. Wang, J. Yang, Z. Bing, Y. Gao, T. Yang, Q. Liu, et al., Refining surface copper species on Cu/SiO_2 catalysts to boost furfural hydrogenation to furfuryl alcohol, Molecules 30 (2) (2025) 225.

[73] Z. Mukadam, S. Liu, A. Pedersen, J. Barrio, S. Fearn, S.C. Sarma, et al., Furfural electrovalorization using single-atom molecular catalysts, Energy Environ. Sci. 16 (7) (2023) 2934–2944.

[74] Y. Zhao, S. Wang, C. Li, B. Che, X. Chen, H. Chen, et al., Regulating deposition kinetics via a novel additive-assisted chemical bath deposition technology enables fabrication of 10.57%-efficiency Sb2Se3 solar cells, Energy Environ. Sci. 15 (12) (2022) 5118–5128.

CHAPTER

12

Polymer-based electrocatalysts for biomass conversion

Mehak Sharma[a], *Kuljit Kaur*[a], *Isha Arora*[a], *Anmol*[a], *Ramasamy Ganesamoorthy*[b], *Muthu Senthil Pandian*[c], and *Rajender Boddula*[d]

[a]School of Natural Sciences, GNA University, Sri Hargobindgarh, Phagwara, Punjab, India, [b]Department of Chemistry, Vinayaka Mission's Kirupananda Variyar Arts and Science College, Vinayaka Mission's Research Foundation (Deemed to be University), Salem, Tamil Nadu, India, [c]Department of Physics, SSN Research Centre, Sri Sivasubramaniya Nadar College of Engineering, Kalavakkam, Chennai, Tamil Nadu, India, [d]School of Sciences, Woxsen University, Hyderabad, Telangana, India

12.1 Introduction

Global warming and climate change are the prime factors that are affecting human life and its sustainability in the future [1]. With these crises, replacement of the conventional fossil fuels with green energy alternatives is the need of the hour. The electrochemical energy conversion process uses cheap and easily available biomass resources that enable the development of efficient catalysts to bridge the energy barrier of the electrochemical reactions [2,3]. Many electrochemical energy conversion technologies are currently in use, including water electrolysis, metal–air

Electrochemical Biomass Conversion: Sustainable Processes and Applications for Bioenergy, Biofuels, and Bioproducts
DOI: https://doi.org/10.1016/B978-0-443-26757-4.00013-1

batteries, metal-ion batteries, and fuel cells [4,5]. Besides the fact that these technologies are safer and have sustainable attributes, some challenges are being faced regarding the efficiency of these technologies. Various electrochemical reactions involved in these technologies are the reactions that are inherently nonspontaneous, and some external energy is required for such reactions to occur. Hence, the role of electrocatalytic materials comes into play. Traditional metals that come under the category of platinum-group metals include rhodium, ruthenium, palladium, iridium, osmium, and platinum. These metals are being used as the most popular and effective catalysts [6]. It is, however, their high cost, scarcity, and sensitivity to Earth's geography that do not promote their use in the new energy sector. To overcome this problem, different research groups shifted their focus toward PGM –free electrocatalysts. In this regard, carbon-based materials are found to offer an effective alternative to support a variety of applications [7]. Currently, commercially available carbon materials are obtained from petrochemical products that require sophisticated synthesis equipment [8]. Since these petrochemical resources are expensive and not available in abundance, their large-scale commercialization has not been possible yet.

Biomass is a low-cost, renewable material that primarily consists of proteins (e.g., hemoglobin, collagen, etc.) and polysaccharides like chitin and cellulose, which can be effectively used to synthesize carbon materials that can act as electrocatalysts [9]. Biomass serves as an important source of synthesizing carbon materials and has many advantages, like preservation of the characteristic features of the structural morphology of the native material and the porous morphology of the carbon carriers [10]. Biomass-derived carbon compounds are highly popular for their role as catalysts, owing to their abundance and low cost, renewability, and porous morphologies. This biomass further needs to be processed to convert it into a valuable biofuel. The process of conversion of biomass into biofuel and biochemical requires valorization of the biomass, and in that context, electrolytic valorization of the biomass is very crucial. The process is made to proceed using the renewable sources of energy, such as solar and wind energy, promoting green chemistry. Many scientific groups are working toward advancing the electro-catalytic biomass conversion process that includes electrochemical hydrogenation and electrooxidation [11]. More recent studies have witnessed the advances in organic polymer-based materials to serve as electrocatalysts. These materials form an interesting category of materials because of their pyrolysis-free synthesis, a more tunable molecular composition, adjustable electrical conductivity, inexpensiveness, and stability. Biomass-derived carbon materials can also serve as polymer-based electrocatalysts [10,12]. Developing such materials as the active components and their efficiency enhancement is the need of the hour, as well as a global interest.

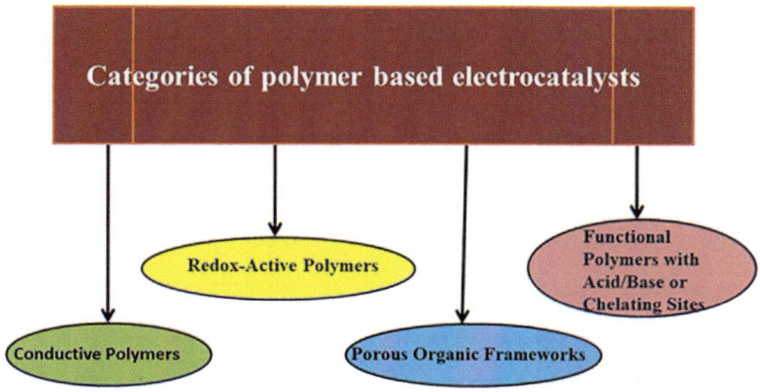

FIGURE 12.1 Categories of polymer-based electrocatalysts.

12.2 Categories of polymer-based electrocatalysts

Over the past several years, polymer-based electrocatalysts have gained popularity as sustainable and diverse alternatives to inorganic catalysts for energy transformation and ecological applications. The advantage of polymer-based electrocatalysts is the tunability of polymer chemistry, allowing researchers to customize conductivity, porosity, redox activity, and functional groups. Additionally, these can offer a combination of conductivity with structural adaptability, chemical stability, and mechanical flexibility. Polymer-based electrocatalysts can be comprehensively classified (Fig. 12.1) into four primary types, which are conductive polymers, redox-active polymers that catalyze intrinsically, porous polymers, and functional polymers that are acidic, basic, or chelating.

12.2.1 Conductive polymers (e.g., polyaniline, polypyrrole, and poly(3,4-ethylenedioxythiophene))

The discovery of conducting polymers introduced a new dimension to the global research market. These materials provide a wide-ranging applications in biomedical, sensors, photovoltaics, Light emitting diodes (LEDs), transducers, supercapacitors, and superconductors, etc. Conductive polymers are organic materials that have the ability to conduct electricity due to the extent of their π-conjugation across the polymer backbone. These materials are lightweight, environmentally friendly, and stable, and have exceptional electrical, optical, and thermal properties. Due to these attractive properties, conductive polymers have the tendency to replace metal components in electronic instruments. The electrical conductivity of these polymers depends primarily on the synthesis route chosen and the reaction conditions. Common examples of conductive polymers

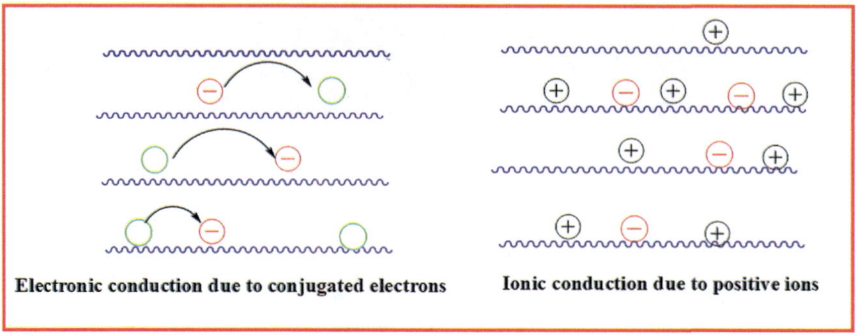

FIGURE 12.2 Mechanism of conductivity and catalysis.

are polyacetylene, polyaniline (PANI), polypyrrole (PPy), polyparaphenylene, polythiophene, and PEDOT. Polyacetylene is the first electroconductive polymer that has $(C_2H_2)n$ as the repeating unit. It is also represented as $(CH)_x$. The most well-studied conductive polymers utilized for electrocatalysis are PPy, PANI, and poly(3,4-ethylenedioxythiophene) (PEDOT) due to their environmental stability, redox-reversible processes, and relative ease of synthesis through chemical or electrochemical means.

Conductive polymers conduct by way of delocalized electrons (Fig. 12.2) moving along the polymer backbone, and electrons are often delocalized because of incorporating acid or counterions to dope the conductive polymer, thereby creating polarons and bipolarons—quasiparticles that are responsible for charge transport in the oxidized state of the polymer [13]. When utilized for electrocatalysis, conductive polymers can function in two ways: as an electron conductor as well as an active catalytic matrix, which is especially useful for conducting redox reactions like electrochemical sensing. the oxygen reduction reaction (ORR), hydrogen evolution reaction (HER).

Applications and examples

Polyaniline (PANI): PANI exhibits pH-unit sensitive redox behavior and serves various uses in the fields of supercapacitors and sensors (Fig. 12.3). PANI in the emeraldine salt form has been found to exhibit conductivity levels in excess of 10^0 S/cm. PANI also offers exciting possibilities in electrocatalysis, where it functions as a support for metal nanoparticles, which benefits the metal by supporting electron transfer and transport, and in some cases, it can even improve the catalytic performance [14] (Table 12.1).

Polypyrrole (PPy): PPy Fig. 12.2 offers a more favorable electropolymerization and higher mechanical flexibility when compared with PANI. PPy is particularly advantageous for use in the electrooxidation of alcohols, and it has developed suitable applications as a support in the

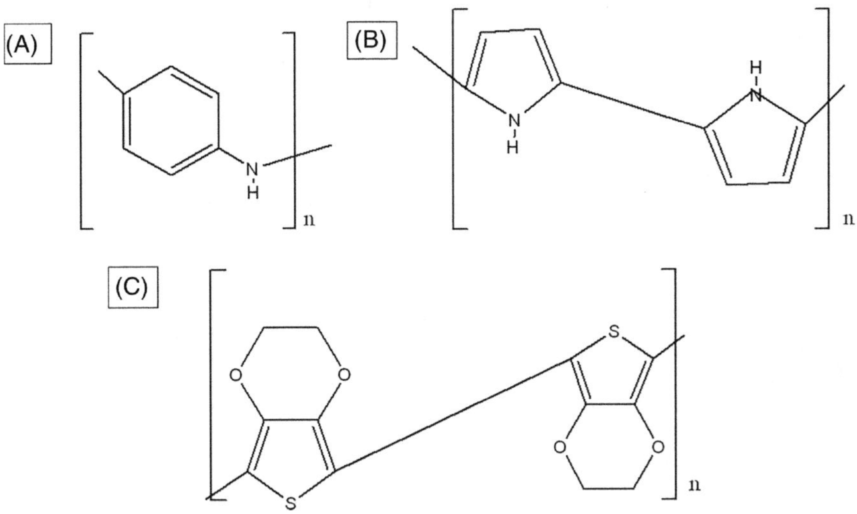

FIGURE 12.3 Structural formulae of (A) PANI, (B) PPy, and (C) PEDOT.

TABLE 12.1 Comparison table for conducting polymers.

S. No	Polymer	Catalyst Type	Potential	FE (Faradiac Efficiency)	Conversion Efficiency	References
1.	PANI	Film on carbon paper	1.96 V	-	76%	[15]
2.	PEDOT	Electrodeposited PEDOT doped with ClO_4^-, PSS^- or Aquivion	≈1.2–1.4 V	≈80 %	>90%	[16]
3.	PPy	Conductive support for NiO	1.45– 1.65 V	60%–90 %	70%– 92%	[17,18]

electrooxidation of alcohols with metal catalysts utilized in fuel cells, such as Pt, Pd, and Au [19].

PEDOT: PEDOT (Fig. 12.3) is noted for high conductivity (up to 1000 S/cm when doped), transparency, and greater long-term environmental stability. PEDOT composites have been documented for use in electrochemical water splitting, supercapacitors, and the creation of biosensors as a result of their great ability to transport charge [20].

12.2.2 Redox-active polymers with intrinsic catalytic action

Redox-active polymers differ from conductive polymers, which are simply a means of conduction of electrons. They have some sort of catalytic units (such as transition metal complexes, quinones, or nitrosyl radicals) incorporated into, or pendant from the polymer backbone, that participate

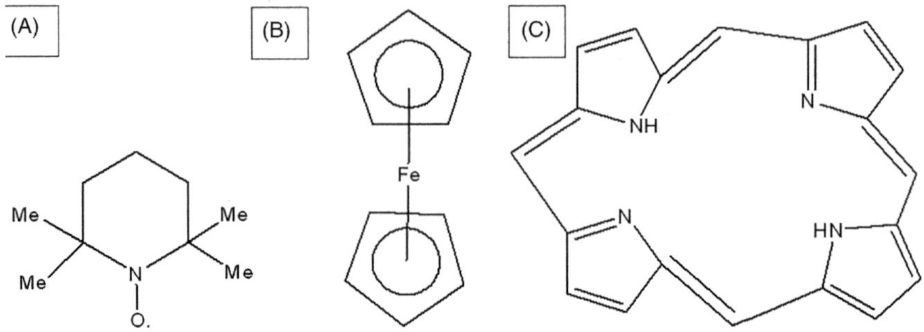

FIGURE 12.4 Structural formulas of (A) TEMPO radical, (B) ferrocene, and (C) porphyrin.

in a redox reaction and therefore involve either the transfer of electrons or protons in some sort of electrocatalytic process.

Structural characteristics and mechanism

Many redox-active polymers are chemically synthesized by attaching organometallic complexes or redox-active organic functional groups to polymer chains. Examples of such polymers include those containing TEMPO (2, 2, 6, 6-tetramethylpiperidine-1-oxyl) radicals, ferrocene, or metal porphyrins. The reaction mechanisms typically involve electron shuttling from the active sites to the electrode through the polymer matrix [21].

Applications

TEMPO-based polymers: There is a broad application of TEMPO-based polymers (Fig. 12.4A), which have been largely applied for the electrochemical oxidation of alcohols and biomass-derived polyols [22].

Metallopolymers: Redox-active polymers containing Co, Fe, Ru, or Mn (Fig. 12.4B and C) complexes are active in reactions like CO_2 reduction, O_2 evolution, and nitrate reduction. For example, cobalt-based poly(4-vinylpyridine) complexes have shown stunning performance for electrochemical hydrogen production [23,24].

Organic redox catalysts: Quinone-functionalized polymers exhibit significant potential for oxygen reduction and bioelectrocatalysis owing to redox processes being analogous to natural enzyme systems [25].

12.2.3 Porous organic frameworks

Porous organic frameworks, which include the classes of covalent organic frameworks (COFs) and porous organic polymers (POPs), represent a new and rapidly evolving class of polymer electrocatalysts. As with inorganic frameworks such as Metal-Organic frameworks (MOFs), POFs provide high surface area and porosity along with the mechanical stability and processability of organic polymers.

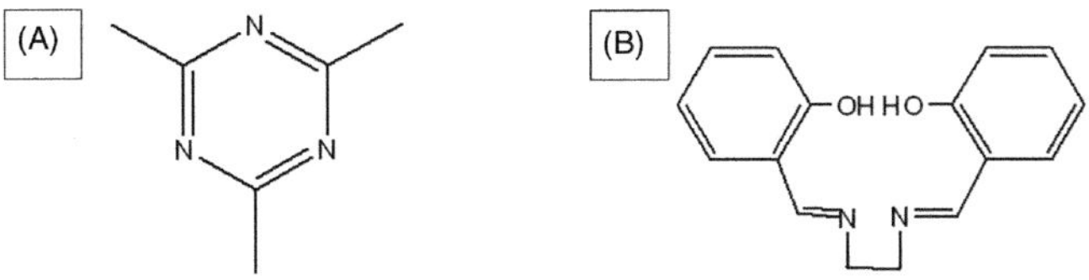

FIGURE 12.5 Structural formulae of (A) triazine (B) salen ligand.

Structural properties

COFs: COFs are crystalline, highly ordered polymers produced by covalent bonds between organic building blocks, and pore size, functionality, and crystallinity can be easily controlled [26].

POPs: POPs are sometimes amorphous, but still have very high levels of porosity and thermal/chemical resistance. They are produced using polymerization methods such as Friedel–Crafts, Sonogashira coupling, or Schiff base condensation [27].

Both COFs and POPs can have electroactive units such as triazine or porphyrin, or metal-binding ligands such as bipyridine or salen (Fig. 12.5).

Electrocatalytic applications

CO_2 reduction: Ni-porphyrin COFs have shown selective conversion of CO_2 to CO in electrochemical conditions [28].

OER and HER: Fe- or cofunctionalized COFs exhibit promising activity and durability for water splitting, and may offer enhanced kinetics of oxygen evolution reaction (OER).

Energy storage: Several POPs have been utilized as active layers in batteries and supercapacitors because they have the ability to host several redox-active sites or cations/anions of metals [29].

12.2.4 Functional polymers with acid/base or chelating sites

A secondary class of polymeric electrocatalysts includes functional polymers bearing acidic, basic, and chelating moieties, which may contribute to catalysis via chemical coordination, proton shuttling, or microenvironment modification. These polymers are often found acting as solid electrolytes, proton conductors, or simply as hosts to immobilize metal ions.

Examples of functionalization

Functionalization of polymers sulfonated polystyrene (SPS), poly(ethyleneimine) (PEI), poly(acrylic acid) (PAA) occurs with (Fig. 12.6):

Acidic moieties (e.g., $-SO_3H$, –COOH) provide proton conduction, useful for HER and OER.

Basic moieties (e.g., $-NH_2$, imidazole) provide for improved adsorption and activation of CO_2.

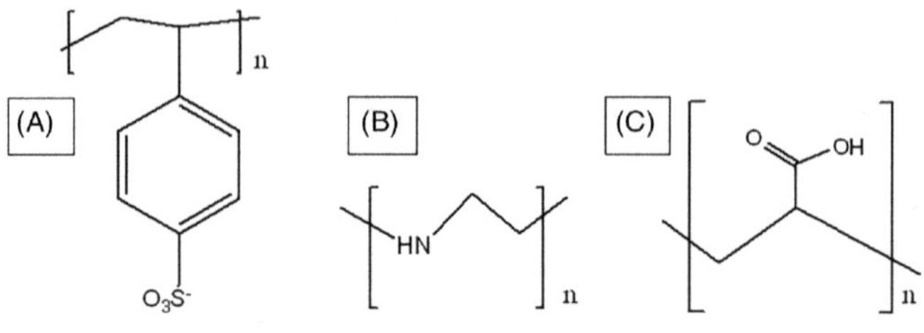

FIGURE 12.6 Structural formulas of (A) sulphonated polystyrene, (B) PEI, and (C) PAA.

Chelating ligands (e.g., bipyridine, EDTA-type structures) provide stabilization for the transition metal center for redox reactions.

Applications

Proton exchange membranes (PEMs): Sulfonated polymers are appropriate for PEM fuel cells and electrolyzers for proton conduction at low cost [30].

Metal complex immobilization: PEI or PAA matrices can chelate metals (Fe, Cu, or Mn) and provide individual electro reduction of NO_3^-, CO_2, or O_2 [31,32].

Electrochemical sensors: Functional polymers can selectivity and stability of sensing electrodes used to monitor the environment and biosensing [33].

12.2.5 Polymer degradation, dopant leaching, and recyclability

While conducting polymers have exhibited high catalytic activity and tunable electronic properties, they continue to be plagued by a lack of chemical stability and recyclability in the context of electrochemical biomass conversion.

Degradation of the polymer backbone: Repeated redox cycling and exposure to oxidizing potentials can lead to irreversible structural degradation of the conjugated backbone. For example, PANI and PPy can be overoxidized at high anodic potentials, generating quinone-type or ring-opened species that reduce charge conductivity and accessibility of active sites. Although PEDOT is more robust to degradation, ether cleavage and oxidation of sulfur can still occur under extended bias. Degradation reduces catalytic lifetime but also has consequences on charge transport uniformity and product selectivity [34,35].

Dopant leaching: The electrochemical environment can help drive the migration or leaching of mobile dopants (ClO_4^-, PSS^-, SO_4^{2-}, BF_4^-), particularly when solvents such as water and alkaline electrolytes are used. When dopants are lost, polymers experience changes in oxidation state, conductivity, and morphology. Selection of the counter ion, such as PSS(bulky sulfonates) or Aquivion, and crosslinking of polymers have been shown

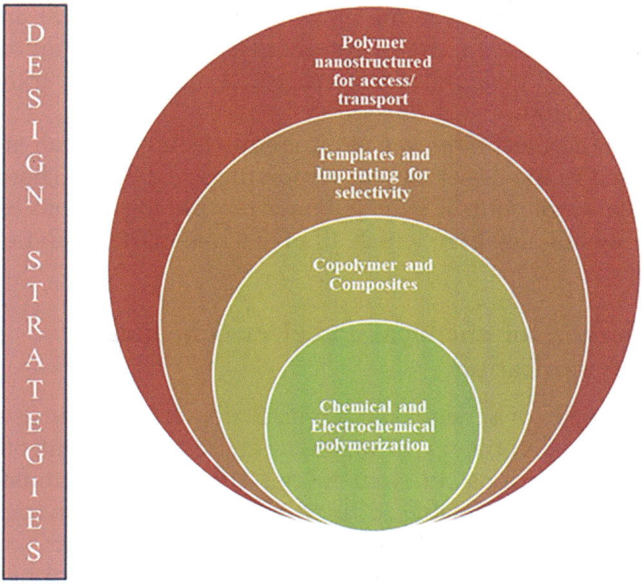

FIGURE 12.7 Strategy for design and synthesis of a catalyst.

to enhance dopant retention and mechanical stability during long-term electrolysis [36,37].

Ability to be recycled and regenerated: The reusability of conducting polymer electrodes also suffers from these degradation mechanisms. Strategies for improving recyclability include: (1) utilizing self-healing dynamic bonds (imine, disulfide, and supramolecular hydrogen bonds) that will allow for partial recovery of the structure. (2) designing covalently attached redox mediators (TEMPO units on EDOT backbones) to reduce mediator leaching from the electrode. (3) employing redeposition or electrochemical redoping, which would potentially restore conductivity and surface activity following each catalytic cycle [38,39].

12.3 Strategies for the design and synthesis of a catalyst

Researching the design and synthesis of catalysts that are efficient, selective, and robust is an important aspect of modern chemistry with applications ranging from energy conversion to remediation of environmental toxins to chemical production. Due to the versatile structural behavior, simple functionalization, and modulation of the physical properties of polymers, polymer-based catalysts are of particular interest. The strategic design and synthesis of polymer-based catalysts (Fig. 12.7) rely on several foundational approaches—some of which include chemical and electrochemical polymerization, use of copolymers and composites, templating for selectivity, and design of nanostructured materials for transport and access to active sites.

Polymerization techniques, chemical and electrochemical, are primary methods used in the development of polymeric catalysts. These methods provide significant control over polymer morphology, chain length, and degree of functionalization, which all play a role in the catalytic activity.

12.3.1 Method for chemical and electrochemical polymerization

The mechanism of a chemical polymerization involves the use of radical initiators, cationic initiators, or anionic initiators. With controlled/living polymerization, various different modes such as atom transfer radical polymerization and reversible addition-fragmentation chain transfer polymerization can produce polymers with definite molecular weights and structure [40]. These methods allow for various catalytic functionalities, such as metal-binding ligands or acid/base sites, to be incorporated through the polymer backbone or side chains.

Electrochemical polymerization has additional benefits, especially in regard to the formation of conductive polymers, such as PANI, PPy, and polythiophene. Accordingly, conductive polymers have historically been used in electrocatalysis applications, primarily due to their redox properties and electrical conductivity. In situ polymerization on the electrode surface also generates thin film deposition, which is easy to control in terms of thickness, morphology, and porosity, which is favorable for sensor and energy device applications [41].

12.3.2 Copolymer and composites

The incorporation of polymers and polymer composites increases the functional diversity and structural stability of catalysts. Copolymerization provides the opportunity to combine different monomeric units within the same polymer chain, creating domains that have different physical or chemical properties; for example, block copolymers can undergo microphase separation with one domain exhibiting the catalytic groups independently, and the other actively hosting transport elements. Amphiphilic block copolymers contain both hydrophilic and hydrophobic components, such as poly(styrene)-block-poly(acrylic acid), that can self-assemble into either vesicles or micelles in aqueous media. These nanostructures are ideal for hosting catalytic species within their hydrophobic cores or hydrophilic surfaces, allowing for improved substrate access and product departure [42].

Composites, on the other hand, consist of polymers with inorganic components such as metal oxides, carbon nanotubes, or graphene that improve stability/conductivity or introduce mechanical properties. An example includes poly(ionic liquid)-based composites containing graphene oxide that provide improved CO_2 reduction catalysis due to their ionic conductivity and large surface area [43,44].

Recent studies showed considerable advancements in the use of conductive polymers, particularly PEDOT and its derivatives, alongside nickel–iron (NiFe) layered double hydroxides (LDHs), for improved electrocatalytic performance. NiFe/PEDOT composites showed improved conductivity, charge-transfer efficiency, and electrochemical stability compared with bare NiFe electrocatalysts. For example, NiFe-LDH modified with poly(3-thiophenemalonic acid) had improved OER activity due to strong electronic coupling between polymer and NiFe active sites to enhance electron mobility while limiting degradation during long-term operation. Another study continued to explore strategies with conductive polymers with PEDOT: PSS cradles on NiFe-LDH showed improved performance in alkaline and seawater electrolysis applications. The polymer treatments created an electrostatic barrier to repel chloride ions to enhance corrosion resistance and maintain catalysts for longer operation [45–47].

Aside from NiFe systems, polymer-based bifunctional catalysts have emerged as multifunctional platforms for both the OER and the ORR that are critical for metal–air batteries and overall water splitting. Conductive polymers such as PEDOT, PANI, and PPy offer high electronic conductivity and provide structural flexibility that stabilizes transition-metal sites. Recent reviews have noted that polymer matrices can restrict one or two metal centers, modify their electronic structure, and prevent agglomeration or leaching during catalytic cycles [48]. For example, polymer-derived carbon frameworks with Fe and Co species produce promising bifunctional performance and high stability in alkaline solutions.

The synergistic behaviors of polymer composites are largely derived from the unique attributes of both components, which present complementary properties. Polymers can provide processability and flexibility, while the inorganic fillers attribute rigidity, thermal stability, and catalytic activity. This is especially relevant in terms of heterogeneous catalysis, electrocatalytic and electrochemical energy sources, such as supercapacitors and fuel cells.

12.3.3 Templates and imprinting for selectivity

A significant challenge in catalyst design is achieving high selectivity, particularly in complex chemical environments. Molecular imprinting and templating strategies are viable methods to design polymeric high-selectivity catalysts for substrate recognition.

A template molecule is required for the synthesis of molecularly imprinted polymers (MIPs), around which functional monomers and crosslinking monomers are polymerized. After the template is removed, the polymer contains cavities complementary in shape, size, and chemical functionality to the template, to selectively bind or catalyze [49].

This templating approach is particularly useful for the design of enzyme-mimetic catalysts. In this case, the transition state or intermediate can be imprinted, and MIPs can be designed to stabilize the transition state or intermediate to lower the activation energy and subsequently raise the reaction rate. Recent reports have also confirmed the utility of imprinted polymers for the asymmetric catalysis of prochiral substrates, and the method has potential for enantioselective applications [50].

Polymers are templated using hard or soft templating to manipulate the morphology and porosity of catalytic structures. Silica or anodized alumina can serve as a hard template in order to produce polymers with mesoporous geometries, which are removed by etching after polymerization. Surfactants or block copolymer micelles can serve as soft templates, facilitating the formation of nanostructured polymers with high surface areas and tunable pore volumes [51]. Control over polymer structure is critical for limiting mass transport limitations, as well as ensuring catalytic active sites are accessible, and will dictate whether catalytic efficiencies are high.

12.3.4 Polymer nanostructured for access/transport

A nano-structured polymer is a valuable approach for overcoming diffusion limitations and improving catalytic behavior. Specifically, engineering polymers on the nanoscale improves accessibility to active sites, increases surface area, and controls the local microenvironments.

Many different forms of nanostructured polymers have been studied in regard to their catalytic application, including nanofibers, nanotubes, and hollow spheres. For instance, polymer fibers created from electrospinning offer high surface-to-volume ratios with dynamic pore interconnected pore structures, which promote diffusion of reactants and discharge products [52,53].

Core-shell structures are another example of extremely efficient catalytic systems that can be designed so that the core is catalytically active and represents the catalytic core, while the outer shell provides additional stabilization/protection via a polymeric, permeable shell. This design permits the selective access of desired reactants while still blocking poisons/inhibitors. For example, core–shell architectures of a catalytically active metal nanoparticle with a cross-linked polymer shell have been shown to provide excellent rates of productivity and recyclability for hydrogenation reactions [54].

TABLE 12.2 Comparison table of strategies for the design and synthesis of the catalyst.

Sr.no	Strategy	Feature	Advantage	Application
1.	Chemical and electrochemical polymerization	A process where polymer chain growth can be controlled. Polymers that are electroactive or redox-active; Films (the surface that a polymer is deposited on)	Control over polymer architecture; Control over thickness and morphology; Potential for being directly integrated with electrodes	Catalyst for electrosynthesis, sensors, energy storage, or energy harvesting
2.	Copolymers and composites	Mixed monomer unit; Hybridization with inorganic components; Amphiphilic or functional groups	Functional diversity increase mechanical and thermal stability; Greater catalyst synergy—	Utilize hybrid catalytic systems, environmental remediation, and fuel cells
3.	Templates and imprinting for selectivity	Molecular imprinting; Hard/soft templating; Shape/size selectivity;	High substrate specificity; Stabilization of the transition state; Enzyme-mimic behavior	Applicable in biosensing, selective catalysis, and molecular recognition
4.	Polymer; Nanostructure for access/transport	Nanofibers, nanotubes, and mesoporous structures; Core-shell structures; High surface area designs	Enhanced diffusion of reactants; More active sites are available for the reaction; Responsive control	Very suited for high-efficiency heterogeneous catalysis and in multistep reactions

Stimuli-responsive nanostructures, for example, temperature- or pH-sensitive polymers, can also allow for pseudo-dynamic control of catalyst access by adjusting the material properties [55]. Smart materials can change their conformation in response to an external stimulus, thereby opening or closing access to catalytic sites as required.

The availability of porous polymer networks such as COFs and hypercrosslinked polymers (HCPs) adds yet another layer to catalyst development. The porosity, well-defined pores, and tunable functional groups/chemistry provided from porous polymer networks are excellent characteristics for heterogeneous catalysis, primarily their applications in gas-phase reactions and interfacial catalysis such as photocatalysis [56]. All the strategies for the design and synthesis are compared in Table 12.2.

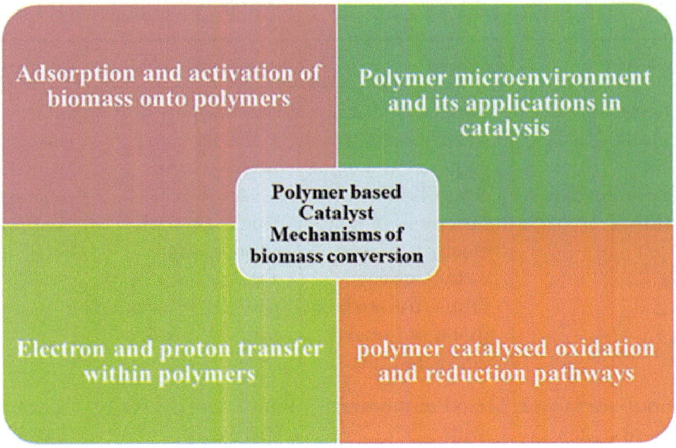

FIGURE 12.8 Types of polymer-based catalyst mechanisms of biomass conversion.

12.4 Mechanisms of biomass conversion by polymer-based catalyst

Converting biomass to value-added chemicals and fuels is an important approach to making advancing a sustainable, circular economy. Biomass is a highly oxygenated, heterogeneous, and very recalcitrant feedstocks that need multifunctional catalysts that can facilitate operating steps such as hydrolysis, oxidation, hydrogenation, dehydration, and reforming. Polymer-based catalysts have entered the stage as suitable candidates due to their tunable microenvironments, tailor-made architectures, and ability to host multiple catalytic sites. Usually, these polymers serve as heterogeneous catalysts or supports to substrate recognizing, activating, and facilitating reactions. Different categories are explained in Fig. 12.8.

12.4.1 Adsorption and activation of biomass onto polymers

In biomass conversions, an optimal reaction yield begins with the adsorption and activation of biomass or biomass molecules on absorption surfaces. Polymer-based catalysts can be synthesized with certain functional groups (e.g., sulfonic acid, amines, carboxylates, and phosphonic acids) that react with biomass-derived substrates (e.g., glucose, cellulose, or lignin fragments), hydroxyl, carboxylic, and ether donation functionalities.

Hydrophilic polymers provided easier hydrogen bonding with hydroxyl-rich components of the biomass, such as poly(acrylic acid) and poly(vinyl alcohol), which aids in capturing and correctly orienting the substrate for reaction [57]. In addition to the Brønsted acid sites (e.g., sulfonated), which can activate the glycosidic linkages in cellulose and hemicellulose for hydrolysis of the substrate to yield sugars [58].

In the case of hydrophobic lignin-derived fractions, polymeric catalysis with added aromatic or hydrophobic domains can aid in adsorption through π–π stacking or van der Waals forces. Molecular imprinting or polymeric imprinting can further the process so that polymerization occurs around the biomass fractions, providing even more recognition/discrimination spaces for the biomass and selectivity during complex feedstocks [59].

12.4.2 Electron and proton transfer within polymers

Polymer-based catalysts often perform the very important function of mediating processes involving electron and proton transfers, which are key to redox transformations (reductions and oxidations), such as oxidizing polyols or reducing furfural and levulinic acid. In many of these processes, polymers are capable of incorporating redox-active centers, such as transition metal complexes, organic radicals, or conjugated backbones, which can shuttle electrons during catalysis.

Conducting polymers (such as PANI and PPy) are particularly good for transporting electrons, owing to their delocalized π- electron systems, and have been employed in electrochemical biomass conversion, in which electron transfer from an electrode is mediated through the polymer to oxidize or reduce biomass-derived molecules [60].

In acid- or base-catalyzed reactions, proton transfer is the predominant mechanism. In these cases, sulfonated polymers (e.g., SPS-divinylbenzene resins) incorporate the accessible Brønsted acid sites to allow for efficient donation of protons to initiate dehydration or isomerization steps [61]. The density or accessibility of these sites ensures the polymer will maintain effective catalytic performance.

In addition, poly(ionic liquids) (PILs) leverage ionic conductivity with a tunable chemical environment that favors concurrent proton transfer and electron transfer. The high ionic mobility of PILs and ability to tailor the cation/anion combinations make them highly effective materials for promoting electrochemical oxidation of biomass-derived substrates, including 5-hydroxymethylfurfural (HMF) to 2,5-furandicarboxylic acid (FDCA) [62].

12.4.3 Polymer microenvironment and its applications in catalysis

The microenvironment of polymers—the local polarity, hydrophilicity, charge distribution, pore dimensions, and steric arrangement—significantly affects reaction kinetics, intermediate stability, and product selectivity. For example, hydrophilicity can solubilize and efficiently process aqueous biomass feedstocks for biorefining, while hydrophobicity domains can stabilize transition states with nonpolar reactive intermediates. Most polymers have both hydrophilic and hydrophobic domains, which can spatially compartmentalize discrete reaction steps, allowing for organization comparable to an enzyme [63].

Microporous and mesoporous polymers can also provide confined spaces to enhance catalysis through stabilizing reactive intermediates with a confinement effect that lowers the activation barrier [64]. COFs and HCPs are platforms with well-defined porous structures and functional surfaces shown to selectively catalyze sugar derivative and lignin fragment transformations [65]. Also ionic microenvironments, as with PILs or sulfonated polymers, can affect local pH and modulate acid/base equilibria that are fundamental to dehydration and hydrolysis reactions. Ionic fields can stabilize charged intermediates, as in the conversion of glucose to HMF in acidic polymer gels [66].

12.4.4 Comparison of polymer-catalyzed oxidation and reduction pathways

Biomass conversions generally involve oxidative and reductive transformations, and polymer catalysts can be tailored to facilitate either reaction or both, depending on polymer functionalization and reaction conditions.

Oxidation pathways:

The oxidation reactions, such as converting HMF to FDCA or glycerol to glyceric acid, thermodynamically require an electron acceptor, which typically means aerobic or electrochemical conditions. Functionalized polymer-supported metal complexes, such as poly(acrylamide)-functionalized TEMPO or poly(4-vinylpyridine)-chelated Pd or Ru, have been shown to be effective for these oxidation reactions [67].

Conductive polymers, such as PPy, can shunt electrons from the biomass substrate to the electrode for electrocatalytic oxidation. When metal nanoparticles (Au, Pt) also occupy the polymer space, it typically enhances catalytic performance due to some type of synergistic effect [68].

Reduction process:

Reduction reactions, for example, the hydrogenation of furfural to furfuryl alcohol or levulinic acid to γ-valerolactone, will require hydrogen donors or electrons sourced from the working electrode. Polymer-based hydrogenation catalysts include porous polymers with metal, for example, Pd or other metal alloys, and redox-active COFs that can activate H_2 or formic acid for using them as reducing agents.

Pd nanoparticles in a polymer micelle have been shown to selectively hydrogenate furfural to furfuryl alcohol, with the polymer stabilizing the nanoparticle and preventing agglomeration [69]. In electrochemical reductions, polymer-modified electrodes have been used to facilitate the PCET steps.

Oxidation generally requires an aerobic or oxidative potential, and reductions, a protic environment, and reducing equivalents. The design of polymer oxidation catalysts focuses on stability when subjected to oxidative stress and electronic conductive properties, while reduction catalysts can benefit from using proton conductivity and electron donor supplies.

Hybrid systems that combine both capabilities, for example, polymers with both acidic and redox-active sites, can catalyze cascade reactions and give rise to overall transformations, which biology performs so effortlessly, allowing for one-pot transformations of biomass-derived substrates [70]. All the mechanisms are compared in Table 12.3.

12.5 Electrochemical pathways in biomass valorization

Electrochemical biomass valorization is a fast-growing area that uses renewable electricity to drive chemical transformations, enabling biomass-derived feedstocks to be converted into high-value chemicals and fuels under milder and greener conditions. As opposed to traditional thermochemical processes that demand heat, pressure, and corrosive reagents, electrochemical processes operate at or near room temperature, with the ability to link renewable energy sources and follow the principles of green chemistry and sustainable development [71,72]. More broadly, valorization of biomass via electrochemistry can be grouped into oxidative, reductive, and paired electrolysis pathways.

12.5.1 Electrochemical oxidation pathways

Electrochemical oxidation is a popular means to upgrade biomass-derived molecules such as glycerol, HMF, and lignin-derived phenolic into value-added chemicals. For example, HMF oxidation to FDCA has been a focus of recent studies, as FDCA can serve as a renewable monomer for bio-based plastics [73]. In these cases, polymer-based catalysts-such as conducting polymers (PANI, PPy) or redox-active polymers with metal complexes-were utilized to facilitate electron transfer, stabilize reaction intermediates, and improve selectivity. Furthermore, oxidation was promoted with anodic potentials while in aqueous or biphasic media, which enabled selective C–H and C–OH bond activation, while avoiding overoxidation to CO_2 [72]. The most abundant and renewable feedstock used is lignocellulosic biomass, comprising cellulose, hemicellulose, and lignin,

TABLE 12.3 Comparison of mechanisms in polymer-based catalysts for biomass conversion.

Sr.no	Mechanism	Feature	Advantages	Challenges	Relevance to Biomass conversion
1.	Adsorption and activation of biomass	Functional groups ($-SO_3H$, $-NH_2$, $-COOH$); Hydrogen bonding, $\pi-\pi$ stacking, and van der Waals forces Molecular imprinting	Improved substrate-catalyst interaction; Improved selectivity in complicated feedstocks; Preorientation of the substrate for reaction	Limited selectivity in mixed biomass; Deactivation by fouling or nonspecific adsorption	Essential for the initial recognition of the substrate and directing the pathway of conversion
2.	Electron and proton transfer	Redox-active surfaces (e.g., TEMPO, transition metals); Conductors (e.g., polyaniline) Ionic associates (e.g., PILs)	Supports redox reactions (oxidation/reduction); Facilitates electrochemical conversions; Mediates dual electron/proton pathways	Electron transfer efficiency is affected by the conductivity and shape of the polymer; Stability under electrochemical challenges	Core to oxidation (e.g., glucose to HMF) and reduction (e.g., furfural to alcohols) reactions
3.	Polymer microenvironment	Adjustable porosity and polarity; Amphiphilic domains; Confined catalytic spaces; Local pH modulation	Mimics enzyme-like catalysis; Increases stability of intermediates; Provides site isolation for tandem/cascade reactions	Balancing hydrophilic/hydrophobic domains; Maintaining pore accessibility in viscous biomass solutions	Influencing rate, selectivity, and yield of transformations via local microenvironment conditions modulation
4.	Oxidation vs reduction pathways	Oxidation: TEMPO, metal ions, aerobic conditions; Reduction: Pd, Ru nanoparticles, H_2 donors; Cascade/dual-functionality possible	Customizable for desired transformation; Fusion of dual-function pathways; Reduces the need for external reagents	Oxidative degradation of polymer; Relevant to maintain the appropriate redox potential within one system	Central to valorizing biomass into chemicals/fuels (e.g., oxidation of sugars or reduction of platform molecules like levulinic acid)

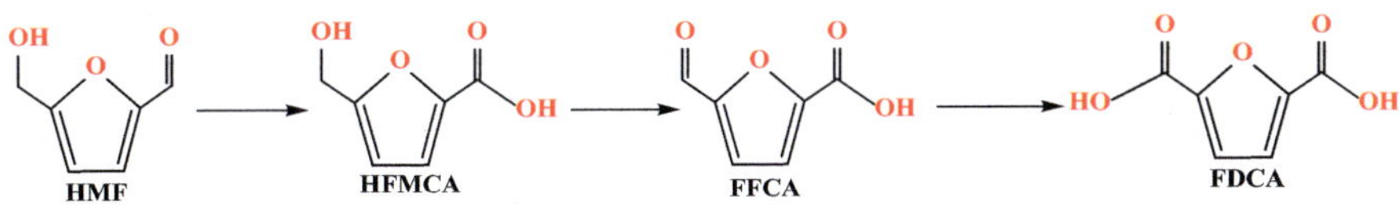

SCHEME 12.1 Chemical conversion of FDCA using TiO_2 supported gold catalyst.

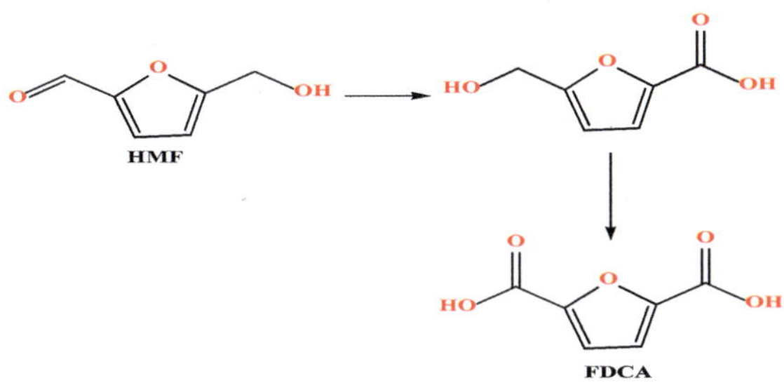

SCHEME 12.2 Chemical conversion of FDCA using Merrifield resin and cobalt porphyrin complex.

used for promising fuel production. The U.S. Department of Energy has released the most capable bio-derived platform molecules, such as HMF, furfural (FF) [3], levulinic acid, sorbitol, and FDCA [74]. The electrolytic conversion of some of the molecules has been explained later:

12.5.1.1 *Electrocatalytic oxidation of 5-hydroxymethylfurfural*

HMF is prepared via the dehydration of C_6 carbohydrates, whereas FDCA is the counterpart to terephthalic acid, which can easily replace PET [75]. HMF is considered the most promising precursor for FDCA production. FDCA can be produced via two methods, chemical conversion and bioconversion. Chemical conversion includes High-pressure air oxidation of HMF with $Co(OAc)_2$ and $Mn(OAc)_2$, oxidation by pure oxygen or by using oxidizing agents such as hydrogen peroxide and potassium permanganate, using metal catalysts like platinum (Pt), gold (Au), palladium (Pd), etc. Oxidation can also be carried out using metal-free nano-porous graphite carbon catalysts [76]. Albonetti et al [77] performed the oxidation of HMF via a TiO_2-supported gold catalyst for conversion to FDCA (Scheme 12.1). The intermediates, such as 5-hydroxymethyl-2-furancarboxylic acid and 5-formyl-2-furancarboxylic acid, are formed during the oxidation process. The reaction pathway is illustrated in Scheme 12.1.

Gao et al. [78] reported the conversion of HMF to FDCA using Merrifield resin and a cobalt porphyrin complex. A total of 95.6% of HMF was successfully converted to FDCA (Scheme 12.2) with an obtained yield of 90.4%.

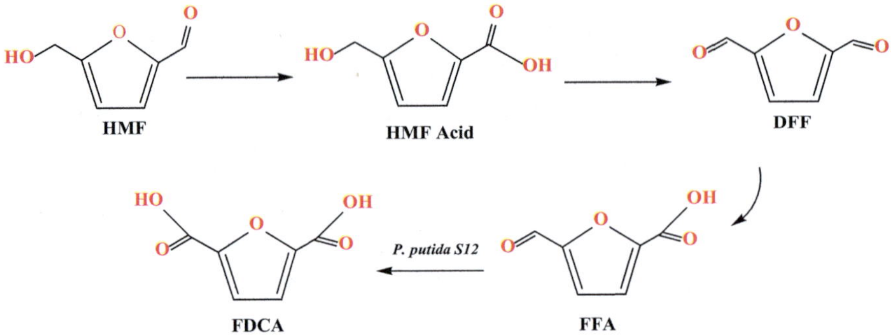

SCHEME 12.3 Bioconversion of FDCA using *P. putida S12* enzyme.

The proposed catalytic route for the conversion is shown in Scheme 12.2.

Bioconversion of HMF to FDCA is achieved by modifying microorganism strains such as *Cupriavidus basilensis, Methylovorus sp. strain MP688, C. basilensis, Synechococcus elongatus, Pseudomonas putida S12,* and *Escherichia coli.*[79]. Hsu et al. [80] employed genetically engineered *P. putida S12* as a biocatalyst to achieve the conversion. The whole-cell microorganism produced 545 mM of FDCA after 72 h. The following scheme of conversion of HMF to FDCA follows three successive steps, and generates different intermediates in between, where in the first step, HMF undergoes oxidation to form the corresponding aldehyde to generate diformylfuran, then the compound formed undergoes oxidation by an enzyme, converting it to formylfuroic acid (Scheme 12.3), and finally it is oxidized to form FDCA [80].

Sheng et al. [81] converted HMF to FDCA using *Acinetobacter calcoaceticus NL14.* A concentration of 5 g/L of HMF was converted 100% to FDCA using the enzyme within 48 h. Similarly, Cunha et al. [82] performed the bioconversion using *Saccharomyces cerevisiae* strain, and the strain was engineered in such a way that the yield of product increased 15 times of the conventional methods (Scheme 12.4).

Papanikolaou et al. [83] used an Au/ZTC (Zeolite-templated carbon) catalyst for the selective oxidation of furfural to furoic acid. The preparation of the catalyst was achieved by deposition of Au nanoparticles onto β-ZTC carbon. The process avoids decarboxylation and esterification, leading to a 90% conversion of furfural to furoic acid (Scheme 12.5).

Different electrochemical conversion technologies have been used in developing high-performance catalytic materials that reduce the energy barrier of electrochemical processes. The techniques involve metal–air batteries, water electrolysis, and fuel cells, along with reactions such as oxidation reactions, reduction reactions, HERs, electrochemical carbon dioxide reaction, and OER [84].

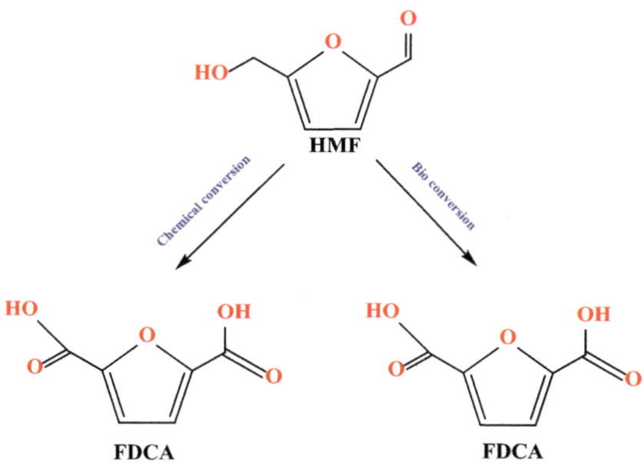

SCHEME 12.4 Conversion of HMF using different methods.

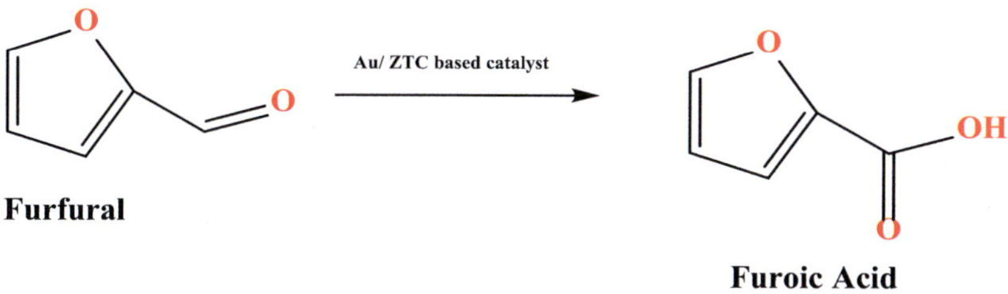

SCHEME 12.5 Conversion of furfural to furoic acid using Au/ZTC catalyst.

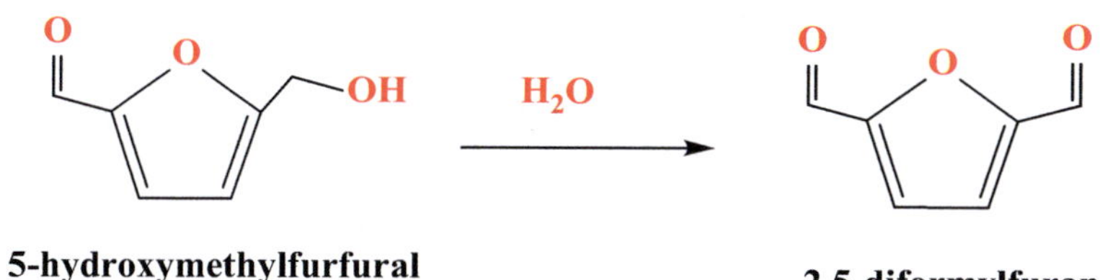

SCHEME 12.6 Oxidation of 5-hydroxymethylfurfural to 2,5-diformylfuran using water.

Xue et al. [85] converted HMF into 2,5-diformylfuran using water as a green solvent. Using water as a solvent increased the kinetics and activation barrier of the reaction. The reaction pathway is illustrated in Scheme 12.6:

12.5.1.2 Electrocatalytic oxidation of furfural

Furfural, derived from lignocellulosic biomass, has drawn interest as a sustainable material used as a precursor to produce hydrocarbons of

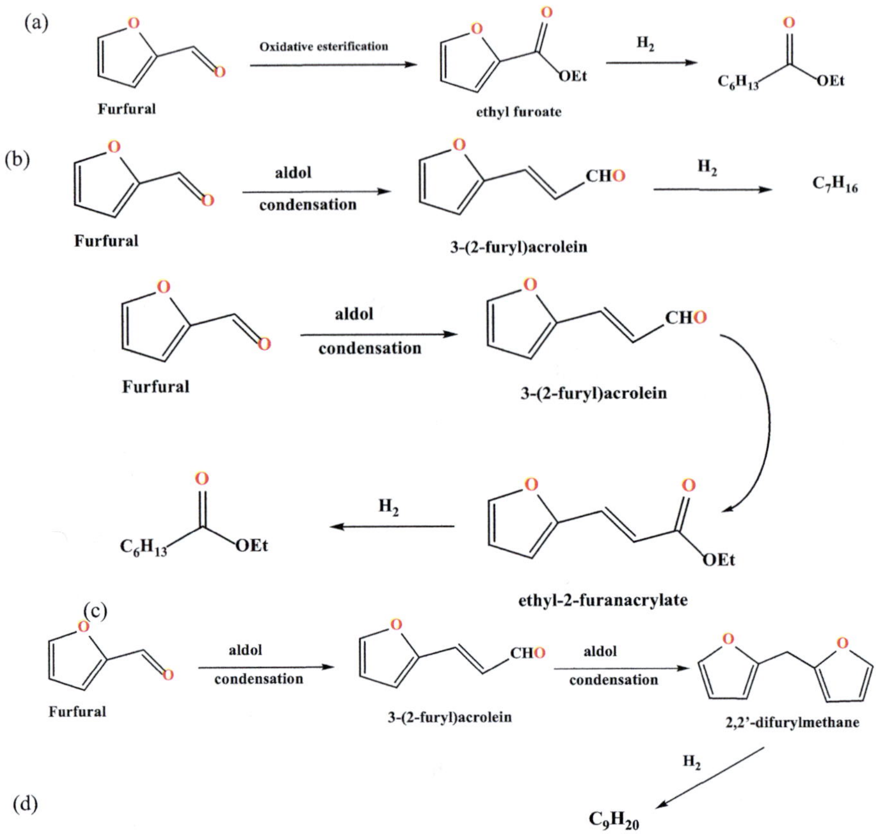

SCHEME 12.7 (A) Conversion of furfural to a C9 compound via oxidative esterification, (B) conversion to C7 hydrocarbon via aldol condensation, (C) conversion to C9 compound via aldol condensation, (D) formation of the final products 3-(2-furyl)acrolein and C9 compound.

C5–C7 length. Fang et al. [86] produced C7–C9 hydrocarbons via a one-pot synthesis (oxidative esterification and cross-aldol condensation) using gold and nickel oxide nanoparticles as catalysts. Four products, namely, 3-(2-furyl)acrolein (FA), ethyl furoate (EF), ethyl-2-furanacrylate (EFA), 2,2′-difurylmethane (DFM), and ethyl acetate (EAc) [86]. A detailed overview of the reaction pathway is illustrated in Scheme 12.7.

Gupta et al. [87] used 1,3-bis(2,4,6-trimethylphenyl) imidazolium chloride with a strong organic base (1,8-diazabicyclo [5.4.0]undec-7-ene) as a homogeneous catalyst for the formation of furoic acid (Scheme 12.8) via oxidation.

Kanan et al. [88]performed electrooxidation of glycerol, methanol, ethylene glycol, and xylitol using PPy/manganese and oxyhydroxide/palladium nanocomposite electrodes. Among all of these, ethylene

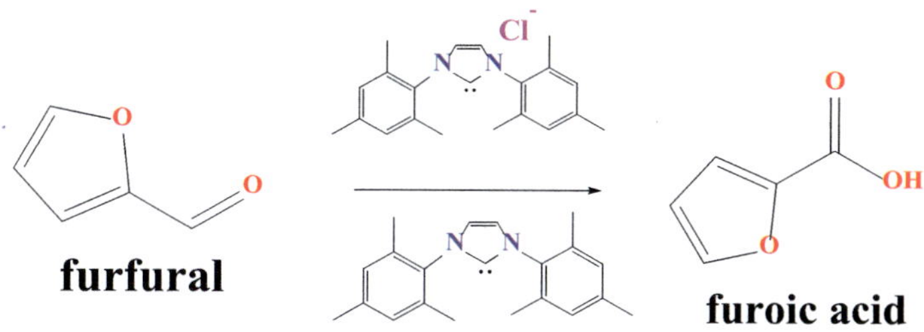

SCHEME 12.8 Conversion of furfural to furoic acid using 1,3-bis(2,4,6-trimethylphenyl)imidazolium chloride and (1,8-diazabicyclo[5.4.0]undec-7-ene).

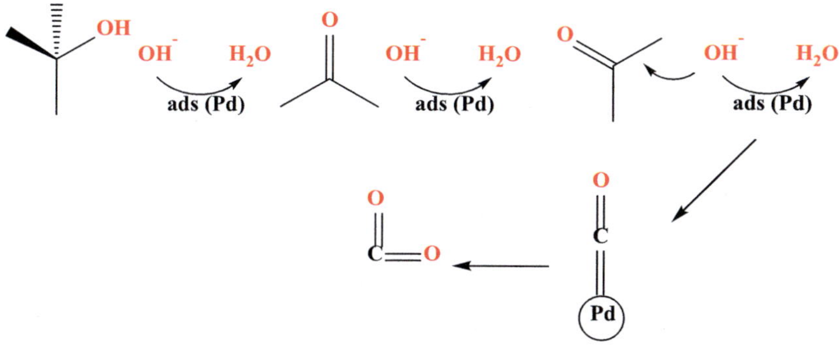

SCHEME 12.9 Mechanism for electrooxidation of ethylene glycol using a Pd-based catalyst.

glycol represented the maximum power density of 430 mA/cm^2. The detailed mechanism of the reaction has been shown in Scheme 12.9.

12.5.1.3 *Electrocatalytic oxidation of glycerol*

Glycerol is produced as a byproduct during the production of biofuels via the transesterification reaction. It is also known as glycerine and has the molecular formula $C_3H_8O_3$. Due to the extensive production of biodiesel, the production of glycerol has also increased, resulting in the production of high-value chemicals using glycerol as a precursor. The oxidation of glycerol results in the production of high-value products, such as glyceric acid and dihydroxyacetone [89]. Dihydroxyacetone is essentially used in sunless tanning products, as well as in pharmaceutical, medical, and food applications [90]. The most common method used for the conversion of glycerol into useful products is oxidation. For this purpose, different catalysts have been used. For instance, Mendoza et al. [91]. used iron-pillared clays for the selective production of dihydroxyacetone (Scheme 12.10) from

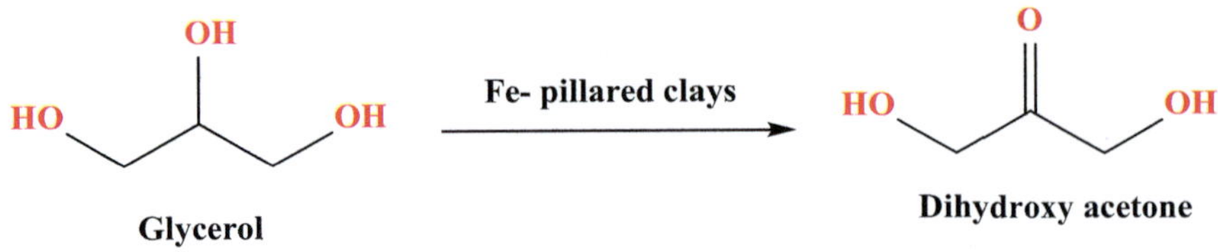

SCHEME 12.10 Conversion of glycerol to dihydroxyacetone using Fe-pillared clays.

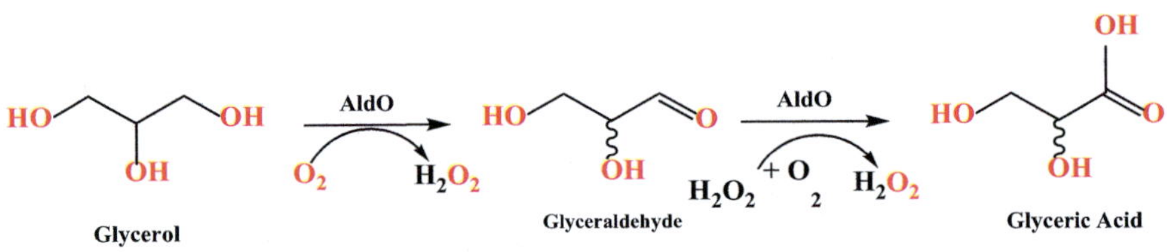

SCHEME 12.11 Conversion of glycerol to glyceric acid using alditol oxidase.

glycerol via a photo-assisted oxidation reaction. The product yield was found to be more than that of a conventional TiO_2-based catalyst.

Similarly, Hu et al. [92] used a Pt-Bi/C catalyst to selectively produce dihydroxyacetone from glycerol via oxidation. The secondary hydroxyl group of glycerol is oxidized, and dihydroxyacetone is obtained as a product in the presence of this catalyst. The catalyst was optimized under different conditions, such as temperature, pressure, and pH. While the most optimized reaction conditions for high product yield were at a temperature of 80°C, pressure of 30 psi, and pH 2. 80% of glycerol was converted to the desired product using semi batch reactor [92]. Along with the formation of dihydroxyacetone, glycerol also forms glyceric acid (Scheme 12.11) using different heterogeneous and homogeneous catalysts. Gerstenbruch et al. [93] used alditol oxidase in Scheme 12.11.

12.5.2 Electrochemical reduction pathways

Reduction pathways in biomass valorization are characterized by the transformation of oxygen-rich compounds into reduced products with a higher energy density. Examples include the electrochemical hydrogenation of levulinic acid (LA) to γ-valerolactone (GVL) and the reduction of furfural to furfuryl alcohol [72]. In both cases, polymer-supported metal nanoparticles or nitrogen-doped conductive polymers serve as the cathode materials to enable efficient proton-coupled electron transfer (PCET). The employment of water as the proton source in these reduction reactions mitigates toxic chemical usage, thus aligning with the design principles of

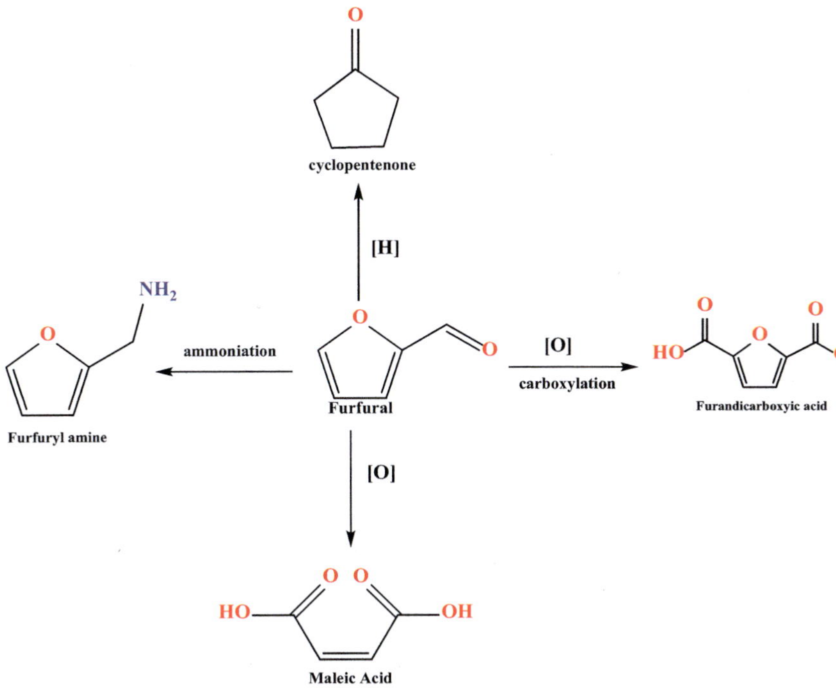

SCHEME 12.12 Production of different compounds using furfural.

sustainable chemistry. Furthermore, the polymeric matrix helps tune the local hydrophilicity/hydrophobicity, ultimately affecting product distribution through the control of access of the reactants to the catalytic sites. Furfural is a versatile precursor used to synthesize high-value products such as furandicarboxylic acid, furfurylamine, furfuryl alcohol (furoic acid), and cyclopentanone (Scheme 12.12). It is produced from corn cobs and wood waste, and its main producers are China, the Dominican Republic, and South Africa [94]. Furfural is reduced electrocatalytically to furfuryl alcohol under acidic and basic conditions, and it was found that the yield was maximum under basic conditions [11].

Levulinic acid is produced by acid-catalyzed dehydration and hydrolysis of hexose sugars. It is used to produce a number of biochemicals, such as succinic acid, polymers, resins, herbicides, pharmaceuticals, and flavoring agents, biofuels/oxygenated fuel additives [95]. Acid-catalyzed dehydration of cellulose produces levulinic acid and formic acid as products, and further hydrogenation produces γ-valerolactone as shown in Scheme 12.13 [96].

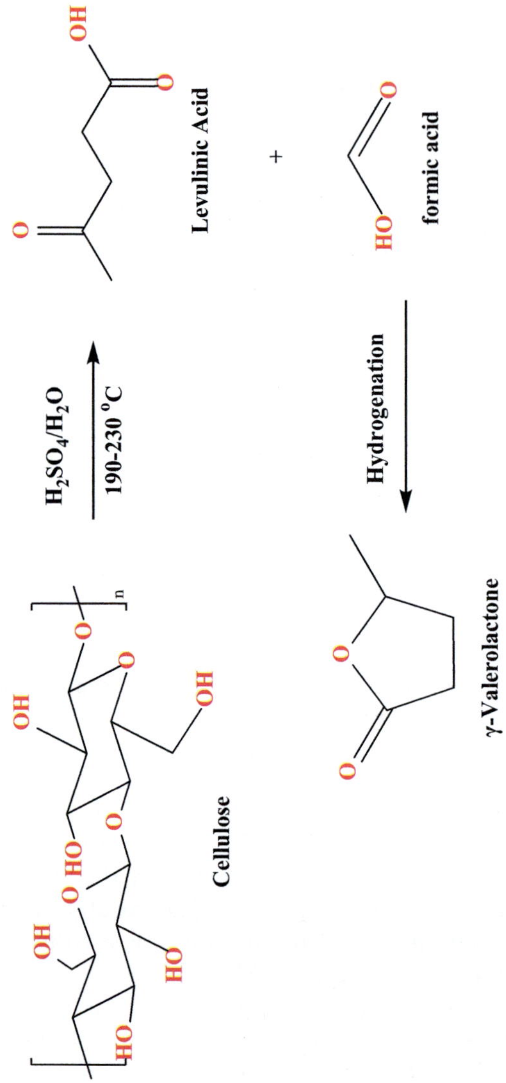

SCHEME 12.13 Production of Levulinic acid using cellulose and conversion into γ-valerolactone.

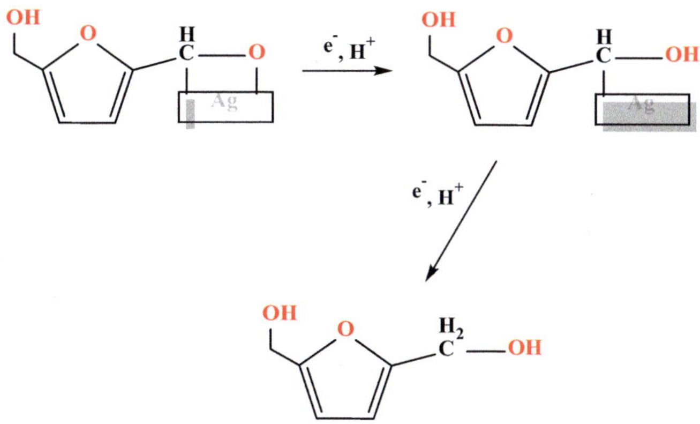

SCHEME 12.14 Conversion of 5-hydroxymethylfuran to 2,5-bis-(hydroxymethyl)furan (BHMF) using Ag electrodes.

12.5.3 Paired electrolysis and integrated oxidation–reduction

Paired electrolysis is an energy-efficient way to perform oxidative and reductive processes in one cell, thus allowing both electrodes to generate valuable products from biomass-derived waste [97]. For instance, anodic oxidation of glycerol can utilize cathodic reduction of CO_2 to yield high-value chemicals at both electrodes. Polymer-based electrocatalysts are particularly well-positioned for paired electrolysis systems due to their intrinsic tunability in electronic structure, stability across differing electrochemical environments, and ability to operationalize both catalytic processes within a single structure. Electrochemical hydrogen evolution and the oxygen reduction reaction (HER and ORR) are important in this category. The most scalable and clean approach for hydrogen production is the HER. A highly efficient electrocatalyst is essential for hydrogen production. Sekar et al. [98] synthesized activated carbon nanosheets derived from eucalyptus biomass. It was proven that the synthesized catalyst showed promising reaction performance suitable for hydrogen evolution. HERs have been widely used in the synthesis of hydrogen peroxide. A wide range of electrocatalysts have been employed in the synthesis process, such as carbon-based materials, metals or alloys, nonnoble transition metals, and molecular complexes [99]. Roylance et al. [100] performed the conversion of 5-hydroxymethylfurfural to 2,5-bis-(hydroxymethyl)furan (BHMF) using Ag catalytic electrodes. Water was used as the source of hydrogen. Scheme 12.14 shows the mechanism of the reaction involved in the conversion.

Li et al. [101] synthesized an $Mn_2Mo_3O_8$-Ni-based electrocatalyst in order to improve HER performance. The prepared electrocatalyst was integrated with biomass-derived lignin to further improve the performance.

The synthesized sample was proven to be highly stable for 140 h at −500 mA/cm^2 of current density [101]. Energy systems are evaluated on the basis of performance and efficiency using ORR. Because the ORR kinetics are inherently slow, sturdy electrocatalysts are used to make the process more feasible. Pt-based catalysts have gained much attention as promising electrocatalysts for the ORR [102]. Oezalsan et al. [103] synthesized Pt-based nanoparticles and used them as an electrocatalyst for ORR in a fuel cell system. Ca´rdenas-Arenas et al. [104] synthesized an Fe-N-C-based electrocatalyst derived from *Posidonia oceanica* biomass. The most important property to improve ORR kinetics, the electrochemically active surface area, was improved using these catalysts. The catalyst was highly stable even after 8000 cycles in a rotating disk electrode under O_2-saturation conditions [104]. Yabu et al. [105] synthesized a carbon-alloy-based electrocatalyst mixed with cellulose nanofibers and blood meal from a pig. The ORR performance was evaluated on the basis of properties such as onset potential, half-wave potential, and maximum current density.

12.6 Benefits and drawbacks of electrochemical methods relative to thermal or catalytic routes

The conversion of biomass into chemicals or fuels can occur in a variety of ways, with electrochemical methods representing an alternative to thermal or heterogeneous catalytic processes. Although these methods are not without limitations, they do have benefits that can make electrochemical routes more attractive depending on the application. In particular, electrochemical approaches are advantageous in terms of sustainability, operating conditions, and selectivity of the final product. Nevertheless, electrochemical processes have limitations that prohibit them from large-scale implementation.

12.6.1 Benefits of electrochemical biomass upgrading

Mild reaction conditions: Electrochemical reactions typically occur at or near ambient temperatures and pressures – very different from thermal processes like pyrolysis or gasification, which are done at temperatures typically above 400°C–800°C and high pressures [106]. Operating under milder conditions will result in reduced energy requirements and the lifetime of the catalysts or parts of the reactor.

Selectivity and tunability: In electrochemical systems, the applied potential provides a unique control parameter and can aid in gradually adjusting the reaction delivery location to the desired products [107]. In a reaction, like the electrooxidation of HMF to FDCA, the anodic reaction potential can be optimized to lower the potential for overoxidation to CO_2. Other catalytic systems generally rely on changing temperature and pressure to modulate reaction pathways, which is less controllable.

Integrating with renewable energy: Electrochemical methods can be driven directly by renewable electricity, such as from solar, wind, or hydropower sources, thus separating chemical production from a fossil energy source [108]. Although thermal methods can be electricity-driven, they still require considerable heat generation, which is less efficient.

Decreased usage of hazardous reagents: Water is the solvent and proton source for many electrochemical biomass valorization reactions, allowing us to avoid the toxic use of harsh oxidants like chlorine gas or hydrogen peroxide and reductants like high-pressure hydrogen [71]. This also improves the safety and sustainability of the process.

Paired electrolysis and process intensification: Paired electrolysis means that anodic oxidation reactions and cathodic reduction occur simultaneously, using the same positively and negatively poised electrodes within one vessel. This can lead to a single reaction vessel producing multiple high-value products [97]. This is not typically possible in thermal reactions, where you are either doing oxidation or reduction separately.

12.6.2 Limits regarding electrochemical biomass conversion

Scalability and capital cost

Scaling electrochemical systems for biomass processing is not straightforward due to the large amount of electrode surface area that is needed in many cases, and capital costs associated with advanced reactor designs and manufacture [60]. In addition, thermal catalytic reactors are mature technologies for which substantial industrial investments have been made to build the industrial-scale production.

Catalyst stability and fouling

Biomass-derived feedstocks will often have many impurities, such as lignin fragments, salts, and proteins, which would be able to adsorb on the surface of the electrode, resulting in deactivation of the catalyst and continued loss of performance over time [72]. Thermal processes tend to be more tolerant with regard to feedstock heterogeneity.

Current efficiency and energy demand

While electrochemical processes can utilize renewable electricity, it is difficult to achieve high current efficiencies due to side reactions (e.g., OER, HER) competing with the conversion routes of biomass [73]. Thermal processing methods are typically capable of higher mass throughput per energy input, but sometimes at the expense of selectivity.

Limited industrial application and knowledge gap

The commercialization of electrochemical biomass valorization is still in the early phases of development, with very few standardized methods to measure, standardized reactor designs, or techno-economic assessments of performance, compared with the considerable knowledge and methodology for catalytic upgrading of bio-oils, sugars, and lignin via thermal processing [109].

Mass transport limitations

An electrochemical system can convert biomass to form carbon material by allowing the reactions to occur at the electrode–electrolyte interface, which also places constraints on product yields since they depend on effective mass transport of reactants to these active sites. In a thermal catalytic processing system where the reactions occur in the bulk phase, product yields are not dependent on any active site but as a function of the total mass transport aspect of the process, which is less impacted by interfacial transport processes [110].

12.7 Electrocatalyst performance assessment

12.7.1 Performance metrics—Activity, selectivity, and faradaic efficiency

The activity of a catalyst evaluates the spontaneity of an electrochemical reaction. The activity of a reaction is commonly evaluated by calculating overpotential at required current density, exchange current density, or turnover frequency [111]. While selectivity is termed as the selection of a desired reaction pathway over other undesired side reactions. Chen et al. [112] discussed that the traditional noble metal-based catalysts are less selective and have poor stability. In order to solve this issue, a pulsed potential electrolysis strategy has been opted for higher selectivity of the products and is applied to the conversion of glycerol to glyceric acid [112]. Faradic efficiency specifies the percentage of electrons contributing to the formation of the desired product. It can be calculated as follows [113]:

$$FE(\%) = \frac{n \times F \times No.\ of\ moles\ of\ product}{Total\ charge\ passed} \times 100$$

12.7.2 Stability and sustainability under realistic electrochemical conditions

Stability of a catalyst refers to the resistance of the catalyst to degradation over long-term use. The more stable the catalyst, the more efficient it is. The stability of a catalyst is evaluated using Accelerated degradation testing and long term chronopotentiometry [114]. The use of conducting polymer-based electrocatalysts in continuous-flow electrochemical reactors represents a feasible strategy for improving productivity and process scalability, which is essential for biomass valorization. Continuous flow ensures rapid mass transport of substrates and products, minimizes local concentration gradients, as well as the ability to control of residence time. Conducting polymers such as PANI, PEDOT, and PPy can be uniformly coated on porous electrodes, as well as incorporated into flow-through supports (e.g., carbon felt, nickel foam), where their high electrical connection is maintained under flow conditions. However, the stability of the electrode remains a key challenge—polymeric matrices are prone to undergo redox cycling, swelling, or dopant loss through bias over extended periods [46]. Recent reports have demonstrated that enhanced stability in current density and product selectivity over >20 h of continuous electrolysis through covalent attachment of redox mediators (e.g., TEMPO to PEDOT) or hybridization of the conducting polymer-based frameworks with metal oxides, resulting in improved performance stability. These results suggest that polymer-based catalysts can be mixed and incorporated into flow electrochemical reactors, depending on addressing mechanical stability and charge transport across the polymer/catalytic interface [115]. The term sustainability refers to the resistance that leads to fouling, morphology changes, and leaching of active sites. It is specially implemented in polymer systems due to their swelling characteristics in liquid media. An ideal electrocatalyst should have high turnover frequency, high exchange current density, and high standard rate constant.

12.8 Conclusion

Polymer-based electrocatalysts are a new generation of catalysts that have gained popularity because they are structurally tunable, functionally diverse, and compatible with green catalytic processes. By the strategic inclusion of conductive polymers (e.g., PPy, PANI, PEDOT), redox-functional groups (e.g., TEMPO radicals and ferrocene groups), and functional ligands (e.g., bipyridine, salen, and sulfonic acid groups), scientists have designed catalysts with improved electron/proton transfer, substrate selectivity, and environmental sustainability. These systems find good applications in biomass conversion and electrochemical energy technologies due to their modularity, high surface areas, and designed microenvironments. Porous structures such as COFs and polymer composites also enhance mass transport and catalytic efficiency. Polymer-based electrocatalysts as a whole play the intermediary role between molecular-level accuracy and macroscopic performance, providing a sustainable pathway for addressing intricate chemical conversions.

12.9 Future scope and outlook

Looking to the future, further progress will benefit from deeper mechanistic insight into catalyst-substrate interactions, preferably with in situ spectroscopy and computational simulation. Stimuli-responsive and biomimetic architectures based on enzyme active sites hold promise for next-generation precision catalysis. Additionally, scalable synthesis methods, integration into electrochemical devices, and the utilization of biodegradable or bio-derived monomers will enable industrial applicability and environmental sustainability. Newly developing tools like machine learning and high-throughput screening are likely to catalyze the discovery and optimization of these systems. Polymer-based electrocatalysts are thus poised to play a crucial role in pushing forward green chemistry, renewable energy conversion, and circular materials science.

Electrocatalysts based on polymers offer a promising platform for the sustainable conversion of biomass feedstocks because of their tunable electronic properties, flexibility in design, and potential for low-cost substrate integration. However, for long term use, durability, recyclability, and techno-economic viability must be carefully assessed.

- **Durability**
  Chemical and electrochemical durability: Conducting polymers can be susceptible to oxidative degradation and chain scission from severe redox conditions. Stabilizing the π-conjugated backbone, adding heteroatoms, and covalent crosslinking can potentially improve the lifetime.
  Mechanical and interfacial durability: Structural fatigue, swelling, and delamination at the interface between the electrode and electrolyte can contribute to a loss in durability due to electrode materials becoming detached or insoluble residues caused by polymer chain scission at the surface. Reinforcing the material with other 2D materials such as MXenes or graphene, or by embedding dynamic covalent strong hydrogen bonds or supramolecular chemistry for self-healing capabilities, can also improve operational reliability.
  Durability in an operational sense: Dynamic covalent bonds or supramolecular chemistry can be utilized to prove self-healing properties to recover active sites over extended cycling periods.
- **Recyclability**
  Closed-loop material design: The use of depolymerizable or recyclable polymer frameworks allows for the recovery and regeneration of catalysts after deactivation.
  Component separation: Systems with modular designs and freely detachable catalytic centers, for example, metal-ligand or enzyme-polymer conjugates, allow for selective recycling.
- **Techno-economic considerations**
  Cost effectiveness of materials: Partially to fully replacing precious metal with lower-cost earth-abundant metals (Fe, Co, and Ni) in conductive polymers decreases raw material costs significantly.

Scalability of manufacturing processes: Controlled Processes for electrochemical polymerization, roll-to-roll coating, or additive manufacturing have the potential for low-cost, large-area manufacture.
Trade-off between high and low cost: Performance should be optimized to maximize the catalytic turnover while minimizing the cost associated with polymer synthesis and processing.
Lifecycle and economic consideration: Assessing these factors in techno-economic assessment supports long-term sustainability by considering recyclability, stability, and material recovery.
Green synthesis: The use of bio-based monomers, reagent-free polymerization, and mild routes to polymerization will minimize the overall environmental impact of the synthetic route.
End of life processing: Use of selective solvent-based recovery methods will allow for recovery and reclamation of polymeric components without compromising structural integrity.

References

[1] V.T. Nguyen, K. Cho, Y. Choi, B. Hwang, Y.K. Park, H. Nam, D. Lee, Biomass-derived materials for energy storage and electrocatalysis: recent advances and future perspectives, Biochar 6 (2024) 1–30.
[2] Y. Ramli, V. Chaerusani, Z. Yang, R. Yang, J. Zhang, A. Abudula, G. Guan, Electrochemical conversion of biomass derivatives to value-added chemicals: a review, Green Carbon 3 (2025) 113–129.
[3] P. Zhu, M. Shi, Z. Shen, X. Liao, Y. Chen, Electrocatalytic conversion of biomass-derived furan compounds: mechanisms, catalysts and perspectives, Chem. Sci. 15 (2024) 4723–4756.
[4] T. Liu, H. Yabu, Biomass-derived electrocatalysts: low-cost robust materials for sustainable electrochemical energy conversion, Adv. Energy Sustain. Res. (2023), doi:10.1002/aesr.202300168.
[5] M. Heydari Gharahcheshmeh, K. Chowdhury, Fabrication methods, pseudocapacitance characteristics, and integration of conjugated conducting polymers in electrochemical energy storage devices, Energy Adv. 3 (2024) 2668–2703.
[6] X. Yu, S. Ye, Recent advances in activity and durability enhancement of Pt/C catalytic cathode in PEMFC. Part II: degradation mechanism and durability enhancement of carbon supported platinum catalyst, J. Power Sources 172 (2007) 145–154.
[7] J. Wang, H. Zhang, X. Wang, Recent methods for the synthesis of noble-metal-free hydrogen-evolution electrocatalysts: from nanoscale to sub-nanoscale, Small Methods 1 (2017) 1700118.
[8] K. Lee, J. Zhang, H. Wang, D.P. Wilkinson, Progress in the synthesis of carbon nanotube- and nanofiber-supported Pt electrocatalysts for PEM fuel cell catalysis, J. Appl. Electrochem. 36 (2006) 507–522.
[9] S. Gao, X. Wei, H. Fan, L. Li, K. Geng, J. Wang, Nitrogen-doped carbon shell structure derived from natural leaves as a potential catalyst for oxygen reduction reaction, Nano Energy 13 (2015) 518–526.
[10] Y. Cao, Y. Sun, R. Zheng, Q. Wang, X. Li, H. Wei, L. Wang, Z. Li, F. Wang, N. Han, Biomass-derived carbon material as efficient electrocatalysts for the oxygen reduction reaction, BioMass BioEnergy 168 (2023) 106676.

[11] R.S. Sherbo, A. Kurimoto, C.M. Brown, C.P. Berlinguette, Efficient electrocatalytic hydrogenation with a palladium membrane reactor, J. Am. Chem. Soc. 141 (2019) 7815–7821.

[12] S. Dutta, A. Bhaumik, K. C-W.Wu, Hierarchically porous carbon derived from polymers and biomass: effect of interconnected pores on energy applications, Energy Environ. Sci. 7 (2014) 3574–3592.

[13] T.A. Skotheim, Handbook of Conducting Polymers, 2nd rev. and expanded ed., CRC Press, New York, 1997.

[14] D.A. Fungaro, A. Oliveira-Neto, M. Linardi, Electrocatalytic oxidation of ethanol on platinum nanoparticles modified polyaniline electrode, Lat. Am. Appl. Res. 37 (2007) 223–226.

[15] X. Lu, K. Qi, X. Dai, Y. Li, D. Wang, J. Dou, W. Qi, Selective electrooxidation of 5-hydroxymethylfurfural to 5-formyl-furan-2-formic acid on non-metallic polyaniline catalysts: structure-function relationships, Chem. Sci. 15 (2024) 11043–11052.

[16] S. Carli, Electrocatalytic poly(3,4-ethylenedioxythiophene) for electrochemical conversion of 5-hydroxymethylfurfural, Langmuir. 40 (2024) 10115–10128.

[17] H. Ai, L. Fan, Y. Wang, Z. Wang, H. Zhang, J. Zhao, M. Jiao, B. Lv, X. Han, OER catalytic performance of a composite catalyst comprising multi-layer thin flake Co3O4 and PPy nanofibers, RSC Adv. 13 (2023) 32045–32053.

[18] Z. Ma, L. Wang, G. Li, T. Song, Recent advances in electrocatalytic oxidation of 5-hydroxymethylfurfural to 2,5-furandicarboxylic acid by heterogeneous catalysts, Catalysts 14 (2024) 1–32.

[19] A. Sierra-Padilla, J.J. García-Guzmán, D. López-Iglesias, J.M. Palacios-Santander, L. Cubillana-Aguilera, E-tongues/noses based on conducting polymers and composite materials: expanding the possibilities in complex analytical sensing, Sensors 21 (2021) 4976.

[20] Y. Liang, A. Offenhäusser, S. Ingebrandt, D. Mayer, PEDOT: PSS-based bioelectronic devices for recording and modulation of electrophysiological and biochemical cell signals, Adv. Healthc. Mater. 10 (2021) 1–24.

[21] J. Kim, J. Ling, Y. Lai, P.J. Milner, Redox-active organic materials: from energy storage to redox catalysis, ACS Mater. Au 4 (2024) 258–273.

[22] A. Das, S.S. Stahl, Noncovalent immobilization of molecular electrocatalysts for chemical synthesis: efficient electrochemical alcohol oxidation with a pyrene–TEMPO conjugate, Angew. Chemie—Int. Ed. 56 (2017) 8892–8897.

[23] D.A. Corbin, D.M. Shircliff, B.J. Reeves, B.M. Boardman, Metallopolymers from direct polymerization of functionalized cobalt chalcogenide clusters and thiophene comonomers, Polym. Chem. 8 (2017) 3801–3809.

[24] T. Shimomura, N. Oyama, F.C. Anson, Polarography of a cobalt(III) complex of poly(4-vinylpyridine), J. Electroanal. Chem. 112 (1980) 265–270.

[25] T.L. Silva, M.de L.S.G. de Azevedo, F.R. Ferreira, D.C. Santos, C. Amatore, M.O.F. Goulart, Quinone-based molecular electrochemistry and their contributions to medicinal chemistry: a look at the present and future, Curr. Opin. Electrochem. 24 (2020) 79–87.

[26] C.S. Diercks, O.M. Yaghi, The atom, the molecule, and the covalent organic framework, Science 355 (2017) eaal1585.

[27] C. Xu, W. Zhang, J. Tang, C. Pan, G. Yu, Porous organic polymers: an emerged platform for photocatalytic water splitting, Front. Chem. 6 (2018) 1–12.

[28] Y. Fan, M. Chen, N. Xu, K. Wang, Q. Gao, J. Liang, Y. Liu, Recent progress on covalent organic framework materials as CO_2 reduction electrocatalysts, Front. Chem. 10 (2022) 1–8.

[29] H. Wang, Z. Li, Z. Meng, X. Guo, Y. Du, H. Yang, An easily obtained hypercrosslinked pyrene-based porous organic polymer as a high performance electrode material for lithium-ion batteries, New J. Chem. 45 (2021) 7060–7064.

[30] F. Wang, M. Hickner, Y. Seung, T.A. Zawodzinski, J.E. Mcgrath, Direct polymerization of sulfonated poly(arylene ether sulfone) random (statistical) copolymers: candidates for new proton exchange membranes, J. Memb. Sci. 197 (2002) 231–242.
[31] D. Balatskiy, I. Tkachenko, I. Malakhova, N. Polyakova, Polyethylenimine as a non-innocent ligand for hexacyanoferrates immobilization, Molecules 27 (2022) 8489.
[32] T. Osako, A. Ohtaka, Y. Uozumi, Development of polymer-supported transition-metal catalysts and their green synthetic applications, Catalyst Immobilization: Methods and Applications, 2019, pp. 325–368, doi:10.1002/9783527817290.ch10.
[33] M. Nemiwal, T.C. Zhang, D. Kumar, Pectin modified metal nanoparticles and their application in property modification of biosensors, Carbohydr. Polym. Technol. Appl. 2 (2021) 100164.
[34] R. Holze, Overoxidation of intrinsically conducting polymers, Polymers 14 (2022) 1584.
[35] T.H. Le, Y. Kim, H. Yoon, Electrical and electrochemical properties of conducting polymers, Polymers 9 (2017) 150.
[36] S.Y. Kim, H.-K. Song, Conducting polymers with functional dopants and their applications in energy, environmental technology, and nanotechnology, Clean Technol. 21 (2015) 12–21.
[37] S.H.K. Paleti, Y. Kim, J. Kimpel, M. Craighero, S. Haraguchi, C. Müller, Impact of doping on the mechanical properties of conjugated polymers, Chem. Soc. Rev. 53 (2024) 1702–1729.
[38] C. Zhang, L. Yao, M. Pu, C. Zhou, Recent material development and applications of conjugated polyelectrolytes by leveraging electronic and ionic transport properties, RSC Appl. Polym. 3 (2025) 549–573.
[39] X.P. Zhou, J. Kateb, J. Fan, G.A. Kim, B. Lodygensky, D.P. Amilhon, et al., Conducting polymer films and bioelectrodes combining high adhesion and electro-mechanical self-healing, J. Mater. Chem. C. 12 (2024) 5708–5717.
[40] K. Matyjaszewski, J. Xia, Atom transfer radical polymerization, Chem. Rev. 101 (2001) 2921–2990.
[41] S.-A. Chen, Y.-C. Tsai, Electrochemical polymerization of pyrrole on a fabric, Die Angew. Makromol. Chemie. 169 (1989) 153–157.
[42] J. Eichhorn, A.K. Mengele, C. Neumann, J. Biskupek, A. Turchanin, U. Kaiser, S. Rau, F.H. Schacher, Block copolymer micelles as colloidal catalysts for photocatalytic NAD+ reduction, Polym. Chem. 15 (2024) 4810–4823.
[43] Y.D. Sintayehu, H.C. Ananda Murthy, Polypyrrole based biofunctional composite layer for bioelectrocatalytic device system, Adv. Mater. Lett. 10 (2019) 524–532.
[44] C. He, S. Sun, H. Peng, C.P. Tsui, D. Shi, X. Xie, Y. Yang, Poly(ionic liquid)-assisted reduction of graphene oxide to achieve high-performance composite electrodes, Compos. Part B Eng. 106 (2016) 81–87.
[45] Lv, C.Y. Yanqin, PEDOT: PSS-modified NiFe layered double hydroxide enables efficient and durable seawater electrolysis at high current density, J. Mater. Chem. A. 13 (2025) 25329–25334.
[46] H. Noh, H. Yeo, B.W. Boudouris, B.M. Tackett, Conducting polymer transforms hydrophobic porous membranes into robust gas diffusion layers in electrochemical applications, Energy Environ. Sci. 18 (2024) 1272–1281.
[47] A. Pacuła, J. Gurgul, R.P. Socha, P. Pietrzyk, M. Ruggiero-Mikołajczyk, D. Duraczyńska, M. Krzan, M. Nattich-Rak, G. Mordarski, J. Żukrowski, M. Sadowska, B.D. Napruszewska, Electrocatalysts and components for polymer-based composite films: pyrolytic carbon deposited on metal and metal oxides derived from hydrotalcite-like materials, J. Mater. Sci. 60 (2025) 10997–11034.
[48] Z. Terzopoulou, A. Zamboulis, N.D. Bikiaris, E. Xanthopoulou, R.O. Ioannidis, D.N. Bikiaris, A decade of innovation: synthesis, properties and applications of PLA copolymers, Prog. Polym. Sci. 167 (2025) 101991.

[49] B. Sellergren, A.J. Hall, Molecularly imprinted polymers, In Supramolecular Chemistry (2012) 172–182, doi:10.1002/9780470661345.smc137.

[50] C. Alexander, H.S. Andersson, L.I. Andersson, R.J. Ansell, N. Kirsch, I.A. Nicholls, et al., Molecular imprinting science and technology: a survey of the literature for the years up to and including 2003, J. Mol. Recognit. 19 (2006) 106–180.

[51] D. Zhao, J. Feng, Q. Huo, N. Melosh, G.H. Fredrickson, B.F. Chmelka, et al., Triblock copolymer syntheses of mesoporous silica with periodic 50 to 300 angstrom pores, Science 279 (1998) 548–552.

[52] D. Li, Y. Xia, Electrospinning of nanofibers: reinventing the wheel? Adv. Mater. 16 (2004) 1151–1170.

[53] H. Soderlund, L. Kaariainen, C.H. Von Bonsdorff, Properties of Semliki Forest virus nucleocapsid, Med. Biol. 53 (1975) 412–417.

[54] Y. Lu, Y. Mei, M. Ballauff, M. Drechsler, Thermosensitive core-shell particles as carrier systems for metallic nanoparticles, J. Phys. Chem. B 110 (2006) 3930–3937.

[55] D. Schmaljohann, Thermo- and pH-responsive polymers in drug delivery, Adv. Drug Deliv. Rev. 58 (2006) 1655–1670.

[56] S.Y. Ding, W. Wang, Covalent organic frameworks (COFs): from design to applications, Chem. Soc. Rev. 42 (2013) 548–568.

[57] L. Wu, J. Song, B. Zhang, B. Zhou, H. Zhou, H. Fan, Y. Yang, B. Han, Very efficient conversion of glucose to 5-hydroxymethylfurfural in DBU-based ionic liquids with benzenesulfonate anion, Green Chem. 16 (2014) 3935–3941.

[58] Z. Zhang, G.A. Tompsett, S. Granados-Focil, C.R. Lambert, M.T. Timko, et al., Rational design of solid-acid catalysts for cellulose hydrolysis using colloidal theory, Phys. Chem. Chem. Phys. 23 (2021) 10236–10243, doi:10.1039/D1CP00198A.

[59] G. Vasapollo, R.Del Sole, L. Mergola, M.R. Lazzoi, A. Scardino, S. Scorrano, et al., Molecularly imprinted polymers: present and future prospective, Int. J. Mol. Sci. 12 (2011) 5908–5945.

[60] P. Prabhu, Y. Wan, J.M. Lee, Electrochemical conversion of biomass derived products into high-value chemicals, Matter 3 (2020) 1162–1177.

[61] M. Hara, M. Okamura, M. Toda, J.N. Kondo, A. Takagaki, K. Domen, T. Tatsumi, S. Hayashi, 2007. Sulfonated incompletely carbonized glucose as strong Brønsted acid catalyst. doi: 10.1016/B978-0-444-53202-2.50090-7

[62] B. Qiu, B. Lin, F. Yan, Ionic liquid/poly(ionic liquid)-based electrolytes for energy devices, Polym. Int. 62 (2013) 335–337.

[63] K. Kuperkar, D. Patel, L.I. Atanase, P. Bahadur, Amphiphilic block copolymers: their structures, and self-assembly to polymeric micelles and polymersomes as drug delivery vehicles, Polymers. 14 (2022), doi:10.3390/polym14214702.

[64] S. Dutta, A. Bhaumik, K.C.-W. Wu, Hierarchically porous carbon derived from polymers and biomass: effect of interconnected pores on energy applications, Energy Environ. Sci. 7 (2014) 3574–3592.

[65] A. López-Magano, B. Ortín-Rubio, I. Imaz, D. Maspoch, J. Alemán, R. Mas-Ballesté, Photoredox heterobimetallic dual catalysis using engineered covalent organic frameworks, ACS Catal. 11 (2021) 12344–12354.

[66] X. Qi, M. Watanabe, T.M. Aida, R.L. Smith, Catalytical conversion of fructose and glucose into 5-hydroxymethylfurfural in hot compressed water by microwave heating, Catal. Commun. 9 (2008) 2244–2249.

[67] R. Das, D. Chakraborty, I 2-TEMPO as an efficient oxidizing agent for the one-pot conversion of alcohol to amide using FeCl 3 as the catalyst, Catal. Commun. 26 (2012) 48–53.

[68] Z. Liu, L. Zhang, S. Poyraz, X. Zhang, Conducting polymer—metal nanocomposites synthesis and their sensory applications, Curr. Org. Chem. 17 (2013) 2256–2267.

[69] M.M. Dell'Anna, M. Gagliardi, P. Mastrorilli, G.P. Suranna, C.F. Nobile, Hydrogenation reactions catalyzed by a supported palladium complex, J. Mol. Catal. A Chem. 158 (2000) 515–520.
[70] J.A. Glaser, Green chemistry with nanocatalysts, Clean Technol. Environ. Policy 14 (2012) 513–520.
[71] R.L. Arechederra, S.D. Minteer, Complete oxidation of glycerol in an enzymatic biofuel cell, Fuel Cells 9 (2009) 63–69.
[72] F.W.S. Lucas, R.G. Grim, S.A. Tacey, C.A. Downes, J. Hasse, A.M. Roman, et al., Electrochemical routes for the valorization of biomass-derived feedstocks: from chemistry to application, ACS Energy Lett. 6 (2021) 1205–1270.
[73] K. Li, Y. Sun, Electrocatalytic upgrading of biomass-derived intermediate compounds to value-added products, Chem.—A Eur. J. 24 (2018) 18258–18270.
[74] S. Takkellapat, T. Li, M. A. Gonzalez, An overview of biorefinery-derived platform chemicals from a cellulose and hemicellulose biorefinery, Clean Technologies and Environmental Policy 20 (2018) 1615–1630.
[75] G. Totaro, L. Sisti, P. Marchese, M. Colonna, A. Romano, C. Gioia, M. Vannini, A. Celli, Current advances in the sustainable conversion of 5-hydroxymethylfurfural into 2,5-furandicarboxylic acid, ChemSusChem 15 (2022) 1–13.
[76] M. Sajid, X. Zhao, D. Liu, Production of 2,5-furandicarboxylic acid (FDCA) from 5-hydroxymethylfurfural (HMF): recent progress focusing on the chemical-catalytic routes, Green Chem. 20 (2018), doi:10.1039/C8GC02680G.
[77] S. Albonetti, T. Pasini, A. Lolli, M. Blosi, M. Piccinini, N. Dimitratos, J.A. Lopez-Sanchez, D.J. Morgan, A.F. Carley, G.J. Hutchings, F. Cavani, Selective oxidation of 5-hydroxymethyl-2-furfural over TiO 2-supported gold-copper catalysts prepared from preformed nanoparticles: effect of Au/Cu ratio, Catal. Today 195 (2012) 120–126.
[78] L. Gao, K. Deng, J. Zheng, B. Liu, Z. Zhang, Efficient oxidation of biomass derived 5-hydroxymethylfurfural into 2, 5-furandicarboxylic acid catalyzed by merrifield resin supported cobalt porphyrin, Chem. Eng. J. 270 (2015) 444–449.
[79] J.T. Cunha, A. Romaní, L. Domingues, Whole cell biocatalysis of 5-hydroxymethylfurfural for sustainable biorefineries, Catalysts 12 (2022) 1–15.
[80] C. Hsu, Y. Kuo, Y. Liu, S. Tsai, Green conversion of 5-hydroxymethylfurfural to furan- 2, 5-dicarboxylic acid by heterogeneous expression of 5-hydroxymethylfurfural oxidase in *Pseudomonas putida* S12, Microbial Biotechnol. 13 (2020) 1094–1102.
[81] Y. Sheng, X. Tan, X. Zhou, Y. Xu, Bioconversion of 5-hydroxymethylfurfural (HMF) to 2,5-furandicarboxylic acid (FDCA) by a native obligate aerobic bacterium, *Acinetobacter calcoaceticus* NL14, Appl. Biochem. Biotechnol. 192 (2020) 455–465.
[82] J.T. Cunha, A. Romaní, L. Domingues, Production of HMF-derivatives from wine residues using *Saccharomyces cerevisiae* as whole-cell biocatalyst, Bioresour. Bioprocess. 12 (2025) 8, doi:10.1186/s40643-025-00840-5.
[83] G. Papanikolaou, P. Lanzafame, S. Perathoner, G. Centi, D. Cozza, G. Giorgianni, et al., High performance of Au /ZTC based catalysts for the selective oxidation of bio-derivative furfural to 2-furoic acid 1, Catal. Commun. 149 (2021) 106234.
[84] T. Liu, H. Yabu, Biomass-derived electrocatalysts: low-cost, robust materials for sustainable electrochemical energy conversion, Adv. Energy Sustain. Res. 5 (2024), doi:10.1002/aesr.202300168.
[85] W. Xue, J. Ye, Z. Zhu, R. Kumar, J. Zhao, Harnessing trace water for enhanced photocatalytic oxidation of biomass-derived alcohols to aldehydes, Energy Environ. Sci. 18 (2024) 214–226.
[86] Q. Fang, Z. Qin, Y. Shi, F. Liu, S. Barkaoui, H. Abroshan, et al., Au /NiO composite: a catalyst for one-pot cascade conversion of furfural, ACS Appl. Energy Mater. 2 (2019), doi:10.1021/acsaem.9b00001.

[87] N.K. Gupta, K. Fukuoka, K. Nakajima, Metal-free and selective oxidation of furfural to furoic acid with an N-heterocyclic carbene catalyst, ACS Sustainable Chemistry & Engineering 6 (2018) 6134–6142. https://doi.org/10.1021/acssuschemeng.7b03681.

[88] R. Kannan, A. Rhan, Enhanced electrooxidation of methanol, ethylene glycol, glycerol, and xylitol over a polypyrrole /manganese oxyhydroxide / palladium nanocomposite electrode, J. Appl. Electrochem. 44 (2014) 893–902.

[89] J. Li, Y. Tu, K. He, C. Chen, L. Liang, C. Ruan, Mechanistic insights into glycerol oxidation to high-value chemicals via metal-based catalysts, Molecules 30 (6) (2025) 1310.

[90] B. Synthon, L. Bricotte, K. Chougrani, V. Ladmiral, Dihydroxyacetone: a user guide for a challenging bio-based synthon, Molecules 28 (6) (2023) 2724.

[91] A. Mendoza, R. Romero, G.P. Gutiérrez-Cedillo, G. López-Tellez, O. Lorenzo-González, R.M. Gómez-Espinosa, et al., Selective production of dihydroxyacetone and glyceraldehyde by photo-assisted oxidation of glycerol, Catal. Today 358 (2020) 149–154.

[92] W. Hu, D. Knight, B. Lowry, A. Varma, Selective oxidation of glycerol to dihydroxyacetone over Pt-Bi/C catalyst: optimization of catalyst and reaction conditions, Ind. Eng. Chem. Res. 49 (2010) 10876–10882.

[93] S. Gerstenbruch, H. Wulf, N. Mußmann, T.O. Connell, K. Maurer, Asymmetric synthesis of D -glyceric acid by an alditol oxidase and directed evolution for enhanced oxidative activity towards glycerol, Appl. Microbiol. Biotechnol. 96 (5) (2012) 1243–1252.

[94] X. Zhang, S. Xu, Q. Li, G. Zhou, H. Xia, Recent advances in the conversion of furfural into biochemicals through chemo- and bio-catalysis, RSC Adv. 11 (2021) 27042–27058.

[95] D.W. Rackemann, W.O.S. Doherty, T. Crops, The conversion of lignocellulosics to levulinic acid, Biofuel. Bioprod. Biorefin. 5 (2011) 198–214.

[96] W.R.H. Wright, R. Palkovits, Development of heterogeneous catalysts for the conversion of levulinic acid to γ-valerolactone, ChemSusChem 5 (9) (2012) 1657–1667.

[97] N.P. Martínez, M. Isaacs, K.K. Nanda, Paired electrolysis for simultaneous generation of synthetic fuels and chemicals, New J. Chem. 44 (2020) 5617–5637.

[98] S. Sekar, A. Shanmugam, G. Senthilkumar, K. Thangasami, H. Jung, Y. Lee, et al., Enhanced hydrogen evolution reaction using biomass-activated carbon nanosheets derived from eucalyptus leaves, Materials 18 (2025), doi:10.3390/ma18030670.

[99] N. Wang, S. Ma, P. Zuo, J. Duan, B. Hou, Recent progress of electrochemical production of hydrogen peroxide by two-electron oxygen reduction reaction, Adv. Sci. 8 (2021) 1–26.

[100] J.J. Roylance, T.W. Kim, K.S. Choi, Efficient and selective electrochemical and photoelectrochemical reduction of 5-hydroxymethylfurfural to 2,5-bis(hydroxymethyl)furan using water as the hydrogen source, ACS Catal. 6 (2016) 1840–1847.

[101] L. Li, G. Qian, L. Ge, Y. Wang, X. Cao, Embedding $Mn_2Mo_3O_8$/Ni heterojunctions in self-supported lignin-derived carbon frameworks for enhanced ampere-level hydrogen evolution reaction under pH-universal conditions, Chem. Eng. J. (2025) 166556, doi:10.1016/j.cej.2025.166556.

[102] S. Das, S. Ghosh, T. Kuila, N.C. Murmu, A. Kundu, 2022. Biomass-derived advanced carbon-based electrocatalysts for oxygen reduction reaction 155–177. doi: 10.3390/biomass2030010

[103] M. Oezaslan, F. Hasché, P. Strasser, Pt-based core-shell catalyst architectures for oxygen fuel cell electrodes, J. Phys. Chem. Lett. 4 (2013) 3273–3291.

[104] A. Cárdenas-Arenas, A. Pedersen, F.J. Recio, J. Barrio, M.M. Titirici, V. Montiel, et al., Magnesium-templated Fe-N doped carbon materials from biomass waste as electrocatalysts for oxygen reduction reaction in alkaline media, Electrochim. Acta 538 (2025) 146899.

[105] H. Yabu, K. Ishibashi, M.S. Grewal, Y. Matsuo, Bifunctional rare metal-free electrocatalysts synthesized entirely from biomass resources, Sci. Technol. Adv. Mater. 23 (2022) 31–40.

[106] R. Latsuzbaia, R. Johannes, M. Bisselink, M. Crockatt, J.C. van der Waal, E.L.V. Goetheer, Biomass Processing via Electrochemical Means, Wiley, 2021, doi:10.1002/9783527825028.ch8.

[107] M. Yang, Z. Yuan, R. Peng, S. Wang, Y. Zou, Recent progress on electrocatalytic valorization of biomass-derived organics, Energy Environ. Mater. 5 (2022) 1117–1138.

[108] S.R. Kubota, K.S. Choi, Electrochemical oxidation of 5-hydroxymethylfurfural to 2,5-furandicarboxylic acid (Fdca) in acidic media enabling spontaneous FDCA separation, ChemSusChem 11 (2018) 2138–2145.

[109] Y.B. Huang, F.Y. Jiang, J.Y. Luo, Y.H. Zhou, Ag nanocomposites in biomass utilization: from tailored catalytic materials to chemical conversions, Chem. Catal. 3 (2023) 100676.

[110] A. Bogaerts, G. Centi, V. Hessel, E. Rebrov, Challenges in unconventional catalysis, Catal. Today 420 (2023) 114180.

[111] G. Kibria, J.P. Edwards, C.M. Gabardo, C. Dinh, A. Seifitokaldani, D. Sinton et al., 2019. Electrochemical CO2 reduction into chemical feedstocks : from mechanistic electrocatalysis models to system design. Adv. Mater. 31(31):1807166.

[112] W. Chen, L. Zhang, L. Xu, Y. He, H. Pang, S. Wang, et al., Pulse potential mediated selectivity for the electrocatalytic oxidation of glycerol to glyceric acid, Nat. Commun. 15 (2024) 1–11.

[113] M. Ismael, M. Wark, A recent review on photochemical and electrochemical nitrogen reduction to ammonia: strategies to improve NRR selectivity and faradaic efficiency, Appl. Mater. Today 39 (2024) 102253.

[114] A.S. Bandarenka, E. Ventosa, A. Maljusch, J. Masa, W. Schuhmann, Techniques and methodologies in modern electrocatalysis: evaluation of activity, selectivity and stability of catalytic materials, Analyst 139 (2014) 1274–1291.

[115] M.I. Pilo, G. Sanna, N. Spano, Conducting polymers in amperometric sensors: a state of the art over the last 15 years with a focus on polypyrrole-, polythiophene-, and poly(3,4-ethylenedioxythiophene)-based materials, Chemosensors 12 (2024) 81.

CHAPTER

13

Electrocatalysts for biomass conversion beyond inorganic materials

Liya Babu, Prathyusha P.V., Rukhsana Badarudeen, Taniya Rose Abraham, and Sabu Thomas

School of Energy Materials, Mahatma Gandhi University, Kottayam, Kerala, India

13.1 Introduction

Biomass has the potential to replace fossil fuels because it contains ample carbon, hydrogen, and oxygen content that is stored in massive natural polymers, such as lignocellulose (comprising cellulose, hemicellulose, and lignin). Biomass serves as a renewable feedstock for electrocatalytic conversion, transforming widely available lignocellulosic materials, agricultural residues, and bio-alcohols into valuable chemicals such as fuels, platform molecules, and fine chemicals via sustainable electrochemical processes [1–3]. A number of complementary electrocatalytic processes, such as selective alcohol oxidation, lignin depolymerization, electroreduction of platform molecules (furans), and paired (anodic/cathodic) tandem processes that increase energy efficiency, can electrochemically upgrade biomass to value-added chemicals and fuels. Distributed and low-temperature valorization methods for lignocellulosic biomass and its derivatives are enabled by these techniques, which operate in mild conditions, offer precise potential-controlled selectivity, and can directly convert renewable energies into chemical bonds [4–6].

Although effective biomass oxidation and reduction are made possible by noble metal and transition oxide electrocatalysts, their high cost,

Electrochemical Biomass Conversion: Sustainable Processes and Applications for Bioenergy, Biofuels, and Bioproducts
DOI: https://doi.org/10.1016/B978-0-443-26757-4.00009-X

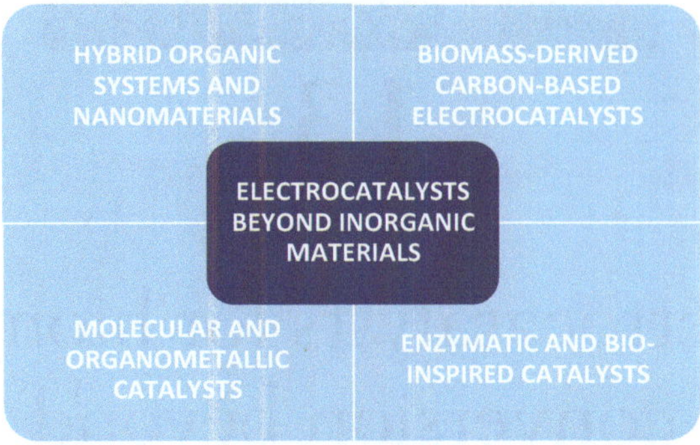

FIGURE 13.1 Biomass electrocatalysts beyond inorganic materials.

elemental scarcity, and environmental impact restrict their commercialization in decentralized bio-refineries. By utilizing renewable feedstocks for efficient and sustainable performance in alcohol oxidation, lignin depolymerization, and platform molecule valorization, organic or bio-derived catalysts (Fig 13.1), such as pyrolysis-derived N/P/S-doped carbons, bio-inspired enzymes, and heterogenized molecular systems, address these limitations [7,8].

13.2 Fundamentals of electrocatalysts beyond inorganic materials

To attain selectivity and efficiency on par or even better than noble metals, noninorganic electrocatalysts for biomass conversion rely on three interrelated design paradigms: heteroatom doping, porosity engineering, and active sites design.

Heteroatom doping: Through pyrolysis of biomass precursors or post-synthetic alteration, heteroatom doping introduces nitrogen, phosphorus, sulfur, or boron to carbon matrices. By tuning the Fermi level and activating nearby carbon atoms, nitrogen doping (pyridinic, graphitic, and pyrrolic forms) improves oxygen reduction and alcohol oxidation. These dopants also modify electron density and produce polarized active sites. While sulfur doping increases hydrophilicity and promotes mass transport of hydrophobic biomass substrates, phosphorus doping enhances Lewis acidity for C–O bond activation in lignin oxidation [7–9].

Porosity engineering: In order to balance accessible surface area and diffusion kinetics, porosity engineering regulates micro-, meso-, and macropore distributions. In high-concentration biomass slurries, hierarchical pores facilitate quick substrate or product movement, micropores offer a large surface area for active site density, and tailored pore size distributions (usually 2–50 nm) avoid pore-blocking by biomass polymers and intermediates. Biomass-derived carbons are more affordable since they naturally display variable porosity through activation protocols (KOH, CO_2, and steam) [8].

Active sites design: In enzymatic and organic systems, active site design is different from inorganic systems. In carbon materials, phosphorus-containing moieties and pyridinic-N sites function as catalytic centers, frequently strengthened by synergistic interactions with heteroatoms or residual metal impurities. While anchoring on carbon supports via covalent bonds, electrostatic interactions, or physical adsorption to maintain redox accessibility, immobilization protects the natural coordination sphere (copper centers, heme groups) for enzyme-based catalysts (laccases, peroxidases). Molecular catalysts use tunable organic ligands, such as phenanthroline coordinating Ni/Co and bipyridine, to form distinct, electrochemically accessible active sites with exquisite steric/electronic control over substrate binding and product selectivity [8,10].

Standard metrics are necessary for rigorous evaluation of noninorganic catalysts in order to facilitate comparison with inorganic benchmarks and direct optimization. Overpotential (η), measured in mV, which is the applied potential above the thermodynamic need for a reaction, is the fundamental metric. Overpotentials of 200 to 400 mV at 10 mA/cm^2 are typical for optimal bio-derived carbons in biomass oxidation reactions (e.g., glycerol to glucuronic acid), often competitive with or better than Pt/IrO_2 couples. To account for geometric and roughness variables, overpotential is plotted using linear sweep voltammetry (LSV) and compared at iso-current densities [7,8].

By normalizing current to the number of active sites, turnover frequency (TOF), which is obtained from electrochemical characterization or physical approaches, evaluates catalytic efficiency. It is measured in s^{-1}. The kinetics of C–C/C–O cleavage are immediately reflected in TOF for molecular catalysts; typical values for Ni-complexes in glycerol oxidation range from 0.1 to 1 s^{-1}. To detect rate-limiting stages, such as electron transfer, substrate transport, or chemical change, TOF is crucial [2].

In biomass electrolysis, where multiple byproducts (CO_2, H_2, side-oxidized species) compete, Faradaic efficiency (FE), measured in percentage (%), is crucial since it quantifies the percentage of electrons delivered that result in the desired product. FE is calculated as:

$$FE\ (\%) = 100\% \times \frac{\textit{mole of product}}{\textit{total charge passed}/(n \times 96485\ \textit{Cmol}^{-1})}$$

Because mild, enzyme-like routes are more selective than aggressive inorganic oxidizers, bio-derived catalysts frequently obtain 60% to 90% FE in selective alcohol oxidation or furfural reduction [5,6].

13.3 Biomass-derived carbon-based electrocatalysts

13.3.1 Precursors and synthesis

Lignocellulosic wastes like corncob and cattail are ideal, sustainable, and low-cost precursors for electrocatalytic carbons because their intrinsic multiscale structure and biochemical makeup yield hierarchical porosity and abundant heteroatom anchoring sites upon thermal conversion. This process creates high-surface-area, defect-rich carbons suitable for advanced electrocatalysis, including ORR/OER/HER and metal–air battery cathodes [11].

Three interconnected processes make up a typical processing route:

1. Pretreatment and drying (size reduction, washing, and optional hydrothermal pretreatment),
2. pyrolysis/carbonization under an inert environment (typically 500°C–1000°C) and
3. activation and/or chemical doping to open pores and introduce heteroatoms.

The lignocellulose is physically pyrolyzed into a carbonaceous matrix whose microstructure (degree of graphitization, defect density) is highly temperature dependent. Larger graphitic domains are preferred at higher carbonization temperatures, but surface functional groups that act as doping sites may be reduced. High specific surface areas and a hierarchical pore network are often produced by chemical activation (KOH, H_3PO_4, $ZnCl_2$ are common). For corncob-derived carbons, KOH activation at 700°C to 900°C with alkaline, biochar ratios in the range ~3:1 to 5:1 frequently yields large BET areas and well-developed micropore/mesopore structures [11].

Heteroatom doping (N, P, and S) can be accomplished ex situ (post-treatment, like ammonia annealing) or in situ (mixing biomass with precursors like urea or melamine prior to pyrolysis). Melamine and urea are common N-sources that add catalytically active species that alter electronic structure, such as pyridinic and graphitic -N. Phosphoric acid (H_3PO_4) is frequently used to add phosphorus, whereas elemental S or thiourea is frequently used to introduce sulfur. Importantly, codoping techniques (N+P, N+S, or N+P+S) are very successful because they offer additional defect sites and synergistic electronic effects that greatly increase electrocatalytic activities (ORR/CO2RR/OER) [12].

13.3.2 Structure–property relationships

Three important components contribute to the improved electrocatalytic performance of carbons obtained from biomass: heteroatom dopants, defect sites/edge-rich domains, and high accessible surface area/hierarchical porosity. Nitrogen configurations have distinct roles: graphitic-N increases conductivity, whereas pyridinic-N and pyrrolic-N change intermediate adsorption to increase ORR activity. Furthermore, as confirmed by experimental and DFT studies, codoping with P or S asymmetrically disrupts the electronic network, generating polarized sites that reduce reaction barriers and increase selectivity [13].

13.3.3 Performance in key reactions

Carbons generated from biomass, such as wood, corncobs, and fruit shells, are a desirable and environmentally friendly type of electrocatalyst support and catalyst. These materials acquire high surface area, variable surface chemistry, and hierarchical porosity through controlled pyrolysis, activation, and heteroatom doping (N, P, S, and B). They are very useful for a variety of electrochemical processes because of these characteristics, which guarantee plenty of active sites, enhanced mass transport, and superior electrical conductivity [14].

Oxygen reduction (ORR) and oxygen evolution reaction (OER): Carbons generated from biomass serve as conductive substrates for transition metals and oxides, as well as metal-free electrocatalysts when doped with N/P/S. In alkaline environments, properly doped carbons exhibit ORR performance competitive with Pt/C, exhibiting good tolerance and durability. By forming intimate metal–carbon contacts and accessible active sites, decorating the porous carbon framework with single atoms or metal oxides improves performance for bifunctional ORR/OER activity. Even if they are quite successful, Pt-level activity usually still needs to be further optimized utilizing metal-containing hybrid structures, especially in acidic media [15].

Nitrate reduction (eNO$_3$RR)/biomass electrolysis: In the electrochemical nitrate-to-ammonia conversion NO_3^- - NH_3, biomass-derived carbons predominantly serve as high-surface-area, porous substrates for metal nanoparticles (e.g., Cu, Co) or single-atom centers that facilitate the selective multielectron reduction. These carbon-supported metal catalysts have benefits such as increased metal dispersion, mass transport, and variable adsorption characteristics, resulting in promising NH_3 production rates and selectivity in lab tests. However, the process confronts several hurdles, including competition from the hydrogen evolution reaction (HER) and low Faradaic efficiency. As a result, carefully developed metal/bio-carbon hybrids are currently outperforming pristine metal-free carbons [9].

13.3.4 Case studies

Vertically aligned nitrogen-containing carbon nanotube arrays (VA-NCNTs) were found to be a superior metal-free catalyst for the oxygen reduction reaction (ORR) in alkaline medium. The VA-NCNTs demonstrated significantly greater steady-state ORR performance than the traditional platinum-on-carbon (Pt/C) catalyst, achieving a higher current density at comparable potentials. This was a groundbreaking discovery because it was the first high-profile showing that an N-doped carbon material might outperform Pt/C in alkaline ORR [16].

VA-NCNTs' dense vertical alignment and substitutional nitrogen doping result in highly accessible, positively polarized carbon sites with excellent mass transport, permitting an efficient 4-electron ORR pathway and giving greater stability and CO-tolerance compared to platinum [17].

Nanoporous carbons and nitrogen-rich carbon frameworks developed from low-cost graphitic carbon nitride (g-C_3N_4) precursors have demonstrated outstanding performance as metal-free catalysts for the ORR in alkaline media. Their ORR activity is very competitive, with reported half-wave potentials (E 1/2) that frequently approach or even exceed those of commercial platinum-on-carbon (Pt/C). This makes g-C_3N_4 -templated carbons an attractive, low-cost alternative to platinum for energy conversion technologies such as fuel cells and metal-air batteries [18]. g-C_3N_4 is a nitrogen-rich scaffold that forms graphitic domains with abundant N-functional sites and hierarchical porosity, making it ideal for ORR active-site exposure and fast mass transfer [19].

13.4 Enzymatic and bio-inspired catalysts

Enzymatic and bio-inspired systems have emerged as potential solutions for selective biomass upgrading under mild conditions, and the investigation of noninorganic electrocatalysts has greatly expanded. These systems offer substrate specificity, environmental compatibility, and low-energy pathways that are ideal for valorizing complex feedstocks such as lignin and glycerol, in contrast to conventional metal-based catalysts.

13.4.1 Natural enzyme-based electrocatalysts

Laccases and peroxidases, widely found in fungi and plants, have demonstrated high efficiency in C–O bond cleavage, especially in lignocellulosic substrates. Under ambient conditions, depolymerization is enabled by laccases, multicopper oxidases that catalyze the oxidation of phenolic units in lignin while reducing O_2 to H_2O [20]. Similarly, through peroxide-driven processes, horseradish peroxidase (HRP) and manganese peroxidase (MnP) facilitate the oxidative breakdown of alcohol and ether bonds in glycerol and lignin-derived intermediates [21].

13.4.2 Immobilization strategies and electrode engineering

These enzymes are often immobilized on conductive substrates, such as carbon nanotubes, graphene, and polymer matrices, to enhance operating stability in electrochemical conditions. Immobilization increases the lifetime of the catalyst, improves electron transport, and stops denaturation [22]. Immobilization methods (adsorption, covalent attachment, entrapment) provide an important framework for designing enzyme-electrode interfaces suitable for biomass electrocatalysis [23]. In bioelectrocatalytic reactors, carbon felt and screen-printed electrodes have also been utilized as reliable substrates for enzyme-electrode integration. These methods can be used directly in biomass conversion systems, where steady, effective electron mediation is necessary for selective oxidation and depolymerization.

13.4.3 Bio-inspired molecular catalysts

Quinones and metalloporphyrins are examples of bio-inspired chemical catalysts that mimic the active sites of enzymes and function in comparably mild environments. For the low-potential electrooxidation of glycerol and alcohols produced from lignin, quinone-based mediators work well. Inspired by heme enzymes, porphyrin complexes provide metal-centered redox sites that can be adjusted for hydrogen abstraction, C–O activation, and controlled oxidation of biomass derivatives [7,24]. Their modular chemistry enables robust anchoring to conductive supports and accurate redox property adjustment.

As sustainable photocatalysts that mimic bio-inspired redox pathways, organic dyes are becoming more widely acknowledged. Tacchi et al. [25] review of the photocatalytic oxidation of alcohols with organic dyes—moving from homogeneous systems to supported dye-sensitized materials—has made a substantial contribution. Their research demonstrates how photo-induced electron transfer allows for moderate, selective oxidation of organic dyes like eosin Y, rose bengal, and porphyrin-based chromophores.

Enzymes, enzyme mimics, and organic dye photocatalysts are examples of biological and biomimetic catalytic systems that present intriguing avenues for selective, energy-efficient biomass conversion. The shift from laboratory-scale demonstrations to sustainable industrial processes will be expedited by ongoing advancements in enzyme immobilization, electrode design, and molecular catalyst engineering, together with discoveries from organic dye photocatalysis [25].

13.5 Molecular and organometallic catalysts

Molecular and organometallic catalysts, particularly those based on earth-abundant transition metals such as nickel and cobalt, play a vital role in modern sustainable chemistry. Through ligand design, particularly with bipyridine frameworks, their activity and selectivity can be precisely adjusted, allowing for effective C–C bond activation and functionalization. In order to enhance stability, recyclability, and compatibility with continuous flow systems, recent developments also concentrate on heterogenizing these homogenous complexes by anchoring them onto solid supports. In order to produce value-added compounds from renewable biomass platforms like furfural while retaining high efficiency and selectivity, such catalyst advances are essential.

13.5.1 Homogeneous Ni/Co complexes with bipyridine ligands for C–C activation

Nickel–bipyridine catalysts: Homogeneous organometallic catalysts featuring earth-abundant transition metals such as nickel (Ni) and cobalt (Co) have seen rapid development owing to their cost-effectiveness, tunability via ligand design, and ability to activate otherwise inert C–C bonds under mild conditions. In contemporary cross-coupling and C–C activation chemistry, nickel complexes bound by bipyridine and its derivatives have been thoroughly investigated. Ni(II) precursors, such as NiX_{2h}bpy, are used in a significant class of catalysts. These precursors undergo redox cycling during catalysis, allowing for both radical and two-electron routes, depending on substrates and circumstances. Ni(bpy) systems are essential to photoredox Ni catalysis and cross-electrophile coupling techniques, in which oxidative addition and reductive elimination cycles are coordinated by Ni(I)/Ni(III) intermediates [26].

Bipyridyl ligands affect the speciation and reactivity of Ni in critical phases such as oxidative addition of aryl halides and radical capture processes, according to a growing body of mechanistic research. Variants of bpy with sterically bulky or electron-withdrawing substituents have been utilized to control catalytic intermediates and modify reduction potentials, which directly affect the selectivity and efficiency of C–C bond forming [26].

Cobalt–bipyridine catalysts: Another first-row metal that has similar reactivity in C–C and C–X activation processes is cobalt. It has been demonstrated that cobalt catalysts containing nitrogen-donor ligands, such as bipyridine analogs, influence single-electron transfer (SET) and concerted metalation–deprotonation (CMD) mechanisms in activations such as cycloadditions, C–H functionalizations, and cross-couplings. Although the breadth of Co(bpy) complexes for direct C–C activation currently lags behind Ni and Pd systems, their capacity to participate in radical pathways and promote C–H/C–X activation emphasizes the usefulness of adjusting ligand electronics to access novel transformations [27].

13.5.2 Heterogenization: Anchoring homogeneous complexes onto supports for recyclability and flow catalysis

Molecular Ni/Co catalyst heterogenization has become a potent tactic to combine the practical benefits of heterogeneous systems with the accuracy of homogeneous catalysis, especially for continuous-flow and recyclable catalytic processes. For industrial and sustainable applications, immobilizing traditional homogeneous catalysts on solid supports is crucial because these catalysts frequently face problems such as metal leaching, problematic separation, and short operating lives. Covalent anchoring of ligands or metal–ligand complexes onto inorganic or polymeric substrates, such as silica, mesoporous oxides, or functional polymers, is one often used technique. Covalent grafting guarantees low leaching and great reusability in flow reactors, in addition to stabilizing the active site. Reviews show that Ni/Co complexes can be directly immobilized onto organic–inorganic hybrid surfaces to maintain molecular-level activity and facilitate catalyst recycling [28].

The integration of catalytic complexes into ordered porous frameworks, such as periodic mesoporous silica materials with integrated ligand functions, metal–organic frameworks (MOFs), and covalent organic frameworks (COFs), is an even more sophisticated approach. Postsynthetic metalation with Ni or Co can provide spatially isolated but molecularly well-defined active sites in frameworks containing preinstalled bipyridine or pyridyl units [29]. These heterogenized catalysts improve diffusion, selectivity, and turnover number by utilizing the high surface area, adjustable pore environment, and chemical resilience of porous materials. For instance, because of their consistent binding pockets and inhibited catalyst aggregation, bipyridine-functionalized organosilica nanotubes and MOFs have shown exceptional activity in C–H and C–C activation processes [30].

These strategies-covalent anchoring, supramolecular immobilization, and framework incorporation are complementary means to preserve the complex reactivity of molecular Ni/Co catalysts while significantly enhancing their stability, recyclability, and flow system applicability. A significant step toward scalable and sustainable catalytic reactions is represented by these heterogeneous systems.

13.5.3 Selectivity advantages and electronic tuning for platform chemicals

Since furfuryl alcohol (FA), tetrahydrofurfuryl alcohol (THFA), and other derivatives are important intermediates for polymers, fuels, and fine chemicals, selective catalytic conversion of biomass-derived platform chemicals like furfural is a crucial goal in sustainable chemistry. Directing the reaction pathway to a desired product while inhibiting side reactions like over-hydrogenation, decarbonylation, or ring saturation is the main problem in furfural upgrading. This is where supported and programmable molecular catalysts show a big benefit.

The electrical and structural characteristics of the active metal sites, which can be modified by ligand design, alloying, and support interactions, have a significant impact on catalyst selectivity. For instance, Ni–Cu alloy systems supported on modified carbon or oxide supports show excellent selectivity toward THFA with nearly complete conversion under mild conditions (e.g., 100% conversion and ~99% selectivity to THFA over Ni_2Cu_1/RHPC catalysts at 50°C and 1 MPa H_2) [31].

Similar to this, it has been demonstrated that varying the metal composition of Cu–Ni catalysts can affect the equilibrium between the hydrogenation and C–O bond cleavage stages, allowing for great selectivity toward furfuryl alcohol or, depending on the intended use, deeper hydrogenated products [32]. Combinations of metal active sites can enhance H_2 activation and intermediate stability in Co–Ni and other multimetallic doped systems, resulting in selectivity for furfuryl alcohol exceeding 99% in specific compositions [33].

These findings highlight the significance of electronic tuning in regulating adsorption modes and reaction pathways, whether by metal alloying, ligand effects in molecular catalysts, or support-induced electronic modulation. Because furfural can be efficiently valorized into high-value compounds with excellent selectivity through strategic catalyst design, these systems are becoming more and more appealing for biomass upgrading in both batch and continuous processes.

13.6 Hybrid organic systems and nanomaterials

For biomass upgrading, hybrid organic systems combine polymeric scaffolds, covalent frameworks, and MOF-derived structures with catalytically active sites to provide synergistic advantages over pure organics or inorganics. These materials achieve increased activity in complex feedstocks such as lignocellulosic slurries by utilizing the mechanical resilience and tunability of organics with high surface areas and hierarchical porosity [7].

13.6.1 Polymer composites

Polymer composites are flexible, conducting scaffolds that facilitate mass transport and electron transfer while immobilizing molecular catalysts or enzymes. When polyaniline (PANI) is doped with heteroatoms or sulfonic acids, it creates carbon-supported nanocomposites that offer π-conjugated routes for quick charge transport during alcohol oxidation. Redox-active states (~0.4–0.8 V vs RHE) in emeraldine salt forms of PANI work in conjunction with anchoring Ni/Co complexes to selectively break C–O in glycerol or furfural, resulting in overpotentials <350 mV and FEs >80%[2].

Covalent organic frameworks (COFs) are crystalline porous polymers composed of light elements (C, H, N, O, and B) by strong covalent bonds like imine, boronate, or β-ketoenamine linkages. They offer ultrahigh surface areas (>2000 m^2/g), precise pore size control (1–5 nm), and tunable electronics through organic building blocks [34,35]. Redox-active linkers such as bipyridine, porphyrin, phthalocyanine, or triazine units are incorporated during solvothermal synthesis to create ordered 2D/3D frameworks with semiconducting properties (band gaps 1.5–2.5 eV) and π-conjugation for charge transport, which can often be enhanced by graphene or carbon nanotube hybrids. Linkers like tetrathiafulvalene (TTF) or thieno[3,2-b]thiophene optimize electron routes for biomass electrocatalysis, while amino/imidazolium groups increase substrate capture (such as furans and alcohols) and lower activation barriers [34,36].

Because of their size-selective pores and well-defined active sites, COF-based electrocatalysts are excellent at selectively upgrading biomass platform molecules. Ni/Co-porphyrin COFs, such as NiPor-CTF, use porphyrin metal centres for C–O cleavage to catalyze 5-Hydroxymethylfurfural (HMF) oxidation to 2,5-Furandicarboxylic acid (FDCA) or furfural reduction with FEs >90% and TOFs >0.5 s^{-1}. Dioxin linkers improve stability for more than 24 h, and phthalocyanine-linked COFs obtain >96% FE for CO generation from CO_2. When it comes to glycerol or lignin oxidation, hierarchical macro or microporous COFs outperform amorphous polymers in facilitating mass transport in viscous biomass feeds. Through electronic manipulation, postsynthetic metalation, such as CoPc or Ni-Bpy COFs, adjusts selectivity. Electron-donating groups move d-band centers to improve the binding of oxygenated intermediates [34,37].

13.6.2 MOF-derived organics

After being pyrolyzed or postsynthetically altered, MOFs produce carbon scaffolds that are rich in nitrogen while maintaining organic linker motifs for durable catalysis.

To create hybrid nodes that resemble natural active sites, amine-functionalized UiO-66 or ZIF-8 are grafted with porphyrins or enzymes. Cu-porphyrin@UiO-66 selectively breaks β-O-4 bonds for lignin depolymerization by controlled redox cycling, with vanillin yields >40% in real kraft lignin feeds [10].

Single-atom metal-nitrogen sites within organic matrices are preserved during high-temperature carbonization of MOFs. Because of the protective carbon encapsulation, these catalysts perform exceptionally well in oxygen evolution during paired biomass electrolysis, maintaining an overpotential of less than 300 mV over 200 h [7].

13.7 Challenges, applications, and outlook

13.7.1 Challenges

Electrocatalytic biomass conversion has made significant strides, but there are still a number of important obstacles in the way of its widespread practical application. Achieving long-term stability and durability of catalysts under actual, complicated biomass feed circumstances is one of the main challenges; changes in pH, contaminants, and variable reactant compositions all hasten degradation and shorten catalytic lifespan. This problem is similar to more general electrocatalysis problems with deterioration in severe electrochemical conditions, which shorten the life of equipment like fuel cells and electrolyzers [38]. Additionally, many biomass electrocatalytic reactions still lack a thorough mechanistic understanding, especially when it comes to multielectron, multistep transformations like the oxidation or reduction of lignocellulosic derivatives. Rational catalyst design and predicted selectivity adjustment are hampered by unclear structure-activity connections [2].

13.7.2 Real-world applications

Proton exchange membrane (PEM) electrolyzers and other electrochemical devices that incorporate sophisticated electrocatalysts offer interesting avenues for biomass upgrading applications. Electrocatalytic proton exchange membrane designs can enable electrochemical refinement akin to traditional PEM water splitting technologies by efficiently hydrogenating and deoxygenating biomass intermediates at low temperatures and pressures (electrochemical conversion of biomass) [39]. Whether for the conversion of biomass feed or the evolution of hydrogen, electrocatalysts designed for PEM systems must strike a balance between stability and activity under the very acidic or oxidative conditions typical of these systems [40]. Furthermore, photovoltaic-coupled electrochemical systems are becoming more popular. Although this increases the complexity of system integration and controls, directly connecting renewable electricity to biomass electrocatalysis can improve the sustainability and energy efficiency of converting biomass to fuels or chemicals.

13.7.3 Future direction

By anticipating the ideal compositions, morphologies, and active site settings for intricate biomass conversion processes, AI-driven catalyst design and optimization provide a mechanism to speed up discovery and enhance performance in the future. By combining machine learning with multiscale modelling and high-throughput experimentation, existing mechanistic gaps might be filled, and materials selection could be more precisely guided. In order to assess the economic feasibility and environmental impact of biomass electrocatalytic technologies in comparison to existing thermochemical processes, it will be essential to develop techno-economic analysis (TEA) frameworks and life cycle assessments. These analyses must take into account feedstock sourcing, catalyst manufacturing costs, separation and purification requirements, and system integration complexities [41].

References

[1] M. Mujtaba, L. Fernandes Fraceto, M. Fazeli, S. Mukherjee, S.M. Savassa, G. Araujo De Medeiros, et al., Lignocellulosic biomass from agricultural waste to the circular economy: a review with focus on biofuels, biocomposites, and bioplastics, J. Cleaner Prod. 402 (2023) 136815, doi:10.1016/j.jclepro.2023.136815.

[2] Y. Ramli, V. Chaerusani, Z. Yang, R. Yang, J. Zhang, A. Abudula, et al., Electrochemical conversion of biomass derivatives to value-added chemicals: a review, Green Carbon 3 (2) (2025) 113–129, doi:10.1016/j.greenca.2024.10.004.

[3] D. Zhao, T. Su, Y. Wang, R.S. Varma, C. Len, Recent advances in catalytic oxidation of 5-hydroxymethylfurfural, Mole. Catal. 495 (2020) 111133, doi:10.1016/j.mcat.2020.111133.

[4] F.W.S. Lucas, R.G. Grim, S.A. Tacey, C.A. Downes, J. Hasse, A.M. Roman, et al., Electrochemical routes for the valorization of biomass-derived feedstocks: from chemistry to application, ACS Energy Lett. (2021) 1205–1270, doi:10.1021/acsenergylett.0c02692.

[5] H. Zhou, Y. Ren, B. Yao, Z. Li, M. Xu, L. Ma, et al., Scalable electrosynthesis of commodity chemicals from biomass by suppressing non-Faradaic transformations, Nat. Commun. 14 (1) (2023) 5621, doi:10.1038/s41467-023-41497-y.

[6] P. Zhu, M. Shi, Z. Shen, X. Liao, Y. Chen, Electrocatalytic conversion of biomass-derived furan compounds: mechanisms, catalysts and perspectives, Chem. Sci. 15 (13) (2024) 4723–4756, doi:10.1039/D4SC00546E.

[7] T. Liu, H. Yabu, Biomass-derived electrocatalysts: low-cost, robust materials for sustainable electrochemical energy conversion, Adv. Energy Sustain. Res. 5 (1) (2024) 2300168, doi:10.1002/aesr.202300168.

[8] M. Wang, S. Wang, H. Yang, W. Ku, S. Yang, Z. Liu, et al., Carbon-based electrocatalysts derived from biomass for oxygen reduction reaction: a minireview, Front. Chem. 8 (2020) 116, doi:10.3389/fchem.2020.00116.

[9] T. Xie, X. Li, J. Li, J. Chen, S. Sun, Y. Luo, et al., Co-nanoparticles decorated corncob-derived biomass carbon as an efficient electrocatalyst for nitrate reduction to ammonia, Inorg. Chem. 61 (35) (2022) 14195–14200, doi:10.1021/acs.inorgchem.2c02499.

[10] X. Liu, Y. Wang, H. Duan, Recent progress in electrocatalytic conversion of lignin: from monomers, dimers, to raw lignin, Prec. Chem. 2 (9) (2024) 428–446, doi:10.1021/prechem.4c00024.

[11] B. Escobar, D.C. Martínez-Casillas, K.Y. Pérez-Salcedo, D. Rosas, L. Morales, S.J. Liao, et al., Research progress on biomass-derived carbon electrode materials for electrochemical energy storage and conversion technologies, Int. J. Hydrog. Energy 46 (51) (2021) 26053–26073, doi:10.1016/j.ijhydene.2021.02.017.

[12] J.C. Martínez-Loyola, M.A. Carrasco-Cordero, I.L. Alonso-Lemus, F.J. Rodríguez-Varela, P. Bartolo-Pérez, B. Escobar-Morales, et al., Systematic study of the N concentration effects on metal-free ORR electrocatalysts derived from corncob: less is more., Electrochem. Commun. 166 (2024) 107792, doi:10.1016/j.elecom.2024.107792.

[13] X. Hao, X. An, A.M. Patil, P. Wang, X. Ma, X. Du, et al., Biomass-derived N-doped carbon for efficient electrocatalytic CO_2 reduction to CO and Zn–CO_2 batteries, ACS Appl. Mater. Interfaces 13 (3) (2021) 3738–3747, doi:10.1021/acsami.0c13440.

[14] M. Wang, S. Wang, H. Yang, W. Ku, S. Yang, Z. Liu, et al., Carbon-based electrocatalysts derived from biomass for oxygen reduction reaction: a minireview, Front. Chem. 8 (2020) 116, doi:10.3389/fchem.2020.00116.

[15] S. Park, J. Kim, K. Kwon, A review on biomass-derived N-doped carbons as electrocatalysts in electrochemical energy applications, Chem. Eng. J. 446 (2022) 137116, doi:10.1016/j.cej.2022.137116.

[16] K. Gong, F. Du, Z. Xia, M. Durstock, L. Dai, Nitrogen-doped carbon nanotube arrays with high electrocatalytic activity for oxygen reduction, Science 323 (5915) (2009) 760–764, doi:10.1126/science.1168049.

[17] M. Scardamaglia, C. Bittencourt, Metal-free catalysis based on nitrogen-doped carbon nanomaterials: a photoelectron spectroscopy point of view, Beilstein J. Nanotechnol. 9 (2018) 2015–2031, doi:10.3762/bjnano.9.191.

[18] Y. Zheng, Y. Jiao, J. Chen, J. Liu, J. Liang, A. Du, et al., Nanoporous graphitic-C_3 N_4 @carbon metal-free electrocatalysts for highly efficient oxygen reduction, J. Am. Chem. Soc. 133 (50) (2011) 20116–20119, doi:10.1021/ja209206c.

[19] H. Piao, G. Choi, X. Jin, S.-J. Hwang, Y.J. Song, S.-P. Cho, et al., Monolayer graphitic carbon nitride as metal-free catalyst with enhanced performance in photo- and electro-catalysis, Nano-Micro Lett. 14 (1) (2022) 55, doi:10.1007/s40820-022-00794-9.

[20] S. Riva, Laccases: blue enzymes for green chemistry, Trends Biotechnol. 24 (5) (2006) 219–226.

[21] M. Dashtban, H. Schraft, T.A. Syed, W. Qin, Fungal biodegradation and enzymatic modification of lignin, Int. J. Biochem. Mol. Biol. 1 (1) (2010) 36–50.

[22] R.A. Sheldon, S. van Pelt, Enzyme immobilisation in biocatalysis: why, what and how, Chem. Soc. Rev. 42 (15) (2013) 6223–6235.

[23] C.Y. Chen, A.A. Bojang, D. Damayanti, H.S. We, Immobilization of enzymes on electrodes and electrode design in biofuel cells, Catalysts 15 (3) (2025) 253, doi:10.3390/catal15030253.

[24] K.S. Kadish, The Porphyrin Handbook, Academic Press, San Diego, CA, 2000.

[25] E. Tacchi, G. Rossi, A. Sartorel, Photocatalytic oxidation of alcohols with organic dyes: from homogeneous systems to sensitized materials, Chemistry 31 (2025) e202501124.

[26] A.D. Cagan, D. Bim, N.P. Kazmierczak, R.G. Hadt, Mechanisms of photoredox catalysis featuring nickel–bipyridine complexes, ACS Catal 14 (11) (2024) 9055–9907.

[27] M.M. Tohidi, B. Paymard, S.R. Vasquez-García, D. Fernández-Quiroz, Recent progress in applications of cobalt catalysts in organic reactions, Tetrahedron 136 (2023) 133352.

[28] B. Zhang, J.N.H. Reek, Supramolecular strategies for the recycling of homogeneous catalysts, Chemistry 16 (23) (2021) 3851–3863.

[29] S. Kumar, B. Mohan, Z. Tao, H. You, P. Ren, Incorporation of homogeneous organometallic catalysts into metal–organic frameworks for advanced heterogenization: a review, Catal. Sci. Technol 11 (17) (2021) 5734–5771.

[30] S. Zhang, H. Wang, M. Lei, J. Han, X. Liu, J. Gong, Molecular heterogeneous catalysts derived from bipyridine-based organosilica nanotubes for C–H bond activation, Chem. Sci 8 (2017) 4489–4496.

[31] Z. Ding, Y. Gao, L. Hu, Highly efficient and selective hydrogenation of biomass-derived furfural using interface-active rice husk-based porous carbon-supported NiCu alloy catalysts, Molecules 29 (11) (2024) 2638.

[32] Y. Gao, J. Wang, Z. Liu, S. Li, Q. Liu, Boosting furaldehyde hydrogenation to furfuryl alcohol: role of Ni in Cu5Nix/SiO_2 bimetallic catalysts, Catalysts 15 (12) (2025) 1151.

[33] S. Li, Y. Fan, C. Wu, C. Zhuang, Y. Wang, X. Li, et al., Selective hydrogenation of furfural over the co-based catalyst: a subtle synergy with Ni and Zn dopants, ACS Appl. Mater. Interfaces 13 (2021) 8507–8517.

[34] Y. Fan, M. Chen, N. Xu, K. Wang, Q. Gao, J. Liang, et al., Recent progress on covalent organic framework materials as CO_2 reduction electrocatalysts, Front. Chem. 10 (2022) 942492, doi:10.3389/fchem.2022.942492.

[35] J. Feng, W. Zhang, Z. Gu, Covalent organic frameworks for electrocatalysis: design, applications, and perspectives, ChemPlusChem 89 (10) (2024) e202400069, doi:10.1002/cplu.202400069.

[36] S. Tao, D. Jiang, Covalent organic frameworks for energy conversions: current status, challenges, and perspectives, CCS Chem. 3 (1) (2021) 2003–2024, doi:10.31635/ccschem.020.202000491.

[37] X. Zhao, P. Pachfule, S. Li, T. Langenhahn, M. Ye, C. Schlesiger, et al., Macro/microporous covalent organic frameworks for efficient electrocatalysis, J. Am. Chem. Soc. 141 (16) (2019) 6623–6630, doi:10.1021/jacs.9b01226.

[38] S.M. Golekar, Electrocatalysis in green energy systems: from hydrogen production to CO_2 reduction, Int. J. Adv. Acad. Studies 7 (2025) 125–131.

[39] *Electrochemical conversion of biomass.* ChemCatBio. (n.d.). https://www.chemcatbio.org/capabilities/evaluation/electrochemical-conversion.

[40] Y. Zhou, H. Zhong, S. Chen, G. Wen, L. Sehn, Y. Wang, et al., Proton exchange membrane-based electrocatalytic systems for hydrogen production, Carbon Energy 7 (1) (2024) e629.

[41] J. Barrio, A. Pedersen, S. Favero, H. Luo, M. Wang, S. Sharma, et al., Bioinspired and bioderived aqueous electrocatalysis, Chem. Rev. 123 (2022) 2311–2348.

CHAPTER

14

Integrated microbial and electrochemical biomass conversion

Mrutyunjay Padhiary[a], *Ramyakrishna Pothu*[b,c], *and Amit Kumar Rajak*[d]

[a]Department of Agricultural Engineering, Assam University, Silchar, Assam, India, [b]Centre for Research and Development Cell, Sharda University, Greater Noida, Uttar Pradesh, India, [c]School of Physics and Electronics, College of Chemistry and Chemical Engineering, Hunan University, Changsha, Hunan, China, [d]Department of Energy & Environmental Engineering, CSIR-Indian Institute of Chemical Technology, Hyderabad, Telangana, India

14.1 Introduction

14.1.1 Overview of biomass as a renewable resource

Biomass is a renewable energy source originating from organic materials, including plants, agricultural byproducts, and animal excrement [1]. These materials can be directly incinerated for heat or converted into biofuels, such as ethanol and biodiesel, which serve as substitutes for fossil fuels in transportation. Biomass captures solar energy through photosynthesis and may be converted into biofuels, chemicals, and energy using various biological, thermal, and electrochemical methods [2]. Primary sources of biomass encompass agricultural wastes, algae, municipal and industrial waste, and animal and microbial biomass. Biomass is sustainable, carbon-neutral, and may be stored and changed as required, making it an efficient renewable resource.

Electrochemical Biomass Conversion: Sustainable Processes and Applications for Bioenergy, Biofuels, and Bioproducts
DOI: https://doi.org/10.1016/B978-0-443-26757-4.00015-5

Charcoal, enzymatic hydrolysis, pyrolysis, and combustion are some of the biochemical methods that can be used to turn biomass into valuable chemicals and energy products [3]. Other electrochemical methods include bioelectrochemical systems (BES), electrocatalytic enhancement, and hybrid microbial-electrochemical methods. Applications for biomass conversion include bio-based compounds like organic acids, solvents, bioplastics, medicines, bioenergy, waste valorization, and biofuels like ethanol, biodiesel, biogas, and hydrogen. Challenges and prospects comprise feedstock supply, logistics, conversion efficiency, financial sustainability, and environmental factors [4]. Challenges cover seasonal fluctuations, transportation expenses, preprocessing necessities, technical advancements for enhanced yields and reduced costs, rivalry with fossil fuels, and environmental factors such as land use issues, water usage, and emissions control [5].

14.1.2 Role of integrated microbial and electrochemical systems

Microbial and electrochemical processes are practical options for enhancing biomass conversion efficiency [6]. These hybrid technologies enable sustainable biofuel, biochemical, and energy production by integrating microbial metabolism with electrochemical control. Major benefits include optimized biomass decomposition, specific processes of metabolism, enhanced electron transfer, and waste valorization. BES uses both microbial and electrochemical processes. Some examples are bioelectrochemical biorefineries, microbial electrolysis cells (MECs), microbial fuel cells (MFCs), and microbial electrosynthesis (MES) [7]. The sustainable production of biofuels, biochemicals, and bioelectricity becomes feasible with these systems.

Integrated systems offer advantages in increased productivity, process adaptability, sustainability, and waste reduction. However, scalability, maintaining microbiological stability, optimizing the use of external power for energy-efficient procedures, and industrial system integration are among the difficulties [8]. Integrating microbial and electrochemical processes enables an innovative approach to biomass conversion that produces effective and long-lasting bioeconomy solutions. Enhancing system performance, developing new bioelectrochemical processes, and advancing commercial applications should be the primary objectives of future research.

14.1.3 Scope of the chapter

The integration of microbial and electrochemical systems for efficient biomass conversion is reviewed in this chapter, with a focus on their potential for changing the production of bioenergy and bioproducts. It highlights the advantages of combining electrochemical and biological processes for increased productivity and their role in a circular bioeconomy. MFCs, electro-fermentation (EF), MECs, MES, and hybrid bioreactors are examples of fundamental technologies in integrated systems [9].

Some long-term and scalable uses are turning waste into useful products, making bioenergy, creating new bioproducts, and integrating biotechnology into large biorefineries and distributed bioenergy systems [10]. Improving electron flow efficiency, maximizing microbial consortia, and proposing accessible resources are among the difficulties. Adaptability, cost-effectiveness, and regulatory support for commercialization are examples of economic and policy considerations. Advances in synthetic biology, bioelectrochemical engineering, and sustainable energy techniques are among the research areas. The integrated microbial and electrochemical systems for biomass conversion are carefully reviewed in this chapter, with a focus on how they can support scalable and sustainable improvements in bioenergy and bioproducts [11].

14.1.4 Objectives of the chapter

With a focus on their technological, financial, and environmental significance, this chapter studies the potential of integrated microbial-electrochemical systems for biomass conversion. It explores the interaction between electrochemical processes and microbial metabolism. These methods for converting biomass into valuable resources include electro-fermentation, MFCs, and MESs. The viability of using BES to produce sustainable energy and bioproducts is reviewed in this chapter. Case studies on the production of biofuel, bioelectricity, and biochemical synthesis are discussed, along with innovative uses in bioenergy and biorefineries. Cost-effectiveness, ecological benefits, and operational advantages are all evaluated in this chapter, along with the systems' economic and environmental benefits. It also discusses technological issues like microbial stability and electron transfer efficiency, and offers microbial-electrochemical system research and development. This chapter explains how these systems convert biomass sustainably and their bioeconomic implications.

14.2 Fundamentals of microbial and electrochemical biomass conversion

14.2.1 Microbial conversion processes

Microbial conversion techniques are essential for biomass valorization to change organic feedstocks into energy, biofuels, and biochemicals. These

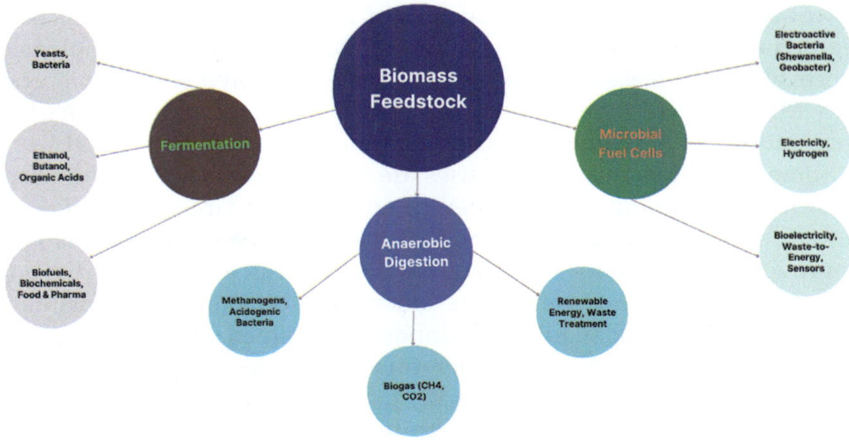

FIGURE 14.1 Comparison of microbial conversion processes.

procedures use microorganisms' metabolic abilities to reduce complex biomass components into useful byproducts. MFCs, anaerobic digestion, and fermentation are the main microbial conversion processes. In the metabolic process known as fermentation, microorganisms in anaerobic or microaerophilic environments process organic substrates into chemicals and biofuels [12]. Ethanolic fermentation, acetone-butanol-ethanol (ABE) fermentation, lactic acid fermentation, and electro-fermentation are different forms of fermentation. Applications include the manufacture of organic acids, bioethanol, and bioplastics. Anaerobic digestion, which breaks down organic materials without oxygen, produces biogas. Hydrolysis, acidogenesis, acetogenesis, and methanogenesis are the phases of anaerobic digestion. Applications include treating organic waste and producing biogas for fuel, heating, and electricity. By oxidizing organic substrates and moving electrons to an electrode, microorganisms create electricity in MFCs, which are bioelectrochemical devices [13]. The proton exchange membrane (PEM), cathode chamber, and anode chamber are the necessary elements. Applications include the production of hydrogen in MECs, biosensors, wastewater treatment, and bioelectricity [14]. Waste valorization, renewable energy production, and long-term biomass utilization are all feasible with microbial conversion techniques. These methods are more effective for large-scale bioenergy and bioproduct applications when combined with electrochemical systems.

The efficiency and scalability of these systems may still create obstacles to their mainstream adoption in large-scale applications, even though microbial conversion techniques might provide long-term advantages [15]. Also, MECs may not be feasible for some companies or geographical areas due to their complexity and the expense of maintenance. Fig. 14.1

compares key microbial conversion processes, including MFCs, MECs, and fermentation. It visualizes differences in energy recovery, conversion efficiency, and scalability. This comparison helps highlight the advantages of electroactive microbial systems in biomass valorization.

14.2.2 Electrochemical conversion processes

Electrochemical conversion methods, which integrate electrochemistry with biological systems to improve the production of biofuels, chemicals, and energy, include EF, MECs, and electrochemical reduction [16]. These methods use electric potential to increase the activity of microbes, speed up specific biochemical reactions, and make the conversion of biomass more efficient.

Improved production of ethanol, butanol, and organic acids; control of the metabolic processes for specific bioproduct synthesis; and reduction of harmful substances are some of the applications of EF, which uses an electric potential to control the metabolic processes of microbes during fermentation, maintaining redox balance, increasing product yields, and directing metabolic energy toward specific molecules. Microorganisms are used in MECs to help with electrolysis, which turns organic materials into hydrogen or methane when an external voltage is applied [17]. They operate under anaerobic conditions with an applied voltage to promote reactions. Biomethane and biohydrogen generation, wastewater treatment with simultaneous energy recovery, and the conversion of organic waste into renewable fuels are some examples of applications [18].

Electrochemical reduction is a process that converts biomass intermediates into useful compounds and fuels. Electrocatalysts help break down organic molecules, selectively driving certain redox processes using both microbial and nonmicrobial catalysts [19]. Examples of electrochemical reduction include the conversion of CO_2 to fuels, the enhancement of biomass molecules, and the production of biochemicals. Carbon-neutral energy sources, enhanced biomass processing, and recyclable chemical production from renewable biomass are all possible with BES [20]. These technologies enhance microbial activity, reduce energy requirements, and facilitate selective bioproduct synthesis, making them viable for sustainable biomass utilization and industrial-scale applications. This process is expensive, requires more infrastructure and research, and must address the environmental impact of creating rare earth metals for electrodes.

14.2.3 Synergy in integrated systems

Energy production, waste management, and resource recovery can all be enhanced by the integration of microbial and electrochemical processes. MES involves microorganisms interacting with electrodes to transfer electrons, improving energy efficiency and sustainability [21]. MFCs provide electricity for power generation and wastewater treatment by using microorganisms to break down organic materials. When an external voltage is applied to waste materials, MECs produce hydrogen gas. In wastewater and contaminated areas, low-voltage electricity will speed up the breakdown of contaminants by stimulating microbial metabolism. Metals may be dissolved by microbiological processes and recovered by electrochemical techniques. Also, microorganisms have the potential to either prevent or induce corrosion, which may have an effect on industrial applications. As a result, it is crucial to thoroughly monitor and regulate microbial activity in these systems to guarantee optimal efficiency and performance. Researchers may develop designs to improve resource recovery and energy production in MFCs and MECs by studying how microorganisms and electrochemical processes interact [22]. To reduce damage to industrial equipment and infrastructure, further research is needed on the pros and cons of microbial corrosion. For wastewater treatment and energy generation, MFCs and MECs have significant potential, but effective microbial control is essential. Fig. 14.2 illustrates the integration of microbial and electrochemical systems for biomass conversion. It shows how electron flow between microbes and electrodes enhances reaction efficiency. This synergy enables simultaneous energy generation, waste treatment, and chemical production.

14.3 Study methodology

14.3.1 Research design

An innovative method for enhancing the production of biofuel and biochemicals is the integrated microbial and electrochemical biomass conversion [23]. This technique improves biomass breakdown and product creation by utilizing electrochemical technologies and microbial metabolism. A well-organized computational and experimental framework is indispensable in this field. We use computer methods, bioelectrochemical analysis, metabolomics, choosing the right microbial strain, processing and converting biomass feedstock, bioelectrochemical reactor design, and bioproduct development [24]. Genetic engineering, adaptive evolution, screening, and separation are all included in the selection and engineering of microbial strains. Dual-chamber and single-chamber configurations, carbon-based electrodes, and optimal operating conditions are all part of the design of electrochemical reactors. Characterization of metabolism and bioproducts includes 16S rRNA sequencing, fluorescence microscopy, and HPLC/GC-MS [25,26]. A few examples of computational approaches are flux balance analysis, genome-scale

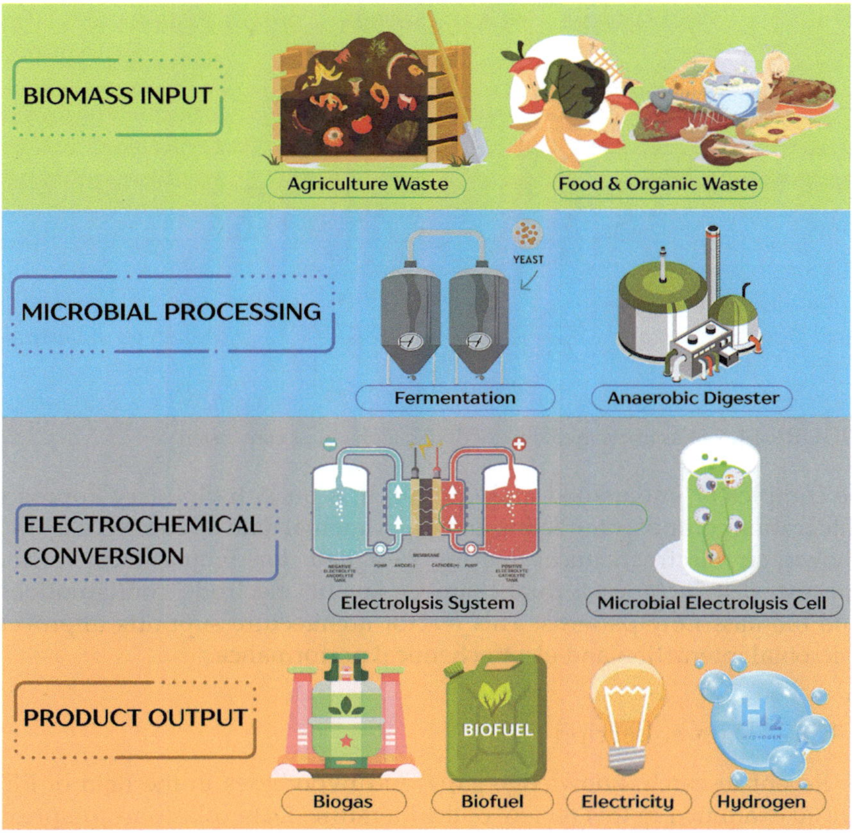

FIGURE 14.2 Schematic of integrated microbial-electrochemical biomass conversion system.

metabolic models, dynamic energy balancing models, computational fluid dynamics, density functional theory, and finite element analysis. Neural networks and artificial intelligence are examples of machine learning and data analytics that forecast the interactions between microorganisms and electrodes and integrate proteins, metabolomics, and genomes to improve system performance [27]. Combining computational and experimental approaches is essential; this includes automated bioprocess control, techno-economic analysis, and hybrid modeling. The objective of these methods is to enhance the production of biofuels and biochemicals by optimizing biomass decomposition and product synthesis [28]. Computer models can help break down biomass and generate new products more effectively, but they frequently fail to illustrate complex biological systems. Using mainly machine learning and data analytics may miss essential factors or interactions, resulting in efficiency failures. For these systems to

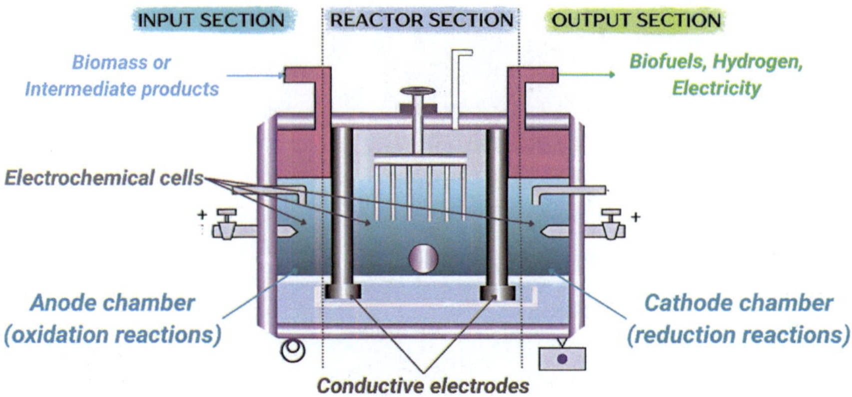

FIGURE 14.3 Electrochemical reactor design for biomass conversion.

be completely understood and optimized, a multidisciplinary approach integrating computer modeling and experimental data is required. Fig. 14.3 shows an electrochemical reactor designed for biomass conversion processes. It includes a dual-chamber layout, electrode configurations, and biomass flow patterns. This reactor architecture supports improved microbial interaction and electrochemical performance.

14.3.2 Data collection

Recent research, knowledge gaps, and hypotheses in the field of BES are all examined in this review of the literature. Data collection comprises sources like peer-reviewed journals, conference proceedings, patents, and technical reports pertaining to biofuel conversion technology. The article talks about different types of microbes and metabolic processes that are important for electrochemical biomass conversion [29]. It also talks about electrode materials and configurations, techniques for pretreatment of biomass, process parameters, computer models, and how the technology can be scaled up. Experiments on a laboratory scale provide controlled environments that maximize system efficiency and validate theoretical models. Microbial cultures, electrodes and materials, biomass feedstock, and bioelectrochemical reactors constitute the experimental apparatus. Impedance spectroscopy, chronoamperometry, and cyclic voltammetry are used in electrochemical analysis. Parameters measured include electrochemical performance, biomass breakdown efficiency, metabolite analysis, and microbial community composition [30].

Pilot-scale experiments assess system performance in real-world scenarios, reducing the gap between laboratory research and industry applications. The design of the pilot reactor incorporates automation, real-time monitoring, performance evaluation, interaction with industrial biomass flows, and scaling of BES. Process optimization, feasibility analysis for commercial-scale applications, and comparison of laboratory-scale results are all part of data collection and validation [31]. Pilot-scale experiments might underestimate the intricacy and challenges of full-scale commercial implementations, as they may supply significant information about system performance [32]. The scale of BES and their integration with biomass streams may add more variables and challenges missing in laboratory studies.

14.3.3 Analytical tools

Bioinformatics techniques are crucial for the analysis of microbial populations, metabolic processes, and genetic modifications in BES. The technologies comprise 16S rRNA sequencing, whole genome sequencing, shotgun metagenomics, metabolic and proteomic analysis, computational systems biology, electrochemical analysis, and process modeling [33]. Metagenomic and genomic analysis uses tools like 16S rRNA sequencing, WHOT, MEGA, PhyloSuite, BLAST, KEGG, MetaCyc, and metabolomics to find out the different kinds of microbes and which ones are most common. Metabolomics studies gene expression and protein interactions in electroactive microorganisms to better understand how biomass is converted metabolically. Computational systems biology uses flux balance analysis (FBA) and genome-scale metabolic models (GEMs) to forecast metabolic flux distribution in microbial processes [34].

As part of electrochemical analysis, tools like cyclic voltammetry (CV), electrochemical impedance spectroscopy (EIS), chronoamperometry (CA), and differential pulse voltammetry (DPV) are used to find out how fast electrons move, how difficult it is for charges to move, and how well biofilms conduct electricity [35,36]. Coulombic efficiency (CE), power density, and current output are used to check how well electrons are recovered in BES and keep an eye on how well bioreactors are working [37]. Process modeling integrates experimental and computational data to optimize system performance. Computational fluid dynamics (CFD) models how matter moves, how fluids move, and how electrodes interact with each other [38]. Kinetic and thermodynamic modeling uses Michaelis-Menten kinetics and density functional theory (DFT) to look at how electrons move at the interface between the electrode and the microbe [39]. Machine learning and AI for optimization forecast microbial-electrode interactions and optimize bioelectrochemical process parameters.

14.3.4 Ethical and sustainability considerations

The integration of ethical practices with green chemistry principles is crucial for sustainable biofuel and biochemical production. Waste biomass, including agricultural residues, industrial waste, and municipal waste, offers a renewable feedstock and addresses waste management challenges [40]. Sustainability benefits include waste reduction, resource efficiency, and promoting resource recycling and bioeconomy development. Ethical considerations include avoiding competition with food supply, collaborating with local waste producers, and ensuring fair access to technology. Green chemistry principles guide the design of safer, more sustainable chemical processes, including prevention and waste minimization, energy efficiency, renewable feedstock selection, and reducing toxicity and environmental impact [41]. Biocompatible electrode materials and green solvents are used in bioelectrochemical reactors, while biocatalysts and microbial consortia enhance reaction specificity and reduce the need for chemical catalysts.

Lifecycle assessment (LCA) is essential for evaluating the entire lifecycle of the biomass conversion process, including carbon footprint, energy balance, and water footprint [42]. Tools for LCA include SimaPro, GaBi, and TRACI. Social and economic dimensions include job creation, support for rural economies, and transparency and accountability. Integrating ethical practices with green chemistry principles is crucial for sustainable biofuel and biochemical production [43, 44]. Utilizing waste biomass, reducing environmental effects, and encouraging ethical sourcing are methods to produce biofuels and biochemicals in an efficient way.

14.4 Applications of integrated systems

14.4.1 Biofuel production

Lignocellulosic biomass, derived from agricultural wastes, forestry waste, and energy crops, serves as a promising feedstock for biofuel generation. It comprises cellulose (35%–60%), hemicellulose (25%–41%), and lignin (15%–20%), which can be converted into bioethanol, biogas, bioalcohols, and high-value compounds [45]. Biomass may be derived from multiple sources, including corn stover, wheat straw, sugarcane bagasse, sawdust, wood chips, bark, and switchgrass. Integrated systems for biofuel production utilize microbiological and electrochemical methods to improve the conversion efficiency of lignocellulosic biomass. Pretreatment consists of physical methods such as milling, grinding, and steam explosion; chemical approaches including acid and alkali treatment; and biological treatments including enzymatic hydrolysis using cellulases and hemicellulases [46]. Microbial conversion involves microbial consortia or genetically modified strains decomposing hydrolyzed biomass, including the fermentation of C6 sugars by *Saccharomyces cerevisiae* and *Zymomonas mobilis* for bioethanol production and the utilization of C5 sugars by engineered *Escherichia coli* and *Pichia stipitis* for improved conversion efficiency [47].

BES uses MFCs, MECs, and EF to increase the activity of microbes and the amount of product they make [48]. Integrated systems may produce multiple forms of biofuels, such as bioethanol, biogas (methane and hydrogen), biodiesel, biohydrogen, and bio-oil. To enhance biofuel production and process efficiency, strategies involve microbial genetic engineering, electrochemical optimization, real-time monitoring, and data-driven optimization. Pretreatment and enzymatic hydrolysis are expensive, feedstock is inconsistent, electrode fouling and biofilm management are difficult, and the process is not financially viable [49]. Hybrid systems, by-product value, and scalability from lab to pilot and commercial applications are examples of future promise.

14.4.2 Chemical production

Platform chemicals, including acetate, ethanol, and hydrogen, act as essential intermediates in the synthesis of biofuels, bioplastics, and fine chemicals [50]. Integrated microbial and electrochemical systems provide an effective and sustainable method for synthesizing these compounds from renewable biomass resources. Acetate is utilized in food preservatives, solvents, textiles, bioethanol, pharmaceuticals, and as a precursor for MES [51]. Ethanol is utilized in biofuels, pharmaceuticals, and as a precursor in the chemical industry [52]. Hydrogen serves as a clean energy carrier, a feedstock for ammonia synthesis, and a component in hydrogen fuel cells. Microbial synthesis processes use microbes that turn carbon sources or biomass into platform chemicals through acetogenic fermentation. These microbes change CO_2 and organic acids into acetate [53]. *Acetobacterium woodii* and *Clostridium aceticum* are the main microorganisms that break down feedstocks like lignocellulosic hydrolysates and syngas [54]. Electrochemical enhancement methods, including BES, can improve efficiency and product yield by including electrodes as electron donors or acceptors.

Proposed solutions for existing difficulties include microbial genetic engineering, redox potential management, real-time monitoring, and process integration. For genetic engineering to work, the key enzymes needed to make acetate and ethanol must be overexpressed. For electrochemical control, the electrode voltage must be precisely adjusted to control the metabolic pathways of microbes. Process integration encompasses effective biomass pretreatment and feedstock conditioning, while downstream processing entails the recovery and purification of acetate, ethanol, and hydrogen via distillation, gas stripping, and membrane separation [55]. Challenges comprise electrode fouling, product inhibition, and economic feasibility resulting from the elevated costs of improved electrode materials and microbial engineering. Proposed solutions consist of dynamic electrode coatings, adaptive microbial strains, cost reduction techniques, sustainability and economic effect assessments, circular economy methodologies, low carbon footprint initiatives, and economic feasibility through the valuation of by-products and integration with existing biorefineries [56,57].

14.4.3 Wastewater treatment and resource recovery

A sustainable method for controlling waste streams while producing useful resources is the simultaneous treatment of wastewater and energy recovery [58]. This approach integrates microbiological and electrochemical technologies to clean wastewater efficiently and recover energy in the form of electricity and biogas. Conventional wastewater treatment techniques often require substantial energy and resources; however, incorporating microbial and electrochemical systems presents a transformative method, converting treatment plants into resource recovery facilities. MFCs are BES that harness the metabolic processes of microorganisms to oxidize organic matter in wastewater, producing electrons to generate electricity [59]. MFCs have been used to treat various wastewater types, including domestic sewage, industrial effluents, and agricultural runoff, effectively facilitating pollutant removal while enabling energy recovery. Osmotic microbial fuel cells (OsMFCs) combine forward osmosis (FO) with MFC technology, enhancing wastewater treatment and energy generation [60].

High-rate algal ponds (HRAPs) employ microalgae for wastewater treatment and generate biomass suitable for bioenergy conversion. HRAPs combined with biogas production are ecologically advantageous, especially in warm climates. Case studies indicate that the Philadelphia Water Department established biogas cogeneration at the Northeast Water Pollution Control Plant, fulfilling nearly 85% of the plant's energy requirements by harnessing methane generated during sewage treatment [61]. Technical issues involve enhancing system efficiency, scaling technology, and managing operational complications. Further research is necessary to improve biocatalyst efficiency, develop cost-effective materials, and integrate systems into the existing structure. Supportive policies and economic incentives are crucial for promoting the adoption of these integrated systems.

14.4.4 Carbon capture and utilization

Bioenergy with carbon capture and storage (BECCS) is a promising approach to achieve sustainable energy production while mitigating atmospheric CO_2 levels [62]. BECCS involves using biomass as an energy source and capturing the CO_2 produced during its conversion, storing it to prevent its release into the atmosphere. This integration not only provides renewable energy but also contributes to carbon dioxide removal

(CDR), addressing climate change challenges. BECCS may produce neutral emissions by capturing and storing CO_2 along with producing electricity, heat, and biofuels from biomass [63]. It similarly promotes waste valorization and the principles of the circular economy. The challenges of BECCS include the sustainability of biomass supply, economic viability, and technological development; policy incentives such as carbon credits and subsidies may enhance economic feasibility; and BECCS systems require improvements in carbon capture efficiency, biomass conversion technologies, and storage solutions.

The UK's Drax Power Station, which switched from coal to biomass and is spending up to $12.5 billion on BECCS projects in the United States over the next 10 years, is one example of a case study of recent progress [64]. Studies on the combination of CCS and hydrothermal liquefaction of biomass have demonstrated promise for creating advanced liquid biofuels with zero CO_2 emissions. Universal carbon reduction initiatives expect to benefit greatly from the combination of carbon capture technologies and biomass conversion. To solve current issues and expand BECCS implementations, it is imperative to conduct continuous studies, develop new technologies, and create supportive policy frameworks [65]. BECCS provides a workable alternative for atmospheric CO_2 reduction and sustainable energy production as part of a diversified strategy for reaching emission-free operation. BECCS can offset emissions from other industries and contribute to climate change mitigation by absorbing and storing CO_2 emissions from biomass conversion processes [66]. The development of advanced liquid biofuels may also assist with the transition to a more sustainable energy system and less dependency on fossil fuels. This approach has the potential to significantly advance the transition to a carbon-neutral future with sustained investment and focus on developing BECCS technology. Fig. 14.4 presents a process flow diagram for converting biomass into biofuel through microbial-electrochemical integration. The flow outlines feedstock input, pretreatment, microbial conversion, and CO_2 capture. It reflects a closed-loop, carbon-efficient bioenergy system with resource recovery.

14.5 Integration with other advanced technologies

14.5.1 Artificial intelligence and machine learning for process optimization

Artificial Intelligence (AI) and machine learning (ML) are utilized in integrated microbial and electrochemical biomass conversion systems to improve efficiency and sustainability [67]. These technologies enhance predictive modeling and real-time monitoring, leading to better outcomes in biomass conversion processes. Predictive modeling plays a crucial role

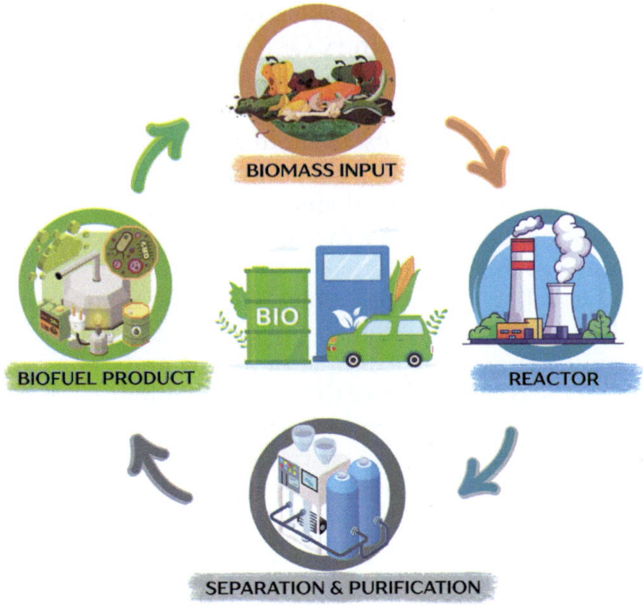

FIGURE 14.4 Process flow diagram of biomass to biofuel conversion.

in optimizing parameters and anticipating challenges related to biomass conversion. ML techniques have been applied to forecast outcomes in thermochemical processes, such as pyrolysis and gasification. AI-driven models are used to predict biomass characteristics, process performance, and biofuel properties, thus enhancing the efficiency of bioenergy production by providing insights into complex biochemical interactions [68].

Real-time monitoring is crucial for the continuous monitoring and control of biomass conversion systems. Combining ML with microbial electrochemical sensors makes it possible to test for multiple toxicants at the same time and in real time. This protects the integrity of the system and stops conditions that might stop microbes from working. When IoT devices are combined with AI, they make it easier to collect and analyze data in real time. This lets predictive maintenance and dynamic changes happen, which makes biomass conversion processes more reliable and effective. Dynamic flux balance analysis (dFBA) and computational modeling platforms, like the computation of microbial ecosystems in time and space (COMETS), are two strategic ways to use AI and ML [69,70]. The biomass conversion study group needs to standardize data collection and sharing processes to build confidence and promote its use in industrial applications. Data quality, accessibility, and model interpretability remain issues.

14.5.2 Internet of things enabled smart bioreactors

The integration of Internet of Things (IoT) technology into bioreactors facilitates remote monitoring and control of microbial-electrochemical systems [71]. This increases data collection and improves the efficiency of the procedures used to convert biomass into valuable commodities. Internet of Things-enabled bioreactors use sensors and microcontrollers to monitor crucial parameters like temperature, pH, viscosity, and electrical potential. Real-time data is uploaded to cloud platforms for remote access and control. Sensors for temperature and viscosity control are incorporated into embedded IoT devices, like the ESP32, which enable continuous data transfer for real-time monitoring and process improvement [72]. As IoT systems provide constant data collection and transmission, remote monitoring and control are possible, enabling operators to remain focused on bioreactor conditions from a distance. By using automated control systems, advanced IoT-enabled bioreactors may reduce manual involvement and increase process dependability.

IoT integration in microbial-electrochemical systems provides major advantages for bioelectrochemical process monitoring and improvement. MFCs enable real-time data collection and analysis by operating as biosensors for process monitoring [73]. Advanced analytics become feasible with the constant flow of data from IoT-enabled sensors, which improves microbial-electrochemical interactions and increases system performance in general. But problems like integration complexity, system scalability, and data security need to be rectified. Improving sensor accuracy, developing standardized communication protocols, and installing comprehensive cybersecurity measures to protect sensitive data may be the primary objectives of future developments [74,75]. When IoT technology is used in microbial-electrochemical systems, remote monitoring and control capabilities are enhanced, which improves the flexibility and effectiveness of biomass conversion processes.

14.5.3 Hybrid renewable energy systems

In biomass conversion processes, hybrid renewable energy systems (HRES) provide a practical way to increase sustainability and energy efficiency [76]. To provide a sustainable and uninterrupted power supply, these systems combine multiple energy sources, such as solar energy during the day and wind energy at sunset. This is especially useful in off-grid or rural regions where stable electricity access is challenging. Numerous benefits, including continuous operation, increased efficiency, and environmental benefits, come from using HRES in microbiological and electrochemical biomass conversion processes [77]. In Spain, a hybrid facility combines biomass and solar thermal energy to produce energy continuously, even when there is no sunlight. According to case studies, biomass is used at night or in dark conditions, whereas solar thermal energy is used at Termosolar Borges in Spain during sunny hours. Husk Power Systems uses solar energy during the day, biomass in the evening, and batteries at night to provide rural areas in India with continuous, sustainable electricity [78].

The availability of resources, economic considerations, and technical complexity are some of the obstacles that must be overcome. Resilient infrastructure and advanced control mechanisms are necessary for technical complexity. Conversely, economic considerations require a substantial initial cost. Long-term benefits, such as lower operational costs and environmental effects, offset these expenses. Because the availability of resources depends on the local availability of renewable resources, comprehensive analyses are required to determine the proper energy composition for a given area. Using hybrid renewable energy sources combined with integrated microbiological and electrochemical biomass conversion systems presents an exciting opportunity for efficient and sustainable energy production [79]. Integrated systems may be able to get around the limitations of distinct energy sources by combining the strengths of solar, wind, and biomass energy, ensuring a reliable and continuous power supply.

14.6 Case studies and experimental results

14.6.1 Case study 1: Biofuel production from agricultural waste

The manufacture of biofuel from agricultural waste has the potential to substitute fossil fuels and offer a sustainable energy source. This case study focuses on the methods, advantages, difficulties, and practical uses of producing biofuel from agricultural waste. Crop wastes, animal manure, and industrial byproducts are just a few examples of the agricultural waste that typically burns or decomposes to produce methane emissions [80]. Biofuel production from agricultural waste reduces waste and provides a sustainable energy source. Corn combustion, rice husks, wheat straw, and sugarcane bagasse are some of the feedstocks collected from the agricultural and processing sectors as part of the biofuel production process [81]. Thermolysis, gasification, and hydrothermal liquefaction are a few pretreatment methods that increase the effectiveness of upcoming conversion procedures [82]. Thermolysis, gasification, hydrothermal liquefaction, and biochemical conversion using enzymes and microbes are typical conversion procedures.

Raw biofuel is processed and improved to meet fuel regulations, remove impurities, and raise fuel quality. It may be used in appropriate engines and power plants or mixed with standard fuels for distribution and use.

The bioethanol program in India is an example of a program that converts agricultural waste into bioethanol, reducing carbon emissions and improving the rural economy [83]. High transportation costs, seasonal supply, technological obstacles, and competition from traditional fuels are some of the difficulties. Potential developments in the future include integration with the ideas of the circular economy and improvements in biotechnology and process engineering. An integrated solution to the problems of energy, waste management, and the environment is provided by the production of biofuel from agricultural waste. Table 14.1 details the performance metrics of different biomass types used for biofuel production. It includes yield (L/ton), energy efficiency, carbon emission reduction, and cost savings. This comparative data supports strategic feedstock selection for optimized bioenergy output.

14.6.2 Case study 2: Electro-fermentation for chemical production

Microbial fermentation and electrochemical systems are combined in the biotechnological process known as electro-fermentation (EF), which increases the yield of useful compounds [92]. EF provides a sustainable and effective way to produce chemicals, biofuels, and bioplastics by controlling microbial metabolism with electrical inputs. Electrochemical stimulation, increased fermentation, substrate feeding, and product recovery are all aspects of the process. Redox management, operating flexibility, and integration with renewable energy sources are among EF's primary benefits. When compared with standard fermentation techniques, a case study conducted at the University of Tübingen demonstrated the use of EF to convert glycerol into acetate using *Clostridium pasteurianum*, which produced a 30% increase in acetate production [93].

The synthesis of organic acids, alcohols, and bioplastics, as well as the synthesis of biofuels and bioremediation, are examples of applied uses. The advantages of EF include sustainability, increased yields, and energy efficiency because it may utilize low-grade or renewable electricity. The challenges include system complexity, microbiological compatibility, and growth constraints. Future potential for EF includes advancements in microbial engineering, electrode materials, and process control [94]. Advancements in these areas might lead to the possibility of integrating EF with renewable energy systems, which would usher in a new era of green chemical production technologies. By combining microbiology and electrochemistry, EF provides a viable method for the efficient and sustainable production of chemicals, supporting a circular and low-carbon economy. Table 14.2 outlines chemical production metrics for various compounds synthesized via EF. It compares production rates, energy inputs, purity levels, and economic benefits. These insights help evaluate the efficiency and viability of electro-biochemical pathways.

TABLE 14.1 Performance metrics of biofuel production system.

Biomass type	Biofuel yield (L/ton)	Energy efficiency (%)	Carbon emission reduction (%)	Cost savings (%)	References
Corn stover	380–410	83–85	38–42	22–25	[84]
Sugarcane bagasse	340–360	88–91	43–46	28–31	[85]
Wheat straw	270–300	78–82	33–36	18–21	[45]
Rice husk	240–260	73–76	28–31	13–16	[86]
Wood chips	310–330	85–89	40–43	26–29	[49]
Algae biomass	390–420	91–93	48–52	33–36	[87]
Food waste	250–280	75–79	30–34	16–19	[88]
Municipal waste	220–240	68–71	25–29	10–13	[89]
Switchgrass	300–320	84–87	36–39	20–23	[90]
Miscanthus	235–340	87–90	42–45	24–27	[91]

TABLE 14.2 Chemical production metrics.

Chemical	Production rate (g/L/h)	Energy input (kWh/kg)	Purity (%)	Cost reduction (%)	References
Acetate	0.4–0.6	1.2	95	30	[51]
Ethanol	0.4	1.5	90	25	[95]
Butanol	0.3	1.8	85	20	[96]
Hydrogen	0.6	1.0	98	35	[97]
Lactic acid	0.4	1.4	92	28	[98]
Succinic acid	0.35	1.6	88	22	[99]
Propionate	0.25	1.7	84	18	[100]
Formate	0.45	1.1	96	32	[101]
Methane	0.55	1.3	97	30	[102]
Butyrate	0.3	1.6	86	20	[103]

14.6.3 Case study 3: Wastewater treatment with resource recovery

The goal of wastewater treatment has changed over time, moving from pollution control to resource recovery through a coordinated approach. Modern technologies have made it possible to extract valuable resources from wastewater, such as clean water, electricity, and fertilizers, which facilitates a circular economy [104]. This case study examines cutting-edge techniques for treating wastewater while recovering resources, highlighting practical uses and advantages. Membrane filtration, advanced oxidation, energy recovery, nutrient recovery, material recovery, and chemical recovery are examples of resource recovery methods. Advanced oxidation eliminates pathogens and micropollutants, while membrane filtering produces high-quality reclaimed water. Anaerobic digestion, MFCs, thermal hydrolysis, struvite precipitation, ammonia stripping, cultivation of algae, utilization of biosolids, and chemical recovery are all components of energy recovery [105].

Using natural treatment technologies, the Water Hub at Emory University in the United States runs a decentralized wastewater treatment facility that handles up to 400,000 gallons of campus wastewater every day. Up to 40% less potable water is utilized at the institution when the treated water is recycled for heating, cooling, and toilet flushing [106]. Climate change mitigation, sustainability, economic value, and environmental effects are all advantages of resource recovery in wastewater treatment. High initial and ongoing maintenance costs, legal restrictions, issues with public opinion, and process effectiveness are some of the difficulties. Future opportunities like "sewer mining," decentralized treatment, and intelligent monitoring systems are increasing the potential for resource recovery. Operational carbon footprints may be further decreased by combining renewable energy sources with wastewater treatment. The shift to resource recovery in wastewater treatment is consistent with the circular economy's principles and aims of global sustainability. Innovative treatment techniques that convert waste streams into useful resources support resource security and economic resilience in addition to protecting the environment [107].

14.6.4 Economic and environmental benefits

Integrated microbial and electrochemical biomass conversion is a hybrid technique that turns biomass into useful products, including chemicals, biofuels, and bioplastics, by combining biotechnology with electrochemical processes [108]. Cost-effectiveness, value addition, energy efficiency, market potential, and government incentives are some of the many economic and environmental advantages that this technology provides.

Increased yields from electrochemical control, lower chemical inputs, and the creation of high-value goods like organic acids, specialized chemicals, hydrogen, and bioplastics are among the financial advantages [109]. The use of renewable energy is also possible because excess renewable electricity may power electrochemical processes. Market potential includes the integration of the circular economy, which allows companies to access markets that are centered on green technologies and sustainability. Financial incentives are frequently offered by governments to technologies that lower emissions and promote the objectives of renewable energy.

Environmental benefits include waste reduction, greenhouse gas mitigation, pollution control, resource conservation, and synergies with a circular and green economy. A pilot project at the University of Queensland in Australia showed how to use microbial fermentation and electrochemical systems together to turn waste biomass into acetate [110]. The project increased the yield of the product by 40% and powered the electrochemical process with clean solar energy. Being a part of biorefineries that turn different types of biomass into different outputs and allowing local biomass processing, lowering transportation emissions, and helping rural economies are all things that work well with a circular and green economy. Ultimately, combining microbial and electrochemical biomass conversion technologies offers a potent strategy for attaining financial success and environmental sustainability.

14.7 Challenges and limitations

14.7.1 Scalability and economic viability

The integration of microbial and electrochemical biomass conversion technologies faces several technical, economic, regulatory, and market barriers. Process complexity, a lack of microbial options, high capital expenditures (CAPEX), problems with operational stability, market volatility, a lack of funding and investment, complex approvals from regulators, environmental compliance, certification difficulties, uneven policy support, and social barriers are some of these [111]. Complicated approval procedures, environmental compliance, lack of standardization, certification difficulties, uneven policy support, and subsidy gaps are signs of regulatory and policy challenges. The development of a proficient workforce, supply chain constraints, industry resistance, and consumer awareness all hamper market acceptability. Resource intensity, water and chemical consumption, waste management, and environmental effects are aspects of technological and environmental constraints [112].

Innovation and research, concentrating on less expensive, more reliable electrode materials, optimizing microbial strains, utilizing government incentives, encouraging consumer education, and forming alliances with industry are some of the methods used to overcome these obstacles. While carbon credits and green bonds may improve project economics, government incentives may reduce CAPEX and OPEX [113]. The environmental advantages of bio-based products may be presented to consumers through education, and collaborations may develop to incorporate innovations into current systems. Integrated microbial and electrochemical biomass conversion technologies may be taken from pilot projects to mainstream use in industry by figuring out how to deal with technical, financial, regulatory, and market issues and by using new ideas, government support, and strategic partnerships [114]. This will significantly contribute to sustainable development and the circular economy.

14.7.2 Microbial and electrochemical stability

The long-term performance of integrated biomass conversion systems depends on microbial and electrochemical stability. Variations in pH, temperature, and substrate availability may lead to changes in the composition of the microbial community, competition and predation, mutational movement, biofilm management, and maintenance requirements [115]. Development of biofilms on electrodes may increase resistance and decrease electron transfer efficiency, while shear stress problems may cause instability in microbial performance. The requirement for regular biofilm management may increase operational complexity and costs.

Electrode toxicity, nutritional constraint, and stress from the electrochemical environment are the causes of the decrease in microbial activity [116]. Electrode fouling and degradation may also affect the growth of microbes. Material wear, surface fouling, corrosion, decreased redox capacity, passivation layers, and changes in current density can all lead to unstable performance. Maintaining ideal conditions such as pH and temperature variability, substrate supply management, sensor drift, control system failures, and energy input consistency requires process control and operational stability [117]. Some things that may be done to make things more stable are genetic engineering, managing microbial consortia, biofilm control methods, long-lasting electrode materials, antifouling coatings, regular maintenance, advanced monitoring systems, predictive maintenance, and energy management systems.

The METland project in Spain serves as an illustration of this, having effectively attained long-term stability and steady wastewater treatment effectiveness with minimal maintenance [118]. Overcoming these obstacles and promoting the scalability of these technologies will require the use of advanced monitoring technology, creative materials, and effective microbial control techniques. For integrated biomass conversion systems to function well over the long run, microbial and electrochemical stability must be maintained. Addressing biological, material, and process control issues improves system resilience, maintenance, and economic viability.

14.7.3 Regulatory and standardization issues

Various regulatory sectors, such as biotechnology, chemical production, environmental standards, and energy laws, have fractured the current regulatory landscape for integrated microbial and electrochemical biomass conversion systems [119]. The fact that this differs between nations makes international initiatives more challenging. The absence of a structure for these operations creates regulatory gaps and complicated approval processes. The absence of regulations creates problems for safety and compliance, including waste management, electrochemical dangers, and possible genetically modified organism (GMO) issues [120]. Challenges also arise from market and investment barriers, including investor reluctance, difficulties with product certification, and operational and technical constraints.

System design and operation, bioreactor and electrode standards, process control requirements, microbial management, disinfection and containment, electrochemical systems, environmental impact and waste management, and resource recovery certification are all areas of regulation that need to be standardized. Regulatory challenges include mixed regulations in the European Union, regulatory overlap in the United States, and bio-based product certification [121]. To address these issues, potential solutions include developing integrated standards through cross-sector collaboration, working with international standards bodies, proactive regulatory engagement through pilot projects with regulators, and industry-led initiatives [122]. These approaches may help reduce regulatory risks and facilitate smoother market entry for innovative bio-based technologies. The lack of regulatory guidelines for integrated microbial and electrochemical biomass conversion systems presents significant challenges to commercialization and scalability. A coordinated approach involving regulatory bodies, industry stakeholders, and research institutions is needed to develop standardized frameworks, promote pilot projects, and advocate for supportive policies. By developing standardized frameworks, promoting pilot projects, and advocating for supportive policies, the industry may reduce regulatory risks and facilitate smoother market entry for innovative bio-based technologies [123].

14.7.4 Technical expertise and infrastructure

Advanced infrastructure and educated workers are critical to the success of integrated microbiological and electrochemical biomass conversion technologies. Professional staff members have expertise in environmental and regulatory compliance, automation and control systems, electrochemical engineering, microbiology and biotechnology, LCA, and environmental impact analysis [124]. A restricted group of workers, cross-disciplinary skills gaps, and training requirements make workforce development difficult. Establishing educational programs, forming industry connections, and promoting continuous education are all necessary to develop competence. Biotechnology labs with microbial culture, strain development, and bioprocess optimization capabilities; electrochemical testing facilities with electrode fabrication, electrochemical performance testing, and material durability assessment capabilities; integrated pilot plants for waste treatment, resource recovery, and energy generation; sensor technology for real-time monitoring of pH, redox potential, microbial activity, and electrochemical parameters; data management systems like SCADA; ML tools for predictive analytics; and remote monitoring via IoT devices are illustrations of the advanced laboratory and pilot-scale facilities that infrastructure requirements include [125].

Energies such as reliable power supplies, water supply and treatment systems, waste management facilities for managing byproducts, wastewater treatment, and the secure disposal of spent materials, and supply chain logistics for the collection, preprocessing, and transportation of biomass are illustrations of supporting infrastructure. The German BioElectroMed project produced a pilot-scale microbial electrochemical device for pharmaceutical wastewater treatment [126]. Financial investment, teamwork, and cross-industry learning are needed to overcome expertise and infrastructure development obstacles. Government funding for infrastructure development and training initiatives may be obtained through government grants, corporate investment, training incentives, and international research collaborations. International research collaborations, open-source projects, and cross-industry learning are other methods to collaborate and share information. Advanced infrastructure and trained workers are critical to the success of integrated microbiological and electrochemical biomass conversion technologies [127]. While investing in modern facilities will enable safe, effective, and scalable operations, it is imperative to address skills gaps through education, training, and industry alliances.

14.8 Future directions and opportunities

14.8.1 Advanced biocatalysts and electrodes

High-performance materials play a crucial role in the efficiency and scalability of integrated microbial and electrochemical biomass conversion systems. Advanced biocatalysts and electrode materials are essential for enhancing biological activity, electron transfer, and chemical stability. Biocatalysts come in many forms, such as electroactive microorganisms, engineered enzymes, biofilm-based systems, genetic engineering, synthetic biology approaches, nano-enriched biocatalysts, and conductive nanomaterials [128]. Advanced electrode materials require high conductivity, chemical stability, biocompatibility, and structural durability. Carbon-based electrodes offer high conductivity and large surface areas, while metal and metal oxide electrodes provide high conductivity but can be costly [129]. Conductive polymers offer flexibility and biocompatibility, and composite electrodes combine carbon materials with metal oxides or polymers for tailored properties.

Surface functionalization, chemical coatings, nano-patterning, redox mediators, better structural design, and synergy between biocatalysts and electrodes are all ways to improve electrode performance. Creating biohybrid systems means creating electrodes that combine living cells with inorganic materials, biofilms that stick together as biocatalytic layers, and redox-active compounds that help move electrons around [130]. Some ways to improve performance are to speed up the transfer of electrons, watch the electrode potential and microbial activity in real time, and use genetic techniques to make biofilms more conductive.

The ElectroBioCatalyst Initiative in the Netherlands, which developed high-efficiency electrodes and altered microorganisms for biochemical production, is an illustration of high-performance materials in operation [131]. The project succeeded in increasing current density by 30% and achieving higher target chemical yields. Costs of materials, scalability, and long-term stability are a few technical difficulties. Lifecycle analysis, hybrid techniques, and innovative materials should be the primary fields of research and development. High efficiency, stability, and scalability in integrated microbial and electrochemical systems depend on the development of advanced biocatalysts and electrodes [132]. It is possible to improve electron transmission, increase material durability, and lower costs by utilizing genetic engineering, material science advancements, and biohybrid technologies. Technological advancements and economic adoption will be facilitated by sustained research and development as well as cooperation among academics and industry.

14.8.2 Integration with renewable energy sources

Carbon neutrality, financial savings, operational stability, and increased efficiency are only a few advantages of combining renewable energy sources with microbial and electrochemical biomass conversion systems. Solar, wind, hydropower, biogas systems, and geothermal energy are a few renewable energy sources that might be integrated [133]. While wind energy may be utilized for off-grid and grid integration applications, solar energy may be immediately converted into electricity for electrochemical operations. For large-scale operations, hydropower offers consistent and reliable power, whereas biogas systems employ organic waste to generate heat or energy. Geothermal energy powers electrochemical processes and regulates temperature by supplying heat and electricity. Immediate powering of electrochemical systems, adaptive control systems, battery storage, hydrogen production, power-to-gas (P2G) technology, and thermal storage are technical methods for integrating renewable energy [134]. Economic advantages include resource efficiency, reduced carbon emissions, more profitability, and cheaper energy prices. Benefits to the environment include resource efficiency, support for circular economy models, and lower emissions.

BioElectroSun, a successful integration project in Spain, combined solar energy with microbial electrochemical systems to produce biogas and clean wastewater. By providing excess power, the project helped stabilize the local grid, achieved energy self-sufficiency, and reduced operating costs by 25%[135]. Technical compatibility, early investment expenses, and inconsistent power supply are obstacles to renewable energy integration. Using power electronics and smart grid technologies to even out fluctuations in the energy supply, integrating battery storage and hybrid systems with different renewable inputs, and including the use of government incentives, subsidies, and green financing options are some solutions. Blockchain technology for energy trading, AI-driven systems for forecasting energy production and optimizing consumption in biomass conversion processes, and hybrid renewable systems are examples of future developments and directions [136]. Facilities may improve process stability, attain energy independence, and support carbon reduction initiatives by utilizing solar, wind, and other renewable energy sources. Future developments in hybrid systems, smart grid technologies, and energy storage will improve operational effectiveness and facilitate widespread adoption.

14.8.3 Artificial intelligence-driven process optimization

Biomass conversion is undergoing an advancement because AI and ML are making real-time monitoring, predictive control, and optimization feasible. These technologies lower operating costs, increase efficiency, and facilitate dynamic decision-making. Important uses include data collecting and processing employing IoT devices for high-frequency data collection and analysis, as well as real-time monitoring using smart sensors to detect pH, temperature, redox potential, substrate concentration, and biomass growth [137]. Another important use of AI and ML in biomass conversion systems is forecasting. By identifying departures from normal operating circumstances, ML algorithms may promote early intervention, reducing equipment breakdowns and process interruptions. AI-driven processes in automated process control systems automatically adjust feed rates, pH levels, temperature, and electrical parameters [138]. Model predictive control maximizes system performance by anticipating changes in processes [139].

Different types of ML are used in biomass conversion systems. These include supervised learning, unsupervised learning, reinforcement learning, adaptive feedback loops, self-tuning systems, process parameter optimization, and simulation-based methods like digital copies of biomass conversion systems. Cost-effectiveness, resource efficiency, and waste reduction are all advantages of AI-driven process optimization [140]. The BioSmartTech project in Germany, which aimed to develop an AI-driven platform for real-time optimization of microbial electrochemical systems, is a case study of AI in biomass conversion. Predictive maintenance systems were put in place to minimize downtime, and ML algorithms were integrated with sensor networks to monitor bioreactor performance. As a result of continuously changing process parameters, biogas production efficiency increased by 30%, and energy consumption was significantly reduced [141].

AI integration faces challenges such as data availability, complexity, and skilled workers. Solutions include comprehensive data collection and storage systems, hybrid modeling techniques, training courses, and collaboration with academic institutions. Future developments include AI algorithms that adapt without human intervention, autonomous systems for process efficiency, and predictive maintenance for equipment [142]. Real-time simulations of biomass conversion systems enable safe testing of novel approaches without affecting actual operations. AI-driven process optimization offers opportunities to improve sustainability, stability, and efficiency in microbial and electrochemical biomass conversion systems [143]. Facilities can increase yields, reduce costs, and reduce environmental impact by utilizing ML, real-time data analysis, and predictive control. Fig. 14.5 demonstrates how AI-driven models enhance biomass conversion system performance. It depicts sensor networks, predictive analytics, and automated control loops in real-time optimization. Such integration improves reliability, reduces energy input, and maximizes output in smart bioreactors.

14.8.4 Global adoption and impact

Energy security, waste management, and climate change may be solved by increasing the adoption of combined microbial and electrochemical biomass conversion systems. These technologies promote sustainable development objectives by producing biofuels, biochemicals, and clean water from organic waste and renewable resources [144]. Waste management, environmental sustainability, energy security, and regional opportunity are some of the primary challenges. By using local biomass resources, such

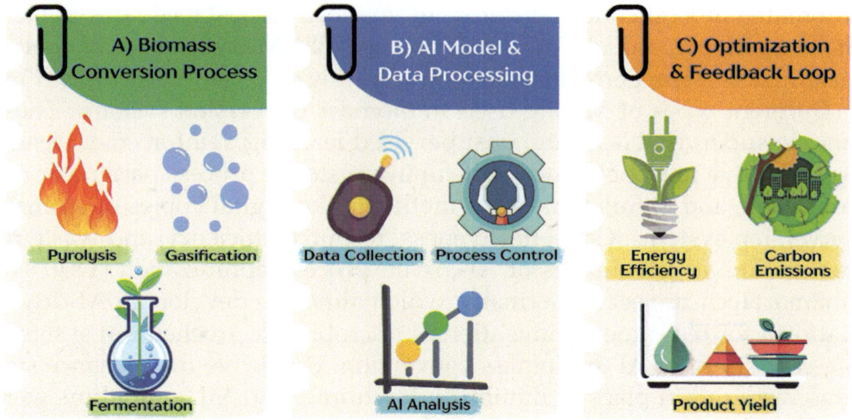

FIGURE 14.5 AI-driven optimization of biomass conversion processes.

as municipal and agricultural waste, biomass conversion systems provide energy independence through localized energy generation through biogas, biohydrogen, and bioelectricity [145]. By recycling waste into useful goods, promoting circular economy models, and lowering emissions through carbon-neutral biofuels, biomass conversion systems may help address environmental sustainability issues. With the world's waste creation expected to exceed 3.4 billion tons by 2050, waste management is a major concern. Developing economies, developed countries, consumer demand, and the biomass conversion sector are proof of regional potential [146]. Technologies for biomass conversion may promote agriculture, enhance industrial operations, generate jobs in waste collection, bioprocessing, and technology maintenance, improve health results, and supply inexpensive energy in developing countries.

Increased demand for bio-based goods and sustainable energy is expected to propel the biomass conversion industry to reach $98.5 billion by 2030. Bio-based technologies offer investment prospects as well as career opportunities in supply chains, plant operation, maintenance, and technology development. The Paris Agreement, the European Green Deal, China's Circular Economy Promotion Law, renewable energy requirements, and government subsidies are some of the legislative and policy factors that are driving biomass conversion [147]. Expanding access to technology, encouraging information exchange in underdeveloped nations, creating modular solutions, and removing adoption obstacles through creative funding models and regulatory harmonization are some instances of future societal effects. Technologies for the combined microbiological and electrochemical conversion of biomass may improve industrial productivity, assist in rural development, and assist in the worldwide carbon reduction campaign.

14.9 Conclusion

The integrated microbial-electrochemical biomass conversion systems, which combine electrochemical accuracy with biological innovation, provide a new approach to sustainable technology. Issues such as waste management, water shortages, climate change, and energy security may be resolved by these systems. They illustrate the principles of a circular economy by recycling organic waste into useful biofuels, biochemicals, and clean water. Researchers must continue developing new ideas, legislators must develop regulations that encourage them, industry participants must fund pilot projects, and communities and organizations must promote sustainable habits if these technologies are to reach their full potential.

Integrated microbial-electrochemical systems may reduce carbon emissions in industries, enhance rural economies, and improve urban sustainability, all of which contribute to the vision of a sustainable future. We may develop a strong, sustainable, and equitable entire ecosystem by utilizing the power of microbes and electrochemistry. Using biomass conversion technology, we may develop a healthier, cleaner, and more sustainable society by acting immediately, investing boldly, and working widely.

References

[1] T. Kalak, Potential use of industrial biomass waste as a sustainable energy source in the future, Energies 16 (4) (2023) 1783, doi:10.3390/en16041783.

[2] R. Boddula, Y.-Y. Lee, S. Masimukku, G.-P. Chang-Chien, R. Pothu, R.K Srivastava, et al., Sustainable hydrogen production: solar-powered biomass conversion explored through (photo)electrochemical advancements, Proc. Saf. Environ. Prot. 186 (2024) 1149–1168, doi:10.1016/j.psep.2024.04.068.

[3] M.K. Awasthi, T. Sar, S.C. Gowd, K. Rajendran, V. Kumar, S. Sarsaiya, et al., A comprehensive review on thermochemical, and biochemical conversion methods of lignocellulosic biomass into valuable end product, Fuel 342 (2023) 127790, doi:10.1016/j.fuel.2023.127790.

[4] H. Mortadha, H.B. Kerrouchi, A. Al-Othman, M. Tawalbeh, A comprehensive review of biomass pellets and their role in sustainable energy: production, properties, environment, economics, and logistics, Waste Biomass Valor. (2025), doi:10.1007/s12649-024-02873-x.

[5] M. Padhiary, A. Hoque, G. Prasad, K. Kumar, B. Sahu, Precision agriculture and AI-driven resource optimization for sustainable land and resource management, in: A.J. Ruiz-Vanoye, O. Díaz-Parra (Eds.), Smart Water Technology for Sustainable Management in Modern Cities, IGI Global, 2025, pp. 197–232, doi:10.4018/979-8-3693-8074-1.ch009.

[6] B. Chen, B. Rattan, S. Gan, A. Garg, S. Sreedeep, Bioelectricity from pyrolyzed biomass-aided plant microbial fuel cell considering impact of different particle sizes of biochar, Biomass Conver. Biorefin. (2025), doi:10.1007/s13399-025-06650-1.

[7] Hussain, Sajjad, H. Khan, A. Maqbool, M. Usama, S. Gul, Microbial electrolysis: an emerging biotechnology for energy production and wastewater treatment, in: A.M. Yatoo, M. Sillanpää (Eds.), Smart Waste and Wastewater Management by Biotechnological Approaches, Interdisciplinary Biotechnological Advances, Springer Nature Singapore, Singapore, 2025, pp. 435–461, doi:10.1007/978-981-97-8673-2_21.

[8] A.I. Osman, Z. Chen, A.M. Elgarahy, M. Farghali, I.M.A. Mohamed, A.K. Priya, et al., Membrane technology for energy saving: principles, techniques, applications, challenges, and prospects, Adv. Energy Sustain. Res. 5 (5) (2024) 2400011, doi:10.1002/aesr.202400011.

[9] R.A. Timmers, D. Hidalgo, M.A. Sánchez-Gatón, Novel applications of microbial electrolysis cells in anaerobic digestion systems: trends and perspectives, in: V. Kumar, P. Verma (Eds.), Biotechnological Applications in Industrial Waste Valorization, Interdisciplinary Biotechnological Advances, Springer Nature Singapore, Singapore, 2025, pp. 367–405, doi:10.1007/978-981-96-2302-0_11.

[10] S. Vickram, S.S. Infant, B.S. Balamurugan, P. Jayanthi, M. Sivasubramanian, Techno-economic and life cycle analysis of biorefineries: assessing sustainability and scalability in the bioeconomy, Environ. Qual. Manage. 34 (4) (2025) e70077, doi:10.1002/tqem.70077.

[11] M. Shanthi, P. Sivashanmugam, Sustainable carbon management: exploring innovative bio-green technologies for CO_2 sequestration and energy utilisation, in: S. Kandasamy, M.P. Shah, K. Subbiah, N. Manickam (Eds.), *Microbial Niche Nexus Sustaining Environmental Biological Wastewater and Water-Energy-Environment Nexus*, Environmental Science and Engineering, Springer Nature Switzerland, Cham, 2025, pp. 541–563, doi:10.1007/978-3-031-62660-9_21.

[12] A.V. Jadhav, S.S. Phadtare, Mangrove microbiome as an alternative source of bioenergy, in: S. Muthusamy, R. Manikkam, G. Venugopal (Eds.), Mangrove Microbiome, Springer Nature Singapore, Singapore, 2025, pp. 151–170, doi:10.1007/978-981-96-2602-1_7.

[13] E.D. Penteado, Y.J.O. Asencios, A.A.R. Diniz, L.H.D. da Silva Martins, J.A.R. De Oliveira, A. Komesu, Microbial fuel cells and sustainable electricity generation from lignocellulosic waste biomass, in: G. Mukherjee, S. Dhiman (Eds.), Value Addition and Utilization of Lignocellulosic Biomass, Springer Nature Singapore, Singapore, 2025, pp. 139–165, doi:10.1007/978-981-96-2786-8_6.

[14] J. Arun, P.S SundarRajan, K.G Pavithra, P. Priyadharsini, S. Shyam, R. Goutham, et al., New insights into microbial electrolysis cells (MEC) and microbial fuel cells (MFC) for simultaneous wastewater treatment and green fuel (hydrogen) generation., Fuel 355 (2024) 129530, doi:10.1016/j.fuel.2023.129530.

[15] L. Samylingam, N. Aslfattahi, C.K Kok, K. Kadirgama, M. Schmirler, T. Yusaf, et al., Underlying developments in hydrogen production technologies: economic aspects and existent challenges, Korean J. Chem. Eng. 41 (11) (2024) 2961–2984, doi:10.1007/s11814-024-00264-5.

[16] Y.u-C. Liang, K. Li, X.-Q. Zhao, F.-W.u Bai, C.-G. Liu, Enhanced glucose utilization and succinic acid production in *Corynebacterium glutamicum* by integration of metabolic engineering and electro-fermentation, Chem. Eng. J. 505 (2025) 159522, doi:10.1016/j.cej.2025.159522.

[17] N. Deb, T. Rahman, Z. Alam, M.S. Jami, M.F.B.t. Mansor, Efficient technology for food waste valorization: an integrated anaerobic digestion—microbial electrolysis cell for biomethane production, Int. J. Environ. Sci. Technol. 22 (7) (2025) 5609–5622, doi:10.1007/s13762-024-06011-x.

[18] N. Goyal, M. Padhiary, I.A. Karimi, Z. Zhou, Flux measurements and maintenance energy for carbon dioxide utilization by *Methanococcus maripaludis*, Microb. Cell Fact. 14 (1) (2015) 146, doi:10.1186/s12934-015-0336-z.

[19] Tripathi, V. Kumar, K.M Tripathi, M. Shrivastava, N. Dhiman, Degradation of textile waste for environmental protection, in: A.K. Mishra (Ed.), Nano-Engineered Materials for Textile Waste Remediation, Environmental Footprints and Eco-design of Products and Processes, Springer Nature Singapore, Singapore, 2023, pp. 201–225, doi:10.1007/978-981-19-7978-1_10.

[20] P. Kumar, J. Singh, Innovative bio-green technologies for sequestration of carbon dioxide and utilisation for sustainable energy, in: S. Kandasamy, M.P. Shah, K. Subbiah, N. Manickam (Eds.), Microbial Niche Nexus Sustaining Environmental Biological Wastewater and Water-Energy-Environment Nexus, Environmental Science and Engineering, Springer Nature Switzerland, Cham, 2025, pp. 121–148, doi:10.1007/978-3-031-62660-9_5.

[21] Y. Li, Y. Zong, C. Feng, K. Zhao, The role of anode potential in electromicrobiology, Microorganisms 13 (3) (2025) 631, doi:10.3390/microorganisms13030631.

[22] W. Cui, S. Yin, Microbial electrolysis cells for H2 generation by treating acid mine drainage: recent advances and emerging trends, Fuels 6 (1) (2025) 14, doi:10.3390/fuels6010014.

[23] A. Saravanan, P.R. Yaashikaa, P.S. Kumar, A.S. Vickram, S. Karishma, R. Kamalesh, et al., Techno-economic and environmental sustainability prospects on biochemical conversion of agricultural and algal biomass to biofuels, J. Cleaner Prod. 414 (2023) 137749, doi:10.1016/j.jclepro.2023.137749.

[24] S. Dhiman, B. Thakur, S. Kaur, M. Ahuja, S. Gantayat, S. Sarkar, et al., Closing the loop: technological innovations in food waste valorisation for global sustainability, Discov. Sustain. 6 (1) (2025) 258, doi:10.1007/s43621-025-01073-4.

[25] N.A.N. Johnson, S.Y-S.S. Adade, J.-N. Ekumah, B.A. Kwadzokpui, J. Xu, Y.i Xu, et al., A comprehensive review of analytical techniques for spice quality and safety assessment in the modern food industry, Crit. Rev. Food Sci. Nutr. (2025) 1–26, doi:10.1080/10408398.2025.2462721.

[26] K. Saveetha, C.S. Somala, T. Anand, D. Balamurugan, V. Vasudevan, K.M. Saravanan, et al., Impact of soil microbiomes on mung bean cultivation: insights from 16S rRNA metagenomics, Mol. Biotechnol. (2025), doi:10.1007/s12033-025-01425-5.

[27] M. Naveed, Z. Memon, M. Abdullah, S.I. Makhdoom, A. Azeem, S. Mehmood, et al., Use cases and future aspects of intelligent techniques in microbial data analysis, in: A. Khamparia, B. Pandey, D.K. Pandey, D. Gupta (Eds.), Microbial Data Intelligence and Computational Techniques for Sustainable Computing, Microorganisms for Sustainability, Springer Nature Singapore, Singapore, 2024, pp. 259–280, doi:10.1007/978-981-99-9621-6_17.

[28] S. Poornima, S. Manikandan, R. Prakash, S.R. Deena, R. Subbaiya, N. Karmegam, et al., Biofuel and biochemical production through biomass transformation using advanced thermochemical and biochemical processes—a review, Fuel 372 (2024) 132204, doi:10.1016/j.fuel.2024.132204.

[29] F. Alqarzaee, MdA. Al Bari, S.A. Razzak, S. Uddin, Biomass-based hydrogen production towards renewable energy sources: an advance study, Emerg. Mater (2024), doi:10.1007/s42247-024-00931-7.

[30] D. Li, Q. Ping, R. Mo, W. Guo, S. Zhang, L. Wang, et al., Revealing synergistic mechanisms of biochar-assisted microbial electrolysis cells in enhancing the anaerobic digestion performance of waste activated sludge: extracellular polymeric substances characterization, enzyme activity assay, and multi-omics analysis, Water Res. 267 (2024) 122501, doi:10.1016/j.watres.2024.122501.

[31] D. Dominic, S. Baidurah, A review of biological processing technologies for palm oil mill waste treatment and simultaneous bioenergy production at laboratory scale, pilot scale and industrial scale applications with technoeconomic analysis, Energy Conver. Manag. 26 (2025) 100914, doi:10.1016/j.ecmx.2025.100914.

[32] V. Senatore, T. Zarra, M.G. Galang, G. Oliva, A. Buonerba, C.-W. Li, et al., Full-scale odor abatement technologies in wastewater treatment plants (WWTPs): a review, Water 13 (24) (2021) 3503, doi:10.3390/w13243503.

[33] K.T. Nachammai, S. Ramachandran, C. Nagarajan, L. Kulanthaivel, G.K. Subbaraj, K. Chandrasekaran, et al., Exploration of bioinformatics on microbial fuel cell

technology: trends, challenges, and future prospects, J. Chem. 2023 (2023) 1–8, doi:10.1155/2023/6902054.

[34] S. Wu, Z. Qu, D. Chen, H. Wu, Q. Caiyin, J. Qiao, Deciphering and designing microbial communities by genome-scale metabolic modelling., Comput. Struct. Biotechnol. J. 23 (2024) 1990–2000, doi:10.1016/j.csbj.2024.04.055.

[35] S. Ameer, H. Ibrahim, M.U Yaseen, F. Kulsoom, S. Cinti, M. Sher, Electrochemical impedance spectroscopy-based sensing of biofilms: a comprehensive review, Biosensors 13 (8) (2023) 777, doi:10.3390/bios13080777.

[36] D. Sushma, J. Das, S. Dave, P. Mohanty, G.M. Priya, et al., Contribution of cyclic voltammetry and electrochemical impedance spectroscopy in deciphering the electron transport system in biofilm, in: M. Nag, D. Lahiri (Eds.), Analytical Methodologies for Biofilm Research, Springer Protocols Handbooks, Springer US, New York, NY, 2021, pp. 129–152, doi:10.1007/978-1-0716-1378-8_6.

[37] A.é Gemünde, J. Gail, D. Holtmann, Anodic respiration of *Vibrio Natriegens* in a bioelectrochemical system, ChemSusChem 16 (16) (2023) e202300181, doi:10.1002/cssc.202300181.

[38] S. Wolf, M. Lippke, A. Schoo, A. Kwade, C. Schilde, A computational fluid dynamics–discrete element method model for physics-based simulation of structure formation during battery electrode drying, Energy Technol. 12 (4) (2024) 2301004, doi:10.1002/ente.202301004.

[39] M. Sun, L. Zhao, T. Liu, Z. Lu, G. Su, C. Wu, et al., Construction of CuO/Fe_2 O_3 nanozymes for intelligent detection of glufosinate and chlortetracycline hydrochloride, ACS Appl. Mater. Interfaces 15 (47) (2023) 54466–54477, doi:10.1021/acsami.3c12157.

[40] B. Koul, M. Yakoob, M.P. Shah, Agricultural waste management strategies for environmental sustainability, Environ. Res. 206 (2022) 112285, doi:10.1016/j.envres.2021.112285.

[41] M. Padhiary, R. Kumar, Assessing the environmental impacts of agriculture, industrial operations, and mining on agro-ecosystems, in: M. Azrour, J. Mabrouki, A. Alabdulatif, A. Guezzaz, F. Amounas (Eds.), *Smart Internet of Things for Environment and Healthcare*, Studies in Computational Intelligence, Springer Nature Switzerland, Cham, 2024, pp. 107–126, doi:10.1007/978-3-031-70102-3_8.

[42] A.I. Osman, N. Mehta, A.M. Elgarahy, A. Al-Hinai, A. Al-Muhtaseb, D.W. Rooney, Conversion of biomass to biofuels and life cycle assessment: a review, Environ. Chem. Lett. 19 (6) (2021) 4075–4118, doi:10.1007/s10311-021-01273-0.

[43] N. Ghosh, D. Rhithuparna, S.L Rokhum, G. Halder, Ethical issues pertaining to sustainable biodiesel synthesis over trans/esterification process, Sustain. Chem. Pharm. 33 (2023) 101123, doi:10.1016/j.scp.2023.101123.

[44] A. Hoque, M. Padhiary, S. Roy, Precision agriculture meets sustainable chemistry: innovations for eco-friendly farming, Sustainable Chemistry and Pioneering Green Engineering Solutions, IGI Global, 2025, pp. 233–270, doi:10.4018/979-8-3373-1409-9.ch009.

[45] N. Dutta, M. Usman, M.A Ashraf, G. Luo, M.G. El-Din, S. Zhang, Methods to convert lignocellulosic waste into biohydrogen, biogas, bioethanol, biodiesel and value-added chemicals: a review, Environ. Chem. Lett. 21 (2) (2023) 803–820, doi:10.1007/s10311-022-01511-z.

[46] T.S.P. De Souza, H.Y. Kawaguti, Cellulases, hemicellulases, and pectinases: applications in the food and beverage industry, Food Bioprocess Technol. 14 (8) (2021) 1446–1477, doi:10.1007/s11947-021-02678-z.

[47] Pooja, S.K. Yadav, Diversity and use of genetically modified microorganisms for second-generation ethanol production, in: C.R. Soccol, A.G. Pereira, C.-G. Dussap, L.P. De Souza Vandenberghe (Eds.), Liquid Biofuels: Bioethanol, Biofuel and

Biorefinery Technologies, Springer International Publishing, Cham, 2022, pp. 187–207, doi:10.1007/978-3-031-01241-9_9.

[48] M. Quraishi, K. Wani, S. Pandit, P.K. Gupta, A.K. Rai, D. Lahiri, et al., Valorisation of CO_2 into value-added products via microbial electrosynthesis (MES) and electro-fermentation technology, Fermentation 7 (4) (2021) 291, doi:10.3390/fermentation7040291.

[49] S. Bhushan, U. Jayakrishnan, B. Shree, P. Bhatt, S. Eshkabilov, H. Simsek, Biological pretreatment for algal biomass feedstock for biofuel production, J. Environ. Chem. Eng. 11 (3) (2023) 109870, doi:10.1016/j.jece.2023.109870.

[50] K. Kumar, M. Padhiary, The scope of hydrogen production and future in agricultural automation in India, Cureus J. Engin. (2025), doi:10.7759/s44388-025-05899-6.

[51] D.A. Jadhav, K. Gunaseelan, G.T.H. Le, T. Eisa, S.-G. Park, S. Gajalakshmi, et al., Sustainable vision toward development of microbial electrosynthesis for diverse resource recovery: circular economy, J. Environ. Chem. Eng. 12 (5) (2024) 114027, doi:10.1016/j.jece.2024.114027.

[52] S. Rajeswari, D. Baskaran, P. Saravanan, M. Rajasimman, N. Rajamohan, Y. Vasseghian, Production of ethanol from biomass—recent research, scientometric review and future perspectives, Fuel 317 (2022) 123448, doi:10.1016/j.fuel.2022.123448.

[53] B.S. Crandall, S. Overa, H. Shin, F. Jiao, Turning carbon dioxide into sustainable food and chemicals: how electrosynthesized acetate is paving the way for fermentation innovation, Acc. Chem. Res. 56 (12) (2023) 1505–1516, doi:10.1021/acs.accounts.3c00098.

[54] B.M Harahap, B.K. Ahring, Acetate Production from syngas produced from lignocellulosic biomass materials along with gaseous fermentation of the syngas: a review, Microorganisms 11 (4) (2023) 995, doi:10.3390/microorganisms11040995.

[55] S. Beluhan, K. Mihajlovski, B. Šantek, M.I. Šantek, The production of bioethanol from lignocellulosic biomass: pretreatment methods, fermentation, and downstream processing, Energies 16 (19) (2023) 7003, doi:10.3390/en16197003.

[56] K. Karoń, E. Zabłocka-Godlewska, K. Krukiewicz, Recent advances in the design of bacteria-based supercapacitors: current limitations and future opportunities, Electrochim. Acta 430 (2022) 141068, doi:10.1016/j.electacta.2022.141068.

[57] H.Y. Leong, C.-K. Chang, K.S Khoo, K.W. Chew, S.R. Chia, J.W. Lim, et al., Waste biorefinery towards a sustainable circular bioeconomy: a solution to global issues, Biotechnol. Biofuels 14 (1) (2021) 87, doi:10.1186/s13068-021-01939-5.

[58] V. Bohra, K.U. Ahamad, A. Kela, G. Vaghela, A. Sharma, B.J. Deka, Energy and resources recovery from wastewater treatment systems, Clean Energy and Resource Recovery, Elsevier, 2022, pp. 17–36, doi:10.1016/B978-0-323-90178-9.00007-X.

[59] M. Choudhary, P. Verma, S. Ray, A comprehensive review on bio-electrochemical systems for wastewater treatment: process, electricity generation and future aspect, Environ. Dev. Sustain. (2024), doi:10.1007/s10668-024-05866-x.

[60] H. Zhang, L. Duan, S. Li, Q. Gao, M. Li, F. Xing, et al., Simultaneous wastewater treatment and resources recovery by forward osmosis coupled with microbial fuel cell: a review., Membranes 14 (2) (2024) 29, doi:10.3390/membranes14020029.

[61] M. Ha, G. Gutenberger, L. Ou, H. Cai, and T. Hawkins. 2022. *Opportunities for recovering resources from municipal wastewater*. doi:10.2172/1876441.

[62] P. Ganeshan, V.S. Vigneswaran, S.C. Gowd, R. Mishra, E. Singh, A. Kumar, et al., Bioenergy with carbon capture, storage and utilization: potential technologies to mitigate climate change, Biomass Bioenergy 177 (2023) 106941, doi:10.1016/j.biombioe.2023.106941.

[63] M.A. Hayat, K. Alhadhrami, A.M. Elshurafa, Which bioenergy with carbon capture and storage (BECCS) pathways can provide net-negative emissions? Int. J. Greenhouse Gas Control 135 (2024) 104164, doi:10.1016/j.ijggc.2024.104164.

[64] K.S. Herman, B.K. Sovacool, Applying the multi-level perspective to climate geoengineering: sociotechnical bottlenecks for negative emissions and cloud seeding technologies., Energy Res. Soc. Sci. 115 (2024) 103637, doi:10.1016/j.erss.2024.103637.

[65] F.D Yeung, S. Sammarchi, E. Wang, Q.i Gao, J. Li, Interdisciplinary challenges in bioenergy carbon capture utilization & storage deployment: a review., Carbon Capture Sci. Technol. 13 (2024) 100283, doi:10.1016/j.ccst.2024.100283.

[66] D.C. Makepa, C.H. Chihobo, Sustainable pathways for biomass production and utilization in carbon capture and storage—a review, Biomass Convers. Biorefin. (2024), doi:10.1007/s13399-024-06010-5.

[67] S. Sahu, A. Kaur, G. Singh, S.K. Arya, Harnessing the potential of microalgae-bacteria interaction for eco-friendly wastewater treatment: a review on new strategies involving machine learning and artificial intelligence, J. Environ. Manage. 346 (2023) 119004, doi:10.1016/j.jenvman.2023.119004.

[68] Mafat, I. Hussain, S. Palla, D.V. Surya, Machine learning and artificial intelligence for algal cultivation, harvesting techniques, wastewater treatment, nutrient recovery, and biofuel production and optimization, in: S. KumarArya, M. Khatri, G. Singh (Eds.), Value Added Products From Bioalgae Based Biorefineries: Opportunities and Challenges, Springer Nature Singapore, Singapore, 2024, pp. 463–487, doi:10.1007/978-981-97-1662-3_18.

[69] S. Wu, Z. Qu, D. Chen, H. Wu, Q. Caiyin, J. Qiao, Deciphering and designing microbial communities by genome-scale metabolic modelling, Comput. Struct. Biotechnol. J. 23 (2024) 1990–2000, doi:10.1016/j.csbj.2024.04.055.

[70] M. Zulfiqar, V. Singh, C. Steinbeck, M. Sorokina, Review on computer-assisted biosynthetic capacities elucidation to assess metabolic interactions and communication within microbial communities, Crit. Rev. Microbiol. 50 (6) (2024) 1053–1092, doi:10.1080/1040841X.2024.2306465.

[71] V.K. Sandhwar, S. Saxena, D. Saxena, A. Tiwari, S.M. Parikh, Future trends and emerging technologies in water quality management, Computational Automation for Water Security, Elsevier, 2025, pp. 229–249, doi:10.1016/B978-0-443-33321-7.00012-3.

[72] D. Witczak, S. Szymoniak, Review of monitoring and control systems based on internet of things, Appl. Sci. 14 (19) (2024) 8943, doi:10.3390/app14198943.

[73] T. Kumar, S. Naik, S.E. Jujjavarappu, A critical review on early-warning electrochemical system on microbial fuel cell-based biosensor for on-site water quality monitoring, Chemosphere 291 (2022) 133098, doi:10.1016/j.chemosphere.2021.133098.

[74] S. Daousis, N. Peladarinos, V. Cheimaras, P. Papageorgas, D.D. Piromalis, R.A. Munteanu, Overview of protocols and standards for wireless sensor networks in critical infrastructures, Fut. Int. 16 (1) (2024) 33, doi:10.3390/fi16010033.

[75] D. Rajababu, S. Surya, M. Padhiary, H. Modi, Blockchain for Cybersecurity_ Securing data transactions and enhancing privacy in digital systems, 2025 First International Conference on Advances in Computer Science, Electrical, Electronics, and Communication Technologies (CE2CT), IEEE, 2025, pp. 1426–1430, doi:10.1109/CE2CT64011.2025.10939863.

[76] R. Kumar, H.K. Channi, A PV-Biomass off-grid hybrid renewable energy system (HRES) for rural electrification: design, optimization and techno-economic-environmental analysis, J. Cleaner Prod. 349 (2022) 131347, doi:10.1016/j.jclepro.2022.131347.

[77] M. Sharma, E.l-S. Salama, N. Thakur, H. Alghamdi, B.-H. Jeon, X. Li, Advances in the biomass valorization in bioelectrochemical systems: a sustainable approach for microbial-aided electricity and hydrogen production., Chem. Eng. J. 465 (2023) 142546, doi:10.1016/j.cej.2023.142546.

[78] S. Husain, N.A. Shrivastava, Novel power management strategy for a solar biomass off-grid power system, Residential Microgrids and Rural Electrifications, Elsevier, 2022, pp. 183–213, doi:10.1016/B978-0-323-90177-2.00002-5.

[79] S. Das, R. Raj, S. Das, M.M. Ghangrekar, A sustainable approach for the production of green energy with the holistic treatment of wastewater through microbial electrochemical technologies: a review, Front. Sustain. 2 (2021) 792028, doi:10.3389/frsus.2021.792028.

[80] M.N. Khan, T.A. Sial, A. Ali, F. Wahid, Impact of agricultural wastes on environment and possible management strategies, in: A. Núñez-Delgado (Ed.), *Frontier Studies in Soil Science*,, Springer International Publishing, Cham, 2024, pp. 79–108, doi:10.1007/978-3-031-50503-4_4.

[81] D. Saha, M. Padhiary, N. Chandrakar, AI vision and machine learning for enhanced automation in food industry: a systematic review, Food Humanity 4 (2025) 100587, doi:10.1016/j.foohum.2025.100587.

[82] N. Dutta, M. Usman, G. Luo, S. Zhang, An insight into valorization of lignocellulosic biomass by optimization with the combination of hydrothermal (HT) and biological techniques: a review, Sustain. Chem. 3 (1) (2022) 35–55, doi:10.3390/suschem3010003.

[83] P. Bhattacharyya, J. Bisen, D. Bhaduri, S. Priyadarsini, S. Munda, M. Chakraborti, et al., Turn the wheel from waste to wealth: economic and environmental gain of sustainable rice straw management practices over field burning in reference to India, Sci. Total Environ. 775 (2021) 145896, doi:10.1016/j.scitotenv.2021.145896.

[84] S. Beluhan, K. Mihajlovski, B. Šantek, M. Ivančić Šantek, The production of bioethanol from lignocellulosic biomass: pretreatment methods, fermentation, and downstream processing, Energies 16 (19) (2023) 7003, doi:10.3390/en16197003.

[85] S. Baksi, D. Saha, S. Saha, U. Sarkar, D. Basu, J.C. Kuniyal, Pre-treatment of lignocellulosic biomass: review of various physico-chemical and biological methods influencing the extent of biomass depolymerization., Int. J. Environ. Sci. Technol. 20 (12) (2023) 13895–13922, doi:10.1007/s13762-023-04838-4.

[86] H. Bhoomeeswaran, T.V. Visvesh Yasyanth, N.A. Akash, Exploring materials for green hydrogen production: opportunities and obstacles in biological hydrogen production, in: K. Kannan, V. Tari (Eds.), Novel Energy Storage and Conversion Technologies for Two-Dimensional MXenes and MBenes, IGI Global, 2025, pp. 349–376, doi:10.4018/979-8-3693-9546-2.ch014.

[87] L.A.G. Fernández, V. Castillo Ramos, M. Sánchez Polo, N.A.M. Castillo, Fundamentals in applications of algae biomass: a review, J. Environ. Manage. 338 (2023) 117830, doi:10.1016/j.jenvman.2023.117830.

[88] K. Dhalsamant, A.K. Dalai, Optimization and characterization of biocrude produced from hydrothermal liquefaction of food waste, Sustain. Energy Fuels 9 (8) (2025) 2119–2136, doi:10.1039/D5SE00136F.

[89] S. Kasiński, M. Dębowski, Municipal solid waste as a renewable energy source: advances in thermochemical conversion technologies and environmental impacts, Energies 17 (18) (2024) 4704, doi:10.3390/en17184704.

[90] J.K. Kenny, S.R. Neefe, D.G. Brandner, M.L. Stone, R.M. Happs, I. Kumaniaev, et al., Design and validation of a high-throughput reductive catalytic fractionation method, JACS Au 4 (6) (2024) 2173–2187, doi:10.1021/jacsau.4c00126.

[91] A. Kazmi, T. Sultana, A. Ali, A. Nijabat, G. Li, H. Hou, Innovations in bioethanol production: a comprehensive review of feedstock generations and technology advances, Energy Strat. Rev. 57 (2025) 101634, doi:10.1016/j.esr.2024.101634.

[92] M.J. Salar-García, V.M. Ortiz-Martínez, S. Sánchez-Segado, R. Valero Sánchez, A. Sáez López, L.J. Lozano Blanco, et al., Sustainable production of biofuels and biochemicals via electro-fermentation technology, Molecules 29 (4) (2024) 834, doi:10.3390/molecules29040834.

[93] B. Virdis, R.D. Hoelzle, A. Marchetti, S.T. Boto, M.A. Rosenbaum, R. Blasco-Gómez, et al., Electro-fermentation: sustainable bioproductions steered by electricity, Biotechnol. Adv. 59 (2022) 107950, doi:10.1016/j.biotechadv.2022.107950.

[94] B. Kim, G. Baek, C. Kim, S.Y Lee, E. Yang, S. Lee, et al., Progress and prospects for applications of extracellular electron transport mechanism in environmental biotechnology, ACS ES&T Engin. 4 (7) (2024) 1520–1539, doi:10.1021/acsestengg.4c00077.

[95] H.-Y. Lee, T.-S. You, C.-L. Chen, Energy efficient design of bio-butanol purification process from acetone butanol ethanol fermentation, J. Taiwan Inst. Chem. Eng. 130 (2022) 104015, doi:10.1016/j.jtice.2021.08.003.

[96] R.F. Darmayanti, M. Muharja, T. Zhao, M. Gao, Y. Tashiro, K. Sakai, et al., Techno-economic analysis of extractive butanol fermentation by immobilized cells with large extractant volume, Jurnal Teknik Kimia dan Lingkungan 6 (2) (2022) 99–111, doi:10.33795/jtkl.v6i2.337.

[97] M. Moon, W.-K. Park, S.Y. Lee, K.-R. Hwang, S. Lee, M.-S. Kim, et al., Utilization of whole microalgal biomass for advanced biofuel and biorefinery applications, Renew. Sustain. Energy Rev. 160 (2022) 112269, doi:10.1016/j.rser.2022.112269.

[98] P. Thanahiranya, Process Design and Evaluation of Value-Added Chemical Production from Crude Glycerol Obtained from Biodiesel Industry, Chulalongkorn University, 2023, doi:10.58837/CHULA.THE.2023.72.

[99] L. Mitrea, B.-E. Teleky, S.-A. Nemes, D. Plamada, R.-A. Varvara, M.-S. Pascuta, et al., Succinic acid—a run-through of the latest perspectives of production from renewable biomass, Heliyon 10 (3) (2024) e25551, doi:10.1016/j.heliyon.2024.e25551.

[100] Y. Chen, N. Shen, T. Wang, F. Zhang, R.J. Zeng, Ammonium level induces high purity propionate production in mixed culture glucose fermentation, RSC Adv. 7 (1) (2017) 518–525, doi:10.1039/C6RA25926J.

[101] M. Sakarika, B. Delmoitié, E. Ntagia, I. Chatzigiannidou, X. Gabet, R. Ganigué, et al., Production of microbial protein from fermented grass, Chem. Eng. J. 433 (2022) 133631, doi:10.1016/j.cej.2021.133631.

[102] A.B.T. Nelabhotla, C. Dinamarca, Electrochemically mediated CO_2 reduction for biomethane production: a review, Rev. Environ. Sci. Bio/Technol. 17 (3) (2018) 531–551, doi:10.1007/s11157-018-9470-5.

[103] Q. Liu, S. Lv, Microbial growth and ecological and metabolic characteristics in bioelectrosynthesis systems, in: A. Wang, W. Liu, B. Zhang, W. Cai (Eds.), Bioelectrosynthesis, Wiley, 2020, pp. 307–332, doi:10.1002/9783527343829.ch13.

[104] M. Mainardis, D. Cecconet, A. Moretti, A. Callegari, D. Goi, S. Freguia, et al., Wastewater fertigation in agriculture: issues and opportunities for improved water management and circular economy, Environ. Pollut. 296 (2022) 118755, doi:10.1016/j.envpol.2021.118755.

[105] M. Zielińska, K. Bułkowska, Sustainable management and advanced nutrient recovery from biogas energy sector effluents, Energies 17 (15) (2024) 3705, doi:10.3390/en17153705.

[106] A.Z. Yaser, N.N. Safie, Sewage treatment in campus for recycling purpose: a review, in: A.Z. Yaser (Ed.), Green Engineering for Campus Sustainability, Springer Singapore, Singapore, 2020, pp. 207–243, doi:10.1007/978-981-13-7260-5_15.

[107] F. Wang, W.-K. Wong, Z. Wang, G. Albasher, N. Alsultan, A. Fatemah, Emerging pathways to sustainable economic development: an interdisciplinary exploration of resource efficiency, technological innovation, and ecosystem resilience in resource-rich regions, Resour. Policy 85 (2023) 103747, doi:10.1016/j.resourpol.2023.103747.

[108] L. Rovira-Alsina, M. Romans-Casas, E. Perona-Vico, A. Ceballos-Escalera, M.D. Balaguer, L. Bañeras, et al., Microbial electrochemical technologies: sustainable solutions for addressing environmental challenges, Advances in Biochemical Engineering/Biotechnology, Springer Berlin Heidelberg, Berlin, Heidelberg, 2024, doi:10.1007/10_2024_273.

[109] R.A. Sheldon, D. Brady, Green chemistry, biocatalysis, and the chemical industry of the future, ChemSusChem 15 (9) (2022) e202102628, doi:10.1002/cssc.202102628.

[110] C.H. Bühlmann, B.S. Mickan, S. Tait, D.J. Batstone, P.A. Bahri, Waste to wealth: the power of food-waste anaerobic digestion integrated with lactic acid fermentation., Front. Chem. Engin. 5 (2023) 1285002, doi:10.3389/fceng.2023.1285002.

[111] H.H. Gandhi, B. Hoex, B.J. Hallam, Private equity renewable energy investments in India, Heliyon 11 (1) (2025) e41098, doi:10.1016/j.heliyon.2024.e41098.

[112] F. Shaheen, M.S Lodhi, J. Rosak-Szyrocka, K. Zaman, U. Awan, M. Asif, et al., Cleaner technology and natural resource management: an environmental sustainability perspective from China, Clean Technol. 4 (3) (2022) 584–606, doi:10.3390/cleantechnol4030036.

[113] L. Zhao, K.Y. Chau, T.K Tran, M. Sadiq, N.T.M. Xuyen, T.T.H. Phan, Enhancing green economic recovery through green bonds financing and energy efficiency investments, Econ. Anal. Policy 76 (2022) 488–501, doi:10.1016/j.eap.2022.08.019.

[114] R. Kandpal, S.W. Ali, S.Z. Ahammad, Commercialization and current market of the microbial electrochemical technologies as biosensor and remediation, in: R. Selvasembian, J. Mal, S. Das, D.K. Verma, I. Anastopoulos (Eds.), Emerging Trends in Microbial Electrochemical Technologies for Sustainable Mitigation of Water Resources Contamination, Springer Nature Switzerland, Cham, 2024, pp. 313–357, doi:10.1007/978-3-031-74636-9_13.

[115] S. Saini, S. Tewari, J. Dwivedi, V. Sharma, Biofilm-mediated wastewater treatment: a comprehensive review, Mater. Adv. 4 (6) (2023) 1415–1443, doi:10.1039/D2MA00945E.

[116] Na Chu, Q. Liang, W. Hao, Y. Jiang, P. Liang, R.J. Zeng, Microbial electrochemical sensor for water biotoxicity monitoring, Chem. Eng. J. 404 (2021) 127053, doi:10.1016/j.cej.2020.127053.

[117] M. Carminati, A. Turolla, L. Mezzera, M. Di Mauro, M. Tizzoni, G. Pani, et al., A self-powered wireless water quality sensing network enabling smart monitoring of biological and chemical stability in supply systems, Sensors 20 (4) (2020) 1125, doi:10.3390/s20041125.

[118] S. Mosquera-Romero, E. Ntagia, D.P.L. Rousseau, A. Esteve-Núñez, A. Prévoteau, Water treatment and reclamation by implementing electrochemical systems with constructed wetlands, Environ. Sci. Ecotechnol. 16 (2023) 100265, doi:10.1016/j.ese.2023.100265.

[119] J.O Unuofin, S.A. Iwarere, M.O. Daramola, Embracing the future of circular bio-enabled economy: unveiling the prospects of microbial fuel cells in achieving true sustainable energy, Environ. Sci. Pollut. Res. 30 (39) (2023) 90547–90573, doi:10.1007/s11356-023-28717-0.

[120] B. Martín-Fernández, C.L. Manzanares-Palenzuela, M. Sánchez-Paniagua López, N. de-los-Santos-Álvarez, B. López-Ruiz, Electrochemical genosensors in food safety assessment, Crit. Rev. Food Sci. Nutr. 57 (13) (2017) 2758–2774, doi:10.1080/10408398.2015.1067597.

[121] S. Bade, O.S. Tomomewo, A. Meenakshisundaram, P. Ferron, B.A. Oni, Economic, social, and regulatory challenges of green hydrogen production and utilization in the US: a review, Int. J. Hydrogen Energy 49 (2024) 314–335, doi:10.1016/j.ijhydene.2023.08.157.

[122] S. Barbhuiya, B.B. Das, D. Adak, Roadmap to a net-zero carbon cement sector: strategies, innovations and policy imperatives, J. Environ. Manage. 359 (2024) 121052, doi:10.1016/j.jenvman.2024.121052.

[123] C. Holland, P. Shapira, Building the bioeconomy: a targeted assessment approach to identifying biobased technologies, challenges and opportunities, Engin. Biol. 8 (1) (2024) 1–15, doi:10.1049/enb2.12030.

[124] C. Matos, M. Castro, J. Baptista, A. Valente, A. Briga-Sá, The use of water in wineries: a review, Sci. Total Environ. 951 (2024) 175198, doi:10.1016/j.scitotenv.2024.175198.

[125] S. Cairone, S.W. Hasan, K.-H. Choo, D.F. Lekkas, L. Fortunato, A.A. Zorpas, et al., Revolutionizing wastewater treatment toward circular economy and carbon neutrality goals: pioneering sustainable and efficient solutions for automation and advanced process control with smart and cutting-edge technologies., J. Water Process Eng. 63 (2024) 105486, doi:10.1016/j.jwpe.2024.105486.

[126] M. Badia-Fabregat, I. Oller, S. Malato, Overview on pilot-scale treatments and new and innovative technologies for hospital effluent, in: P. Verlicchi (Ed.), *Hospital Wastewaters*,, The Handbook of Environmental Chemistry, Springer International Publishing, Cham, 2017, pp. 209–230, doi:10.1007/698_2017_23.

[127] F.W.S. Lucas, R.G. Grim, S.A. Tacey, C.A. Downes, J. Hasse, A.M. Roman, et al., Electrochemical routes for the valorization of biomass-derived feedstocks: from chemistry to application, ACS Energy Lett. (2021) 1205–1270, doi:10.1021/acsenergylett.0c02692.

[128] S. Bajracharya, B. Bian, R. Jimenez-Sandoval, L. Matsakas, K.P. Katuri, P.E. Saikaly, Nature inspired catalysts: a review on electroactive microorganism-based catalysts for electrochemical applications, Electrochim. Acta 488 (2024) 144215, doi:10.1016/j.electacta.2024.144215.

[129] T. Noor, L. Yaqoob, N. Iqbal, Recent advances in electrocatalysis of oxygen evolution reaction using noble-metal, transition-metal, and carbon-based materials, ChemElectroChem 8 (3) (2021) 447–483, doi:10.1002/celc.202001441.

[130] Abdullah, K. Krukiewicz, Development of electrically-conducting biohybrid materials based on electroactive bacteria and conjugated polymers: review and perspectives, Electrochim. Acta 468 (2023) 143191, doi:10.1016/j.electacta.2023.143191.

[131] X. Tan, J. Nielsen, The integration of bio-catalysis and electrocatalysis to produce fuels and chemicals from carbon dioxide, Chem. Soc. Rev. 51 (11) (2022) 4763–4785, doi:10.1039/D2CS00309K.

[132] S.T Boto, L. Cristiani, M.A. Rosenbaum, Biochemical production with microbial bioelectrochemical systems, Curr. Opin. Biotechnol. 93 (2025) 103291, doi:10.1016/j.copbio.2025.103291.

[133] E.T Sayed, T. Wilberforce, K. Elsaid, M.K.H. Rabaia, M.A Abdelkareem, K.-J. Chae, et al., A critical review on environmental impacts of renewable energy systems and mitigation strategies: wind, hydro, biomass and geothermal, Sci. Total Environ. 766 (2021) 144505, doi:10.1016/j.scitotenv.2020.144505.

[134] A. Escamilla, D. Sánchez, L. García-Rodríguez, Techno-economic study of power-to-power renewable energy storage based on the smart integration of battery, hydrogen, and micro gas turbine technologies, Energy Convers. Manag. 18 (2023) 100368, doi:10.1016/j.ecmx.2023.100368.

[135] A. Kartalidis, K. Atsonios, N. Nikolopoulos, Enhancing the self-resilience of HIGH-renewable energy sources, interconnected islanding areas through innovative energy production, storage, and management technologies: grid simulations and energy assessment, Int. J. Energy Res. 45 (9) (2021) 13591–13615, doi:10.1002/er.6691.

[136] K. Ukoba, K.O. Olatunji, E. Adeoye, T.-C. Jen, D.M. Madyira, Optimizing renewable energy systems through artificial intelligence: review and future prospects, Energy Environ. 35 (7) (2024) 3833–3879, doi:10.1177/0958305X241256293.

[137] K. Paul, S.S. Chatterjee, P. Pai, A. Varshney, S. Juikar, V. Prasad, et al., Viable smart sensors and their application in data driven agriculture, Comput. Electron. Agric. 198 (2022) 107096, doi:10.1016/j.compag.2022.107096.

[138] M. Padhiary, P. Roy, P. Dey, B. Sahu, Harnessing AI for automated decision-making in farm machinery and operations: optimizing agriculture, in: S. Hai-Jew (Ed.), Advances in Computational Intelligence and Robotics, IGI Global, 2024, pp. 249–282, doi:10.4018/979-8-3693-6230-3.ch008.

[139] M. Schwenzer, M. Ay, T. Bergs, D. Abel, Review on model predictive control: an engineering perspective, Int. J. Adv. Manuf. Technol. 117 (5–6) (2021) 1327–1349, doi:10.1007/s00170-021-07682-3.

[140] A.N. Abuzaid, AI-driven business processes: efficiency gains, cost reductions, and implications for economic performance, 2024 International Conference on Knowledge Engineering and Communication Systems (ICKECS), IEEE, 2024, pp. 1–5, doi:10.1109/ICKECS61492.2024.10617370.

[141] A.I. Kouzi, M. Puranen, M.H. Kontro, Evaluation of the factors limiting biogas production in full-scale processes and increasing the biogas production efficiency, Environ. Sci. Polluti. Res. 27 (22) (2020) 28155–28168, doi:10.1007/s11356-020-09035-1.

[142] A. Ucar, M. Karakose, N. Kırımça, Artificial intelligence for predictive maintenance applications: key components, trustworthiness, and future trends, Appl. Sci. 14 (2) (2024) 898, doi:10.3390/app14020898.

[143] X. Liu, K. Fan, X. Huang, J. Ge, Y. Liu, H. Kang, Recent advances in artificial intelligence boosting materials design for electrochemical energy storage, Chem. Eng. J. 490 (2024) 151625, doi:10.1016/j.cej.2024.151625.

[144] V. Ashokkumar, G. Flora, R. Venkatkarthick, K. SenthilKannan, C. Kuppam, G.M. Stephy, et al., Advanced technologies on the sustainable approaches for conversion of organic waste to valuable bioproducts: emerging circular bioeconomy perspective, Fuel 324 (2022) 124313, doi:10.1016/j.fuel.2022.124313.

[145] M.J.B. Kabeyi, O.A OlanrewajuA. Messineo (Ed.), Biogas production and applications in the sustainable energy transition, J. Energy 2022 (2022) 1–43, doi:10.1155/2022/8750221.

[146] N.E. Benti, G. Gamachis Sakata, A. Tegenu, A. Abreham Berta, G. Solomon, K. Gashaw Beyene, et al., The current status, challenges and prospects of using biomass energy in Ethiopia, Biotechnol. Biofuels 14 (1) (2021) 209, doi:10.1186/s13068-021-02060-3.

[147] L. Genovaitė, A. Miceikienė, Contribution of the European bioeconomy strategy to the green deal policy: challenges and opportunities in implementing these policies., Sustainability 15 (9) (2023) 7139, doi:10.3390/su15097139.

CHAPTER

15

Biodiesel production through electrochemical biomass conversion

Ramyakrishna Pothu[a], *Amit Kumar Rajak*[b], *Ramakrishna Madaka*[c], *Santhosh Kumar Thatikonda*[d], *Mrutyunjay Padhiary*[e], *Rajesh K. Srivastava*[f], *and Rajender Boddula*[c]

[a]School of Physics and Electronics, College of Chemistry and Chemical Engineering, Hunan University, Changsha, Hunan, China, [b]Department of Energy & Environmental Engineering, CSIR-Indian Institute of Chemical Technology, Hyderabad, Telangana, India, [c]School of Sciences, Woxsen University, Hyderabad, Telangana, India, [d]School of Technology, Woxsen University, Hyderabad, Telangana, India, [e]Department of Agricultural Engineering, Assam University, Silchar, Assam, India, [f]Department of Biotechnology, Gandhi Institute of Technology and Management (GITAM), Visakhapatnam, Andhra Pradesh, India

15.1 Introduction

The quick growth in global needs for energy owing to urbanization and reliance on fossil fuels has increased due to the development of cities and increasing infrastructure demands, producing major environmental problems such as climate change emissions of greenhouse gases. When the quantity of oil and gas reserves continues to decline and raises concerns

Electrochemical Biomass Conversion: Sustainable Processes and Applications for Bioenergy, Biofuels, and Bioproducts
DOI: https://doi.org/10.1016/B978-0-443-26757-4.00005-2

regarding the security of energy and the move to sustainable energy, it is now crucial to employ energy sources that are low-carbon [1]. Biofuels, especially biodiesel, have become a viable substitute because they are biodegradable, have reduced emissions characteristics, and flexibility using current diesel engines. Additionally, the manufacture of biodiesel from unused biomass and oils that are not edible conforms to sustainable development objectives by lessening reliance on food-related materials and encouraging the use of circular resources, thus encouraging a more durable and sustainable energy source in the future.

Biodiesel provides numerous considerable environmental benefits compared to traditional petro-diesel, principally owing to its renewable source, better combustion properties, and reduced emissions throughout the life cycle [2]. Produced from ecological substrates including cooking oil waste, inedible oils, and agricultural wastes, biodiesel helps to recycle carbon instead of adding fresh fossil carbon into the environment. Life-cycle evaluations regularly show that using biodiesel can lower total emissions of carbon dioxide by as much as 60% to 80%, in relation to the raw material and route of manufacture [3]. This decrease is ascribed to the absorption of biogenic carbon while the feedstock is growing, which counteracts emissions produced when combustion occurs. In addition, biodiesel has minimal Sulfur and aromatic chemicals, which produce substantially decreased levels of sulfur oxides (SOx), PAHs, or polycyclic aromatic hydrocarbons, and particle matter. Its enhanced oxygen content promotes efficiency of combustion, resulting in fewer uncompleted combustion byproducts, including CO (carbon monoxide) and unprocessed hydrocarbons. Additionally, biodiesel shows reduced aquatic toxicity and enhanced biodegradability, mitigating hazards related to unintentional leaks versus petro-diesel [4]. Furthermore, biodiesel oxidation creates less soot, which helps lower particle matter-associated health issues and minimizes urban pollutants in the air. When manufactured from waste-generated feedstocks, biodiesel additionally promotes the reduction of waste and valuation of resources. Given the circumstances, these advantages position biodiesel as a purifier, environmentally friendly, and sustainable substitute for diesel fuels made from fossil fuels.

Traditional methods of transesterification, usually catalyzed by uniform bases or acids, experience several constraints that limit their efficiency and ecological responsibility [5]. Base-catalyzed reactions are extremely delicate, performance is inadequate, and feedstock reliability utilizing oils that have significant quantities of moisture or free fatty acid (FFA), leading to decreased production, soap formation, and complicated downstream filtration. Despite acid-catalyzed reactions possessing greater resilience regarding FFAs, notice a slow movement and require higher temperatures, faster response times, and increased use of energy. The two approaches need an excessive amount of cleaning and methods of separation,

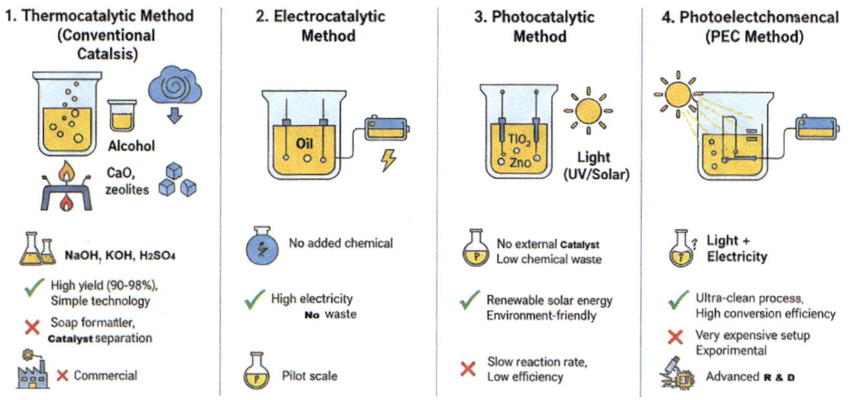

FIGURE 15.1 Biodiesel production by catalysis and energy mode.

producing substantial amounts of wastewater and animal husbandry operating expenditures. In addition, homogeneous catalysts become difficult to recover, leading to waste disposal and reducing circularity in processes [6]. These difficulties demonstrate the need for sophisticated catalytic techniques that promote quicker, more ecologically friendly, and effective biodiesel production.

This section currently offers a thorough analysis of current developments during the use of biodiesel in manufacturing, putting a focus on accomplishments in the creation of environmentally friendly catalytic technology [7]. It does so by talking about how biodiesel plays a major part in global energy challenges and sustainability. The subsequent sections analyze the expansion of sources and fundamental concepts of chemical processes. The development of both traditional and improved catalytic processes (Thermo, photo, electro, and photoelectrochemical pathways) (Fig. 15.1) [8]. Transesterification by electrolysis, the use of catalysts obtained from biomass, and methods for high glycerol are given special attention. Techno-economic issues, industrial development, and sustainable engineering techniques have all been highlighted. The section ends with concerns and goals for future generations, and there are more possibilities to apply cutting-edge catalytic linkages in biorefineries that produce biodiesel in the future.

15.2 Raw materials for biodiesel manufacturing

Biofuel's raw ingredients include oils, which are the primary ones, intermediate, and the third generation, combined with frying oil and any leftover biological pollutants. Oils that are either edible or inedible, such as Helianthus annuus oil, Elaeis guineensis oil, and Oryza sativa bran oil, vary in fatty acid makeup's effects on the efficiency of the response. These chemical characteristics establish potential for transesterification and total production and quality of biodiesel.

15.2.1 Initial, intermediate, and final oil feedstocks

Biodiesel materials are widely grouped into oils that are primary, secondary, and tertiary according to their processing demands and supply. Among the main feedstocks are vegetable oils that are edible, such as sunflower, soybean, and palm oil. They possess minimal impurities and appropriate fatty acid concentrations but express worries about the struggle between food and fuel [9]. Nonedible oils are secondary sources of energy, including neem, Karanja, and jatropha, which are valued for their accessibility, low cost, and substantial FFA amount. Despite this, pretreatment is frequently necessary before transesterification. Tertiary feedstocks comprise waste-produced oils and industrial by-products, such as waste cooking oil, corrosive oil, and leftovers from agro-industrial activities. These components offer substantial economic and environmental advantages that are obtained by promoting waste valorization, though their changing appearance needs more resilient catalytic systems.

15.2.2 Cooking oil waste and leftover biomass pathways

Waste cooking oil (WCO) and sources of residual biomass have been recognized as very sustainable and economical materials for biodiesel manufacturing. WCO, produced by both commercial and household food processing procedures, is high in triglycerides; nevertheless frequently contains increased amounts of FFAs, moisture, and products of breakdown, requiring effective pretreatment or powerful catalysts [10]. Using WCO lowers feedstock prices in addition to but also reduces pollution in the environment related to inappropriate disposal. Remaining derived from biomass oils, including those produced from rice bran, maize sprouts, or various agricultural by-products, increases the supply of biodiesel material. These categories support the ideas of the circular economy transform garbage into fuels with increased value, and lessening reliance on edible oils. Their varied makeup, nevertheless, needs sophisticated processing and catalytic techniques for maximum biodiesel production.

15.2.3 Rice bran oil, sunflower oil, palm oil, and various edible/inedible oils

Sunflower oil, palm oil, rice bran oil, and other edible and inedible oils are common materials for biodiesel because of their worldwide accessibility and varied fatty acid compositions. Rice bran oil, obtained from grinding waste products, includes important bioactive substances, including

FFAs, therefore well suited towards both esterification and transesterification [11]. Sunflower oil, high in linoleic acid, gives great oxidative stability and excellent biodiesel purity. Palm oil, cheap and plentiful, delivers a significant quantity of saturated fatty acids, leading to improved cetane numbers but creating restrictions on cold flow. Nonedible oils such as jatropha, Karanja, and neem avoid food–fuel competition and thrive on marginal lands. Their diverse compositions necessitate specialized catalysts and improved response standards for achieving superior biofuel yields and energy property compliance.

15.2.4 Chemical makeup and transesterification application

The chemical composition of a fluid immediately defines its possibilities in order to transesterify the final result, regarding the quality of biofuel. A description of fatty acids, concentration of FFAs, amount of moisture, and the presence of phospholipids, polysaccharides, and impurities are crucial factors in oils that include saturated fatty acids, which create biofuel with a higher permanence of oxidation and cetane quantity. However, unsaturated fatty acids improve the cold-flow characteristics, which may impact on the consistency of preservation. Increased FFA and too much moisture could interrupt Base-induced damage transesterification through improving the manufacturing of soap, possibly more catalytic techniques, or the esterification process of acid [12]. Furthermore, pollutants like steel or colors could vary the mechanism of interactions and deactivate the catalysts. As a result, understanding the physical-chemical composition of each feedstock is crucial when selecting appropriate catalysts, preprocessing methods, and circumstances of the reaction to encourage efficient conversion of biodiesel.

15.3 Principles of biodiesel manufacturing

Biodiesel synthesis involves converting triglycerides from oils or fats into fatty acid methyl esters through transesterification and ester transformation. A particular kind of catalyst decides on the process, temperature, the ratio of alcohol to oil, and the speed of response. For production to be efficient, suitable reaction conditions, suitable catalysts, and preparing the materials promise quick conversion and the standard of fuel conformity.

15.3.1 Transesterification and esterification in chemical science

Transesterification as well as esterification are both of them fundamental processes of chemical reactions, limiting the manufacture of biodiesel. The process of esterification process involves mixing FFAs with alcohol, typically methanol, whereas an acid catalyst is used for the making of water and fatty acid methylesters (FAMEs). This method is essential for high-FFA materials because it reduces acidity and prevents soap from being produced. The primary method used to produce biodiesel is transesterification, which converts triglycerides that exist in lipids or oils by reacting with methanol to make FAMEs, producing the intermediate glycerol [13]. Typically, this method is sped up by homogeneous or heterogeneous catalysts that are enzymatic, acidic, or basic. The alcohol concentration ratio affects reaction performance, temperature, substrate quality, and catalyst activity; they all affect the quality and quantity of the transformation of biodiesel.

15.3.2 Catalyst category's impact on process performance

The kind of catalyst used is very important in assessing biodiesel manufacturing's overall viability, efficiency, and development rate. Base catalysts like KOH and NaOH provide quick transesterification of short-chain fatty acid oils but are very susceptible to humid conditions and FFAs, which result in the creation of soap and lower yields. Acid catalysts, such as H_2SO_4 and solid acid materials, are more resistant to contaminants and effectively transform utilizing high-FFA feedstocks by concurrently transesterification and esterification, they demand longer responses and greater temperatures [14]. Both basic and acidic heterogeneous catalysts offer benefits like quick separation, reuse potential, and decreased wastewater creation, increasing operational sustainability. Enzyme-derived catalysts provide great selectivity and mild operational conditions but are constrained by price and durability. Therefore, the choice of catalyst has a direct impact on the effectiveness of conversion, complexity of operation, and financial viability.

15.3.3 Procedure parameters include temperature, molar proportion, combination, and time to reaction

Operational factors, including temperature, alcohol-to-oil molar proportion, and combining quantity and time to react, critically impact the effectiveness of the production of biodiesel. Temperature impacts the timing of reactions, catalyst function, and transfer of mass, with ideal values increasing conversion while minimizing methanol leakage 15,16. The molecular proportion of alcohol to oil promotes equilibrium of the reaction toward ester production, where larger ratios can make separation more difficult but often increase yield. Catalyst-reactant interaction is improved by reducing mixing intensity and mass-transfer limitations in complex systems, in particular. Time required for the reaction to have enough time to arrive at equilibrium without producing degradation. High conversion is ensured by better fuel quality, more efficient processes, and appropriate optimization of those elements (Fig. 15.2).

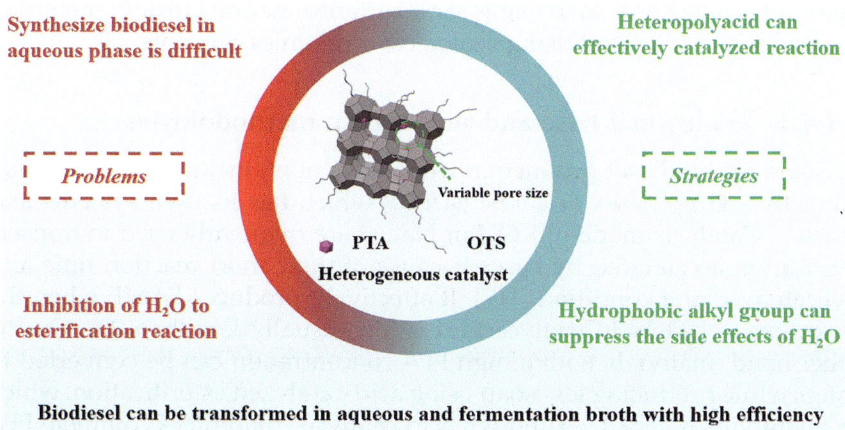

FIGURE 15.2 Challenges in thermocatalytic biodiesel production [16].

15.3.4 The chemistry and mechanisms of biodiesel production

Thermodynamics' connection to and mechanisms of the biodiesel synthesis process, which are crucial to comprehend, improvement of the method of transesterification, and methods for esterification. Depending on the quantity of alcohol ingested, either a first-order pseudo or a second-order response is used to define the process of producing biodiesel. Variables like temperature, kind of catalyst, and reaction times are all impacted by the molar proportion of alcohol to oil and Intense merging [16]. Under ideal circumstances, the process of transesterification in chemical processes is often endothermic and favors higher temperatures, in addition to changes in Gibbs free energy that suggest uncertainty. The rate of conversion is shown by the effects of the activation energy, reaction rate, and equilibrium parameters. Understanding these factors makes it possible to precisely regulate yield, flexibility, and response times, assuring effective biofuel production and limited waste and power usage.

15.4 The development of biodiesel techniques

New developments in manufacturing Simple techniques are boosting biofuel. Alkali-driven transition to sophisticated enzymatic and other catalysts Enhancing efficiency and ecological accountability. In earlier techniques, the oils used were edible. However, non-edible oils, tiny algae, and waste-derived oils are contemporary alternatives. Techniques for improving processes, such as microwave, advanced supercritical, and ultrasonic technology, boost yields and reaction times. Current advancements center on more economical, sustainable, and superior manufacturing, including the creation of biodiesel using ecological economics concepts.

15.4.1 Traditional base and acid-driven methodologies

Standard biodiesel production techniques often include processes that are activated by bases or acids, each of which has its own systems and efficacy. Catalysts made of KOH or NaOH are frequently used in transesterification accelerated by bases because of their short reaction time and delicate working conditions [17]. It effectively produces FAME when the percentage of FFAs in triglycerides is low, usually less than 2%. On the other hand, materials with a high FFA concentration can be converted to esters while producing less soap using acid-catalyzed esterification, which frequently uses H_2SO_4. Although acid catalysis guarantees complete FFA conversion, it is less efficient and requires higher temperatures and longer reaction durations. For both approaches to provide high-quality biodiesel, variables including temperature, catalyst percentage, alcohol-to-oil ratio, and blending must be optimized.

15.4.2 Transesterification by enzymes

Enzymatic conversion is a sustainable method for producing biodiesel that uses lipase enzymes as biocatalysts to convert triglycerides into FAME. Lipases can withstand high FFA levels, in contrast to chemical catalysts, limiting soap formation and permitting the utilization of waste or oils that are not edible [18]. The reaction takes place at a moderate temperature and pressure, lowering the impact on the environment and consumption of energy. Immobilized lipases improve process reliability and reuse potential [19]. Because enzymatic approaches are slower and more expensive than chemical approaches, improvements in enzyme engineering, immobilization methods, and more process optimization techniques are enhancing reaction speeds, yield, and industrial practicality, which makes it an interesting, sustainable biodiesel source.

15.4.3 Supercritical and noncatalyst techniques

Supercritical and noncatalytic techniques provide an alternate way for producing biodiesel, requiring the use of chemical catalysts. When using supercritical alcohol techniques, ethanol or methanol is heated beyond its ideal temperature and pressure, allowing triglycerides to be directly transesterified into FAME, having an excellent conversion rate. Water contaminants and a high FFA level are tolerated by this method, reducing soap development [20–23]. Noncatalytic technologies lessen risks to the environment and decrease downstream treatment. Even though these methods

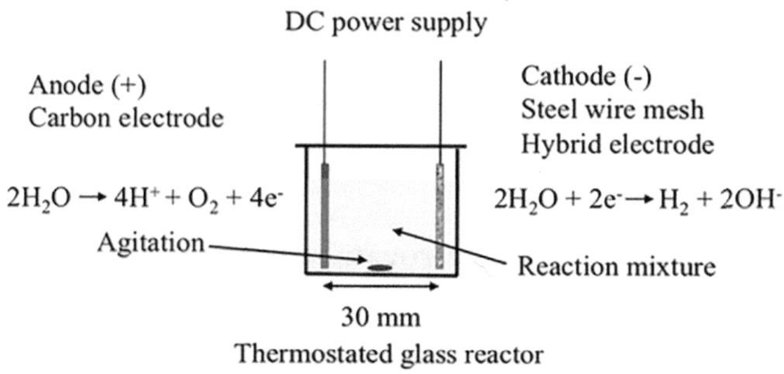

FIGURE 15.3 Schematic diagram of the electrocatalytic reactor system [23].

require increased pressure and temperature, they offer faster responses and greater yields, and adaptability to different feedstocks, providing them with a viable strategy for extensive and economical manufacturing of biodiesel.

15.4.4 New paths for efficient and green manufacturing

New methods for manufacturing biodiesel that are sustainable and green concentrate on limiting the ecological impact while enhancing efficiency. Technologies that include microwave-assisted, ultrasonic-based, and electrolytic transesterification decrease the use of energy and speed up reaction rates. Utilizing recyclable and environmentally friendly heterogeneous catalysts promotes reuse and lowers waste from chemicals (Fig. 15.3) [23]. Edible oils, cooking oil leftovers, and phytoplankton are increasingly being utilized as a raw material, bringing output into conformity with the circular economy's principles. Combining supercritical liquids with enzymatic interactions and continuous circulation systems increases process dependability and yields additional benefits. Collectively, these developments seek to develop biodiesel production techniques that are economically feasible, environmentally friendly, and affordable.

15.5 Electrocatalytic transesterification

An innovative method of producing electrocatalytic transesterification, it generates biodiesel and utilizes an electromagnetic field to enhance the transformation of triglycerides into FAME. By encouraging ion migration, increasing catalytic activity, and frequently lowering the requirement for high temperatures or chemical catalysts, this method speeds up responsiveness. Electrocatalysis can efficiently treat substances with elevated FFA levels while generating less soap. Its energy economics, possible integration with renewable energy sources, and adjustable reaction conditions make it an appealing and environmentally beneficial method of producing biodiesel (Fig. 15.3).

15.5.1 Fundamentals of electrocatalysis for biodiesel manufacturing

This process in biodiesel production involves employing an electric field to stimulate or expedite the process of transesterification and esterification methods of triglycerides, producing FAME. The voltage that is used increases ion migration, increases the transfer of electrons, and accelerates reaction kinetics by increasing catalytic site energy. Electrocatalytic devices may function at mild circumstances, lowering the consumption of energy and minimizing chemical catalyst utilization [21,22]. This method works particularly well with feedstocks having high FFA concentration, limiting soap production. By integrating electrochemical management, electrocatalysis provides an environmentally friendly option using heterogeneous or organic catalysts, successful, and customizable technology for enhanced biodiesel manufacturing.

15.5.2 Arrangements of reactors and electrochemical processes

Furnace designs in electrolytic manufacturing of biodiesel are intended to improve the transfer of mass, electromagnetic field circulation, and reaction effectiveness. Batch, continuously flowing, and microreactor devices are common configurations to improve the transfer of electrons; electrodes with a large surface area are frequently used. When a voltage is supplied, ions migrate as part of the electrochemical action, increasing stimulation of alcohol and molecules of triglycerides at the electrode contact (Fig. 15.4). Electron transport promotes bond breaking and ester synthesis while reducing adverse effects [23]. Precise monitoring of the voltage, current density, and material of electrodes, and furnace structure promotes effective transesterification, Outstanding yield and adaptability of FAME, rendering electrocatalysis a viable method for producing renewable biodiesel.

15.5.3 Varieties of electrocatalysts and electrochemical substances

The production of biodiesel uses three primary kinds of electrocatalysts: metal-based, carbon-based, and hybrid electronic catalysts. Metal-based

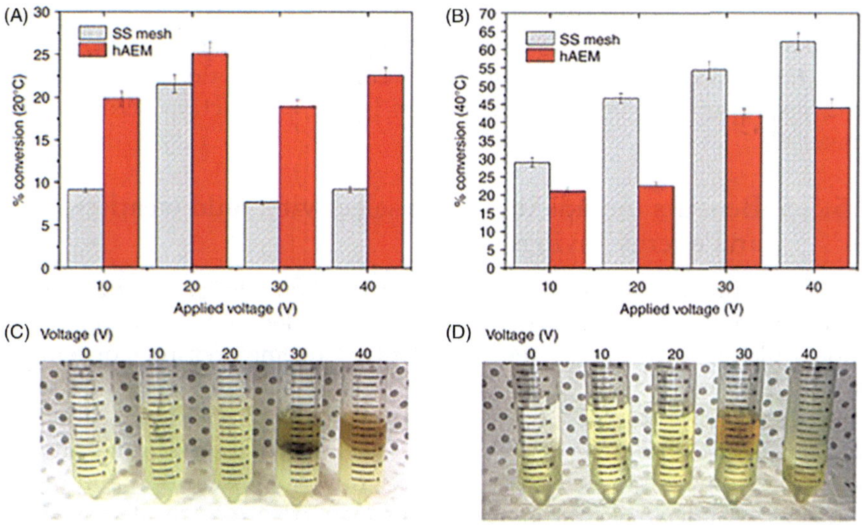

FIGURE 15.4 Influence of electrolysis cell voltages and temperatures on transformation of canola oil in a methanolic reaction mixture at (A) 20°C and (B) 40°C. (C) and (D) Images of the reaction products obtained using stainless steel (SS) Mesh and hAEM electrodes at 40°C [23].

catalysts, which include platinum, nickel, and palladium, give great reactivity and permeability, but they may be expensive. Carbon-derived catalysts, comprising graphene, carbon nanotubes, and biochar, are renewable resources providing enormous surface areas, chemical resistance, and little environmental impact. Composite electrocatalysts incorporate metals and carbon supports to increase electron transport, catalytic effectiveness, and resilience. Electrochemical materials have been selected to enhance surface area, electrical conductivity, and resistance to chemicals. This may involve stainless steel, charcoal, carbon felt, or platinum; the correct choice provides excellent rates of reaction, reliability, and continuous manufacturing of biodiesel [24].

15.5.4 Efficiency considering various feedstocks

The efficacy of electrocatalytic production of biodiesel differs with variable feedstock sources because of changes in triglyceride composition, FFA amount, and contaminants. Inedible oils, leftover cooking oils, and phytoplankton oils, which frequently contain high FFA levels, have advantages of electrocatalysis since it reduces the production of soap. and permits conversion directly under moderate circumstances. Oils that are edible and have a low FFA concentration exhibit faster response speeds and better conversion effectiveness. Raw material viscosity, water content, and levels of impurities affect mass transfer and electrode conductivity [25]. Enhancing process variables, including voltage, density of current, and response time, ensures continuous FAME production across varied materials.

15.5.5 Benefits include improved selectivity, mild weather, and energy savings

Renewable energy is one of the many benefits of electrocatalytic biodiesel production, with managed fields of electricity, reactions move more quickly, lowering the requirement for extreme pressures or temperatures. The method works according to modest circumstances, reducing thermal deterioration of various raw materials and maintaining the high standard of the final output [26]. Sensitive control of the electron transport at the electrode contacts increases FAME, improving selection and reducing negative effects like the generation of soap. Also, electrocatalysis (FFA) materials facilitate the use of enhanced FFA and collaborate with renewable sources of energy to produce biodiesel in a cost-effective, ecologically sound, and sustainable manner.

15.5.6 Challenges and present study gaps

Despite the promising future, there are obstacles to electrocatalytic biofuel synthesis, such as an increase in electrodes and catalysts that react more slowly, obstacles to scale up in comparison to organic approaches, mass transfer limitations, electrochemical corrosion, and energy usage for maintaining electric fields is still a problem. One of the current research goals is to understand mechanistic pathways in order to create inexpensive, long-lasting electrocatalysts, integrating renewable energy sources and improving reactor models for commercial use [27]. Getting over these shortcomings is essential to the electrocatalysis eco-friendly biodiesel process.

15.6 Renewable energy and waste-based catalyst

Biomass-based catalysts and Trash offer sustainable alternatives to the production of biodiesel, using materials such as charcoal, shells, and agricultural waste. They are also affordable, good for the environment, and their surface area is sometimes rather enormous, and places where transesterification is occurring. Such heterogeneous catalysts offer reusability, withstand large FFA substrates, and minimize chemical waste, synchronizing biodiesel manufacturing along with the concepts of green chemistry and the circular economy.

15.6.1 Production of catalysts from industrial and agricultural chemicals

Nutshells, husks, and fly ash are examples of agricultural and industrial waste that may be transformed into effective heterogeneous catalysts for biodiesel manufacturing. The surface area is increased by methods like chemical activation, calcination, and the carbonization process, and creates active locations for transesterification [28]. This transformation of trash not only decreases its ecological footprint but also cuts manufacturing expenses. By combining these environmentally friendly catalysts, the production of biodiesel conforms to the ideas of the circular bioeconomy, encouraging environmentally friendly, economical, and adaptable renewable energy options while making use of wasted biomass and industrial leftovers.

15.6.2 Corncob, coal fly ash, biochar, and shell-based catalysts

Biomass, coal fly ash, maize cob, and products made from shells are all useful, diverse catalysts for the synthesis of biodiesel. Corncob and biochar, formed via carbonization, offer significant surface area and locations for activity; however, following calcination, shells like seashells or eggshells provide alkaline qualities. Because a byproduct of burning coal, fly ash adds porosity and durability [29]. These catalysts are economical, recyclable, and environmentally friendly, permitting effective transesterification of diverse raw materials, containing oils with high levels of FFAs, and encouraging green, circular bioeconomy techniques in biodiesel production.

15.6.3 Doped with metal oxide biochar catalysts (e.g., K_2CO_3, CaO, Nb_2O_5, and MgO)

Doped with metal oxides, biochar catalysts, which include K_2CO_3, CaO, Nb_2O_5, and MgO, indicate a highly sophisticated category of diverse catalysts for biodiesel generation. A large surface area porosity carbon-based support can be provided by biochar, while the inserted oxides of metal provide basic or acidic sites of activity that improve esterification and transesterification processes. Doping enhances catalyst durability, usefulness, and durability against leaching, allowing substances with an elevated FFA concentration to be converted effectively [30]. These catalysts mix the benefits that green biomass feedstocks combine with the catalytic properties of metal oxides, giving sustainable, economical results and environmentally acceptable alternatives for high-performance biodiesel production.

15.6.4 Physicochemical characteristics and structure-activity connections

The essential characteristics of catalysts for biodiesel, including the surface area, pH levels, particle morphology, porosity, and pore size, significantly impact their catalytic effectiveness. Enhanced surface area and Superior porosity increase the availability of chemicals and transfer of mass, whereas the efficiency of esterification and transesterification is influenced by sufficient acidic or basic sites, respectively [31]. Structure-activity correlations show if related reaction kinetics, rate of conversion, and sensitivity are correlated with catalyst arrangement, dopants, and textural features. Comprehending these features enables the development of specific catalysts that enhance the production of FAME, reduce unfavorable reactions, and improve durability and reusability. Comprehending these connections is crucial for creating effective, economical biodiesel catalysts.

15.6.5 Catalytic efficiency in transesterification and esterification

Catalytic efficiency in esterification and transesterification is dependent depending on the catalyst's surface area, locations of activity, and chemical composition. Increased basicity or acidity promotes the transformation of triglycerides and FAME from FFAs, whereas porosity increases mass transport. Biomass and waste-produced heterogeneous catalysts, including those doped with metal oxides and biochar, have excellent stability, reuse potential, and impurity tolerance [32]. Improved catalysts provide rapid reactions and excellent yields and minimum adverse effects, permitting effective biodiesel manufacture from various raw materials in an economical and environmentally friendly manner.

15.6.6 Advantages of sustainability and consequences for the circular economy

Employing biomass- and waste-based catalysts in the manufacture of biodiesel gives Strong environmental advantages and promotes circular-economy concepts. Through utilizing waste from agriculture, waste from industries, and other trash, these catalysts lessen environmental contamination and decrease the demand for virgin chemical supplies [33]. Their reuse potential and sustainability over time decrease the generation of waste and the use of energy. Resource efficiency is promoted by using these kinds of catalysts to make biodiesel, Reduced expenses, and eco-friendly chemical techniques. In summary, this strategy enhances the generation of renewable energy, enabling sustainable growth and establishing a closed-loop system that promotes a circular, regenerative biological economy by turning trash into products with extra value.

15.7 Electrocatalytic glycerol enhancement

Electrolytic development is a creative way to make chemicals and fuels with extra value from glycerol, an important byproduct of the manufacture of biodiesel. Glycerol undergoes specific oxidation or reduction processes using electrocatalysts and regulated electric fields, yielding molecules such as dihydroxyacetone, formic acid, and glyceric acid. This approach provides better selectivity, milder working conditions, and less chemical waste than traditional chemical or thermal procedures. Advances in electrocatalysis link to energy from renewable sources, improving sustainability and preservation. This strategy promotes environmentally friendly, sustainable chemical processes and increases the financial sustainability of the biodiesel business by converting low-value triglycerides into products that are beneficial to the economy.

15.7.1 Glycerol as a highly valued waste product of biodiesel manufacturing

Glycerol is a useful byproduct of transesterification and makes up around 10% of the biodiesel produced. Because of its chemical versatility, it may be transformed into valuable compounds such as renewable energy sources, solvents, acids, and diols. Recognizing glycerol lowers waste, makes biodiesel production more commercially feasible, and promotes sustainable bio-refinery innovation [34]. Enzyme-based electrocatalysis and, under mild conditions, biological upgrading processes enable particular transformation, combining environmentally friendly chemical principles with waste management, and encouraging resource efficiency and overall process sustainability while cultivating a circular environmental economy.

15.7.2 Mechanisms for glycerol electro-reformation and selective reduction

Using particular oxidative reactions and electrolysis to generate glycerol in regulated electric fields, electrochemical catalysts transform glycerol into more useful compounds. Electro-reformation separates C–C and C–O bonds to create hydrogen and very small chemical compounds, while oxidation selectively targets particular groups of function to generate chemicals like the dihydroxyacetone or glycerol acid. These approaches boost glycerol utilization and enhance biofuel process profitability, and perform within mild, ecological circumstances [35].

15.7.3 Manufacturing of value-added chemicals, such as DHA, LA, and HPA

Glycerol during biodiesel manufacturing can be turned into compounds with added value, such as dihydroxyacetone (DHA), lactic acid (LA), and hydroxy pyruvic acid (HPA), by electrocatalytic, biological, and chemical oxidation processes. DHA is commonly utilized in cosmetics and medicines, LA acts as a foundation ingredient for polymers that are biodegradable, and HPA has uses in premium chemicals and medicines [36]. Electrocatalytic techniques give great selectivity, little waste, and gentle conditions for operation. By turning glycerin into these little particles, biodiesel production becomes more profitable, and waste is reduced, promoting the creation of sustainable and environmentally friendly compounds by integrating renewable energy sources with circular biologic refinery ideas.

15.7.4 Catalyst development techniques for specific glycerol transformation

In order to influence customer perceptions of desired goods, this creation of a catalyst for efficient glycerol degradation focuses on enhancing surface characteristics, active areas, and technical components. Among the techniques are treating metals or metal oxides, producing acidic or basic sites, and enhancing mass movement through hierarchy susceptibility. Mixed electrocatalysts boost their immobilization of the enzyme as well as their specificity and endurance [37]. By changing the form, content, and electrode insertion, response monitoring of circuits is made possible, maximizing the amounts of desirable substances such as HPA, LA, or DHA, hence reducing adverse reactions and energy expenditure.

15.7.5 Glycerol assessment integration with biodiesel renewable energy facilities

Glycerol valuation increases overall process efficiency and economic viability in biorefineries that make biodiesel. By using electrolytic, enzyme-based processes to convert glycerol into extremely useful substances like hydrogen, lactic acid, or dihydroxyacetone, or chemical reactions, waste is prevented. There are many different sources of income. [38]. When glycerol is processed and biodiesel is produced, a useless byproduct is transformed into an ecologically favorable, consumable, encouraging green, efficient, and well-rounded renewable energy system.

15.8 Process intensification and sustainable engineering techniques

Green methods of engineering and process improvement in the manufacture of biodiesel attempt to boost efficiency, decrease the consumption of energy, and decrease environmental impact. Highly critical alcohol processes, constant-flow furnaces, supersonic and microwave-assisted transesterification, and other technologies accelerate reaction and improve results. When these solutions are combined with biological or heterogeneous catalysts, the waste chemicals are reduced, allowing for moderate operating conditions and encouraging feedstock variety. Incorporating energy from green sources and waste valuation significantly enhances sustainability, improving the manufacturing of biodiesel. More ecologically friendly, cost-effective, and consistent with the circular bioeconomy and environmentally friendly chemistry ideas.

15.8.1 Microwave-assisted and ultrasound-based biodiesel production

Assisted by ultrasound and microwave, the production of biodiesel is an intensified process, with methods that increase yield and response speeds. Microwaves offer quick, consistent heating, cutting down on reaction time and energy use, whereas cavitation is produced by ultrasonics, enhancing the mixing and bulk transfer of oil and alcohol phases. Both techniques allow for effective transesterification under moderate circumstances employing heterogeneous or bio-based catalysts [39]. These methods improve the synthesis of FAME, decrease the use of chemicals, and encourage the manufacture of sustainable biodiesel by providing high-yield, eco-friendly, and energy-efficient substitutes for traditional thermal techniques.

15.8.2 Hydrothermal, solvent-based, and electrosynthesis techniques

Thermal, solvent-based processing, and catalysts and biodiesel may be produced using advanced electrosynthesis technology. Hydrothermal processes require extremely high temperatures. High-pressure water is used to convert biomass or ignite catalysts, boosting surface activity and permeability [40]. Solvothermal processes use organic solvents under controlled conditions to produce homogenous particles and carry out the functionalization process. During electrosynthesis, electricity sources are included to promote catalyst formation or glycerol boosting, increasing response times and selectivity. These methods increase catalytic efficacy, offer precise control over material properties, and enable high-performance, ecologically sustainable engines and economical methods for producing biodiesel.

15.8.3 Without solvents and water-based systems

Absence of solvents and solutions made of water provides dependable choices in the production of biofuel. Solvent-free techniques get rid of solvents, which are organic, lowering costs, simplifying cleaning, and reducing the impact on the surroundings, considering outstanding conversion results [41]. Water-based mechanisms help with humidity or increased levels of FFAs, lowering the output of cleansing soap and enabling instantaneous transesterification. Combining these techniques with different catalysts derived from plants enhances environmental sustainability, efficacy, and reaction time. Overall, they promote eco-friendly, energy-saving techniques, the process of making biodiesel linked with a circular approach to sustainable development, and the eco-friendly design of process concepts.

15.8.4 Sustainable measurements and techno-economic factors

Environmental metrics and technological and Economic issues are very important for studying the processes used in the production of biofuel. Crucial prerequisites include energy conservation, effects on the environment, water usage, waste generation, and catalytic recycling potential that assesses the impact on the environment and maximizes the use of resources [42]. The technological and economic study examines energy costs, production costs, feedstock accessibility, operational flexibility, and economic viability in terms of money. Utilizing catalysts made from waste or biomass, speeding up the process, and using glycerol enhances maintainability and economic feasibility. An exhaustive assessment guarantees the environmental friendliness of biodiesel solutions. Cost-effective and compatible with circular biological economy concepts for widespread, sustainable adoption.

15.9 Opportunities, difficulties, and prospects for the future

The development and production of biodiesel is hampered by issues including unpredictable feedstock, expensive catalysts, high-energy procedures, and the treatment of glycerol effluent. Nonetheless, there are chances to create heterogeneous catalysts made from biomass or trash, process enhancement, electrocatalytic improvement, and the enhancement of wastes. Potential futures involve combining energy from renewable sources, increasing the use of environmentally friendly technologies, and implementing circular bioeconomy strategies. Developments in catalyst creation and reactor efficiency, and Sustainable design are anticipated to increase productivity, lessen the adverse effects on the natural world, and increase economic viability, promoting biodiesel as an environmentally friendly substitute for fossil fuels.

15.9.1 Catalyst removal and recovery

Removal of the catalyst due to contamination, leaching, or heating decreases the synthesis of biodiesel's effectiveness and specificity. Biomass-based waste-derived diverse catalysts could lose sites of activity as a result of contaminants or through fatty acid-free adsorption. Restoration techniques, such as re-impregnating, washing, and calcination, or enzymatic reactivation, recovering surface area, permeability, and locations of activity, and prolonging catalyst lifetime [43]. Adequate deactivation control enables continuous FAME production, decreases operational expenses, and promotes sustainable development and the financial viability of biodiesel procedures, facilitating long-term utilization in industry.

15.9.2 Expanding advanced electrocatalytic systems

In order to scale up electrolytic biodiesel operations, experimental to pilot projects or large-scale furnaces that preserve efficiency, flexibility, and high-quality goods. The primary obstacles are creating a reliable dispersion of electrical fields, controlling the transfer of mass, the robustness of the electrodes, and the consumption of energy [44]. Improvement of furnaces in electrodes with huge surface regions and systems with constant flow enhances adaptability and communication with procedures for tracking and conservation with regard to energy, guaranteeing dependable performance. Successful growth encourages business sustainability, allowing considering sustainability over time and high-quality production of biofuel when using electrolysis in sustainable energy and bioeconomic solutions that are regenerative.

15.9.3 Coupling with improvement based upon artificial intelligence/machine learning

Artificial intelligence (AI) integration, together with machine learning (ML) in biodiesel production, enables data-driven enhancement between electrolysis and conventional methods. AI and ML techniques analyze the factors of response, catalytic properties, fluctuation of feedstock, and operational parameters of forecasting optimal conditions to optimize the yield, effectiveness of power, and accuracy [45]. The previously identified instruments offer real-time procedure supervision, recognizing faults and required maintenance, lowering the amount of exploratory testing. The application of artificial intelligence also assists in catalytic creation by connecting structural attributes with outstanding efficacy. Adaptability, cost, and Sustainable environmental practices are enhanced by combining AI and ML, allowing more intelligent, more flexible, and superior production efficiency of biofuel consistent with engineering for ecology and the sector of circular bioengineering.

15.9.4 Strategy, waste-utilization frameworks, along with the ecological impact

Government support, waste management structures, and considerations about the environment are important for the efficient production of biodiesel. Regulatory rules, regulations, and subsidies encourage the consumption of nonedible oils, environmental catalysts, as well as waste raw materials. Regulations for organized waste handling and appraisal allow Successful transformation of agricultural waste, transforming industrial waste into added-value goods and catalysts, lowering pollutants in the environment [46]. Considering life-cycle effects and greenhouse gas emissions, eco-friendly operations are ensured via resource utilization. Combining these tactics improves sustainability, promotes circular bioeconomy procedures, and supports widespread, economically viable, and ecologically responsible manufacturing of biodiesel.

15.10 Developing efficient and carbon-neutral renewable fuels: Future prospects in catalytic and electrolytic systems

The development of environmentally friendly and carbon-neutral biodiesel requires innovative enzymatic and electrocatalytic methods. Future research will concentrate on diverse catalysts made from garbage, biomass, renewable fuels, doped with metal oxides, and customized electrocatalysts to improve transesterification and triglyceride degradation. Combination with energy from renewable sources, intensification of process procedures, and AI/ML-based management improves productivity, flexibility, and sustainability. Focused on being free of solvents, water-resistant materials, and circular renewable energy techniques, minimizes environmental effect. Combining these strategies enables cost-effective, high-producing, and eco-friendly production of biodiesel, encouraging eco-friendly and eco-friendly energy objectives.

15.10.1 Approaches toward carbon-free biodiesel manufacturing

Two significant paths of carbon-neutral initiatives in biofuel highlight the management of materials and the mechanism linkage. The first strategy focuses on using non-edible oils, algae, frying oil waste, and agricultural waste, lowering lifetime pollutants, such as carbon dioxide, and relyingon fossil fuels. These starting points of renewable energy also align with the circular bioeconomy ideas by using trash [47]. The next action involves taking action incorporate green process methods, some instances are supercritical transesterification of ethanol, electrocatalysis using sustainable energy, and technology for process improvement, which might involve ultrasonic or microwave assistance. In conjunction with concepts for ecologically friendly reactors, Valuation of triglycerides, recycling, and reusing catalysts, these techniques lessen the generation of waste, pollution, and energy consumption [48]. When combined, these methods increase the production of biodiesel, getting closer to a renewable, environmentally responsible, sustainable, and promising future for the economy.

15.10.2 Future prospects in catalysis and electrocatalytic complexes

Future developments in catalysis and electrically powered systems for biofuels generation concentrate on improving environmental sustainability, adaptability, and effectiveness. Sophisticated heterogeneous catalysts made from waste from industry, biomass, and doped with metals, biochar have been created with customized surface characteristics, increased porosity, as well as maximized basicity or acidity, the effectiveness of esterification, and the transesterification process [49]. Electrocatalytic devices continue to develop to increase glycerol oxidation and reaction efficiency, selectivity, and conservation of energy, frequently coupled with electricity from renewable sources. Optimization of processes utilizing AI and ML allows for catalyst development and proactive management, control, and continuous tracking for improved yields and lower use of resources. Methods of process intensification, such as the microwave, ultrasonic, hydrothermal, and continuously flowing reactors, increase processes and decrease demand for energy [50]. Future studies will also focus on circular biological economy implementation, water-resistant and solvent-free frameworks, and lifecycle-oriented metrics for sustainability. In general, the emphasis is on building inexpensive and strong structures and ecologically sustainable catalytic frameworks that change the manufacturing of biodiesel into a completely sustainable, scalable process and a beneficial manufacturing procedure.

15.11 Conclusion

In summary, the growth of environmentally friendly and environmentally friendly biodiesel technology offers an essential leap in energy efficiency and environmentally friendly chemistry. Developments in diverse, biomass- and waste-generated, and doped with metal oxide catalysts, in combination with electrocatalytic technologies, have greatly boosted the effectiveness, quality, and flexibility of biodiesel synthesis. Blend of methods, accelerated methods, such as ultrasonography, microwave, and coupled supercritical procedures, with a focus on progress about AI and ML, offers outstanding leadership of chemical properties, increasing output when restricting energy consumption and effects on the surroundings. The transformation of residue, like glycerol, into useful substances further strengthens the concept of a circular biological economy. Chemical reactors with continuous flow and no solvents, water-resistant systems, and economically viable and environmentally benign advances are encouraged. Considering the deactivation of catalysts, uncertain feedstock, and challenges with scaling production, ongoing research offers interesting solutions. In summary, these developments in biodiesel's reputation as a reliable, sustainable alternative to fossil fuels promote safe energy sources and sustainability on a global scale.

References

[1] D.F. Dongo, S. Relvas, Evaluating the role of the oil and gas industry in energy transition in oil-producing countries: a systematic literature review, Energy Res. Soc. Sci 120 (2025) 103905. https://doi.org/10.1016/j.erss.2024.103905.

[2] W.N.A.W. Osman, M.H. Rosli, W.N.A. Mazli, S. Samsuri, Comparative review of biodiesel production and purification, Carbon Capture Sci. Technol 13 (2024) 100264. https://doi.org/10.1016/j.ccst.2024.100264.

[3] Monika, S. Banga, V.V. Pathak, Biodiesel production from waste cooking oil: a comprehensive review on the application of heterogenous catalysts, Energy Nexus 10 (2023). https://doi.org/10.1016/j.nexus.2023.100209.

[4] B. Panchal, C-H. Su, C-C. Fu, S-J. Wu, H-Y. Juan, Ecofriendly and cost-effective biodiesel production from water containing feedstocks through electrolysis: a review, Fuel Process. Technol. 276 (2025) 108277. https://doi.org/10.1016/j.fuproc.2025.108277.

[5] H.H. Naseef, R.H. Tulaimat, Transesterification and esterification for biodiesel production: a comprehensive review of catalysts and palm oil feedstocks, Energy Convers. Manag. 26 (2025) 100931. https://doi.org/10.1016/j.ecmx.2025.100931.

[6] M. Mariachiara, F. Patrizia, M. Anastasia, M. Angela, Recovery/reuse of heterogeneous supported spent catalysts, Catalysts 11 (2021) 5. https://www.mdpi.com/2073-4344/11/5/591.

[7] R.M. Ahmad, A.Y. Elnaggar, L.K. Teong, S. Sultana, M. Zafar, M. Munir, et al., Sustainable and eco-friendly synthesis of biodiesel from novel and non-edible seed oil of Monotheca buxifolia using green nano-catalyst of calcium oxide, Energy Convers. Manag 13 (2022) 100142. https://doi.org/10.1016/j.ecmx.2021.100142.

[8] S.M. Farouk, A.M. Tayeb, S.M.S. Abdel-Hamid, R.M. Osman, Recent advances in transesterification for sustainable biodiesel production, challenges, and prospects: a

comprehensive review, Environ. Sci. Pollut. Res. Int 31 (2024) 12722–12747. https://doi.org/10.1007/s11356-024-32027-4.

[9] D. Krisdiyanto, K. Wijaya, A. Syoufian, A.J. Saviola, R.A. Fitria, P. Wahyuningsih, et al. Exploring non-conventional techniques in biodiesel production: a brief review. Mole. Catal. 584 (2025) 115289. https://doi.org/10.1016/j.mcat.2025.115289.

[10] S.S. Patil, D.B. Hulwan, V.S. Kumbhar, A. Pandey, S. Chandrakant, R. Goyal, et al., Influence of waste cooking oil characteristics on the quality of produced biodiesel, Int. J. Thermofluids 30 (2025) 101469. https://doi.org/10.1016/j.ijft.2025.101469.

[11] M.C. Capellini, V. Giacomini, M.S. Cuevas, C.E.C. Rodrigues, Rice bran oil extraction using alcoholic solvents: physicochemical characterization of oil and protein fraction functionality, Ind. Crop. Prod 104 (2017) 133–143. https://doi.org/10.1016/j.indcrop.2017.04.017.

[12] N.P. Vidal, O.A. Adigun, T.H. Pham, A. Mumtaz, C. Manful, G. Callahan, et al., The effects of cold saponification on the unsaponified fatty acid composition and sensory perception of commercial natural herbal soaps, Molecules 14 (23) (2018) 2356. https://doi.org/10.3390/molecules23092356.

[13] L.P. Amador-Gómez, D. Hernández-Romero, J.M. Rivera-Villanueva, S. Rosete-Luna, C.A. Cruz-Cruz, E. Méndez-Bolaina, et al. Transesterification/esterification reaction catalysed by functional hybrid MOFs for efficient biodiesel production. Reactions. 6 (2025) 4. https://www.mdpi.com/2624-781X/6/4/58.

[14] N. Shobhana-Gnanaserkhar, G. Asikin-Mijan, A. Kareem-Alsultan, S.M.I. Sivasangar-Seenivasagam, Y.H. Taufiq-Yap, Biodiesel production via simultaneous esterification and transesterification of chicken fat oil by mesoporous sulfated Ce supported activated carbon, Biomass Bioenergy 141 (2020) 105714. https://doi.org/10.1016/j.biombioe.2020.105714.

[15] R. Nabaterega, R.N. Vesuwe, O.T. Iorhemen, R.W. Thring, Biodiesel production unpacked: data-driven analysis of key operational factors, Biomass Bioenergy 205 (2026) 108457. https://doi.org/10.1016/j.biombioe.2025.108457.

[16] D. Li, Q. Shi, F. Liang, D. Feng, The green synthesis of biodiesel via esterification in water catalyzed by the phosphotungstic acid–functionalized hydrophobic MCM–41 catalyst, Catalysts 14 (2) (2024) 142. https://doi.org/10.3390/catal14020142.

[17] M.K.A. Viraj, U.S.P.R. Arachchige, T.M.M. Marso, S. Gamunu, Biodiesel production through the transesterification of waste cooking oil over typical heterogeneous base or acid catalysts, Catalysts 13 (2023) 3. https://www.mdpi.com/2073-4344/13/3/546.

[18] A. Moschona, A. Spanou, I.V. Pavlidis, A.J. Karabelas, S.I. Patsios, Optimization of enzymatic transesterification of acid oil for biodiesel production using a low-cost lipase: the effect of transesterification conditions and the synergy of lipases with different regioselectivity, Appl. Biochem. Biotechnol 196 (2024) 8168–8189. https://doi.org/10.1007/s12010-024-04941-3.

[19] R. Camila, P. Hackenhaar, D. Abellanas-Perez, J. Carballares, M. Bolivar, C. Rafael, et al., Preparation and characterization of different immobilized and chemically modified preparations of lipase B from *Candida* antarctica: is it the activation energy a good indicator of the biocatalyst expressed activity, Int. J. Biol. Macromol 323 (2025) 147310. https://doi.org/10.1016/j.ijbiomac.2025.147310.

[20] H. Taher, S. Al-Zuhair, A.H. Al-Marzouqi, Y. Haik, M.M. Farid, A review of enzymatic transesterification of microalgal oil-based biodiesel using supercritical technology, Enzyme Res 11 (2011) 468292. https://doi.org/10.4061/2011/468292.

[21] B. Thangaraj, P.R. Solomon, Biodiesel production by the electrocatalytic process: a review, Clean Energy 5 (1) (2021) 19–31. https://doi.org/10.1093/ce/zkaa026.

[22] J. Park, J.H. Lim, J-H. Kang, J. Lim, H.W. Jang, H. Shin, et al., A review of understanding electrocatalytic reactions in energy conversion and energy storage systems via scanning electrochemical microscopy, J. Energy Chem 91 (2024) 155–177. https://doi.org/10.1016/j.jechem.2023.12.015.

[23] F.M. Allioux, B.J. Holland, L. Kong, F.D. Ludovic, Electro-catalytic biodiesel production from canola oil in methanolic and ethanolic solutions with low-cost stainless steel and hybrid ion-exchange resin grafted electrodes, Front. Mater 4 (2017). https://doi.org/10.3389/fmats.2017.00022.

[24] J.L. Schultz, K. Christian, J.N. Brinck, T.M. Hedegaard, C. Tanmay, Biohydrogen production in microbial electrolysis cells utilizing organic residue feedstock: a review, Energies 15 (2022) 22. https://www.mdpi.com/1996-1073/15/22/8396.

[25] H. Becker, J. Murawski, D.V. Shinde, I.E.L. Stephens, G. Hinds, G. Smith, Impact of impurities on water electrolysis: a review, Sustain. Energy Fuels 7 (2023) 1565–1603. https://doi.org/10.1039/D2SE01517J.

[26] K.K. Dash, U. Fayaz, A.H. Dar, R. Shams, S. Manzoor, A. Sundarsingh, et al., A comprehensive review on heat treatments and related impact on the quality and microbial safety of milk and milk-based products, Food Chem. Adv 1 (2022) 100041. https://doi.org/10.1016/j.focha.2022.100041.

[27] S.G. Nnabuife, A.K. Hamzat, J. Whidborne, B. Kuang, K.W. Jenkins, Integration of renewable energy sources in tandem with electrolysis: a technology review for green hydrogen production, Int. J. Hydrogen Energy 107 (2025) 218–240. https://doi.org/10.1016/j.ijhydene.2024.06.342.

[28] S.Z. Naji, C.T. Tye, A review of the synthesis of activated carbon for biodiesel production: precursor, preparation, and modification, Energy Convers. Manag 13 (2022) 100152. https://doi.org/10.1016/j.ecmx.2021.100152.

[29] C. Yirong, L-P. Vaurs, Wasted salted duck eggshells as an alternative adsorbent for phosphorus removal, J. Environ. Chem. Eng 7 (2019) 103443. https://doi.org/10.1016/j.jece.2019.103443.

[30] K. Yeneneh, G. Sufe, Sustainable biodiesel production from cottonseed oil using a nickel-doped eggshell heterogeneous catalyst optimized via response surface methodology, Sci Rep 15 (2025) 32132. https://doi.org/10.1038/s41598-025-17529-6.

[31] X. Li, X. Jia, W. Li, S. Jia, S. Zhang, J. Song, et al., Comparison of high-efficiency MgO/Na_2CO_3 and MgO/K_2CO_3 as heterogeneous solid base catalysts for biodiesel production from soybean oil, Molecules 7 (30) (2025) 2876. https://doi.org/10.3390/molecules30132876.

[32] V. Gadore, S.R. Mishra, N. Yadav, G. Yadav, M. Ahmaruzzaman, Metal oxide-based heterogeneous catalysts for biodiesel production, Next Sustain 2 (2023) 100012. https://doi.org/10.1016/j.nxsust.2023.100012.

[33] M. Arun, D. Barik, S.S. R. Chandran, Exploration of material recovery framework from waste – a revolutionary move towards clean environment, Chem. Engin. J. Adv 18 (2024) 100589. https://doi.org/10.1016/j.ceja.2024.100589.

[34] A.S. Elgharbawy, M. Farghali, A.I. Osman, M.A. Hanafy, A.H. Al-Muhtaseb, Innovative biodiesel production for sustainable energy: advances in feedstocks, transesterification, and cost efficiency, Biomass Bioenergy 201 (2025) 108114. https://doi.org/10.1016/j.biombioe.2025.108114.

[35] H. Wang, H. Li, C.K. Lee, N.S.M. Nanyan, G.S. Tay, A systematic review on utilization of biodiesel-derived crude glycerol in sustainable polymers preparation, Int. J. Biol. Macromol 261 (2024) 129536. https://doi.org/10.1016/j.ijbiomac.2024.129536.

[36] G.Y. Sun, A. Simonyi, K.L. Fritsche, D.Y. Chuang, M Hannink, Z. Gu, et al., Docosahexaenoic acid (DHA): an essential nutrient and a nutraceutical for brain health and diseases, Prostaglandins Leukot Essent Fatty Acids 136 (2018) 3–13. https://doi.org/10.1016/j.plefa.2017.03.006.

[37] S. Ganchao, W. Xiaobo, Z. Dianping, H. Liben, L. Huiyan, F. Haitian, Immobilization of enzyme electrochemical biosensors and their application to food bioprocess monitoring, Biosensors 13 (2023) 9. https://www.mdpi.com/2079-6374/13/9/886.

[38] G. Sun, X. Wei, D. Zhang, L. Huang, H. Liu, H. Fang, Immobilization of enzyme electrochemical biosensors and their application to food bioprocess monitoring, Biosensors 13 (2023) 886. https://doi.org/10.3390/bios13090886.

[39] D.T. Oyekunle, M. Barasa, E.A. Gendy, S.K. Tiong, Heterogeneous catalytic transesterification for biodiesel production: feedstock properties, catalysts and process parameters, Process Saf. Environ. Prot. 177 (2023) 844–867. https://doi.org/10.1016/j.psep.2023.07.064.

[40] J-U. Abu-Taher, A.S. Shakirudeen, D. Animesh, G.Z. Richard, Hydrothermal conversion of waste biomass from greenhouses into hydrochar for energy, soil amendment, and wastewater treatment applications, Energies 15 (2022) 10. https://www.mdpi.com/1996-1073/15/10/3663.

[41] M. Shahbaz, M. Riaz, U. Momal, I.F. Ul. Rasool, H. Naeem, N. Raza, et al., Green solvent extraction and eco-friendly novel techniques of bioactive compounds from plant waste: applications, future perspective and circular economy, Appl. Food Res 5 (2025) 101434. https://doi.org/10.1016/j.afres.2025.101434.

[42] M. Massoud, G. Vega, A. Subburaj, J. Partheepan, Review on recycling energy resources and sustainability, Heliyon 9 (2023) 2405–8440. https://doi.org/10.1016/j.heliyon.2023.e15107.

[43] O.-A. Abiodun, O. Oluwaseun, O.K. Oladayo, O. Abayomi, A.A. George, E. Opatola, et al., Remediation of heavy metals using biomass-based adsorbents: adsorption kinetics and isotherm models, Clean Technol 5 (2023) 934–960. https://doi.org/10.3390/cleantechnol5030047.

[44] M.A.R. Chowdhury, M.A.H. Bulbul, M.R.I. Sarker, M. Mourshed, Slurry flow electrodes towards scalable environmental and energy storage applications: state of the art and technical challenges, Energy Convers. Manag 28 (2025) 101279. https://doi.org/10.1016/j.ecmx.2025.101279.

[45] M. Arif, A.I. Alalawy, Y. Zheng, M. Koutb, T. Kareri, E.-S. Salama, X. Li, Artificial intelligence and machine learning models application in biodiesel optimization process and fuel properties prediction, Sustain. Energy Technol. Assess 73 (2025) 104097. https://doi.org/10.1016/j.seta.2024.104097.

[46] Y.A. Hajam, R. Kumar, A. Kumar, Environmental waste management strategies and vermi transformation for sustainable development, Environ. Challenges 13 (2023) 100747. https://doi.org/10.1016/j.envc.2023.100747.

[47] A.C.F. Carvalho, S. Ghosh, T.G. Hoffmann, E.S. Prudêncio, C. Krebs de Souza, S. Roy, Valuing agro-industrial waste in the development of sustainable food packaging based on the system of a circular bioeconomy: a review, Cleaner Waste Syst 11 (2025) 100275. https://doi.org/10.1016/j.clwas.2025.100275.

[48] N. Aslan, O. Salman, R. Konakci, A.P. Akan, Circular economy approach to promote sustainable energy from waste, Sustain. Futures 10 (2025) 101238. https://doi.org/10.1016/j.sftr.2025.101238.

[49] C. Zhao, Q. Xu, Y. Gu, X. Nie, R. Shan, Review of advances in the utilization of biochar-derived catalysts for biodiesel production, ACS Omega 23 (8) (2023) 8190–8200. https://doi.org/10.1021/acsomega.2c07909. Erratum in: ACS Omega. 2023 Sep 28;8(40):37622doi: 10.1021/acsomega.3c06089.

[50] Á. Martín, A. Navarrete, Microwave-assisted process intensification techniques, Curr. Opin. Green Sustain. Chem 11 (2018) 70–75. https://doi.org/10.1016/j.cogsc.2018.04.019.

CHAPTER

16

Electrochemical conversion of biomass waste

Ponnaiah Sathish Kumar, Sungwon Lee

Gyeongbuk Institute of Science & Technology (DGIST), Daegu, Republic of Korea

16.1 Introduction

Handling biomass waste is a significant global challenge, with 250 to 280 Gt generated annually. Much of this waste ends up in the earth and ocean daily, posingsignificant environmental and public health risks, including land depletion, new diseases, and a potential drinking water crisis [1]. Ongoing waste dumping could result in the loss of 5% to 8% of arable land over the next five decades, contributing to serious health and environmental problems [2]. As the global population is expected to reach 11 billion by 2050, biomass residues have become a major economic and ecological concern [3]. Leveraging the chemical energy stored in biomass feedstocks allows for efficient electrochemical conversion with minimal environmental footprint [4]. The growing demand for energy and bulk chemicals is accelerating the depletion of finite fossil fuels such as coal, crude oil, and natural gas [5]. In the 1970s, progress in biomass waste management was limited by low investment and a lack of skilled personnel [6]. Today, electrochemical conversion technologies present a sustainable solution by transforming biomass waste into useful products like energy, biofuels, and industrial chemicals [7]. Many leading universities and research centers are actively pursuing innovative approaches in biomass valorization [8]. Core processes in this field include water electrolysis, hydrogen evolution reaction (HER), nitrogen reduction reaction (NRR), oxygen reduction reaction (ORR), carbon dioxide reduction reaction (CO_2RR), and metal–ion or metal–air battery systems [9]. Additionally,

Electrochemical Biomass Conversion: Sustainable Processes and Applications for Bioenergy, Biofuels, and Bioproducts
DOI: https://doi.org/10.1016/B978-0-443-26757-4.00014-3

fossil fuels such as oil, coal, and natural gas are being used up faster because of growing energy demands [5]. In the 1970s, the lack of investment and trained staff made it difficult to recycle biomass effectively [6]. Today, electrochemical conversion is emerging as a promising way to turn biomass waste into valuable resources like energy, chemicals, and biofuels [7]. Top universities and research centers are advancing research in this field, focusing on processes like water electrolysis (HER, oxygen evolution reaction [OER]), NRR, CO_2RR, and metal-ion and metal-air batteries [8,9]. Biomass, nature's primary source of carbon, has recently advanced as a way to create valuable carbon materials (BCM) and generate electricity [10]. BCM is especially promising because it has key properties like high porosity, strong structural stability, low background current, and excellent electrocatalytic performance, making it highly versatile for various applications and offering great potential for future development [11,12]. This chapter examines how electrochemistry can be used for both waste valorization and renewable energy production, focusing on how electrochemical biomass conversion works and its potential. By combining innovative electrochemical technologies with sustainable waste management, this approach offers a pathway to a more efficient, circular bioeconomy.

16.2 A bibliometric review of trends in electrochemical biomass waste conversion

Converting biomass waste into valuable products—like powdered carbon materials, oils, and gases—is a complex and multifaceted process [13]. To improve electrochemical conversion efficiency, research has focused on biomass-derived materials that can form interconnected mesoporous and microporous structures [14]. Since industrialization began, scientists have actively exploring ways to transform various waste materials, including biomass, metals, and polymers, into useful and functional commodities [15]. In this context, the process of converting biowaste and associated resources into reusable materials garnered significant interest throughout the 1970s [6]. Records from 1980 to 2008 show that the bulk of research focused on converting biomass waste into coal, biogas, biofuels, as well as organic manures (both macro and micro systems) [16]. The development of superior bulk biowaste conversion technology also occurred during this time. Nevertheless, between 2008 and 2017, the main focus of scientists' efforts was the creation of nanomaterials for sensing, electrochemical energy, and other applications using biowaste substances [17]. The innovative methods for managing biomass waste emphasize its conversion into carbon nanomaterials (BCM) and subsequent use in electrochemical

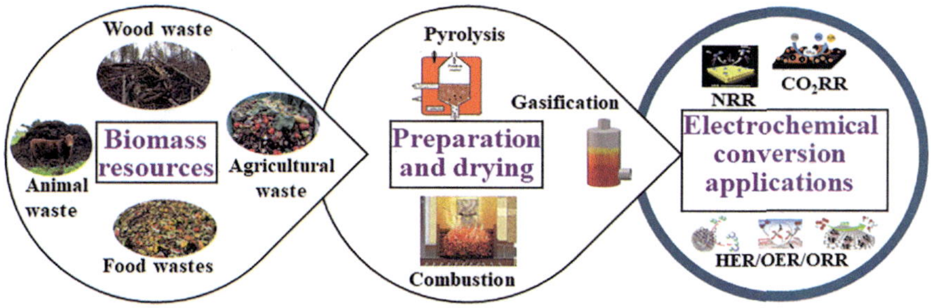

FIGURE 16.1 Illustrates a schematic diagram of innovative approaches for biomass waste management, showcasing its conversion into carbon nanomaterials and subsequent application in electrochemical energy conversion.

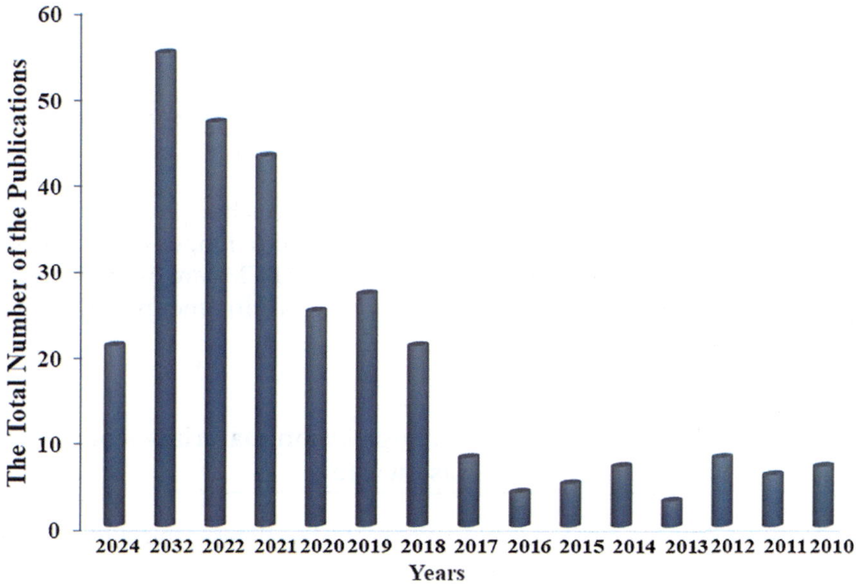

FIGURE 16.2 Total number of publications and the proportions according to the research areas since 2010–2024 (up to April) for biomass-derived carbon in energy storage.

energy conversion, as illustrated in Fig. 16.1. Furthermore, recent research (2010–2024) has largely concentrated on finding biomass materials capable of forming interconnected microporous and mesoporous structures.

A literature search using the keywords "Electrochemical conversion of biomass waste" to explore the method of developing carbon electrode material derived from biomass. The results obtained are depicted in Fig. 16.2. There is now more focus on the uses of environmental biowaste substances, especially when integrated with traditional environmental research and

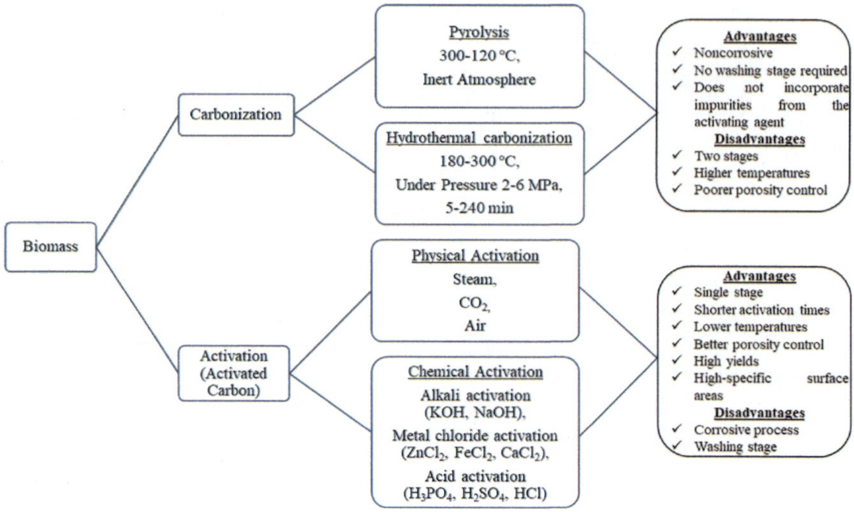

FIGURE 16.3 Illustrate the different methods used to create biomass-derived carbon materials and compares them.

the energy/chemical sector [18]. The unique tunable physicochemical characteristics, strong adsorption capacities, quick ion/electron transit, affordability, and environmental sustainability of BCMs are the reasons for this spike in interest in them as electrode materials for energy conversion research [19].

16.3 Methods of synthesizing carbon materials from biomass waste

Renewability, abundance, environmental friendliness, and high performance are just a few of the exceptional qualities of BCMs, which comprise interconnected porous, multidimensional, and hierarchical structures [20]. BCMs are primarily produced through activation and carbonization processes, as demonstrated in Fig. 16.3. During these processes, carbonization heats organic material without oxygen until it becomes a carbon-rich, coal-like substance [21]. Biochar is a solid byproduct formed during biomass carbonization. Additionally, carbonization parameters such as heating rate, residence time, and temperature impact its quality [22]. In hydrothermal carbonization (HC), also known as wet pyrolysis, biomass is treated under pressures of 2 to 6 MPa while submerged in water or aqueous solutions for 5 to 240 minutes, usually at temperatures between 180°C and 250°C [23]. The main product, known as hydrochar, is characterized by its hydrophobic and brittle nature, which results from a high content of oxygen functional groups [24]. Hydrochar or biochar produced from biowaste through pyrolysis generally exhibits low surface area, and poor porosity [25].

Using a two-step process that combines carbonization with activation can significantly enhance the material's properties. Activation can be done in two ways: physically or chemically [26]. The physical activation promotes the development of pores in biochar and typically takes place between 350°C and 1000°C, using agents like steam, oxygen, air, CO_2, or their combinations. Due to the high working temperature required when utilizing steam as a catalyst for activation, the reaction cannot proceed as intended in this instance [27].

Therefore, to promote porosity growth and increase surface area, a low activation temperature is necessary. In chemical activation, substances like HNO_3, KCl, H_3PO_4, NaOH, $ZnCl_2$, and KOH are mixed with the carbon precursor and heated between 400°C and 900°C [28]. After this process, the material shows increased porosity, and the activating agents are washed away from the final product. The main factors that influence chemical activation include the activation technique, temperature, impregnation ratio, and time [29]. KOH stands out as the preferred activating agent due to its higher efficiency, lower activation temperature requirements, and better improvements in pore volume and surface area [30].

16.4 Electrochemical conversion using carbon compounds obtained from biowaste

Everyday devices, appliances, and various modes of transportation worldwide rely on electricity for power. Research on electric energy production, conversion, and storage is crucial for efficient operations. Fossil fuels, limited and environmentally harmful, cannot meet the growing energy demand due to rising consumption [31]. Thus, alternative energy sources are being investigated to mitigate this issue without disrupting current systems [32]. Energy storage and conversion systems vary, including mechanical, thermal, electrochemical technologies [33]. Recent studies focus on electrochemical devices, but the lack of high-performance electrodes restricts their applications. Therefore, developing novel electrode materials is essential for enhancing the energy conversion device feasibility [34]. Carbon materials derived from biowaste excel in energy conversion and storage, making them suitable for electrodes [35]. Electrocatalytic water-splitting employs anodic and cathodic reactions, OER and HER, to produce H_2 and O_2 without byproducts [36]. The ORR is crucial for fuel cells and metal–air battery development [37]. Recent interest has surged in electrochemically converting NO_2 to ammonia gas (NH_3), due to its efficiency and accessible reaction sources [38]. Additionally, the

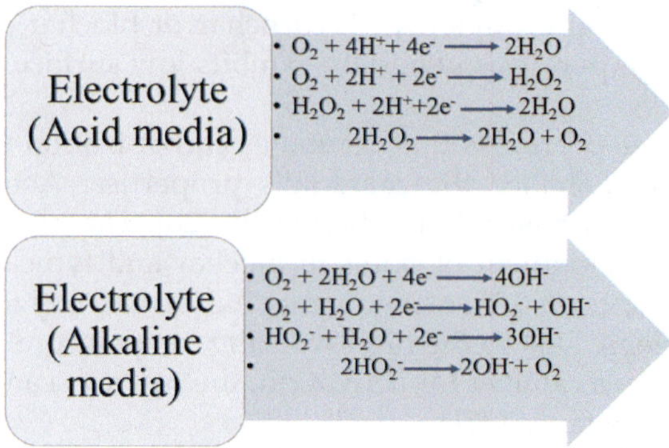

FIGURE 16.4 The ORR occurs in alkaline or acidic electrolyte solutions.

electrocatalytic conversion of CO_2 into valuable chemicals and fuels is gaining attention for its potential to lessen dependence on fossil fuels and mitigate greenhouse gas emissions [39]. This chapter explores engineering strategies to improve the catalytic performance of key electrochemical processes related to biomass-derived carbon compounds.

16.4.1 Using biomass waste in an electrochemical oxygen reduction reaction

O_2 is the most prevalent oxidant in the atmosphere, and its reduction plays a crucial role in corrosion and green energy technologies, such as fuel cells [40]. Fig. 16.4 shows how the ORR reaction pathway differed under acidic and alkaline conditions. However, compared to Pt-based nanoparticles , the overall electrocatalytic capabilities of the examined alternatives are insufficient. Consequently, devising effective strategies to promote catalytic activity is imperative [41]. The ORR is crucial for fuel cell advancements and is present in several energy conversion and storage systems, promoting affordable and clean renewable energy [42]. This efficiency greatly lowers CO_2 emissions and pollutants from fossil fuel use, making ORR vital for sustainable energy development. Effective catalysts are needed to reduce electrochemical overpotential and support high voltages during ORR reactions [43]. Utilizing biomass waste to extract porous carbon enhances material productivity, with sources including wood and corn residue [44]. These carbon materials possess high surface areas and improved porous structures, facilitating better ORR accessibility [45]. Additionally, the type and arrangement of minerals, particularly transition metals, significantly influence the semiconductor's conductivity and catalytic effectiveness for ORR [46].

For example, Zhou et al. [47] produced N-doped carbon nanosheets from walnut shell material (GWS180M800) using melamine activation and hydrothermal treatment, achieving robust stability and electrocatalytic performance for ORR with a near 4-electron route (n of 3.88–3.95). The half-wave potential of GWS180M800 improved from 0.70 to 0.82 V postactivation, outperforming 20 wt% Pt/C in stability and methanol resistance. Ma et al. [48] investigated lignocellulosic waste-derived nitrogen and phosphorus-doped biochars-1000 for ORR/OER applications, demonstrating significant catalytic performance and durability, especially the N/P-biochar-1000, which achieved an $E_{On\text{-}set}$ of 1.095 V and $E_{1/2}$ of 0.862 V at 5.5 mA cm^{-2}, comparable to Pt/C metrics ($E_{On\text{-}set}$ of 1.098 V and $E_{1/2}$ of 0.882 V at 54.8 mA cm^{-2}). Lu et al. [49] used biochars made from soybean straw to create an electrocatalyst doped with Co and N-doped carbon. Despite the presence of Co and N, there was no noticeable effect on the average pore size. CoNASS exhibited the highest ORR execution, with an $E_{1/2}$ of 0.786 V, similar to Pt/C found in commerce ($E_{1/2}$ of 0.827 V vs RHE). Moreover, the transport of electrons value of 3.75 indicated a predominantly single 4-electron ORR route. Considering the excellent performance of ORR electrodes derived from biomass-based carbon materials and nanoparticle-supported structures, these materials are regarded as highly promising candidates for advanced electrode development. Their high surface area, adjustable physicochemical properties, and superior electronic conductivity make them promising candidates for next-generation energy systems [50].

16.4.2 Utilizing biomass waste in an electrochemical oxygen evolution reaction

The OER process is vital for sustainable energy technologies, including water splitting for hydrogen production and metal-air batteries, leading to significant interest in improving the OER performance [51]. The mechanism of the OER is more intricate than that of the HER, involving a 4-electron pathway that subsequently results in slow reaction kinetics. In acidic media, two molecules of water oxidize to form one oxygen molecule. However, most catalysts exhibit instability in acidic media at high working potentials, alkaline media are used for the majority of OER applications [52].In alkaline media, the oxidation of four hydroxide ions can yield one oxygen molecule, as illustrated in Fig. 16.5. Enhancing the performance of renewable energy storage and conversion systems is a significant challenge owing to the slow reaction kinetics of the intricate multistep, 4-electron route of OER [53].

For instance, cedar tree cone-based using a single stage of pyrolysis, hierarchical interconnected porous carbon nanosheets (HPCNS) were synthesized [54].

HPCNS, featuring a pore volume of 1.35 cm^3 g^{-1}, a surface area of 2954 m^2 g^{-1}, and an electrical conductivity of 11.03 S cm^{-1}, showed enhanced

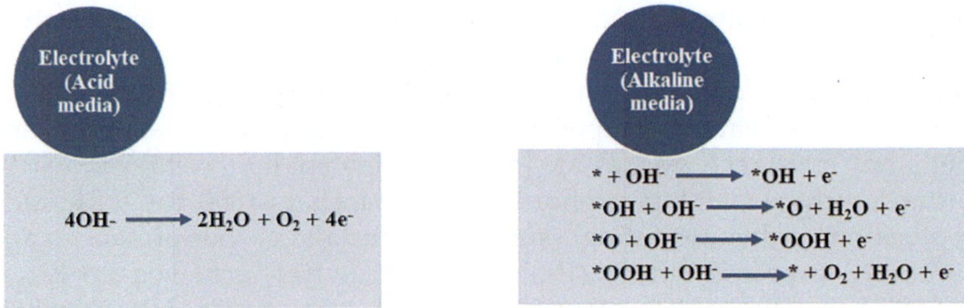

FIGURE 16.5 The OER process occurs in both acidic and alkaline electrolyte solutions.

electrocatalytic activity and rapid ion transport. HPCNS-700 achieved impressive OER performance in 1 M KOH with a Tafel slope of 191 mV dec^{-1} and an overpotential of 106 mV at 10 mA cm^{-2}. Liu et al. [55] used pyrolysis to create nitrogen-doped porous carbon nanosheets (NPCSS) based on corn stalks with a few-layered structure. The resultant NPCSS included numerous active sites, pore sizes (3.77 nm), and a higher active surface area (317.5 m^2 g^{-1}).

The N-doped material (PCNS with a Tafel slope of 239 mV dec^{-1}) showed superior OER activity with an overpotential of 358 mV at 10 mA cm^{-2} and a reduced Tafel slope of 144 mV dec^{-1}, outperforming non-N-doped materials. Wu et al. [56] developed mesoporous carbon microspheres (egg-CMS) doped with nitrogen, phosphorus, and iron, achieving a BET surface area of 970 m^2 g^{-1}, pore volume of 0.5 cm^3 g^{-1}, and low onset potential of 1.51 V with a Tafel slope of 59 mV dec^{-1} at 74.6 mA cm^{-2}, showcasing excellent catalytic activity and stability.

16.4.3 Using biomass waste in an electrochemical hydrogen evolution reaction

Electrolytic water splitting is a sustainable method for hydrogen production, gaining interest as an alternative to conventional techniques [57]. Different reaction pathways apply to electrocatalytic hydrogen production in alkaline and acidic environments. The Volmer–Tafel or Volmer–Heyrovsky reaction processes can be distinguished within the HER based on distinct phases that determine reaction rates [58]. Fig. 16.6 summarizes HER mechanisms in alkaline or acidic electrolytes. When hydrogen or hydrated H ions come into contact with the electrode, they absorb electrons and adsorb them on its surface, becoming active [59]. The Volmer reaction in the electrolyte solution is crucial for mass transport. By adhering to the hydrogen atom in an active state and forming hydrogen molecules using it, the hydrated H ion escapes from the Heyrovsky result; this is a method of electrochemical desorption [60]. Within the intricate desorption process known as the Tafel reaction, to create H_2 molecules, two adsorbed

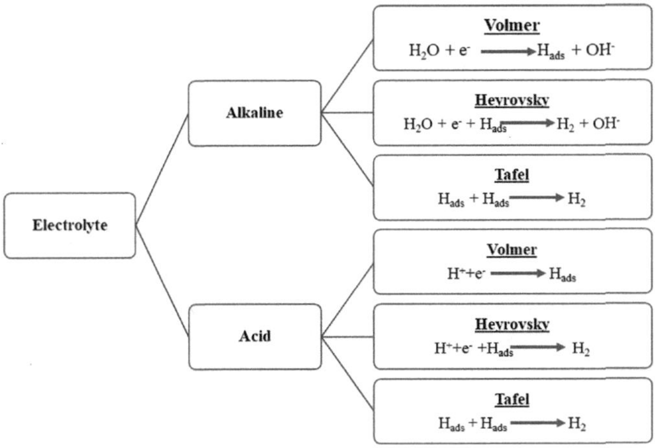

FIGURE 16.6 The HER process occurs in alkaline and acidic electrolytes.

active H atoms come together [61]. The reaction rate on the catalyst electrode serves as an indicator of the overall process efficiency. Although the theoretical water decomposition potential is 1.23 V under standard conditions, a significantly higher voltage is required to drive the reaction [62].

For instance, Zhu et al. [63] employed ball milling to combine KOH with pre-carbonized pine needles, creating N-doped microporous carbon materials for HER electrocatalysis, achieving a Tafel slope of 45.9 mV dec^{-1} and an overpotential of −62 mV o. Fu et al. [64] utilized KOH to activate walnut shells, creating a porous carbon electrocatalyst for HER studies. The structure features a BET surface area of 1037.31 m^2 g^{-1} and a low overpotential of 173 mV, improves ion flow and lowers charge transfer resistance. Cardoso et al. [65] activated grape stems and branches with $ZnCl_2$ to develop carbon composites for Pd nanoparticles, achieving surface areas of 1411 and 1689 m^2 g^{-1}. Upon uniform dispersion on the carbon material of grapevines, Pd nanoparticles showed a low HER overpotential of 210 mV at 10 mA cm^{-2} in 8 M KOH.

Murugan et al. [66] utilized *Platycladus orientalis* tree-cone biomass to fabricate N-doped porous carbon nanosheets with numerous defects, achieving high efficiency in water splitting. These resulting nanosheets exhibit a pore volume of 2.1 cm^3 g^{-1}, surface area of 3369 m^2 g^{-1}, and electrical conductivity of 12.69 S cm^{-1}. They achieved overpotentials of 188 mV and 90 mV for HER and OER, respectively, at 10 mA cm^{-2} in a 0.5 M H_2SO_4 solution. Similarly, Sankar et al. [67] created corrugated graphene nanosheets (RH-CG-700) from rice husk biomass, demonstrating excellent HER activity, substantial surface area (1154 m^2 g^{-1}), modest Tafel slope of 31 mV dec^{-1}, and a low overpotential of 9 mV at 10 mA cm^{-2}. Also, Saravanan et al. [68] created activated carbon nanosheets from peanut shells via pyrolysis and activation, achieving a BET surface area of 2338.56 $m^2\ g^{-1}$, a Tafel slope of 75.7 mV dec^{-1}, and an onset potential of 0.08 V versus RHE, exceeding graphene structures, analogous to transition metal doping.

Deng et al. [69] synthesized nitrogen, phosphorus, and calcium co-doped biochar (PBC-800) through pyrolysis by way of animal bones. achieving a distinct surface area (279.6 $m^2\ g^{-1}$) and excellent conductivity (0.053 S cm^{-1}), exhibiting exceptional HER activity, matching or exceeding that of single-heteroatom-doped carbon electrocatalysts. Additionally, it demonstrates a Tafel slope of 80 mV dec^{-1} at 52.5 μA cm^{-2}, a low onset potential of 80 ± 3 mV, and an overpotential of 162 ± 3 mV at 10 mA cm^{-2}, demonstrating exceptional HER activity comparable to single-heteroatom-doped catalysts.

16.4.4 Using biomass waste in an electrochemical carbon dioxide reduction reaction

An approach to mitigate the impacts of climate change involves utilizing CO_2 electroreduction for energy conversion applications [70]. Converting atmospheric surplus CO_2 into high-value compoundssuch as formate, formic acid, alcohols, hydrocarbons, and CO could enable it to save electrical energy in the same way as chemical energy [71]. A significant quantity of energy is needed for this transition due to the CO_2 molecule's constancy. Many materials have therefore been created to act as catalytic agents, lowering the energy requirement and raising the reaction's efficiency [72]. These precursors are notable for this application owing to their high nitrogen content, enabling the creation of metal-free catalysts with inherent defects that are particularly well-suited for CO_2's electrochemical conversion to CO [73].

Chen et al. [74] prepared catalysts from cocoons of silk through a series of steps that included precarbonization (350°C), $ZnCl_2$ activation, and carbonization (800°C–1100°C) processes. They found a direct correlation between rising temperatures and falling nitrogen concentrations, and current density toward CO generation. Several additional inherent heteroatoms present in biowaste precursors have also been investigated.

For instance, Cai et al. [75] examined heteroatoms in biowaste, creating metal-free catalysts from lignin with urea/melamine and studying chlorine's role during HCl washing, which improved CO faradaic efficiencies. The presence of sulfur and nitrogen enhanced selectivity for CO, benefiting from chlorine's contribution. In 2021, Hao et al. [76] synthesized nitrogen-doped and $FeCl_3$-activated cedar biomass catalysts, achieving efficient electron transport in CO_2RR through active sites, large surface area, and rich micropore structure. Deng et al. [77] used nitrogen-rich municipal sewage to create nitrogen-doped carbon composites via pyrolysis and carbonization, yielding active pyrrolic and pyridinic N species and hierarchical pore architectures that promoted CO_2 reduction and enhanced product selectivity for C_2H_5OH, CH_3OH, and CH_4. Moreover, Lourenco et al. [78] synthesized biochars with ZnO using coffee and chitosan. While both biochars exhibited general activity in CO_2RR and selectivity for CO, the biochar prepared from chitosan outperformed the coffee-derived biochar. The variance in performance was ascribed to the existence of pyridine and pyridinic N species, it enhanced the catalyst's ability to interact with the molecule of CO_2 and made it easier for it to be reduced. Nevertheless, carbonaceous compounds from tobacco cigarette butts have been used as precursors to create Ni nanoparticle supports [79], resulting in high surface area materials and porosity conducive to rapid mass transfer. Additionally, they were nitrogen-doped, sourced from tobacco stem alkaloids and nicotine. In the end, these electrocatalysts demonstrated exceptional faradic efficiencies for CO production and stability for 30 hours. In broad terms, the intrinsic or doped nitrogen content of a material performs a critical role in determining its efficiency, selectivity, and capacity to form the necessary active sites for CO_2 electro-reduction[80]. Moreover, a material type, porosity distribution, and degree of graphitization are crucial for high electrochemical performance, starting with affordable, renewable raw materials [81].

16.4.5 Electrochemical nitrogen reduction reaction utilizing waste biomass

NH_3, hydrazine (N_2H_4), and diazene (N_2H_2) are key NRR products relevant to industry and agriculture [82]. The electrochemical NRR is significant for activating inactive N≡N bonds via an association and dissociation mechanism [83]. During this process, N_2 molecules undergo adsorption and protonation, leading to NH_3 generation [84]. The dissociative processes include adsorption, N atom protonation, and N≡N breaking. The focus is on biomass-derived metal-free electrocatalysts due to their excellent electrochemical properties, affordability, and large surface area [85]. Also, Zeng et al. [86] developed pine tower (PT) -KOH, a metal-free porous carbon electrocatalyst from PT using carbonization and KOH activation, exhibited remarkable selectivity and efficiency in converting nitrogen to NH_3, achieving 3.69% faradaic efficiency (FE) and a production rate of 16.15 $\mu g\ h^{-1}\ mg^{-1}$ catalyst at −0.6 V versus RHE in 0.1 M Na_2SO_4. Additionally, Hollow carbon nanotubes doped with oxygen from kapok fiber showed high selectivity and electrochemical stability, boasting a 9.1% FE at −0.80 V versus RHE and a production rate of 25.12 $\mu g\ h^{-1}\ mg^{-1}$ cat at −0.85 V [87].

Yang et al. [88] produced N-rich carbon (NFC-900) using cicada sloughs through annealing and activation techniques, achieving a high specific surface area that enhanced NH_3 production to 15.7 μg h^{-1} mg^{-1} cat. Increasing catalytic temperatures accelerated N_2 reduction due to improved mass transfer rates. Zhao et al. [89] created N-doped carbon (NPC) from alfalfa via pyrolysis, resulting in NPC material pyrolyzed (500°C) with a porous structure. At −0.4 V versus RHE in H_2SO_4 (0.005 M), it exhibited a FE of 9.98% and an NH_3 production rate of 1.31 mmol h^{-1} g^{-1} cat. (22.3 mg h^{-1} g^{-1} cat.). Pyridine-N incorporation in NPC-500 increased catalytic sites, boosting NH_3 generation during NRR. Feng et al. [90] developed pre-pyrolyzed Fe_2O_3 biochar composites with MoS_2/MoO_2, where the Fe and Mo-based catalysts mimic natural nitrogenase activity. The resulting material shows remarkable N_2 adsorption and activation activity, achieving a maximum FE of 16.5% and 4.87 μg h^{-1} cm^{-2} NH_3 production in Na_2SO_4 (0.1 M).

16.4.6 Alternative energy-conversion and storage systems

The high specific surface area and porous structure of biomass-derived carbon make it effective for energy conversion and storage, particularly in microbial fuel cells (MFCs) that generate electricity and treat wastewater [91]. MFCs utilizing lignocellulosic biomass promote sustainable energy solutions [92]. Research by Li et al. [93] involved extracting cellulose from coconut fiber and purifying it $NaClO_2$, resulting in CON-900, which has a multistage pore architecture comprising oxygen-containing species, graphite, and pyridine nitrogen and a high specific surface area (652 m^2 g^{-1}). This material, when applied as air cathodes, shows a maximum power density of 1014 ± 23 mW m^{-2}. Finally, biowaste-derived functional carbon compounds and their uses in energy conversion applications, as shown in Table 16.1.

16.5 Conclusion and future perspectives

This chapter examines advancements in carbon derived from biomass waste, highlighting its applications in electrochemical conversion and synthesis techniques. The conversion of biomass waste to carbon electrocatalytic resources is influenced by processing parameters such as carbonization temperature, duration, and atmosphere. Activation creates a hierarchical pore structure and enhances conductivity, which is vital for effective electrocatalytic activity in processes like HER, OER, ORR, NRR, and CO_2RR. Significant progress in developing carbon nanostructures based on biomass for energy conversion and electrocatalysis is noted, but challenges remain, including underutilization of biomass and the need for scalable synthesis for eco-friendly materials. Furthermore, the varied morphology and composition of raw biomass support intricate

TABLE 16.1 Carbon-based electrodesderived from diverse biomass wastes and fabricated via various techniques are widely used in energy conversion applications.

S.no	Sources of biomass	Conversion method	Specific surface area (m^2 g^{-1})	Pore size (nm)	Types of nanostructures	Application	Results	References
1.	Sawdust	Carbonization, activation, followed by hydrolysis	3100	0.91	Microporous Carbons	Hydrogen storage	H_2 uptake over 5 wt %; adsorption heat of 4.1–7.5 kJ mol^{-1}	[94]
2.	Boiled coffee bean	Molten-salt treatment	550	–	Hierarchical nanostructures	ORR	Capacitor energy density reaches 22.27 W h kg^{-1}	[95]
3.	Filter, tissue paper + cotton + Iron–tannin framework + dicyandiamide	Basic solution phase reactions, along with shaking	202	–	Carbide/Fe-N-Carbon nanostructure	ORR	High onset potential of 0.98 V vs RHE	[96]
4.	Leather scraps	Regulated pyrolysis	2100	5	Metal-free electrocatalysts	Oxygen Reduction Reaction	On-set potential shifts only 145 mV	[97]

(*continued on next page*)

TABLE 16.1 Carbon-based electrodesderived from diverse biomass wastes and fabricated via various techniques are widely used in energy conversion applications—cont'd

S.no	Sources of biomass	Conversion method	Specific surface area ($m^2 g^{-1}$)	Pore size (nm)	Types of nanostructures	Application	Results	References
5.	Wheat flour combined with KOH	One–step pyrolysis	1269	–	N-doped nanostructured carbon	Electrocatalytic for conversion of CO_2 to CO	Faradaic efficiency of 83.7% and overpotential of 0.71 V	[98]
6.	Lignin waste	Hydrothermal carbonization /KOH activation	3235	–	Carbon materials with well-ordered micropores	CO_2 Storage	CO_2 absorption capacity: 4.6 mmol g^{-1} at 1 bar on 25°C	[99]
7.	Melaleuca bark	Activation using KOH	3170	–	Highly porous carbons nanostructures	Hydrogen storage	H_2 uptake at 77 K and 10 bar. (4.08 wt%)	[100]
8.	Rice husk	Pyrolysis/activation using KOH	1154	2.6	2D graphene nanosheets	HER	Overpotential of 9 mV, Tafel slope 31 mV dec^{-1} at 10 mA cm^{-2}	[67]

(continued on next page)

9.	Cedar tree cone	Pyrolysis/ Activation using KOH	2954	–	Hierarchical porous carbon nanosheets	OER	Overpotential f 106 mV at 10 mA cm^{-2}	[54]
10.	Cornstalks	Chemical vapor deposition	849	–	CNFs	OER	Overpotential 290 mV 10 mA cm^{-2}	[101]
11.	Chitosan	Pyrolysis (Ar atm) at 900°C	–	–	N-doped defective graphene	Transformation of CO_2 into CH_4	Activation energy of 24.3 $kcal.mol^{-1}$ and turnover frequencies of 73.17 s^{-1}	[102]
12.	Chitosan with alginic acid	Pyrolysis (1100°C)	–	–	Graphene with few layers	Hydrogen generation	Positive bias of 1.1 V yields 3.62 μmol H_2	[103]

(continued on next page)

TABLE 16.1 Carbon-based electrodesderived from diverse biomass wastes and fabricated via various techniques are widely used in energy conversion applications—cont'd

S.no	Sources of biomass	Conversion method	Specific surface area ($m^2\ g^{-1}$)	Pore size (nm)	Types of nanostructures	Application	Results	References
13.	Chitosan	Cobalt-catalyzed carbonization	176	–	N-doped carbon nanotube	ORR	Current densities (J) were found to be 5.06 $mA\ cm^{-2}$	[104]
14.	Chlorella	Chemical synthesis	728	6–250	N-doped carbon nanotube	ORR	Half-wave potential shifted 40 mV at 10 $mA\ cm^{-2}$	[105]
15.	Soybean	Templating carbonization and KOH activation	1749	1.2	Nitrogen-doped porous carbon	Supercapacitor	At a power density of 450 $W\ kg^{-1}$, the energy density is 12.5 $Wh\ kg^{-1}$	[106]
16.	Chitosan	KOH activation	1510	–	Carbon horn, tube, black	ORR	Specific capacitances of 106 Fg^{-1} at 0.2 Ag^{-1}	[107]

surface chemistry and porous structures. Uncertainty surrounds certain heteroatom species in the carbon matrix, though, which makes it difficult to comprehend mechanisms, especially those involving porosity and heteroatom doping. Therefore, to accurately control the physical-chemical properties, it is crucial to carefully choose the right sources of biomass. Moreover, extensive *in-situ* analyses are needed to provide a more understandable picture of the connection between carbon compounds made from biomass waste and electrochemical performance.

References

[1] S. Yana, M. Nizar, D. Mulyati, Biomass waste as a renewable energy in developing bio-based economies in Indonesia: a review, Renew. Sustain. Energy Rev. 160 (2022) 112268.

[2] N.M. Clauser, G. González, C.M. Mendieta, J. Kruyeniski, M.C. Area, M.E. Vallejos, Biomass waste as sustainable raw material for energy and fuels, Sustainability 13 (2021) 794.

[3] J. Jjagwe, P.W. Olupot, E. Menya, H.M. Kalibbala, Synthesis and application of granular activated carbon from biomass waste materials for water treatment: a review, J. Bioresour. Bioprod. 6 (2021) 292–322.

[4] L. Zhang, T.U. Rao, J. Wang, D. Ren, S. Sirisommboonchai, C. Choi, H. Machida, Z. Huo, K. Norinaga, A review of thermal catalytic and electrochemical hydrogenation approaches for converting biomass-derived compounds to high-value chemicals and fuels, Fuel Process. Technol. 226 (2022) 107097.

[5] Y. Li, J. Li, S. Zhou, B. Meng, T. Wu, A review on thermogravimetric analysis-based analyses of the pyrolysis kinetics of oil shale and coal, Energy Sci. Eng. 12 (2024) 329–355.

[6] M. Martins, F. Sousa, C. Soares, B. Sousa, R. Pereira, M. Rubal, F. Fidalgo, Beach wrack: discussing ecological roles, risks, and sustainable bioenergy and agricultural applications, J. Environ. Manage. 356 (2024) 120526.

[7] S. Lu, Q. Xiao, W. Yang, X. Wang, T. Guo, Q. Xie, Y. Ruan, Multi-heteroatom-doped porous carbon with high surface adsorption energy of potassium derived from biomass waste for high-performance supercapacitors, Int. J. Biol. Macromol. 258 (2024) 128794.

[8] S.P. Zhou, X. Ke, L.-Q. Jin, Y.P. Xue, Y.G. Zheng, Sustainable management and valorization of biomass wastes using synthetic microbial consortia, Bioresour. Technol. 395 (2024) 130391.

[9] X. Ma, C. Guo, J. Xiang, Y. Qi, J. Yu, K. Li, Z. Tao, J. Wu, Y. Lv, Synthesis and applications of biomass-derived electrocatalysts in water electrolysis, Int. J. Hydrogen Energy 60 (2024) 845–866.

[10] Á. Ramírez, M. Muñoz-Morales, E. López-Fernández, F. Fernández-Morales, J. Llanos, Advancing circular economy: critical insights into waste biomass derived carbon electrodes for (bio) electrochemical water treatment, Curr. Opin. Electrochem. 46 (2024) 101492.

[11] Y. Wang, T. Xu, K. Liu, M. Zhang, X.M. Cai, C. Si, Biomass-based materials for advanced supercapacitor: principles, progress, and perspectives, Aggregate 5 (2024) e428.

[12] A. Plavniece, K. Kaare, D. Simkunaitė, A. Balciunaite, V. Jasulaitiene, G. Niaura, A. Volperts, G. Dobele, L.C. Colmenares-Rausseo, I. Kruusenberg, Manganese-and nitrogen-doped biomass-based carbons as catalysts for the oxygen reduction reaction, Catalysts 14 (2024) 92.

[13] Y. Jiang, Z. Hu, L. Cheng, C. Luo, Linking biomass pyrolysis and biotransformation: a bibliometric review, J. Energy Inst. 114 (2024) 101592.

[14] P. Numee, T. Sangtawesin, M. Yilmaz, K. Kanjana, Activated carbon derived from radiation-processed durian shell for energy storage application, Carbon Resour. Convers. 7 (2024) 100192.

[15] A. Nawaz, S.A. Razzak, Co-pyrolysis of biomass and different plastic waste to reduce hazardous waste and subsequent production of energy products: a review on advancement, synergies, and future prospects, Renew. Energy 224 (2024) 120103.

[16] R. Czubaszek, A. Wysocka-Czubaszek, A. Sienkiewicz, A. Piotrowska-Niczyporuk, M.J. Wassen, A. Bajguz, Possibilities of utilising biomass collected from road verges to produce biogas and biodiesel, Energies 17 (2024) 1751.

[17] V. Molahalli, A. Sharma, K. Bijapur, G. Soman, N. Chattham, G. Hegde, Low-cost biowaste carbon nanocomposites for sustainable electrochemical devices: a systematic review, Mater. Today Commun. 38 (2024) 108034.

[18] V. Ilango, Bioenergy production using biomass wastes: challenges of circular economy, in: A.L. Srivastav, A.K. Bhardwaj, M. Kumar (Eds.), Valorization of Biomass Wastes for Environmental Sustainability: Green Practices for the Rural Circular Economy, Springer International Publishing, Cham, Switzerland, 2024, pp. 171–188.

[19] P. Chaudhary, S. Bansal, B.B. Sharma, S. Saini, A. Joshi, Waste biomass-derived activated carbons for various energy storage device applications: a review, J. Energy Storage 78 (2024) 109996.

[20] S.K. Tiwari, M. Bystrzejewski, A. De Adhikari, A. Huczko, N. Wang, Methods for the conversion of biomass waste into value-added carbon nanomaterials: recent progress and applications, Prog. Energy Combust. Sci. 92 (2022) 101023.

[21] H. Doğan, M. Taş, T.b. Meşeli, G.l.a. Elden, G. Genc, Review on the applications of biomass-derived carbon materials in vanadium redox flow batteries, ACS Omega 8 (2023) 34310–34327.

[22] A. Pistone, C. Espro, Current trends on turning biomass wastes into carbon materials for electrochemical sensing and rechargeable battery applications, Curr. Opin. Green Sustain. Chem. 26 (2020) 100374.

[23] D. Sangaré, M. Moscosa-Santillan, A. Aragón Piña, S. Bostyn, V. Belandria, I. Gökalp, Hydrothermal carbonization of biomass: experimental study, energy balance, process simulation, design, and techno-economic analysis, Biomass Convers. Biorefin. 14 (2024) 2561–2576.

[24] S. Yu, J. He, Z. Zhang, Z. Sun, M. Xie, Y. Xu, X. Bie, Q. Li, Y. Zhang, M. Sevilla, Towards negative emissions: hydrothermal carbonization of biomass for sustainable carbon materials, Adv. Mater. (2024) 2307412.

[25] C. Areeprasert, K. Taweengern, C. Khaobang, T. Thapsamut, S. Aramrak, Circular utilization of sugarcane bagasse for water and nutrient retention in two types of sugarcane cultivation soil via biochar and hydrochar addition, Fuel, 392, 134870.

[26] A. Bardha, S. Prasher, M.-J. Dumont, Waste biomass-derived rubber composite additives: review of current research and future investigations into biowaste tire formulation, Biomass Bioenergy 183 (2024) 107149.

[27] D.A. Peguero, M. Gold, L. Velasquez, M. Niu, C. Zurbrügg, A. Mathys, Physical pretreatment of three biowastes to improve black soldier fly larvae bioconversion efficiency, Waste Manage. 178 (2024) 280–291.

[28] T. Kopac, S. Lin, A review on the characterization of microwave-induced biowaste-derived activated carbons for dye adsorption, Int. J. Environ. Sci. Technol. 21 (2024) 1–32.

[29] S.J. Malode, S. Bilagi, N.P. Shetti, Synthesis of biowaste-derived nanostructured material for the nanomolar detection of glucose: biological sample analysis, J. Pharm. Biomed. Anal. 3 (2024) 100026.

[30] H. Ashraf, B.D. Karahan, Biowaste valorization into valuable nanomaterials: synthesis of green carbon nanodots and anode material for lithium-ion batteries from watermelon seeds, Mater. Res. Bull. 169 (2024) 112492.

[31] L. Rakhymbay, N. Bazybek, K. Kudaibergenov, S.-T. Myung, Z. Bakenov, A. Konarov, Present development and future perspectives on biowaste-derived hard carbon anodes for room temperature sodium-ion batteries, J. Power Sources 602 (2024) 234347.

[32] I. Benchikh, A.O. Ezzat, L. Sabantina, Y. Benmimoun, A. Benyoucef, Investigation of hybrid electrodes of polyaniline and reduced graphene oxide with bio-waste-derived activated carbon for supercapacitor applications, Polymers 16 (2024) 421.

[33] R. Wang, N. Wang, X. Zhang, M. Xie, L. Huang, Q. Zhang, C. Feng, Y. Xu, Y. Jiao, J. Chen, Inhibiting solubility challenges: carbon nanotube encapsulation of stable transition metal oxides for electrochemical energy storage and conversion applications, J. Energy Storage 76 (2024) 109794.

[34] S.M. Alirahmi, T. Gundersen, A. Arabkoohsar, J.J. Klemeš, G. Sin, H. Yu, Process design, integration, and optimization of a novel compressed air energy storage for the coproduction of electricity, cooling, and water, Renew. Sustain. Energy Rev. 189 (2024) 114034.

[35] P.S. Kumar, P. Prakash, Metal free nanocomposite of graphitic carbon nitride, boron nitride and chitosan for efficient evolution of hydrogen: a strategic approach to achieving sustainable and effective electrocatalysis, J. Environ. Chem. Eng. 11 (2023) 109045.

[36] X.-L. Song, L. Chen, J.-T. Ren, L.-J. Gao, Z.-Y. Yuan, Engineering of g-C_3N_4-based composites for photocatalytic and electrocatalytic water splitting: recent progress, challenges and perspective, Coord. Chem. Rev. 507 (2024) 215752.

[37] S. Yang, S. Liu, H. Li, Z. Wang, J. Zhang, M. Liu, J. Ding, S. Qiu, F. Deng, Boosting oxygen mass transfer for efficient H_2O_2 generation via 2e−-ORR: a state-of-the-art overview, Electrochim. Acta 479 (2024) 143889.

[38] J. Theerthagiri, K. Karuppasamy, A.H. Mahadi, C.J. Moon, N. Rahamathulla, S. Kheawhom, S. Alameri, A. Alfantazi, A.P. Murthy, M.Y. Choi, Electrochemical reduction of gaseous nitric oxide into ammonia: a review, Environ. Chem. Lett. 22 (2024) 189–208.

[39] M. Ma, B. Seger, Rational design of local reaction environment for electrocatalytic conversion of CO_2 into multicarbon products, Angew. Chem. Int. Ed. 63 (23) (2024) e202401185.

[40] A.M. Roth-Zawadzki, A.J. Nielsen, R.E. Tankard, J. Kibsgaard, Dual and triple atom electrocatalysts for energy conversion (CO_2RR, NRR, ORR, OER, and HER): synthesis, characterization, and activity evaluation, ACS Catalysis 14 (2024) 1121–1145.

[41] Y. Liu, L. Zhou, S. Liu, S. Li, J. Zhou, X. Li, X. Chen, K. Sun, B. Li, J. Jiang, Fe, N-Inducing interfacial electron redistribution in NiCo Spinel on biomass-derived carbon for bi-functional oxygen conversion, Angew. Chem. Int. Ed. 63 (2024) e202319983.

[42] A. Letona-Elizburu, M. Enterría, A. Aziz, S. Villar-Rodil, J. Paredes, J. Carrasco, N. Ortiz-Vitoriano, Biomimetic nucleotide-graphene hybrids for electrocatalytic oxygen conversion: quantifying biomolecule mass loading, Sustainable Mater.Technol. 39 (2024) e00835.

[43] Y. Zhao, J. Raj, X. Xu, J. Jiang, J. Wu, M. Fan, Carbon catalysts empowering sustainable chemical synthesis via electrochemical CO_2 conversion and two-electron oxygen reduction reaction, Small 21 (2024) 2311163.

[44] C. Ling, W.-Y. Xie, S.-F. He, G.-C. Liang, X.-Y. Xiao, C.-L. Yang, H.-Y. Liu, Si-doped biomass carbon supported Fe-NC catalyst derived from pyrolysis of iron porphyrin for oxygen reduction reaction, J. Alloys Compd. 987 (2024) 174172.

[45] S. Liu, A. Wang, Y. Liu, W. Zhou, H. Wen, H. Zhang, K. Sun, S. Li, J. Zhou, Y. Wang, Catalytically active carbon for oxygen reduction reaction in energy conversion: recent advances and future perspectives, Adv. Sci. 11 (2024) 2308040.

[46] D. Jiang, Y. Chen, K. Xie, S. Li, Y. Wang, Carbon-coated Fe_3O_4@ cotton fiber biomass carbon for the oxygen reduction reaction in air cathode microbial fuel cells, J. Alloys Compd. 982 (2024) 173843.

[47] Y. Zhou, L. Yan, J. Hou, Nanosheets with high-performance electrochemical oxygen reduction reaction revived from green walnut peel, Molecules 27 (2022) 328.

[48] L.-L. Ma, X. Hu, W.-J. Liu, H.-C. Li, P.K. Lam, R.J. Zeng, H.-Q. Yu, Constructing N, P-dually doped biochar materials from biomass wastes for high-performance bifunctional oxygen electrocatalysts, Chemosphere 278 (2021) 130508.

[49] G. Lu, Z. Li, W. Fan, M. Wang, S. Yang, J. Li, Z. Chang, H. Sun, S. Liang, Z. Liu, Sponge-like N-doped carbon materials with Co-based nanoparticles derived from biomass as highly efficient electrocatalysts for the oxygen reduction reaction in alkaline media, RSC Adv. 9 (2019) 4843–4848.

[50] J. Gao, X. Chu, C. He, Z. Yin, H. Lu, X. Li, J. Feng, X. Tan, Biomass-derived carbon for ORR: pine needles as a single source for efficient carbon electrocatalyst, J. Appl. Electrochem. 50 (2020) 1257–1267.

[51] F. Jiang, Y. Li, Y. Pan, Design principles of single-atom catalysts for oxygen evolution reaction: from targeted structures to active sites, Adv. Mater. 36 (2024) 2306309.

[52] X. Ren, Y. Zhai, N. Yang, B. Wang, S. Liu, Lattice oxygen redox mechanisms in the alkaline oxygen evolution reaction, Adv. Funct. Mater. (2024) 2401610.

[53] J. Chen, J. Qian, Insights on MOF-derived metal–carbon nanostructures for oxygen evolution, Dalton Trans. 53 (2024) 2903–2916.

[54] N. Murugan, S. Thangarasu, G.B. Choi, J. Choi, T.H. Oh, Y.A. Kim, Interconnected hierarchical nanoarchitectonics of porous carbon nanosheets derived from renewable biomass for efficient oxygen evolution reaction, J. Alloys Compd. 923 (2022) 166321.

[55] Z. Liu, Q. Zhou, B. Zhao, S. Li, Y. Xiong, W. Xu, Few-layer N-doped porous carbon nanosheets derived from corn stalks as a bifunctional electrocatalyst for overall water splitting, Fuel 280 (2020) 118567.

[56] H. Wu, J. Geng, H. Ge, Z. Guo, Y. Wang, G. Zheng, Egg-derived mesoporous carbon microspheres as bifunctional oxygen evolution and oxygen reduction electrocatalysts, Adv. Energy Mater. 6 (2016) 1600794.

[57] G. Zhao, K. Rui, S.X. Dou, W. Sun, Heterostructures for electrochemical hydrogen evolution reaction: a review, Adv. Funct. Mater. 28 (2018) 1803291.

[58] S. Li, J. Yang, C. Song, Q. Zhu, D. Xiao, D. Ma, Iron carbides: control synthesis and catalytic applications in CO_x hydrogenation and electrochemical HER, Adv. Mater. 31 (2019) 1901796.

[59] M. Sajid, W. Qayyum, A. Farhan, M.A. Qamar, H. Nawaz, Progress in the development of copper oxide-based materials for electrochemical water splitting, Int. J. Hydrogen Energy 62 (2024) 209–227.

[60] A. MariaJoseph, S. Nangan, D. Verma, L. Gnanasekaran, S. Rajendran, T. Natesan, P. Pattananuwat, M. Okhawilai, Rational coupling of selective electrochemical oxidation and reduction reactions for in-situ value-added chemical generation, Fuel 367 (2024) 131408.

[61] S. Bajracharya, B. Bian, R. Jimenez-Sandoval, L. Matsakas, K.P. Katuri, P.E. Saikaly, Nature inspired catalysts: a review on electroactive microorganism-based catalysts for electrochemical applications, Electrochim. Acta 488 (2024) 144215.

[62] J.Y. Kim, S.H. Roh, C. Xia, U. Sim, J.K. Kim, MXene-based hybrid materials for electrochemical and photoelectrochemical H_2 generation, J. Energy Chem. 93 (2024) 111–125.

[63] G. Zhu, L. Ma, H. Lv, Y. Hu, T. Chen, R. Chen, J. Liang, X. Wang, Y. Wang, C. Yan, Pine needle-derived microporous nitrogen-doped carbon frameworks exhibit high performances in electrocatalytic hydrogen evolution reaction and supercapacitors, Nanoscale 9 (2017) 1237–1243.

[64] H.-h. Fu, L. Chen, H. Gao, X. Yu, J. Hou, G. Wang, F. Yu, H. Li, C. Fan, Y.-l. Shi, Walnut shell-derived hierarchical porous carbon with high performances for electrocatalytic hydrogen evolution and symmetry supercapacitors, Int. J. Hydrogen Energy 45 (2020) 443–451.

[65] J. Cardoso, B. Šljukić, M. Erdem, C. Sequeira, D. Santos, Vine shoots and grape stalks as carbon sources for hydrogen evolution reaction electrocatalyst supports, Catalysts 8 (2018) 50.

[66] N. Murugan, S. Thangarasu, S.B. Seo, A. Mariappan, Y.R. Choi, T.H. Oh, Y.A. Kim, N-doped defect-rich porous carbon nanosheets framework from renewable biomass as efficient metal-free bifunctional electrocatalysts for HER and OER application, Renewable Energy 222 (2024) 119801.

[67] S. Sekar, A.T.A. Ahmed, D.H. Sim, S. Lee, Extraordinarily high hydrogen-evolution-reaction activity of corrugated graphene nanosheets derived from biomass rice husks, Int. J. Hydrogen Energy 47 (2022) 40317–40326.

[68] K.A. Saravanan, N. Prabu, M. Sasidharan, G. Maduraiveeran, Nitrogen-self doped activated carbon nanosheets derived from peanut shells for enhanced hydrogen evolution reaction, Appl. Surf. Sci. 489 (2019) 725–733.

[69] L. Deng, Y. Zhang, Y. Wang, H. Yuan, Y. Chen, Y. Wu, In situ N-, P-and Ca-codoped biochar derived from animal bones to boost the electrocatalytic hydrogen evolution reaction, Resour. Conserv. Recycl. 170 (2021) 105568.

[70] P. Uthirakumar, H. Son, V. Dao, Y. Lee, S. Yadav, I.-H. Lee, Accelerating photoelectrochemical CO_2RR selectively of C_{2+} products by integrating Ag/Pd cocatalysts on $Cu/Cu_2O/CuO$ heterojunction nanorods, J. Environ. Chem. Eng. 12 (2024) 112442.

[71] Z. Yang, X. Guo, Y. Chen, L. Gao, R. Wei, G. Xiao, Modulating surface microenvironment based on Ag-adorned CuO flower-liked nanospheres for strengthening C-C coupling during CO_2RR, Surf. Interfaces 48 (2024) 104267.

[72] S. Li, J. Yu, S. Zhang, W. Qiu, X. Tang, Z. Lin, R. Cai, Y. Fang, S. Yang, X. Cai, Operando reconstruction of porous carbon supported copper selenide promotes the C_2 production from CO_2RR, Adv. Funct. Mater. 34 (2024) 2311989.

[73] H. Yang, Y. Wu, Q. Lin, L. Fan, X. Chai, Q. Zhang, J. Liu, C. He, Z. Lin, Composition tailoring via N and S co-doping and structure tuning by constructing hierarchical pores: metal-free catalysts for high-performance electrochemical reduction of CO_2, Angewandte Chemie. 130 (2018) 15702–15706.

[74] M. Chen, S. Wang, H. Zhang, P. Zhang, Z. Tian, M. Lu, X. Xie, L. Huang, W. Huang, Intrinsic defects in biomass-derived carbons facilitate electroreduction of CO_2, Nano Res. 13 (2020) 729–735.

[75] X. Cai, B. Qin, Y. Li, Q. Zhang, G. Yang, H. Wang, Y. Cao, H. Yu, F. Peng, Chlorine-promoted nitrogen and sulfur Co-doped biocarbon catalyst for electrochemical carbon dioxide reduction, Chem. Electro. Chem. 7 (2020) 320–327.

[76] X. Hao, X. An, A.M. Patil, P. Wang, X. Ma, X. Du, X. Hao, A. Abudula, G. Guan, Biomass-derived N-doped carbon for efficient electrocatalytic CO_2 reduction to CO and $Zn–CO_2$ batteries, ACS Appl. Mater. Interfaces 13 (2021) 3738–3747.

[77] L. Deng, H. Yuan, X. Qian, Q. Lu, L. Wang, H. Hu, Y. Chen, Municipal sludge-derived carbon dots-decorated, N-doped hierarchical biocarbon for the electrochemical reduction of carbon dioxide, Resour. Conserv. Recycl. 177 (2022) 105980.

[78] M.A. Lourenço, J. Zeng, P. Jagdale, M. Castellino, A. Sacco, M.A. Farkhondehfal, C.F. Pirri, Biochar/zinc oxide composites as effective catalysts for electrochemical CO_2 reduction, ACS Sustainable Chem. Eng. 9 (2021) 5445–5453.

[79] H. Wang, M. Li, G. Liu, L. Yang, P. Sun, S. Sun, Tobacco stem-derived nitrogen-containing porous carbon with highly dispersed Ni–N sites as an efficient electrocatalyst for CO_2 reduction to CO, New J. Chem. 45 (2021) 1063–1071.

[80] X. Wang, X. Li, S. Ding, Y. Chen, Y. Liu, M. Fang, G. Xiao, Y. Zhu, Constructing ample active sites in nitrogen-doped carbon materials for efficient electrocatalytic carbon dioxide reduction, Nano Energy 90 (2021) 106541.

[81] J. Wang, Y. Chen, S. Zhang, C. Yang, J.Y. Zhang, Y. Su, G. Zheng, X. Fang, Controllable states and porosity of Cu-Carbon for CO_2 electroreduction to hydrocarbons, Small 18 (2022) 2202238.

[82] R. Li, Z. Feng, Y. Li, X. Dai, Y. Ma, Y. Tang, Bioinspired Mo tape-porphyrin as an efficient and selective electrocatalyst for ammonia synthesis, Appl. Surf. Sci. 520 (2020) 146202.

[83] C. Yang, Y. Zhu, J. Liu, Y. Qin, H. Wang, H. Liu, Y. Chen, Z. Zhang, W. Hu, Defect engineering for electrochemical nitrogen reduction reaction to ammonia, Nano Energy 77 (2020) 105126.

[84] X. Zhao, G. Hu, G.F. Chen, H. Zhang, S. Zhang, H. Wang, Comprehensive understanding of the thriving ambient electrochemical nitrogen reduction reaction, Adv. Mater. 33 (2021) 2007650.

[85] S. Zhao, X. Lu, L. Wang, J. Gale, R. Amal, Carbon-based metal-free catalysts for electrocatalytic reduction of nitrogen for synthesis of ammonia at ambient conditions, Adv. Mater. 31 (2019) 1805367.

[86] L. Zeng, A.R. Thiruppathi, J. van der Zalm, X. Li, A. Chen, Biomass-derived amorphous carbon with localized active graphite defects for effective electrocatalytic N_2 reduction, Appl. Surf. Sci. 575 (2022) 151630.

[87] T. Wu, P. Li, H. Wang, R. Zhao, Q. Zhou, W. Kong, M. Liu, Y. Zhang, X. Sun, F.F. Gong, Biomass-derived oxygen-doped hollow carbon microtubes for electrocatalytic N_2-to-NH_3 fixation under ambient conditions, Chem. Commun. 55 (2019) 2684–2687.

[88] X. Yang, K. Li, D. Cheng, W.-L. Pang, J. Lv, X. Chen, H.-Y. Zang, X.-L. Wu, H.-Q. Tan, Y.-H. Wang, Nitrogen-doped porous carbon: highly efficient trifunctional electrocatalyst for oxygen reversible catalysis and nitrogen reduction reaction, J. Mater. Chem. A 6 (2018) 7762–7769.

[89] C. Zhao, S. Zhang, M. Han, X. Zhang, Y. Liu, W. Li, C. Chen, G. Wang, H. Zhang, H. Zhao, Ambient electrosynthesis of ammonia on a biomass-derived nitrogen-doped porous carbon electrocatalyst: contribution of pyridinic nitrogen, ACS Energy Lett. 4 (2019) 377–383.

[90] D. Feng, L. Zhou, C. Liu, H. Li, Y. Zhang, Y. Yao, T. Ma, Enhanced electrochemical reduction of N_2 to NH_3 by interfacial engineering of biomass-derived Fe, Mo-bimetallic composite, Electrochim. Acta 484 (2024) 144096.

[91] J. Arun, P. SundarRajan, K.G. Pavithra, P. Priyadharsini, S. Shyam, R. Goutham, Q.H. Le, A. Pugazhendhi, New insights into microbial electrolysis cells (MEC) and microbial fuel cells (MFC) for simultaneous wastewater treatment and green fuel (hydrogen) generation, Fuel 355 (2024) 129530.

[92] A. Shrivastava, R.K. Sharma, Lignocellulosic biomass based microbial fuel cells: performance and applications, J. Cleaner Prod. 361 (2022) 132269.

[93] R. Li, P. Rao, J. Luo, W. Huang, C. Jia, J. Li, P. Deng, Y. Shen, X. Tian, General method for synthesizing effective and durable electrocatalysts derived from cellulose for microbial fuel cells, ACS Appl. Mater. Interfaces 14 (2022) 13369–13378.

[94] F. Cheng, J. Liang, J. Zhao, Z. Tao, J. Chen, Biomass waste-derived microporous carbons with controlled texture and enhanced hydrogen uptake, Chem. Mater. 20 (2008) 1889–1895.

[95] B. Lu, J. Zhou, Y. Song, H. Wang, W. Xiao, D. Wang, Molten-salt treatment of waste biomass for preparation of carbon with enhanced capacitive properties and electrocatalytic activity towards oxygen reduction, Faraday Discuss. 190 (2016) 147–159.

[96] J. Wei, Y. Liang, Y. Hu, B. Kong, G.P. Simon, J. Zhang, S.P. Jiang, H. Wang, A versatile iron–tannin-framework ink coating strategy to fabricate biomass-derived iron carbide/Fe-N-carbon catalysts for efficient oxygen reduction, Angew. Chem. Int. Ed. 55 (2016) 1355–1359.

[97] I. Alonso-Lemus, F. Rodriguez-Varela, M. Figueroa-Torres, M. Sanchez-Castro, A. Hernandez-Ramírez, D. Lardizabal-Gutierrez, P. Quintana-Owen, Novel self-nitrogen-doped porous carbon from waste leather as highly active metal-free electrocatalyst for the ORR, Int. J. Hydrogen Energy 41 (2016) 23409–23416.

[98] F. Li, M. Xue, G.P. Knowles, L. Chen, D.R. MacFarlane, J. Zhang, Porous nitrogen–doped carbon derived from biomass for electrocatalytic reduction of CO_2 to CO, Electrochim. Acta 245 (2017) 561–568.

[99] W. Sangchoom, R. Mokaya, Valorization of lignin waste: carbons from hydrothermal carbonization of renewable lignin as superior sorbents for CO_2 and hydrogen storage, ACS Sustainable Chem. Eng. 3 (2015) 1658–1667.

[100] Y. Xiao, H. Dong, C. Long, M. Zheng, B. Lei, H. Zhang, Y. Liu, Melaleuca bark based porous carbons for hydrogen storage, Int. J. Hydrogen Energy 39 (2014) 11661–11667.

[101] X. Zhao, X. Yong, Q. Ji, Z. Yang, Y. Song, T. Tian, T. Chen, Z. Yang, L. Xu, X. Shen, Synthesis of all-biomass-derived carbon nanofibers for dual-functional filtration membranes and oxygen evolution reaction electrocatalysts, J. Alloys Compd. 918 (2022) 165600.

[102] B. Jurca, C. Bucur, A. Primo, P. Concepción, V.I. Parvulescu, H. García, N-doped defective graphene from biomass as catalyst for CO_2 hydrogenation to methane, Chem. Cat. Chem. 11 (2019) 985–990.

[103] J. He, A. Anouar, A. Primo, H. García, Quality improvement of few-layers defective graphene from biomass and application for H_2 generation, Nanomaterials 9 (2019) 895.

[104] Y. Zhang, L. Lu, S. Zhang, Z. Lv, D. Yang, J. Liu, Y. Chen, X. Tian, H. Jin, W. Song, Biomass chitosan derived cobalt/nitrogen doped carbon nanotubes for the electrocatalytic oxygen reduction reaction, J. Mater. Chem. A 6 (2018) 5740–5745.

[105] G. Wang, Y. Deng, J. Yu, L. Zheng, L. Du, H. Song, S. Liao, From chlorella to nest-like framework constructed with doped carbon nanotubes: a biomass-derived, high-performance, bifunctional oxygen reduction/evolution catalyst, ACS Appl. Mater. Interfaces 9 (2017) 32168–32178.

[106] G. Lin, R. Ma, Y. Zhou, Q. Liu, X. Dong, J. Wang, KOH activation of biomass-derived nitrogen-doped carbons for supercapacitor and electrocatalytic oxygen reduction, Electrochim. Acta 261 (2018) 49–57.

[107] Z. Zhu, Y. Liu, Z. Ju, J. Luo, O. Sheng, J. Nai, T. Liu, Y. Zhou, Y. Wang, X. Tao, Synthesis of diverse green carbon nanomaterials through fully utilizing biomass carbon source assisted by KOH, ACS Appl. Mater. Interfaces 11 (2019) 24205–24211.

CHAPTER

17

Electrochemical conversion of furfural

Eduardo Magalhães Rodrigues, Sancler da Costa Vasconcelos, Mayane Barbosa dos Santos, Carolina Guimarães de Souza Lima, and Thiago de Melo Lima

Institute of Chemistry, Fluminense Federal University, Niterói, Brazil

17.1 Introduction

Human beings are one of the most significant driving forces on the planet, particularly in terms of their impact on the environment and the detrimental effects of human activities. Industrial development, rising global population, and the increasing number of people living in cities have led to a notable increase in fossil fuel consumption and, consequently, greenhouse gas (GHG) emissions, particularly carbon dioxide (CO_2). Due to high levels of anthropogenic activities, humanity has been experiencing several effects of climate change, such as rising global temperatures that cause unpredictable impacts on natural events. Fossils are expected to worsen if CO_2 emissions continue to increase and if policymakers and climate change panels do not adopt more stringent measures [1,2]. Consequently, mitigation methods for CO_2 (such as carbon capture and utilization), the development of city resilience, and the gradual replacement of fossil fuels with more sustainable sources are consistently discussed in the literature to promote a more sustainable future.

Therefore, biomass is a promising feedstock in the effort to reduce dependence on fossil fuels and nonrenewable energy sources. It is estimated that plants produce between 130 and 170 billion metric tons of organic material annually, and biomass in forests, oceans, and algae may

Electrochemical Biomass Conversion: Sustainable Processes and Applications for Bioenergy, Biofuels, and Bioproducts
DOI: https://doi.org/10.1016/B978-0-443-26757-4.00016-7

absorb around 2.6 to 2.5 billion metric tons of CO_2 each year [3]. When considering biomass as a feedstock, we need to account for the portion that does not compete with the food sector or other vital uses. In this way, residual biomass becomes suitable for a more bio-based society, consisting of organic waste from agriculture, industry, forestry, and urban activities that can be converted into biofuels, biogas, and various other bio-based chemicals. These are essential for a circular bioeconomy and help reduce GHG emissions. Notably, utilizing biomass for these bio-based purposes can be considered a carbon-neutral process: while the burning of biofuel releases CO_2 into the atmosphere, the biomass in nature can reabsorb this gas. Conversely, burning fossil fuels is not considered a carbon-neutral process, as it only releases GHGs into the atmosphere [4–6].

Considering biomass sources and residual biomass, the most abundant to examine is lignocellulosic biomass. As noted in previous chapters, this biomass consists of three main fractions: cellulose, hemicellulose, and lignin, along with minor components such as extractives and ash. It is estimated that this biomass is produced at a rate of 200 to 250 billion tons annually, with residual lignocellulosic biomass reaching at least 75 billion tons each year. Cellulose (35%–50% on a dry basis) is composed of cellobiose units, which form a linear polysaccharide made of β-1,4-linked glucose units [7–11].

Furfural is an essential bio-based platform molecule derived primarily from the hemicellulosic fraction of lignocellulosic biomass (Fig. 17.1), especially from agricultural waste like corn cobs, sugarcane bagasse, oat hulls, and wheat straw. Hemicellulose, which mainly contains pentoses such as xylose and arabinose, is broken down under acidic conditions to release these sugars, which are then dehydrated to produce furfural. The industrial production of furfural typically occurs through acid-catalyzed hydrolysis and dehydration in high-temperature batch or continuous reactors. In this process, biomass is treated with diluted mineral acids, such as sulfuric acid, at temperatures ranging from 150°C to 200°C. Pentoses undergo dehydration to produce furfural, which is continuously removed from the reactor via steam distillation to prevent degradation and side reactions. The overall furfural from pentosan-rich biomass typically ranges from 5% to 8% by weight, depending on the feedstock and process efficiency, with a maximum theoretical yield of up to 80% based on the pentose content [4,7–11].

Globally, furfural production exceeds 300,000 metric tons annually, with China accounting for over 80% of the world's supply due to its abundant agricultural residues and cost-effective processing methods. Other producers include the Dominican Republic, South Africa, and India, although their contributions are relatively small. China's dominance is also driven by the widespread use of corn cobs and bagasse as feedstocks, which are

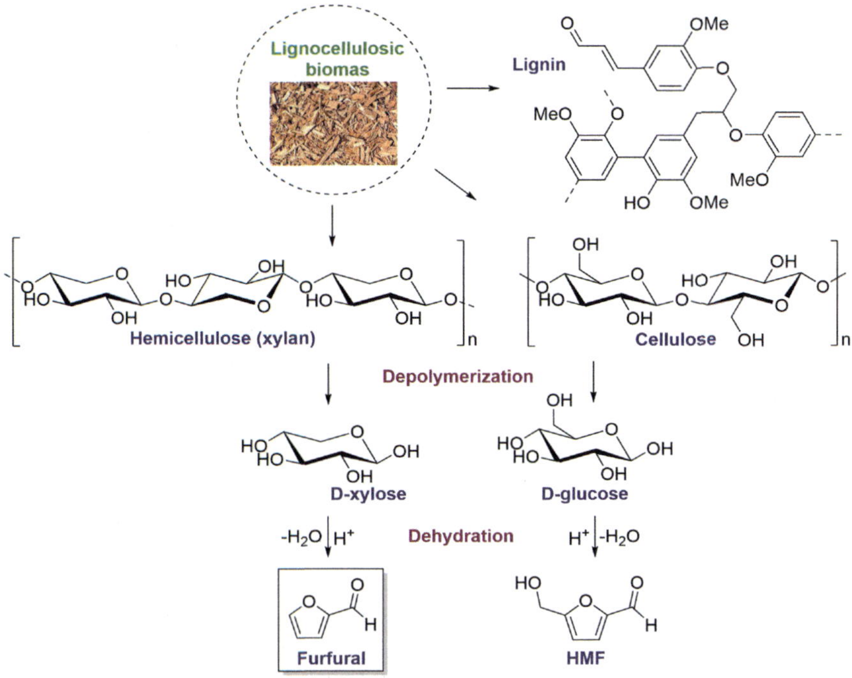

FIGURE 17.1 Overview of the reactions involved in lignocellulosic biomass processing.

processed in traditional batch digesters that, although less efficient, are cost-effective on a large scale [10,11].

Despite its long history of production, dating back to the 1920s, furfural remains a key molecule for the bioeconomy, as it acts as a precursor for several value-added products, including furfuryl alcohol (FA), tetrahydrofurfuryl alcohol, furoic acid (FA), and 2-methylfuran (MF), as well as new polymers and fuels. The move toward greener and more integrated biorefineries has renewed interest in enhancing furfural production through continuous-flow systems, solid acid catalysts, and the utilization of all biomass fractions [10–14].

Conventional catalytic methods for upgrading furfural, such as hydrogenation, hydrodeoxygenation, and oxidation, often require high temperatures, high hydrogen pressures, and noble metal catalysts. Although effective, these processes face challenges including: (1) high energy use and CO_2 emissions; (2) catalyst deactivation caused by fouling or coke buildup; (3) difficulty in controlling product selectivity under harsh conditions; and (4) safety and logistical issues with high-pressure hydrogen systems. These disadvantages motivate the search for alternative, milder, and more sustainable conversion methods [15,16].

While traditional methods for converting furfural primarily rely on thermochemical and heterogeneous catalytic processes, these approaches often necessitate high temperatures, elevated hydrogen pressures, and the use of precious metal catalysts. These conditions not only increase operational costs but also raise concerns about safety, energy use, and the environment. As the world shifts toward electrification and decentralized chemical production, electrochemical conversion becomes an increasingly attractive strategy for the selective upgrading of furfural and other biomass-derived compounds [17–21].

Electrochemical processes operate under mild reaction conditions, typically at ambient temperature and pressure, and can be combined with renewable energy sources, such as solar and wind, thereby reducing the carbon footprint of chemical production. Additionally, electrochemical systems offer unique control over selectivity through the application of potential and current, enabling the adjustment of reaction pathways without altering the catalyst or solvent. These benefits make electrochemical conversion a promising method not only for laboratory-scale synthesis but also for future use in distributed and modular biorefineries.

Within this framework, the electrochemical conversion of furfural has garnered significant attention in recent years through both reductive and oxidative pathways. Furfural electroreduction mainly produces value-added products such as FA, MF, and tetrahydrofurfuryl alcohol, all of which are important in fuel additives and monomer manufacturing. Conversely, electrooxidation can produce FA and other oxygenated compounds used in food preservation and pharmaceuticals. Developing efficient, selective, and stable electrocatalysts, including metal-based electrodes, carbon materials, and innovative hybrid nanostructures, has become a key focus in this area.

Despite increasing research, several challenges still need to be addressed. These include enhancing faradaic efficiency and product selectivity, reducing side reactions (*e.g.*, polymerization or hydrogen evolution), improving catalyst durability, and designing scalable reactors. Additionally, understanding reaction mechanisms at the molecular level, supported by in-situ and operando techniques, as well as density functional theory (DFT) modeling, continues to advance, providing valuable insights into reaction pathways and intermediate species.

This chapter aims to provide a comprehensive and up-to-date overview of the electrochemical conversion of furfural, addressing fundamental principles, reaction mechanisms, electrocatalyst development, and system design. Selected case studies from recent literature will be presented to highlight advances and persisting challenges. Finally, the broader implications of integrating furfural electroconversion into sustainable chemical manufacturing will be discussed, with emphasis on its role in fostering a circular economy and supporting the green energy transition.

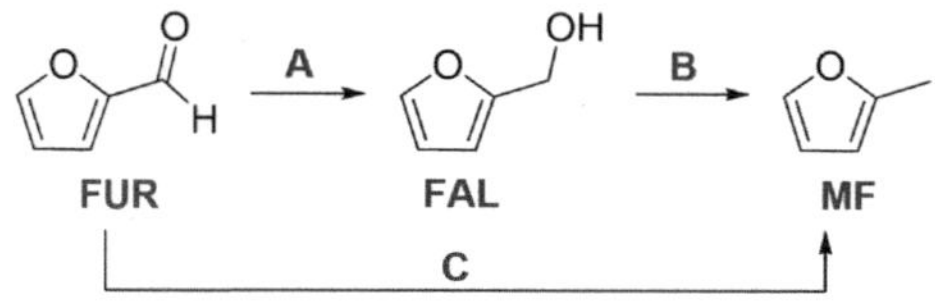

FIGURE 17.2 Typical pathways involved in the electrochemical conversion of FUR to MF.

17.2 Conversion of furfural using electrochemical approaches

17.2.1 Hydrogenolysis to 2-methylfuran

Electrochemical reductions are widely considered a promising alternative to high-pressure, high-temperature catalytic hydrogenation. Among the valuable products obtained from furfural reduction (FUR), MF is an interesting molecule, as it can be applied as an additive for gasoline. Replacing molecular hydrogen with an electric current enables operation under atmospheric pressure and room temperature in aqueous media, thus providing milder reaction conditions.

Notably, the conversion of FUR to MF may take place as a two-step reaction, starting with FUR hydrogenation to FA followed by FA hydrogenolysis to MF (Fig. 17.2, paths A and B, respectively) or directly from FUR to MF (Fig. 17.2, path C); product selectivity can be tailored through catalyst modification and energy input adjustments.

Since electrochemical conversions are heterogeneous processes, the nature of the electrode plays a major role in determining product composition. Accordingly, a pioneering study evaluated various materials (Cu, Ni, Pt, C, Fe, Pb, and Al) as electrodes for MF formation [22]. While Ni and Pt electrodes led to the formation of FA, for C, Fe, Pb, and Al, the pinacol derivative hydrofuroin was the main product. On the other hand, for the Cu catalyst, the major reduction product was MF, with ~80% selectivity. In all cases, the formation of the oxidation product 2-FA was also observed because of the diffusion of FUR, and reaction intermediates to the counter electrode; all reactions were carried out galvanostatically with a current of 200 mA.

Furthermore, it was found that pH and electrolyte solution composition also influence the product distribution. While high selectivity to MF was obtained using 500 mM H_2SO_4 and a water-acetonitrile mixture, by exchanging the medium to a phosphate buffer (pH 7), FA was obtained as the main product [22].

The electrolytic hydrogen evolution reaction (HER) is a competing reaction that occurs in the same potential range as FUR reduction. Using cosolvents such as acetonitrile or ethanol is an alternative to avoid such reactions, and increasing precursor concentration makes hydrogenation the dominating reaction [22]. The drawbacks of HER and harsh pH conditions were also tackled using a phosphorus-doped carbon-supported Cu single-atom catalyst (Cu_1/PC) [23]. Cyclic voltammetry (CV) studies showed that the Cu_1/PC catalyst presented a significantly negative potential [−0.77 V *vs.* reversible hydrogen electrode (RHE)], leading to the suppression of the HER process. After adding FUR to the medium, a decrease in current density was observed, indicating that FUR inhibits the interaction between catalytic sites and hydrogen. Two cathodic peaks were observed, and further tests confirmed that under these conditions, MF is obtained in a two-step process, where FUR is reduced to FA, which is further converted into MF. This may be related to the lack of hydrogen availability in a near-neutral environment; studies performed with a phosphate buffer solution (pH 8) indicated no MF formation.

The tailored catalyst was also tested at a more negative potential, leading to the highest Faraday efficiency (above 60%) at −0.90 V (*vs.* RHE). At more negative potentials than −0.96 V (*vs.* RHE), HER was the dominant reaction, decreasing the Faraday efficiency for MF to approximately 38%. As previously found by other groups, the change in FUR concentration proved to influence the competition between HER and hydrogenolysis [23].

For a more comprehensive understanding of the reaction pathways, other groups have compared thermal and electrochemical methods for the conversion of FUR to MF [24]. Once again, it was observed that FUR conversion and MF selectivity are favored in more negative potentials. Importantly, when the reactions were carried out with a Cu electrode, the evaluation of product distribution in short reaction times indicated that MF is indeed immediately formed at the beginning of the reaction (Fig. 17.2, path C). Under thermal conditions, FA was the major product, and no MF production was identified, indicating that thermal and electrochemical methods present different routes to product formation [24].

17.2.2 Hydrogenation to furfuryl alcohol

As discussed above, the electrochemical conversion of FUR to MF represents one of the seminal transformations involving this substrate. Nevertheless, the electrochemical hydrogenation of furfural can also yield other valuable products, such as FA, 2-hydroxytetrahydrofuran, and 2-methyltetrahydrofuran [25]. Among these, FA is particularly relevant due to the relatively facile nature of its formation and its widespread use as a platform molecule in the production of lubricants, thermostable resins, polymers, and dispersing agents [25,26].

The use of a microchannel flow reactor has enabled the conduction of this reaction in both acidic and basic media, leading to FA with varying yields depending on the reaction conditions (solvent, electrolyte, and electrode materials) [27]. In acidic medium, the best performance (55% yield) was achieved with 0.1 M furfural, 0.1 M HCl, EtOH as solvent, a Cu/stainless steel electrode pair, and Ammoeng 110, an ionic liquid, as supporting electrolyte. However, significantly higher yields (up to 90%) were obtained in basic medium using 0.1 M furfural, 0.05 M potassium ethoxide, EtOH, and Cu/graphite electrodes. Importantly, reaction time optimization revealed an optimal duration of 10 min for both media, markedly shorter than the 6 h required in batch reactors. Regarding operation mode, constant potential provided more stable and tunable selectivity, whereas constant current ensured more controlled conversion. Overall, the Faradaic efficiency was considerably higher in basic medium (90%–92%) compared to acidic medium (18%–19%), confirming the superiority of basic conditions for this transformation.

Given that basic medium has proven to be more suitable for the electrochemical hydrogenation of furfural, several studies have reported the production of FA under such conditions. In one of the early influential studies in the field, several catalysts containing different copper loadings (5%, 15%, 25%) supported on ultra-thin *N*-doped carbon nanosheets were synthesized, deposited in the working electrode, and evaluated in the presence of 1.0 M KOH and 30 mM furfural [28]. The presence of copper led to a shift to a more positive overpotential, reducing the amount of energy required to obtain a current density of 10 mA cm^{-2}, showing the electrocatalytic activity of copper. The catalyst with 15% copper loading exhibited the best performance, achieving 98% FA yield with 99% conversion. In contrast, the 5% Cu catalyst showed lower activity (72% yield and 84% conversion), which can be attributed to the reduced number of active sites available at lower loading. Conversely, the 25% Cu catalyst afforded 82% yield and 87% conversion; the decline compared to the 15% sample was ascribed to pore blockage by copper nanoparticles, which hindered mass transfer and partially blocked active sites. This hypothesis was supported by N_2 adsorption-desorption isotherm analysis.

In another important contribution to the field, Ni and Cu nanoparticles were deposited on nickel foam, generating three different catalysts (NF, NPNi/NF, and Cu–NPNi/NF). Chronoamperometry experiments were carried out in an alkaline medium at different applied potentials (from −1.15 V to −1.45 V) to evaluate furfural conversion. A decrease in potential led to an increase in both formation rate and selectivity toward FA, following the order Cu-NPNi/NF > NPNi/NF > NF, which once again emphasizes the beneficial role of copper in promoting this transformation. Nevertheless, all catalysts exhibited a marked selectivity toward hydrofuroin (Fig. 17.3), underscoring the intrinsic challenge of achieving high selectivity to FA in the electrochemical hydrogenation of furfural [29].

In another significant contribution, copper catalysts with heterojunction nanostructures derived from metal-organic frameworks and supported on a copper foam (CF) were developed, showing enhanced activity toward

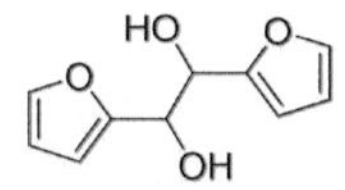

FIGURE 17.3 Chemical structure of hydrofuroin, the pinacol derivative formed as a byproduct during furfural hydrogenation.

furfural electrochemical hydrogenation [30]. Among the five tested materials, the Cu_2O–Cu_2S@CF catalyst stood out, requiring only −93 mV vs RHE to reach a current density of −10 mA cm^{-2}, whereas the other catalysts required more negative potentials (−124 to −185 mV). This superior performance was attributed to the heterojunction between Cu_2O and Cu_2S, which facilitated charge transfer and increased the number of accessible active sites, in addition to the synergistic effect between Cu^0 and Cu^+ species. The electrochemically active surface area (ECSA) analysis confirmed this trend, with Cu_2O-Cu_2S @CF reaching 28 mF cm^{-2}, significantly higher than S-Cu-MOF@CF (17 mF cm^{-2}) or Cu_2O@CF (10 mF cm^{-2}), consistent with its higher catalytic activity.

Expanding on these studies, zinc-manganese bimetallic oxides with surface-derived Zn^{2+} vacancies were investigated, both in the unsupported form (dx-$ZnMn_2O_4$, with etching times of 0.5, 1.0, and 1.5 h) and supported over carbon nanofibers (dx-$ZnMn_2O_4$C) [31]. Linear sweep voltammetry experiments were carried out in alkaline medium (0.5 M NaOH and 20 mM furfural), applying potentials from −0.2 to −0.6 V vs RHE, while keeping the total charge constant (10 C). For all catalysts, the best performance was obtained at −0.5 V vs RHE. Among the unsupported materials, the sample etched for 1 h (d1.0-$ZnMn_2O_4$) delivered the highest performance, reaching a FA yield of 21884 μmol g^{-1}, with a Faradaic efficiency of 42.2% and selectivity of 86.9%. The other materials followed the trend d1.0-$ZnMn_2O_4$ > d1.5-$ZnMn_2O_4$ > d0.5-$ZnMn_2O_4$ > pristine $ZnMn_2O_4$, showing that the introduction of Zn^{2+} vacancies was indeed beneficial, but that there was an optimal etching time to maximize their effect. When the same catalysts were supported by carbon nanofibers, the overall performance increased significantly. Again, the sample etched for 1 h (d1.0-Zn Mn_2O_4-C) was the most efficient, achieving a FA yield of 49461 μmol g^{-1}, with Faradaic efficiency as high as 95.5% and nearly complete selectivity (99.3%). The enhancement was attributed to the carbon nanofiber support, which greatly facilitated hydrogen adsorption and prevented this step from becoming a limiting factor in the reaction. The order of activity among the supported catalysts mirrored that of the unsupported series (d1.0 > d1.5 > d0.5 > pristine), confirming the robustness of the trend. In this sense, it was shown that the combination of defect engineering (vacancies) with conductive supports (carbon nanofibers) can substantially improve activity, selectivity,

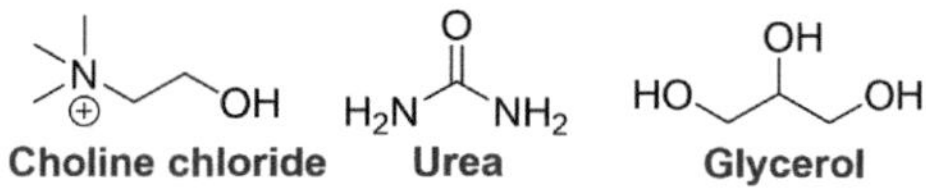

FIGURE 17.4 Chemical structures of DES components used as electrolytes for furfural electrochemical hydrogenation.

and efficiency, thereby providing valuable guidelines for catalyst design in this transformation.

Building on the indication that alkaline electrolytes provide favorable conditions for the electrochemical hydrogenation of furfural, other studies have explored neutral media as a more environmentally benign alternative. In this context, cupric oxide nanowires (CuO NWs) and CF were employed as catalysts to evaluate their performance toward FA production from FUR [32]. LSV experiments were carried out in 1.0 M PBS containing 50 mM FUR, scanning potentials from −0.5 V to 0.0 V vs. RHE. For the CuO NWs, the onset of HER in the absence of FUR occurred near −0.25 V, while in its presence, the current increase shifted positively to −0.19 V, indicating that furfural electroreduction is thermodynamically more favorable than HER. In comparison, the CF showed HER onset at −0.42 V without furfural and at −0.34 V with FUR, evidencing its lower catalytic activity. At −0.2 V, CuO NWs reached a current density of −9.01 $mAcm^{-2}$ for FUR hydrogenation, an order of magnitude higher than CF under the same conditions (−0.95 $mAcm^{-2}$). The catalytic performance was further assessed in terms of conversion and selectivity; at −0.2 V, the CuO NWs delivered 91.5% conversion with 100% selectivity toward FA, while CF reached only 28.7% conversion with 68.8% selectivity. ECSA analysis corroborated these results, with CuO NWs exhibiting 23.55 mF cm^{-2} compared to just 7.51 mF cm^{-2} for CF. Taken together, these results demonstrate that CuO nanowires outperform the CF in both activity and selectivity, confirming the neutral phosphate buffer solution as a viable and greener medium for furfural electrochemical hydrogenation.

Beyond aqueous electrolytes, alternative media have also been investigated for the electrochemical hydrogenation of furfural with the aim of minimizing competing reactions and improving product selectivity. Among these, deep eutectic solvents (DES) have attracted increasing attention. DES are typically composed of a quaternary ammonium salt and a hydrogen-bond donor, and they combine several advantages, such as low vapor pressure, wide electrochemical windows, and tunable physicochemical properties. A key benefit of DES is its intrinsically low water content, which suppresses the competing HER that often decreases the Faradaic efficiency of the FUR process. In one representative study, two eutectic mixtures were examined (Fig. 17.4): choline chloride/urea (ChCl-U, 1:2 molar ratio) and choline chloride/glycerol (ChCl-Gly, 1:2 molar ratio) [33].

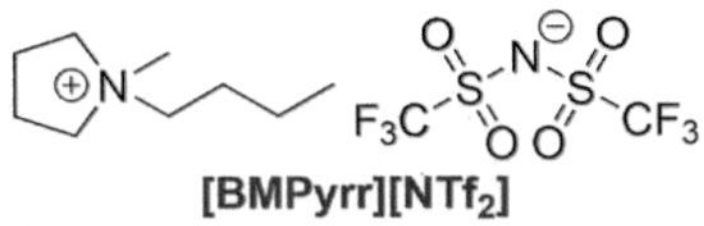

FIGURE 17.5 Chemical structure of an ionic liquid used as an electrolyte for furfural electrochemical hydrogenation.

Prior to electrochemical measurements, the stability of both DES in the presence of 550 mM FUR was tested at 25°C and 50°C for 60 h using ^{1}H NMR spectroscopy; for ChCl-U, FUR stability was poor, with only ~40% of the aldehyde remaining after 60 h at 25°C and complete disappearance at 50°C due to a condensation side reaction with urea. By contrast, ChCl-Gly displayed significantly higher stability; at 25°C, almost no degradation was observed, whereas at 50°C, FUR concentration dropped to ~24% after 60 h. Based on these results, further electrochemical studies were carried out in ChCl-Gly at 25°C to avoid undesired reactions.

CV (−1.8 V to −0.6 V vs. reference) was performed using bulk gold and copper electrodes in solutions containing 200 and 500 mM furfural; for gold, a cathodic peak attributed to FUR appeared at −1.48 V (200 mM) and −1.60 V (500 mM), with higher current density at increased furfural concentration. For copper, no distinct reduction peak was detected, but the onset potential shifted positively with increasing concentration, suggesting catalytic activity toward furfural hydrogenation. To confirm product formation, in situ Fourier-transform infrared spectroscopy was employed, and both electrodes led to the conversion of furfural to FA. These findings point out that DES, particularly ChCl-Gly under mild conditions, can serve as effective and greener alternatives to aqueous electrolytes for this transformation.

On a similar path, room temperature ionic liquids (RTILs) have also been employed as alternative electrolytes for the furfural electrochemical hydrogenation reaction. RTILs are salts composed of an organic or inorganic anion and an organic cation with melting points below 25°C. Their intrinsic ionic conductivity eliminates the need for an additional supporting electrolyte, thereby favoring the selective conversion of furfural. Electrolysis experiments conducted with 500 mM furfural in 1-butyl-1-methylpyrrolidinium bis(trifluoromethanesulfonyl)imide ([BMPyrr] [NTf_2], Fig. 17.5) using different working electrodes (gold, platinum, carbon, and copper) produced the same set of products: FA and hydrofuroin, with relative product distribution being nearly unaffected by the electrode material [34].

These results highlight that RTILs represent a promising alternative to aqueous electrolytes, enabling efficient furfural electrochemical hydrogenation with high selectivity toward FA while minimizing side reactions.

17.2.3 Hydrogenation and hydrogenolysis

Beyond the selective pathways leading exclusively to either 2-MF through hydrogenolysis or to FA via hydrogenation, several studies have demonstrated that both products can be formed within the same reaction system, with their distribution strongly dependent on the reaction conditions. Parameters such as the nature of the catalyst, reaction medium, applied potential, hydrogen source, and temperature play a crucial role in steering the reaction selectivity. Considering that, understanding how these factors govern the balance between C–O bond cleavage and C=O bond reduction is essential to rationalize the dual production of FA and 2-MF, as well as to design strategies for achieving the desired product selectively. In this section, we will discuss representative examples in which the interplay of reaction conditions determines whether furfural is converted preferentially into FA, 2-MF, or a mixture of both.

Computational approaches can provide valuable insights into the reaction mechanism and product distribution of furfural conversion. DFT studies have investigated the pathways leading to FA and 2-MF using silver, lead, and nickel as representative catalysts [35]. The reaction network proceeds in two main steps: hydrogenation of furfural to FA, followed by C–O bond cleavage yielding 2-MF. Free energy diagrams constructed for these metals revealed that, over silver and lead surfaces, both hydrogenation and subsequent hydrogenated intermediate formation are thermodynamically and kinetically favorable. The Brønsted-Evans-Polanyi (BEP) relationship was applied to estimate the energy barrier associated with C–O bond cleavage. Although the formation of 2-MF is thermodynamically more favorable than FA, the high kinetic barrier for C–O bond scission may hinder the overall conversion, therefore impacting product selectivity.

A detailed evaluation of reaction conditions demonstrated that both current density and electrolyte acidity play decisive roles in the electrochemical conversion of furfural, directly affecting product distribution between FA and MF [36]. Experiments carried out in mixed aqueous-acetonitrile solutions (20:80 vol%) with different acidic electrolytes revealed clear trends: the use of 0.2 M NH_4Cl (pH 3.4) led to FA as the major product, with yields of 15% and 23% at −5 and −10 $mAcm^{-2}$, respectively, while with 0.1 M H_2SO_4 (pH 1.1), FA and MF were formed in comparable proportions (11%–14%), and at 0.5 M H_2SO_4 (pH 0), MF became predominant, with yields up to 28% at −10 mA cm^{-2} against 18% at −5 $mAcm^{-2}$. These results indicate that lower acidity favors FA formation, whereas stronger acidic conditions promote MF production, with the yields being strongly dependent on current density. The low values obtained for both FA and MF were associated with competing side reactions, as evidenced by a Faradaic efficiency of 57% for 0.5 M H_2SO_4 at −10 $mAcm^{-2}$, suggesting that hydrogen evolution is significant under such conditions. To assess whether the electrolyte anion affects selectivity, H_2SO_4 and HCl were compared at equivalent concentrations; results showed similar product distributions, confirming that pH, rather than the anion identity, is the determining factor. Since MF formation requires the transfer of four electrons and four protons, low pH enhances the driving force for hydrogen adsorption on the electrode surface, thus promoting MF selectivity. Complementary homogeneous experiments indicated that in acidic media, product concentrations decrease over time, supporting the conclusion that side reactions also occur in solution. Accordingly, fast hydrogenolysis under acidic conditions, coupled with efficient MF separation, has been proposed as a strategy to improve yields while mitigating the impact of side reactions.

Additional investigations were carried out to clarify the influence of reaction parameters on product distribution and the occurrence of side reactions, with the goal of maximizing electrochemical efficiency and avoiding catalyst deactivation [37]. Regarding the applied potential, experiments performed between −0.5 V and −0.95 V (vs. RHE) showed that increasing the potential from −0.5 V to −0.65 V not only enhanced FUR conversion but also more than doubled the combined production rate of FA and MF, while hydrogen evolution remained negligible. A similar trend was observed at −0.8 V, although at this potential, hydrogen evolution became more significant. At −0.95 V, the hydrogen evolution rate surpassed that of furanic product formation, indicating that at higher overpotentials, hydrogen evolution becomes dominant. These results demonstrate that increasing potential accelerates reaction rates but also shifts selectivity, with FA and MF being favored at moderate potential (−0.5 to −0.8 V) and H_2 at higher potentials (−0.95 V). This change was accompanied by a decrease in Faradaic efficiency, from nearly 100% at −0.5 V to about 82% at −0.95 V, suggesting the formation of side products.

The effect of furfural concentration was also examined in the range of 20 to 200 mM. Higher concentrations led to increased yields of furanic products relative to hydrogen evolution, but at the same time caused a reduction in current density, likely due to partial blockage of the electrode surface by adsorbed furanic compounds, which hinders hydrogen adsorption and slows down reaction kinetics. Moreover, higher furfural concentrations were associated with the formation of polymerization and co-polymerization products, involving both furfural and its hydrogenated derivatives. To minimize these undesired pathways, strategies such as reducing the residence time of reactive intermediates in acidic media have been suggested, e.g., by applying nitrogen flow to accelerate evaporation and removal of volatile compounds.

Kinetic aspects of furfural electrochemical hydrogenation have also been investigated by other groups, with particular focus on the competing pathways leading to furfuryl alcohol and 2-MF, as well as the influence of mass transfer, applied potential, and temperature [38]. Mass transport limitations were evident, as magnetic stirring increased the current by approximately 50%. In terms of applied potential, a Butler–Volmer type behavior was observed between −0.517 V and −0.572 V (vs. RHE), with exponential growth in the partial current densities for both FA and MF, followed by a plateau at higher potentials, indicative of a mass-transfer limited regime. Kinetic analysis suggested that the conversion of furfural proceeds through either a one-electron rate-determining step or a two-electron rate-limiting process. Temperature effects were also significant: while increasing the temperature (15°C–45°C) at −0.56 V enhanced FUR conversion, it reduced the yield of MF, revealing the occurrence of parallel homogeneous reactions involving furfuryl alcohol. These reactions were confirmed to be temperature-dependent, since lower-temperature experiments at similar conversions resulted in higher MF yields. Additionally, higher temperatures negatively affected both product selectivity and faradaic efficiency due to the intensification of hydrogen evolution as a side reaction.

In this case, mechanistic studies revealed distinct rate-limiting steps for each product. FUR concentration experiments showed a positive-order dependence at low concentrations, transitioning to a saturation regime consistent with a noncompetitive Langmuir–Hinshelwood mechanism. FA formation was associated with a surface reaction of adsorbed furfural as the limiting step, while MF formation was more strongly governed by the energetically demanding C–O bond cleavage, specifically involving the C_4H_3O–CHOH intermediate.

Complementary mechanistic investigations further disentangled the contributions of electrocatalytic hydrogenation (ECH) and electroreduction pathways [39]. Surface modification of copper electrodes with organothiol self-assembled monolayers (SAMs) revealed that suppression of FAL, MF, and H_2 production occurs when inner-sphere processes are inhibited, whereas hydrofuroin formation remains unaffected, suggesting that it proceeds via an outer-sphere pathway. Isotopic substitution studies provided additional evidence for this mechanistic distinction; When H/D exchange was introduced in the solvent and electrolyte, the production of FA and MF markedly decreased, while hydrofuroin formation increased. These results indicate that the former products arise predominantly from ECH, which requires adsorbed hydrogen (H_ads), whereas hydrofuroin is formed through direct electroreduction. Consistent observations were obtained when replacing Cu with Pb electrodes, where hydrofuroin became the main product, accompanied by negligible H_2 and MF formation, and faradaic efficiencies below 2% for FAL.

Together, these findings confirm that ECH is the dominant pathway for FAL and MF formation on copper electrodes, relying on H_ads, while hydrofuroin originates from electroreduction via outer-sphere electron transfer. Furthermore, the constant selectivity of furfuryl alcohol and MF throughout the reaction time indicates that both are formed in parallel rather than sequentially, in agreement with mechanistic insights from previous studies.

In parallel with experimental mechanistic studies, constant-potential DFT calculations were combined with microkinetic simulations and pH-dependent electrochemical experiments to elucidate the reduction pathways of FUR [40]. Within a grand-canonical framework under acidic conditions, it was demonstrated that formation of the adsorbed intermediate *FCHOH is kinetically preferred over the thermodynamically more stable *FCH_2O, as the activation barrier for its generation through proton-coupled electron transfer (PCET) is significantly lower. The subsequent protonation of *FCHOH, leading either to *FCH_2OH (FAL pathway) or *FCH (MF pathway), was identified as both the selectivity-determining and rate-determining step. Degree of rate control (DRC) analysis further clarified potential-dependent trends. At low overpotentials (0 to −0.2 V vs. RHE), FAL production is limited by protonation of *FCHO to *FCH_2O, while at moderate overpotentials (−0.2 to −0.6 V) the mechanism shifts toward stabilization of the *FCH_2O intermediate. In contrast, MF formation consistently remains constrained by protonation of *FCHOH to *FCH across the entire potential range. At more negative potentials (below −0.6 V vs. RHE), the accumulation of strongly adsorbed *FCHOH species becomes inhibiting, suppressing both pathways. Complementary pH-dependent studies corroborated experimental observations reported by different groups: MF production is favored at low pH and moderate potentials (around −0.5 V vs. SHE), whereas at higher potentials HER increasingly competes with FUR reduction. By comparing PCET and ECH theoretical predictions with experimental data, the authors concluded that PCET is the dominant mechanistic route for FUR reduction on copper electrodes.

Complementary to the mechanistic and theoretical insights previously discussed, divergent electrochemical strategies have also been developed to expand the product scope of FUR conversion [41]. In one such approach, a paired electrochemical system was designed using an ion-exchange membrane to separate anodic and cathodic reactions, enabling the simultaneous generation of 2-furanone and FAL. Preliminary tests established that a cation-exchange membrane offered lower resistance to ion transport compared to its anion-exchange counterpart, and operational cell potentials of 2.4 or 2.9 V were defined.

To address the challenge of long reaction times and current decline with increasing conversion, a flow process was implemented, and this configuration allowed continuous feeding of concentrated FUR solutions while maintaining low concentrations at the reactor outlet, ensuring more uniform current densities along the reactor and significantly reducing reaction times. Using an aqueous NaBr electrolyte, with graphite as the anode and lead as the cathode (to suppress HER), product selectivity could be tuned by adjusting the applied potential. At 2.4 V, a reaction time of only 5 min afforded high yields of FAL (58%), whereas at 2.9 V, 2-furanone (77%) and hydrofuroin (71%) became the dominant products. Importantly, increasing substrate concentration from 0.1 M to 0.6 M did not significantly alter product distribution or efficiency, highlighting the robustness of the system.

Building upon these strategies to control product selectivity, further advances were achieved by modifying the electrode-electrolyte interface [42]. In this case, a Cu electrocatalyst was functionalized with butyl trimethylammonium bromide (BTAB), which profoundly altered both the product distribution and the underlying reaction mechanism. Comparative experiments showed that unmodified Cu electrodes predominantly yielded FAL, with 84% selectivity and 67% FE, while MF accounted for only 13% selectivity (18% FE). In contrast, the Cu-BTAB catalyst inverted this trend, delivering MF as the main product with 80% selectivity and 75% FE, while FAL selectivity decreased to 17% (7.3% FE). Although overall FUR conversion dropped from 30% to 17%, this outcome is consistent with the higher electron demand of the MF pathway (4 e^- per FUR molecule) compared to 2 e^- for FAL.

Kinetic studies revealed that FAL is formed predominantly via a hydrogen atom transfer (HAT) process, in which the aldehyde group of FUR is hydrogenated by adsorbed hydrogen species. Conversely, MF formation proceeds through a proton-coupled electron transfer (PCET) pathway involving $4H^+/4e^-$ steps, strongly dependent on hydronium ion concentration. Experimental evidence suggests that BTAB adsorbed at the Cu surface repels interfacial water and disrupts proton transfer from the bulk electrolyte to the electrode interface, thereby inhibiting the HAT process. As a result, FAL formation is suppressed, while MF production via PCET remains unaffected, leading to the observed reversal in product distribution.

17.2.4 Oxidation reactions

Among the different transformation routes of FUR, its oxidation (Fig. 17.6) has attracted considerable attention, as it allows access to a wide range of value-added products. Depending on the reaction pathway, FUR can be converted not only into FA and 2,5-furandicarboxylic acid (FDCA), but also into 2-furanone, maleic anhydride, maleic acid, fumaric acid, and succinic acid. These oxidation products are of industrial relevance, particularly in the synthesis of polymers, resins, and fine chemicals. Notably, electrochemical oxidation offers a sustainable route, providing precise control over selectivity and efficiency. The following section will discuss the main electrochemical strategies employed for FUR oxidation,

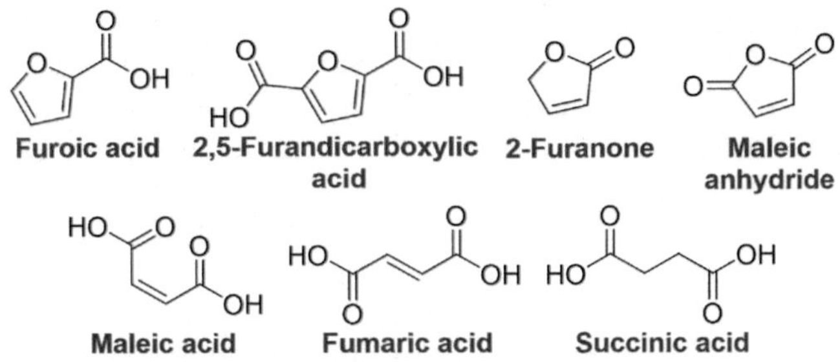

FIGURE 17.6 Furfural oxidation products.

highlighting how catalysts and operational parameters govern product distribution [43–45].

One of the most promising approaches in this field involves the use of metal electrodes such as platinum (Pt), gold (Au), or metal alloys to facilitate furfural oxidation. Some research groups have explored electrooxidation on both monocrystalline and polycrystalline Pt (111) and Au/C surfaces in acidic media (0.1 M H_2SO_4) with an applied potential ranging from 0.2 to 1.2 V (vs. RHE) [43,44,46].

An experimental study using online electrochemical mass spectrometry (OLEMS) and in situ infrared spectroscopy (ATR-SEIRAS) was carried out to (1) monitor reaction selectivity at different potentials, (2) identify oxidation products, and (3) investigate decomposition and overoxidation pathways leading to CO_2. The results revealed that FUR oxidation predominantly yields FA and 5-hydroxyfuroic acid (HFA). Product distribution is strongly potential-dependent: lower potentials favor FA and HFA formation, whereas higher potentials lead to C4 derivatives such as 5-hydroxyfuran-2(5*H*)-one (HFN) and maleic acid (MA). Selectivity can be further tuned by modifying the platinum surface with metals such as Bi or Pb, or by employing SAMs to suppress furan ring adsorption and enhance FA production [46].

A theoretical study based on DFT was performed to model the adsorption of intermediates and to predict reaction pathways and energy barriers in FUR electrooxidation on platinum surfaces [44,46]. The simulations indicated that FUR oxidation preferentially yields FA, with further oxidation being limited by the difficulty of C–C bond cleavage. Two main pathways for FA conversion were proposed: (1) the formation of succinic acid through 2(3H)-furanone as an intermediate, and (2) the formation of maleic acid (MA) and maleic anhydride via 2(5H)-furanone. Thermodynamic and kinetic calculations suggested that product selectivity can be tuned by varying the electrode potential: intermediate potentials favor succinic acid formation, whereas higher potentials promote conversion to MA and maleic anhydride. These theoretical predictions were supported by experimental observations, which showed FA as the predominant product (>80% selectivity) on platinum electrodes.

Subsequent studies further investigated the reaction pathways of FUR oxidation by replacing platinum catalysts with gold. A combination of differential electrochemical reactors, ATR-SEIRAS, and DFT calculations was employed to gain mechanistic insights. Voltammetry analysis revealed that oxidation on Au/C catalysts occurs in two stages, consistent with the oxidation potential of gold. Flow reactor experiments showed that FA is the main product, with higher catalytic activity and faradaic efficiency compared to Pt/C. The results suggested that the rate-determining step is FA desorption and that the arrangement of furoate species on the catalyst surface directly impacts selectivity. Computational analysis supported these findings, confirming that FUR oxidation proceeds favorably on Au(111) and Pt(111) surfaces. However, small amounts of FA (1–10 mM) were found to poison the catalyst surface, significantly inhibiting the reaction. A potential strategy to overcome this limitation is the use of modified alloy catalysts. Overall, Pt-based systems exhibit high selectivity toward FA but are limited by desorption, whereas Au/C catalysts enhance conversion but with lower yield due to byproduct formation [43,46].

Research on catalysts was further expanded with the evaluation of platinum atomic cluster-based catalysts supported on rehydrated layered double hydroxides (Pt/re-Mg_4Al-LDHs). These materials exhibited outstanding performance in the selective electrooxidation of FUR to FA, reaching 99% conversion, 97% yield, and a reaction rate of 676.57 mmol $gPt^{-1}h^{-1}$. This performance surpassed that of conventional platinum catalysts. The superior activity was attributed to the formation of $Pt\delta^-$ species (partially reduced states) on the catalyst surface, arising from localized electron transfer between the support and the platinum atomic clusters [45]. The role of layered double hydroxides (LDHs) was crucial, providing a confinement effect that stabilized and controlled the size of the metallic clusters, directly influencing catalytic performance through metal-support interactions. This effect significantly improved FA selectivity while minimizing side reactions. Several experimental studies confirmed that catalytic activity depends on negatively charged Pt active sites.

It was proposed that the reaction mechanism occurs in two main steps: first, the C=O bond of furfural is adsorbed onto the Ptδ– site and reacts with H_2O, in the so-called aldehyde-water shift process. Then, molecular oxygen is activated, generating reactive oxygen species (O_2^-), which act as hydrogen acceptors in the oxidative dehydrogenation step, leading to FA formation. Experimental characterizations combined with DFT calculations confirmed these conclusions, showing that catalytic furfural oxidation follows these two interconnected processes. The Ptδ– sites play a crucial role in both the selective adsorption of the C=O bond and O_2 activation, as Ptδ– species are bound to the oxygen in the crystalline lattice of the LDH support, reinforcing the electron transfer effect between the support and the metal clusters [45].

An alternative electro-conversion approach for FUR involves molecular complex-catalyzed reactions [47]. For the direct oxidation of FUR and HMF into FA and FDCA, respectively, alkaline water was employed as a formal oxidant in the presence of a ruthenium (Ru)-based acridine PNP catalyst. A key advantage of this process is the generation of high-purity hydrogen gas (H_2), with no detectable carbon monoxide (CO) contamination (detection limit of 15 ppm), making it suitable for direct application in proton-exchange membrane fuel cells. Despite challenges related to process control, optimization, and scalability, this method enabled the conversion of FUR and HMF into their respective oxidation products with yields exceeding 95%. Mechanistic analyses indicated that the reaction is initiated by hydride transfer from FUR to the catalyst ligand, forming a dearomatized complex that drives subsequent oxidation. This step is followed by the Tishchenko coupling of the substrates, which prevents undesirable degradation. The Ru-acridine PNP system demonstrated high selectivity and efficiency, enabling the oxidation of FUR and HMF without substrate decomposition and achieving high yields of FA, FDCA, and H_2.

Beyond molecular complex catalysis, alternative strategies have also been explored. Among them, indirect electrooxidation has emerged as a distinct methodology, as FUR oxidation does not occur directly on the metal electrode but proceeds via intermediates or redox mediators. A paired linear electrolytic system was designed to improve stability, efficiency, and energy savings by integrating valuable electrooxidation and electro-reduction reactions [48]. In this strategy, the conventional half-reaction of water splitting was replaced, consequently reducing energy consumption and enhancing the overall value of the process. This methodology minimizes the dependence of organic electrolysis on electrode potential, as the introduction of redox mediators eliminates the incompatibility between anode and cathode under potentiostatic conditions. In the developed system, FUR was simultaneously and selectively converted into FA at both electrodes, even under high current densities. The anode consisted of a stable three-dimensional PbO_2 structure, while at the cathode, FUR was efficiently oxidized to FA through in-situ H_2O_2 generation via oxygen reduction. A gas-diffusion Pb/RHPC cathode was employed, characterized by a high overpotential for hydrogen evolution; at the anode, FUR oxidation was mediated by 2,2,6,6-tetramethyl-1-piperidinyloxyl (TEMPO), ensuring high selectivity across a wide potential range. The electrochemical evaluation showed excellent compatibility between anodic and cathodic processes, achieving a FA selectivity of 99.3% at the anode and 33.2% at the cathode, with a total electronic efficiency of 127.1% and an energy consumption reduction of about 12.3%.

Notably, both electrodes also demonstrated high stability during prolonged continuous flow operation.

Addressing another critical limitation of furfural electrooxidation, specifically its high sensitivity to alkaline conditions, an anion exchange membrane (AEM) electrolyte was developed and evaluated in both batch and single-pass anodic modes [49]. To mitigate substrate degradation, different feeding strategies were implemented; in batch mode, furfural was introduced continuously over successive electrolysis cycles, reducing its exposure to concentrated alkaline media and thereby suppressing byproduct formation. This approach enabled the accumulation of ~280 mM FA and ~270 mM FDCA, with total yields exceeding 90% across both anolyte and catholyte. In single-pass operation, a pulsed feeding strategy minimized residence time in the electrolyte, ensuring continuous oxidation of 300 mM furfural and 165 mM HMF with product purities above 98% in the anodic effluent. Overall, these results demonstrate that optimizing electrolyte composition and flow dynamics can markedly enhance conversion efficiency and selectivity while simultaneously preventing thermal degradation of biomass-derived aldehydes.

Expanding beyond purely electrochemical strategies, photo(electro) chemical-assisted processes have recently emerged as a promising framework for FUR conversion, as they leverage light-driven pathways to lower oxidation potentials [50]. A Prussian Blue analogue (PBA)-derived NiFe/Ni catalytic electrode, obtained by depositing $Ni_2[Fe(CN)_6]$ onto nickel foam, was shown to undergo in situ reconstruction into high-valence NiOOH under alkaline electrolysis. Oxygen vacancies generated through lattice distortion promoted polaron formation and enhanced reaction kinetics, while the morphological transition from nanocubes to hollow nanostructures increased active surface area, favoring furfural adsorption. The electrode's efficiency was validated in an integrated photoelectrochemical setup, where NiFe/Ni served as the anode and a Ni-TCPP/TiO_2 photocathode facilitated photo-assisted hydrogen evolution, ultimately reducing the required cell voltage by 480 mV at 100 mA cm^{-2} compared to conventional water-splitting systems.

A phenomenon that can enhance photo(electro)chemically assisted processes is localized surface plasmon resonance (LSPR). This occurs when metallic nanoparticles interact with light, generating collective oscillations of conduction electrons. This effect leads to strong light absorption and scattering, influenced by the nanoparticle's size, shape, material, and surrounding medium.

A novel variant of copper nanoparticles containing atomically dispersed nickel was successfully synthesized through a simple one-step electrochemical codeposition process [51]. These nanoparticles were employed as an anode for the electrooxidation of furfural, simultaneously promoting hydrogen gas generation. Voltammetric measurements revealed that the incorporation of small amounts of Ni into the nanoparticles induced a synergistic electronic effect between Cu and Ni, increasing current density by 43% at 0.1 V compared to pure Cu nanoparticles. Furthermore, under visible light irradiation, an extraordinary 505% increase in current density at 0.1 V was observed, attributed to LSPR. This phenomenon, which generates both photothermal and photoelectric effects, significantly enhances the efficiency of furfural electrooxidation and hydrogen production.

Another innovative strategy for furfural electro-conversion is the application of Liquid Flow Fuel Cells (LFFCs), a membrane-free configuration in which microchannels maintain the separation of liquid reactants [52]. This approach reduces costs, avoids membrane degradation, and provides flexibility in fuel utilization. In this system, Ag_2O was employed as the anode catalyst, enabling efficient conversion of furfural into FA, while liquid redox pairs at the cathode promoted oxygen reduction, simultaneously generating H_2, H_2O_2, and electricity. The introduction of an asymmetrical alkaline-acid electrolyte further increased the driving force for electron transfer, significantly improving system efficiency. When vanadyl sulfate $[(VO_2)_2O_4]$ activated by HNO_3 was used as the electron carrier for tetravalent oxygen reduction, the process achieved a maximum power density of 261 mW cm^{-2}, with FA yields of 90% and 0.10 mol/mol of furfural converted to H_2. Alternatively, replacing the carrier with anthraquinone-2-sulfonate (AQS), which favors bivalent oxygen reduction, delivered yields of 0.88 mol/mol for FA, 0.15 mol/mol for H_2, and 0.41 mol/mol for H_2O_2, with a Pmax of 60 mW cm^{-2}.

While membrane-free systems such as LFFCs highlight the potential of innovative cell designs, complementary advances have been achieved through the rational design of metal oxide electrodes, which provide a more straightforward yet highly efficient route for furfural conversion. A nickel-cobalt oxide (Ni-Co) electrode was developed via electrodeposition with a composition of $Ni_{0.48}Co_{0.36}O_{0.16}$, exhibiting a porous, flower-like structure that significantly increased the active area available for the reaction. The coexistence of NiO, Co^{2+}, and Co^{3+} species on the surface was crucial for catalytic activity, ensuring high efficiency and selectivity in furfural oxidation to FA. Electrochemical tests revealed that the electrode achieved a faradaic efficiency of 95.4% for FA formation and a selectivity of 99.6% at a substrate concentration of 20 mM, operating at 1.0 V vs. Hg/HgO and 50°C. Beyond its high electrocatalytic activity, the proposed mechanism suggests that the combination of different metal species facilitates a controlled and efficient oxidation process [53].

Expanding beyond oxide-based catalysts, metal phosphide systems have emerged as a powerful strategy to optimize the adsorption of reaction intermediates and lower the overpotential for furfural oxidation [54,55]. A NiMoP/NF electrode, synthesized from a $NiMoO_4$/NF precursor via a hydrothermal route followed by H_2 reduction and phosphorization, demonstrated outstanding catalytic performance. This material required only 1.46 V vs. RHE to reach a current density of 500 mA cm^{-2}, markedly lower than the value needed for oxygen evolution with RuO_e/NF (1.57 V vs. RHE), underscoring its efficiency. In addition, NiMoP/NF achieved 99.2% selectivity for FA and a Faradaic efficiency (FE) of 96.7%, surpassing conventional catalysts such as RuO_2/NF, NiMo/NF, and Ni_2P/NF. DFT calculations further confirmed that the exceptional performance originates from the high density of oxygenated species and the enhanced adsorption of furfural at the catalyst's active sites.

Recent advances in combining experimental methodologies with density DFT modeling have substantially enhanced the understanding of the electrochemical reduction pathways for furfural. However, significant mechanistic ambiguities persist. DFT studies indicate that the formation of the FCHOH intermediate through proton-coupled electron transfer (PCET) is favored kinetically, whereas the subsequent protonation stages—yielding FCH_2OH (furfuryl alcohol) or *FCH (MF)—are crucial in determining both the rate and selectivity of the reactions. These theoretical predictions correlate well with experimental findings that demonstrate increased formation of MF under low pH conditions and moderate potentials. Despite these advancements, several key mechanistic questions remain unresolved. Critical inquiries include the variations in intermediate stabilities and PCET kinetics under neutral versus alkaline conditions, the impact of interfacial water structure or electrolyte composition on product selectivity, and how the dynamic restructuring of metal surfaces affects catalytic pathways. Current DFT models also lack comprehensive insights into competing HERs and the solvent-specific effects that have been documented in laboratory settings. Overall, the integration of computational and experimental strategies has yielded a more elucidated, albeit still incomplete, understanding of furfural's electroreduction processes. Addressing the remaining uncertainties is essential for advancing rational catalyst design and achieving improved selectivity in electrocatalytic biomass valorization systems.

17.3 Conclusions and perspectives

The electrocatalytic valorization of furfural and HMF through hydrogenation, hydrogenolysis, and oxidation has advanced considerably, with recent studies demonstrating efficient and selective routes to value-added products such as FA, FDCA, alcohols, and alkyl derivatives. Progress in catalyst design, particularly the development of transition-metal oxides, phosphides, and hybrid electrode systems, has significantly reduced overpotentials, enhanced selectivity, and improved faradaic efficiency.

Moreover, innovations in reactor engineering, including flow systems, AEMs, and membrane-free fuel cells, have broadened the operational landscape and enabled integration with cogeneration of hydrogen, electricity, and other energy vectors.

Despite these advances, several challenges remain. Catalyst stability under long-term operation, susceptibility of furanic substrates to degradation in alkaline environments, and the need for scalable, cost-effective synthesis of active materials continue to limit large-scale applications. Furthermore, while mechanistic insights from in situ spectroscopy and DFT modeling have elucidated key reaction pathways, a deeper understanding of structure-activity relationships is still required to enable rational catalyst design and minimize side reactions.

Looking forward, the integration of electrocatalytic furfural conversion with renewable energy inputs presents a promising direction toward sustainable biorefineries. Coupling these processes with photoelectrochemical systems or energy storage devices could further enhance efficiency and reduce dependence on fossil-based feedstocks. The development of multifunctional catalytic systems capable of simultaneously producing high-value chemicals and green energy carriers also offers exciting prospects. Ultimately, bridging fundamental mechanistic studies with reactor-scale optimization will be crucial to translate laboratory-scale achievements into viable industrial technologies for biomass valorization.

Looking ahead, the real impact of electrocatalytic furfural conversion will depend on how effectively these advances can be scaled beyond the laboratory. For this to happen, we need electrode materials that are not only active and selective but also affordable and robust over long operating times. Equally important are electrolytes and reactor designs that can handle high currents without excessive energy demand. Recent progress in flow cells, membrane-free systems, and paired electrolysis shows that continuous and more compact processes are becoming increasingly feasible. However, the economic viability of these technologies will ultimately hinge on factors such as electricity cost, catalyst durability, and the ability to recycle electrolytes. In many regions, especially those rich in agricultural residues, modular electrochemical units powered by renewable energy could offer a practical and competitive alternative to traditional thermo-catalytic routes. To move in this direction, future efforts should combine process simulation, energy analysis, and life-cycle assessment to better understand the real cost and sustainability potential of these emerging systems.

Looking forward, emerging hybrid approaches are likely to reshape the landscape of furfural electro-conversion. Photoelectrochemical (PEC) systems, which combine light harvesting with electrocatalytic upgrading, offer a promising route to lower energy consumption by using photogenerated charge carriers to drive key hydrogenation or oxidation steps. Recent studies using plasmonic metals, semiconductor–metal junctions, and integrated PEC cells illustrate how visible-light activation can reduce applied voltages and enhance selectivity. In parallel, enzyme–electrocatalyst coupling systems are gaining attention for enabling highly selective, mild transformations that complement the broader reactivity of inorganic electrocatalysts. Such bio-hybrid platforms—where redox enzymes operate in tandem with electrochemical hydrogen donors, mediators, or photocathodes—could open new pathways for upgrading furfural into functionalized derivatives that are difficult to access via traditional electrocatalysis alone. Together, these strategies highlight a future in which photon-assisted and bio-assisted electrocatalysis enable more efficient, modular, and sustainable routes for converting biomass-derived molecules into advanced chemicals and fuels.

References

[1] M. Höök, X. Tang, Depletion of fossil fuels and anthropogenic climate change—A review, Energy Policy 52 (2013) 797–809.

[2] S. Suranovic, Fossil fuel addiction and the implications for climate change policy, Glob. Environ. Chang. 23 (2013) 598–608.

[3] A. Kumar, S. Ergas, X. Yuan, A. Sahu, Q. Zhang, J. Dewulf, F.X. Malcata, H. van Langenhove, Enhanced CO_2 fixation and biofuel production via microalgae: recent developments and future directions, Trends Biotechnol. 28 (2010) 371–380.

[4] M.V. Rodionova, R.S. Poudyal, I. Tiwari, R.A. Voloshin, S.K. Zharmukhamedov, H.G. Nam, B.K. Zayadan, B.D. Bruce, H.J.M. Hou, S.I. Allakhverdiev, Biofuel production: Challenges and opportunities, Int. J. Hydrogen Energy 42 (2017) 8450–8461.

[5] T. Sharma, P.S. Chauhan, M. Patel, A. Singh, M. Kaur, G. Chauhan, B.B. Rana, N. Kumar, A. Walia, Carbon negative biofuels: a step ahead of carbon neutrality, Biofuels 16 (2025) 754–774.

[6] E.M. Porqueras, S. Rittmann, C. Herwig, Biofuels and CO_2 neutrality: An opportunity, Biofuels 3 (2012) 413–426.

[7] M.K. Awasthi, T. Sar, S.C. Gowd, K. Rajendran, V. Kumar, S. Sarsaiya, Y. Li, R. Sindhu, P. Binod, Z. Zhang, A. Pandey, M.J. Taherzadeh, A comprehensive review on thermochemical, and biochemical conversion methods of lignocellulosic biomass into valuable end product, Fuel 342 (2023) 127790.

[8] S. Baksi, D. Saha, S. Saha, U. Sarkar, D. Basu, J.C. Kuniyal, Pre-treatment of lignocellulosic biomass: review of various physico-chemical and biological methods influencing the extent of biomass depolymerization, Int. J. Environ. Sci. Technol. 20 (2023) 13895–13922.

[9] J. Cai, Y. He, X. Yu, S.W. Banks, Y. Yang, X. Zhang, Y. Yu, R. Liu, A.V. Bridgwater, Review of physicochemical properties and analytical characterization of lignocellulosic biomass, Renew. Sustain. Energy Rev. 76 (2017) 309–322.

[10] K.J. Yong, T.Y. Wu, C.B.T.L. Lee, Z.J. Lee, Q. Liu, J.M. Jahim, Q. Zhou, L. Zhang, Furfural production from biomass residues: Current technologies, challenges and future prospects, Biomass Bioenerg 161 (2022) 106458.

[11] Y. Bao, Z. Du, X. Liu, H. Liu, J. Tang, C. Qin, C. Liang, C. Huang, S. Yao, Furfural production from lignocellulosic biomass: one-step and two-step strategies and techno-economic evaluation, Green Chem. 26 (2024) 6318–6338.

[12] D. Edumujeze, M.-C. Fournier-Salaün, S. Leveneur, Production of furfural: From kinetics to process assessment, Fuel 381 (2025) 133423.

[13] C. Árvai, A.K. Horváth, K. Komka, L.T. Mika, From molecular level to process design: Co-solvent-free autocatalytic homogeneous hydrogenation of C_5-platform chemical furfural to furfuryl alcohol, Chem. Eng. J. 511 (2025) 161550.

[14] P. Gan, K. Zhang, G. Yang, J. Li, Y. Zhao, J. Chen, Catalytic Production and Upgrading of Furfural: A Platform Compound, Int. J. Mol. Sci. 25 (2024) 11992.

[15] H. Li, Y. Hu, P. Hu, L. Li, D. Wu, Z. Xue, C. Hu, L. Zhu, Sequential extraction and separation of soluble humins from fructose conversion for structural and evolutional understanding, Green Chem. 26 (2024) 5499–5511.

[16] S. Constant, C.S. Lancefield, W. Vogelzang, R.K. Pazhavelikkakath Purushothaman, A.E. Frissen, K. Houben, P. de Peinder, M. Baldus, B.M. Weckhuysen, D.S. van Es, P.C.A. Bruijnincx, Molecular structure and composition elucidation of an industrial humin and its fractions, Green Chem. 26 (2024) 7739–7751.

[17] H. Liang, K. Liu, X. Zhang, Q. Zhang, L. Chen, Y. Chen, X. Zhuang, L. Ma, Recent Advances of the Electrochemical Hydrogenation of Biofuels and Chemicals from Furfural, Energies (Basel) 18 (2025) 3075.

[18] A. Sohail, C. Wattanakit, Electrocatalytic upgrading of furan derivatives, Curr. Opin. Electrochem. 49 (2025) 101628.

[19] J. Deng, R. Li, K. Xiang, L. Zhang, K. Sun, Y.D. Kim, Z. Liu, J. Ge, Z. Peng, Superior matching between electrochemical and non-electrochemical reactions to boost furfural electro-oxidation, Chem. Eng. J. 511 (2025) 162214.

[20] P. Zhu, M. Shi, Z. Shen, X. Liao, Y. Chen, Electrocatalytic conversion of biomass-derived furan compounds: mechanisms, catalysts and perspectives, Chem. Sci. 15 (2024) 4723–4756.

[21] A. Sohail, C. Wattanakit, Electrocatalytic upgrading of furan derivatives, Curr. Opin. Electrochem. 49 (2025) 101628.

[22] P. Nilges, U. Schröder, Electrochemistry for biofuel generation: production of furans by electrocatalytic hydrogenation of furfurals, Energy Environ. Sci. 6 (2013) 2925–2931.

[23] P. Zhou, Y. Chen, P. Luan, X. Zhang, Z. Yuan, S.X. Guo, Q. Gu, B. Johannessen, M. Mollah, A.L. Chaffee, D.R. Turner, J. Zhang, Selective electrochemical hydrogenation of furfural to 2-methylfuran over a single atom Cu catalyst under mild pH conditions, Green Chem. 23 (2021) 3028–3038.

[24] R. Bababrik, D. Santhanaraj, D.E. Resasco, B. Wang, A comparative study of thermal- and electrocatalytic conversion of furfural: methylfuran as a primary and major product, J. Appl. Electrochem. 51 (2021) 19–26.

[25] S. Cheng, H. Zhong, F. Jin, A mini review of electrocatalytic upgrading of carbohydrate biomass—System, path, and optimization, Energy. Sci. Eng. 11 (2023) 2944–2965.

[26] Q. Fu, W. Xie, L. Yang, L. Yan, X. Zhao, Recent advances in electrocatalytic upgrading of biomass-derived furfural, Chem. Eng. J. 485 (2024) 150083.

[27] Y. Cao, T. Noël, Efficient Electrocatalytic Reduction of Furfural to Furfuryl Alcohol in a Microchannel Flow Reactor, Org. Process Res. Dev. 23 (2019) 403–408.

[28] W. Xu, C. Yu, J. Chen, Z. Liu, Electrochemical hydrogenation of biomass-based furfural in aqueous media by Cu catalyst supported on N-doped hierarchically porous carbon, Appl. Catal. B 305 (2022) 121062.

[29] R.J. Dixit, K. Bhattacharyya, V.K. Ramani, S. Basu, Electrocatalytic hydrogenation of furfural using non-noble-metal electrocatalysts in alkaline medium, Green Chem. 23 (2021) 4201–4212.

[30] H.-J. Shen, H.-Y. Guo, S.-Y. Wen, L. Zhang, J.-J. Feng, A.-J. Wang, Metal–Organic Framework-Derived Cu_2O–Cu_2S Heterojunction Nanostructures on Copper Foams for Furfural Electrohydrogenation, ACS Appl. Nano. Mater. 7 (2024) 8113–8120.

[31] M. Qin, S. Fan, X. Li, J. Duan, G. Chen, Electrocatalytic reduction of furfural to furfuryl alcohol using carbon nanofibers supported zinc cobalt bimetallic oxide with surface-derived zinc vacancies in alkaline medium, J. Colloid Interface Sci. 660 (2024) 800–809.

[32] Z. Xia, Y. Li, J. Wu, Y.-C. Huang, W. Zhao, Y. Lu, Y. Pan, X. Yue, Y. Wang, C.-L. Dong, S. Wang, Y. Zou, Promoting the electrochemical hydrogenation of furfural by synergistic Cu^0–Cu^+ active sites, Sci. China Chem. 65 (2022) 2588–2595.

[33] J. Vander Steen, F. Boissou, M. Luhmer, C. Buess-Herman, S. Baranton, C. Coutanceau, T. Doneux, Furfural electroreduction in choline-glycerol deep eutectic solvent, J. Electroanal. Chem. 933 (2023) 117269.

[34] J. Vander Steen, M. Luhmer, C. Buess-Herman, T. Doneux, Electrochemical behavior of furfural, a bio-based building block, in the [BMPyrr][NTf_2] ionic liquid, Electrochim Acta 498 (2024) 144635.

[35] N. Shan, M.K. Hanchett, B. Liu, Mechanistic Insights Evaluating Ag, Pb, and Ni as Electrocatalysts for Furfural Reduction from First-Principles Methods, J. Phys. Chem. C 121 (2017) 25768–25777.

[36] S. Jung, E.J. Biddinger, Electrocatalytic Hydrogenation and Hydrogenolysis of Furfural and the Impact of Homogeneous Side Reactions of Furanic Compounds in Acidic Electrolytes, ACS Sustain Chem. Eng. 4 (2016) 6500–6508.

[37] S. Jung, E.J. Biddinger, Controlling Competitive Side Reactions in the Electrochemical Upgrading of Furfural to Biofuel, Energy Technol. 6 (2018) 1370–1379.

[38] A.S. May, S.M. Watt, E.J. Biddinger, Kinetics of furfural electrochemical hydrogenation and hydrogenolysis in acidic media on copper, React. Chem. Eng. 6 (2021) 2075–2086.

[39] X.H. Chadderdon, D.J. Chadderdon, J.E. Matthiesen, Y. Qiu, J.M. Carraher, J.P. Tessonnier, W. Li, Mechanisms of Furfural Reduction on Metal Electrodes: Distinguishing Pathways for Selective Hydrogenation of Bioderived Oxygenates, J. Am. Chem. Soc. 139 (2017) 14120–14128.

[40] S. Liu, Z. Mukadam, S.B. Scott, S.C. Sarma, M.M. Titirici, K. Chan, N. Govindarajan, I.E.L. Stephens, G. Kastlunger, Unraveling the reaction mechanisms for furfural electroreduction on copper, EES Catalysis 1 (2023) 539–551.

[41] Y. Cao, J. Knijff, A. Delparish, M.F.N. d'Angelo, T. Noël, A Divergent Paired Electrochemical Process for the Conversion of Furfural Using a Divided-Cell Flow Microreactor, Chem. Sus. Chem. 14 (2021) 590–594.

[42] K. Ji, Y. Liu, Y. Wang, K. Kong, J. Li, X. Liu, H. Duan, Steering Selectivity in Electrocatalytic Furfural Reduction via Electrode–Electrolyte Interface Modification, J. Am. Chem. Soc. 146 (2024) 11876–11886.

[43] A.M. Román, N. Agrawal, J.C. Hasse, M.J. Janik, J.W. Medlin, A. Holewinski, Electro-oxidation of furfural on gold is limited by furoate self assembly, J. Catal. 391 (2020) 327–335.

[44] L. Gong, N. Agrawal, A. Roman, A. Holewinski, M.J. Janik, Density functional theory study of furfural electrochemical oxidation on the Pt (1 1 1) surface, J. Catal. 373 (2019) 322–335.

[45] Z. Ren, Y. Yang, S. Wang, X. Li, H. Feng, L. Wang, Y. Li, X. Zhang, M. Wei, Pt atomic clusters catalysts with local charge transfer towards selective oxidation of furfural, Appl. Catal. B 295 (2021) 120290.

[46] A.M. Román, J.C. Hasse, J.W. Medlin, A. Holewinski, Elucidating Acidic Electro-Oxidation Pathways of Furfural on Platinum, ACS Catal. 9 (2019) 10305–10316.

[47] S. Kar, Q.Q. Zhou, Y. Ben-David, D. Milstein, Catalytic Furfural/5-Hydroxymethyl Furfural Oxidation to Furoic Acid/Furan-2,5-dicarboxylic Acid with H_2 Production Using Alkaline Water as the Formal Oxidant, J. Am. Chem. Soc. 144 (2022) 1288–1295.

[48] X. Li, L. Cong, H. Lin, F. Liu, F. Han, N. Lin, Electrochemical conversion of furfural to furoic acid: a more stable, efficient and energy-saving system, Sci. China Chem. 65 (2022) 2576–2587.

[49] Y.Q. Zhou, K. Yang, W.T. Lu, S. Yan, F.F. Cao, G. Zhang, Electrochemical Oxidative Upgrading of Alkali-Sensitive Furfural and 5-Hydroxymethylfurfural in an Anion-Exchange Membrane Electrolyzer with High-Concentration Products, ACS Sustain. Chem. Eng. 12 (2024) 15260–15267.
[50] W. Wang, X. Yu, H. He, Y. Wang, Y. Li, L. Deng, Y.N. Liu, Electrochemical reconstitution of Prussian blue analogue for coupling furfural electro-oxidation with photo-assisted hydrogen evolution reaction, Chem. Eng. J. 465 (2023) 142865.
[51] L. Chen, Y. Xu, L. Su, T. He, L. Zhang, H. Shen, Q. Cheng, L. Liu, S. Bai, S.H. Hong, sible-Light-Enhanced Hydrogen Evolution through Anodic Furfural Electro-Oxidation Using Nickel Atomically Dispersed Copper, Inorg. Chem. 63 (2024) 730–738.
[52] D. Ouyang, D. Gao, J. Hong, Z. Jiang, X. Zhao, Promoting and controlling electron transfer of furfural oxidation efficiently harvest electricity, furoic acid, hydrogen gas and hydrogen peroxide, J. Energy Chem. 79 (2023) 135–147.
[53] Y. Zhang, X. Wang, P. Wu, X. Zhang, Q. Zhou, L. Xing, Y. Fan, Efficient electrochemical oxidation of the biomass platform compound furfural on a $Ni_{0.48}Co_{0.36}O_{0.16}$ electrode, J. Appl. Electrochem. 54 (2024) 2485–2493.
[54] Z. Zeng, S. Wu, X. Huang, Z. Wei, Electrochemical Oxidation of Furfural on NiMoP/NF: Boosting Current Density with Enhanced Adsorption of Oxygenates, Small 20 (2024) 2305462.
[55] J. Deng, R. Li, K. Xiang, L. Zhang, K. Sun, Y.D. Kim, Z. Liu, J. Ge, Z. Peng, Superior matching between electrochemical and non-electrochemical reactions to boost furfural electro-oxidation, Chem. Eng. J. 511 (2025) 162214.

CHAPTER

18

Electrochemical biomass-derived carboxylic acids

Anupam Singha Roy[a], *Amit Kumar*[b]

[a] School of Sciences, Woxsen University, Hyderabad, Telangana, India, [b] Department of Chemistry, School of Allied Sciences, Dev Bhoomi Uttarakhand University, Dehradun, Uttarakhand, India

18.1 Introduction

The global transition to sustainable and circular economies has speed up the hunt for renewable feedstocks to substitute fossil fuels in the manufacturing of chemicals. In this model, platform chemicals made from biomass, particularly 5-hydroxymethylfurfural (HMF), have become crucial linkages. Lignocellulosic biomass, which has received prominent attention in the past ten years, is the most abundant and biorenewable carbon-neutral resource that is widely distributed geographically [1]. Because of HMF's unique bifunctional structure, which consists of an aldehyde and a hydroxyl group, is created by acid-catalyzed dehydration from biomass rich in carbohydrates, and it can be converted into fuels and value-added chemicals [2]. 2,5-furandicarboxylic acid (FDCA), the oxidation product of HMF, is an alternative compound to petroleum-derived terephthalic acid and is used as a precursor to polyethylene terephthalate (PET) polymers. Additionally, FDCA can replace other bio-based polymers like polyamide, polyester, epoxy resin, succinic acid, isophthalic acid, and adipic acid.

Thermocatalytic oxidation as well as traditional techniques for HMF valorization usually depend on high temperatures, noble-metal catalysts (such as Pt, Pd, and Au), and high-pressure oxygen. Catalyst deactivation,

Electrochemical Biomass Conversion: Sustainable Processes and Applications for Bioenergy, Biofuels, and Bioproducts
DOI: https://doi.org/10.1016/B978-0-443-26757-4.00019-2

over-oxidation by-products, and energy-intensive processes can all arise [3]. These drawbacks diluted the idea of green production and prevented its extensive industrial use. The scientific community has recently paid close attention to low-energy and environmentally friendly oxidation processes, like biocatalysis, photocatalysis, and electrocatalysis, of HMF to FDCA. But electrochemical reaction offers a sustainable alternative by utilizing renewable electricity to drive HMF oxidation in ambient conditions. Electrooxidation not only lowers carbon emissions but also provides precise control over reaction pathways through electrode potential variation, potentially improving selectivity for FDCA or other products like 5-hydroxymethyl-2-furancarboxylic acid (HMFCA) or 2,5-diformylfuran (DFF) [4]. This process supports decentralized chemical production by avoiding hazardous oxidants, using water as the oxygen source, lowering reaction temperatures and pressures, and making it easier to combine with renewable energy sources. Noble metal catalysts can be replaced by electrochemical oxidation because many nonnoble metal catalysts work well in alkaline environments and because electrochemical parameters like potential, current, electrolyte, and the electrocatalyst's reactivity can be precisely adjusted to control the reaction rate and the selectivity of the oxidation product.

Despite its potential, the electrooxidation of HMF is full of challenges. Because of the successive electron/proton transfers and challenging pathways, the complex reaction network of HMF requires advanced catalysts that balance stability, selectivity, and activity. Nanostructured carbon materials and nonprecious metal catalysts (like Ni, Co, and Cu oxides) have shown promise, but they often have low durability or slow kinetics [5]. Moreover, the absence of a mechanistic understanding of adsorption, rate-determining mechanisms, and the impact of electrolyte pH hinders sensible catalyst design [6]. Recent advances in in-situ spectroscopy and computational modeling are beginning to address these issues, paving the way for customized electrocatalysts [7].

The electrooxidation of HMF is thoroughly reviewed in this chapter, which also covers reaction engineering, fundamental ideas, catalytic materials, and possible future advancements. We first study HMF in the context of biomass valorization and then analyze the kinetics of the electrochemical transformation. By combining cutting-edge research with commercial opportunities, this chapter aims to guide engineers and scholars toward efficient, sustainable HMF electrooxidation systems.

18.2 Heterogeneous catalysts for electrooxidation of HMF

Here we review the current progress and development of about four classes of heterogeneous electrocatalysts in HMFOR based on their major components (Fig. 18.1).

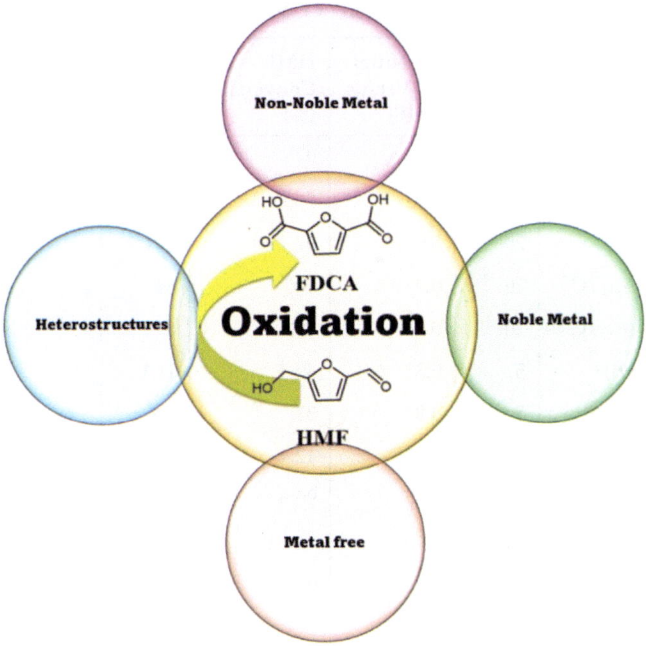

FIGURE 18.1 Different types of heterogeneous electrocatalysts used in HMFOR.

18.2.1 Noble metal catalysts in HMFOR

Because of their low oxidation potentials and very low current densities, certain noble metal electrocatalysts have been discovered to be unsuitable as an effective catalyst in HMFOR. Table 18.1 provides a summary of the most current progress in the electrooxidation of HMF by noble metal catalysts. The primary noble metal to be employed as an HMFOR was platinum. Strasser et al. [8] used Pt/C as a catalyst for the electrooxidation of HMF. However, they were only able to generate a low current density (0.44 mA/cm^2) and a very small amount of FDCA (yield <1%). Li et al. [9] demonstrated that the electrode potential as well as catalyst materials can extensively affect the reaction mechanism by discussing the effects of Pd/C or Au/C catalysts on HMFOR [9]. The oxidation of aldehyde is favored over the oxidation of hydroxymethyl using Au/C catalyst, despite the fact that the yield of FDCA is only 1%. The applied potential has a significant impact on the HMF oxidation pathway. At lower potentials, the DFF pathway succeeds, yielding a 29% FDCA yield with the Pd/C catalyst. In summary, inadequate HMF conversion as well as FDCA yield were the outcomes of employing a sole noble metal as a catalyst. Furthermore, it is difficult to employ noble metals in large-scale industry due to their high cost and low catalytic efficiency. One potential solution to these challenges

TABLE 18.1 Electrochemical oxidation of HMF using noble metal catalysts.

Entry	Catalyst	HMF (mM)	Potential (V Versus RHE)	HMF Conversion (%)	FDCA Yield (%)	FE (%)	References
1	Pt/C	5	-	29	<1		8
2	Au/C	20	0.9	100	1		9
3	Pd/C	20	0.9	97	29		9
4	Pd_2Au_1/C	20	0.9	100	64		9
5	Pd_1Au_2/C	20	0.9	100	83		9
6	$(AuPd)_7$	5	0.82	49.3	11.1		10
7	Au_7/Pd_7	5	0.82	33	6.8		10
8	Pd_7/Au_7	5	0.82	42.4	10.1		10
9	Pd/NiCo	50	1.38	99.6	96.5	95.9	11
10	Ru_1/CoO_x	5	1.45		55	55.2	12
11	Rh-SA/NiFe NMLDH	50	1.3	98	99.8	98.5	13
12	$Ru_{0.3}/NiFe$	5	1.48	99.43	98.68		14

is to design an alloy that can exactly alter the metals' electrical characteristics. This technique lowers the cost of catalyst preparation while enhancing electrocatalytic oxidation activity. This method is beneficial for the transfer of electrons and can alter catalytic activity and selectivity by affecting adsorption and desorption. The unique property of the alloy reduces the drawbacks of single metal catalysts and facilitates the electro-oxidation of HMF to FDCA. According to Li et al. [9], AuPd alloy catalysts demonstrated a collaborative effect during HMFOR. When Au was applied to the Pd surface, it enhanced the hydroxymethyl oxidation and encouraged the oxidation of –CHO groups at lower potentials. Higher FDCA yields were consequently produced by the modified surface composition (Table 18.1, entries 4, 5). Kim et al. [10] developed a 3D amalgam AuPd bimetallic electrode for HMFOR [10]. HMFOR was further advantageous when the $(AuPd)_7$ catalyst had an alternate Pd and Au pattern, as evidenced by the FDCA yield of 10.1% and the HMF conversion of 49.3% (Table 18.1, entries 6–8). Song et al. [11] recently modified Pd on Co-doped $Ni(OH)_2$ to create Pd/NiCo catalysts. Co-doping and Pd loading were found to improve the catalytic performance of $Ni(OH)_2$ for HMFOR. Co upgraded the band structure of Ni and caused lattice distortion. Electron transmission from Pd to Ni sites was facilitated by Pd clusters' electron interactions with

NiCo carriers. The Pd/NiCo catalyst produced a faradaic efficiency (FE) of 95.9%, an FDCA yield of 96.5%, an HMF conversion of 99.6%, and a current density of 50 mA/cm^2 at 1.38 V.

Chen et al. recently created Ru_1/CoO_x catalysts by putting Ru onto cobalt oxide, which showed exceptional activity in HMFOR (Table 18.1, entry 10) [12]. The results showed that adding Ru at a minimum loading of about 0.5 wt% could speed up the electrochemical reduction of $Co^{2+}/Co^{3+}/Co^{4+}$, increase the essential activity of CoO_x, and produce FDCA with 76.5% selectivity, which was higher than in the case of the initial CoO_x electrocatalyst (62.7%). HMF adsorption at the Ru_1/CoO_x interface may be enhanced by a single Ru, which would speed up the rate-limiting step of C–H bond transformation for FDCA yield.

Guo et al. produced $Rh\text{-}O_5/Ni(Fe)$ attached on a nanoporous framework using a NiFe-layered double hydroxide catalyst [13]. At 1.3 V, this catalyst showed remarkable performance for HMFOR with a 50 mA/cm^2 current density. The yield of FDCA was 99.8% with FE 98.5%, and the conversion of HMF was 98% (Table 18.1, entry 11). The barriers to electron transport between the active site and the important intermediates *OH and HMF are lowered by the interaction between the Rh and Ni atoms. By strengthening these synergistic effects, the reaction energy barrier was lowered during HMFOR by facilitating electron exchange and transfer. Li et al. deposited Ru atoms onto NiFe LDH using a two-step electrodeposition technique. The resultant $Ru_{0.3}/NiFe$ catalyst showed outstanding yield (98.68%), FDCA selectivity (99.24%), and HMF conversion (99.43%) (Table 18.1, entry 12) [14]. According to both the theoretical and experimental results, FeNiOOH can more effectively capture protons and electrons during HMFOR by increasing the Ru composition and adjusting the adsorption energy of HMF. Ru was found to improve the yield of FDCA by facilitating the reaction path from HMF to DFF.

18.2.2 Nonnoble metal catalysts in HMFOR

Sun's group was the first to study the use of metal phosphides as well as sulfides in the oxidation of HMF [15]. In these instances, HMF was successfully converted in alkaline media with a 90% FDCA yield. It is crucial to note that the analysis carried out following electrolysis revealed the oxidation of Ni/Co and a fractional loss of P/S, both of which are thought to be active species in HMF oxidation. Zhang et al. combined metal and phytic acid to create a hybrid Ni-PA catalyst. For HMFOR, this catalyst demonstrated exceptional catalytic efficiency with a 99.1% FDCA yield and a FE above 90%[16]. Menezes and his group recently developed a nickel-based NiP/NF catalyst that was made at room temperature using a one-step process. During the HMFOR, this catalyst can change into a highly reactive nickel oxide phase (γ-NiOOH) [17]. It successfully facilitated the electro-oxidation of HMF in an alkaline medium of 1.0 M KOH, with 96% HMF conversion, 96% FDCA yield, and a FE of 96%.

The Ce-Co_2P@NC catalyst was created by Shen et al. using Ce-doped Co_2P supported on N-doped carbon nanosheets [18]. This catalyst demonstrated a FE of 98.5%, 99.3% FDCA yield, and 99.5% HMF conversion. Ce doping changed the electronic band structure and considerably enhanced the electron movement of Co-based active species, which decreased the reaction energy barrier of the HMFOR rate-determining step. A novel Mn-doped Ni_2P electrocatalyst (Mn-5Ni_2P) was developed by Li et al. using microwave-aided DES and phosphorization techniques. It demonstrated remarkable catalytic performance for HMFOR, attaining a 100% HMF conversion and FDCA yield of 98%, with FE of 97.8%[19]. Mn doping, according to DFT calculations, not only successfully modified the electronic structure of Ni, enhancing the adsorption of HMF molecules on Ni sites, but it also created new Mn-based active spots. In order to improve the electrochemical oxidation of HMF in an alkaline environment of 1.0 M KOH, Tao et al. [20] added Mo to CoP to promote lattice expansion. This modification produced an impressive 99.9% HMF conversion with a 95.8% FDCA yield and a FE of 93%. According to DFT theoretical calculations, the CoMoP surface has better adsorption capacity of HMF when Mo is added because it increases the d-band center of Co. Moreover, OH_h can occupy the many oxygen defects on the CoMoP surface, which promotes low-valence metal ions to oxidize.

The hydrothermal sulfuration method was used to create a Ni_2S_3/NF catalyst that reported an approximately 100% HMF conversion, a FE of 94%, and a FDCA yield of 98% [21]. Additionally, Wang and his group produced the S-Ni@C-600 catalyst by hydrothermal reduction reaction using lignin sulfonate (LS), a pulp industry by-product, as a source for sulfur and carbon [22]. With a 96% FDCA and a 96% FE, this catalyst can convert HMF almost entirely. The results exhibited that the electrocatalytic properties were greatly improved by the presence of nickel hydroxide on the outside and the carbon shell. Zheng et al. developed layered Ni-based sulfide nanosheets (NiCo-S) for HMFOR [23], which showed 99.1% HMF conversion with 97.1% yield of FDCA and a FE of 96.4%. The Ni and Co dual sites work together in HMFOR, Co sites assisting the transformation of the –CHO group of HMF into a –COOH group and demonstrating strong adsorption of O and C atoms of the –CHO group, while Ni sites accelerate the rate of FDCA production.

Wang et al. [24] recently created $Mn_{0.2}NiS$/GF, a Mn-doped NiS nanosheet electrocatalyst on a three-dimensional graphite felt (GF). $Mn_{0.2}NiS$/GF catalyst showed exceptional HMFOR performance with 99.3% HMF conversion, 97.6% FDCA yield, and FE of 94.2%. According to the study, Mn and Ni are both efficient for HMF adsorption sites. The addition of Mn facilitates HMFOR by enhancing the effective sites, improving the formation of the nanosheet structure, and introducing more adsorption sites for HMF.

To modify the Ni_3S_2 catalyst, Zheng et al. [25] used a technique that combines surface engineering with heteroatom (Cu, Fe, V, and Zn) [25]. It was found that Cu^{2+} provided many Lewis acid sites for the adsorption of HMF because of its unique electronic structure, with almost complete HMF conversion and FDCA yield, and roughly FE of 100%. Successful surface reconstruction and doping changed the active Ni sites' local coordination environment, improved catalytic activity, and reaction kinetics. Theoretical analysis showed that the superior conductivity of Cu/Ni_3S_2-R and the adsorption of OH and HMF were important elements in increasing the electrocatalytic oxidation activity. Yan et al. used a hydrothermal process and low-temperature calcination and sulfurization to create a W-doped Ni_3S_2 nanocatalyst (W_{20}-Ni_3S_2@NF) [26]. It was discovered that W doping could increase the conductivity of Ni_3S_2 and show more active sites. The adsorption of OH* and HMF was efficiently controlled by W doping. Additionally, the d-band of Ni_3S_2 shifted up with the mixing of high-valence W, which made it easier for water to adsorb and dissociate, producing more OH*. By drawing electrons from Ni, the high-valence W raised the Ni valence state and maximized the HMF adsorption energy. This element produced exceptional HMFOR performance by speeding up the dynamics of HMFOR, and the results showed that the catalyst had 97.3% FE, 99.2% FDCA selectivity, and 100% HMF conversion.

Sun et al. established an electrolysis apparatus that used NiB_x as a cathode for the reduction of p-nitrophenol to p-aminophenol and the anodic HMFOR [27]. According to the experimental results, the generated Ni^{3+} functioned as a reactive intermediate. The conversion, selectivity, efficiency, and FE of both cathodic and anodic sides were all at least 99% when using water as a source of hydrogen and oxygen. Wang et al. produced P-doped nickel boride (NiB_x-P_y) to examine the impact of varying phosphorus quantities on the electrocatalytic efficiency of HMFOR [28]. The results showed that the FDCA yield using NiB_x-P_y catalyst first improved and then decreased as the phosphorus content increased. With the highest 90.6% FDCA yield and 92.5% FE, the NiB_x–$P_{0.07}$ catalyst demonstrated the best electrocatalytic performance. Additionally, investigation has shown that P doping improves electron transference in Ni and promotes the production of NiOOH species, the real catalytic active agents for HMFOR.

The use of transition metal nitrides for HMFOR was first reported by Wang et al. They used a hydrothermal and nitridation process to create a carbon-coated Ni_3N@C catalyst [29]. By adding carbon, it was found that Ni_3N could form nanosheets instead of big, clumped particles, which enhanced the catalytic activity and efficiently controlled the distribution of electrons on the Ni_3N surface. It was found that at 1.35 V, HMF was first oxidized to HMFCA, which was then further oxidized to FDCA as the potential improved to 1.45 V.

Using the co-precipitation technique, Yan et al. produced ultra-thin NiCoFe-LDH nanosheets, which were subsequently applied to HMFOR [30]. NiCoFe-LDH nanosheets were found to have an ultra-thin thickness of only about 1.36 nm compared to NiCo-LDHs and NiFe-LDHs. This suggested that adding Fe could alter the layer between metals. The NiCoFe-LDHs catalyst performed very well in HMFOR. Within an hour, HMF conversion reached 95%, reaching almost an FDCA selectivity of 85%. Using hydrothermal and alkali etching methods, Wang et al. created a d-NiFe LDH/CP catalyst with cation excess defects [31]. The catalytic performance was outstanding with 96.8% FDCA yield and 84.47% FE, at 1.48 V, HMF was nearly fully converted (97.35%). The results explained that the main cause of the increased activity, which considerably altered the electronic configuration of the NiFe LDH, was the abundance of cationic vacancies. While the charge transfer resistance decreased, the active surface area increased significantly. According to recent studies, the simple and efficient way to increase the effective surface area and conductivity of LDHs is to combine them with different conductive nanomaterials to produce independent electrodes. For example, Ni_xSe_y-NiFe LDH@NF catalyst with a three-dimensional layered core–shell structure was created by Fan et al. [32] and used in HMFOR. The Ni_xSe_y nanowire, which is popular for its high conductivity, was found to have strong mechanical strength and to facilitate charge transfer. Numerous active sites were provided by the NiFe LDH that formed the shell, which had a large surface area. Moreover, the main active site was NiOOH, and Fe contributed to changing Ni's electronic structure to increase catalytic activity. Due to these benefits, the Ni_xSe_y-NiFe LDH@NF core-shell catalyst demonstrated outstanding catalytic activity for HMFOR, attaining 99.3% FDCA yield, 98.9% FE, and 96.7% HMF conversion.

NiV layered double hydroxide ultra-thin films (NiV-LDHs) were recently developed by Yan and his group with the goal of efficiently electrocatalytically oxidizing a variety of aldehyde platform molecules [33]. With the smallest thickness of 2.6 nm, the Ni_3V_1-LDHs film was found to have many active sites that allowed aldehydes to be adsorbed and hydroxyl radicals to be produced for the oxidation of various aldehyde molecules, such as benzaldehyde, furfural, HMF, and methoxybenzaldehyde. Experimental and theoretical results proved that NiV-LDHs could produce a large number of OH radicals on their surface, which would promote the FDCA yield. The FDCA yield was found to be 79.5%, whereas the conversion of HMF oxidation was 83.5%.

According to recent studies, phase transitions could result from the incompatible crystal structures of MOOH and $M(OH)_2$, which could make $M(OH)_2$ unstable and reduce its capacity to adsorb and activate HMF. As a result, a single hydroxide is not enough in HMFOR. Guest ion introduction is thought to be a useful tactic for reducing lattice transitions. New active sites are also provided by the doped phase. Fan et al. have created a new Ce-doped Ce-$Ni(OH)_2$ material that can be activated to Ce-NiOOH using an electrochemical remodeling technique [34]. It was observed that adding Ce improved the reaction activity in addition to resolving the challenge of $Ni(OH)_2$ structure collapse during the transformation of $Ni(OH)_2$ to NiOOH. Even after the 50th cyclic voltammetry (CV) test, the 14% Ce-$Ni(OH)_2$ showed outstanding activity, and 27 mA/cm^2 current density. Additionally, the catalyst greatly speeds up the HMFOR kinetics, which raises the yield and FE of FDCA.

Because of their high stability, exceptional catalytic activity, and unique crystal structure, which is defined by the cooperation of tetrahedral and octahedral spinel-type oxides have attracted a lot of interest. In 2021, Lei et al. hydrothermally created sea-urchin-like nanostructures in the spinel $CuMn_2O_4$, which they subsequently dissolved in NH_3 solution to create oxygen vacancies [35]. According to the results, the NH_3-etched $CuMn_2O4$ showed exceptional HMFOR performance. 96% FE of FDCA and an HMF conversion of 100% were observed at 1.31 V in a 1.0 M KOH solution and at a current density of 20 mA/cm^2. Wang et al. investigated the impact of particular blocks on tetrahedral (Co^{2+}) and octahedral (Co^{3+}) sites on the HMFOR activity of spinel cobalt oxides (Co_3O_4, $ZnCo_2O_4$, and $CoAl_2O_4$) [36]. The following is a ranking of the electrochemical performance: $CoAl_2O_4 < ZnCo_2O_4 < Co_3O_4$, suggesting that the catalytic activity of octahedral CoO_6 was superior.

18.2.3 Heterostructures catalysts in HMFOR

Interface engineering has become a successful method for developing effective electrocatalysts since electrocatalytic reactions normally occur at boundaries. In electrocatalytic reactions, heterostructures can modify the catalytic performance by promoting electron transport and affecting the adsorption and desorption energies of active species. Furthermore, the stability and catalytic activity of heterostructures may be further improved by the synergistic interactions between heterogeneous components. A 3D layered nanostructure of NiO-Co_3O_4 electrocatalyst with multiple interfaces was created by Zou et al. [37]. A multitude of cation vacancies were produced by the interface effect, changing Co and Ni's electronic properties and raising the Ni oxidation state. At 1.45 V versus RHE, the NiO-Co_3O_4 catalyst showed exceptional HMFOR activity, attaining nearly 100% HMF conversion, 98% FDCA yield, and a FE of 96%.

To enhance HMFOR, Tao et al. created a porous CoP-CoOOH heterostructure [38]. Electron transfer from CoOOH to CoP occurred at the heterostructure surface, improving its conduction and altering the adsorption of HMF. HMF oxidation was carried out at a concentration of 150 mM at a potential of 1.42 V versus RHE with 98.3% HMF conversion. For HMFOR, Du et al. created $Ni(OH)_2$-NiOOH/NiFeP that showed remarkable catalytic performance, attaining 99% FDCA yield and a FE of more than 94% [39]. The formation of $Ni(OH)_2$ was monitored as a result of the expected oxidation of NiFeP during the anodic process of HMFOR. NiOOH, the actual active site for HMFOR, was created when $Ni(OH)_2$ underwent additional oxidation as the electrode potential increased. The chemical reduction of NiOOH back to $Ni(OH)_2$ occurred after HMFOR, and the transformation of $Ni(OH)_2$ to NiOOH by the applied anodic potential allowed HMFOR to continue cyclically.

Co-Ni_xP@C catalysts on Ni and Co foam were recently created by Wang's group. In addition to improving the free energy of H^* adsorption, the results exhibited that Co-Ni_xP@C had a layered architecture and nanowire-like structure offered many bare active sites to improve the adsorption and activation of the HMF molecule [40]. The conductivity and stability of the catalyst were enhanced by the thin layer of graphite-like carbon. Furthermore, a significant interface lattice mismatch and a large quantity of defects were observed in the cobalt-doped $Ni_{12}P_5/Ni_3P$ heterostructures, which successfully modulated the active sites and electronic structure. The FE was 99.7%, and its FDCA yield and HMF conversion were both close to 100% at 1.38 V versus RHE.

Similarly, Wang et al. created NF/Co_4N@CeO_2 heterostructures [41], and the results showed that CeO_2 acted as an electron source, improving electron movement from Co_4N to CeO_2, leading to a reorganization of charge at the interface. NF/Co_4N@CeO_2 showed exceptional activity in alkaline conditions with 91.1% HMF conversion, 93.6% FDCA yield, and FE of 84.5%. $NiCo_2$@MoO_2/NF heterojunction materials were recently created by Yin et al. utilizing a simple hydrothermal and calcination process [42]. Adding MoO_2 to the $NiCo_2$ could change the electronic arrangement of $NiCo_2$@MoO_2/NF, improve the adsorption properties of intermediates of HMFOR, and make it easier to dehydrogenate the HMFCA to FFCA by deteriorating the C–H and O–H bond. Proton transfer experiments and the kinetic effects of deuterium isotopes showed that the interaction between MoO_2 and $NiCo_2$ improved proton transfer and C–H/O–H bond breaking during the HMFOR. Additionally, the catalyst demonstrated exceptional HMFOR activity, converting HMF almost entirely at 1.40 V.

18.2.4 MOFs, COFs, and non metal-based catalysts for HMFOR

There are few studies that concentrate on the HMFOR applications, even though there are numerous reports describing the use of MOF-based catalysts for transforming biomass into useful products like fine chemicals, biofuels, and functional materials. The first-time metal-doped 2D Ni-MOFs were used as electrocatalysts for HMFOR when Li et al. created a variety of these MOFs using the solvothermal technique. Regarding HMFOR, the Co-doped NiCoBDC-NF was found to exhibit outstanding activity, selectivity, and stability [43]. The FDCA yield (99%) in an electrolyte with a pH of 13 was 20.1 μmol cm^{-2} h^{-1} at 1.55 V versus RHE. The outcomes explained that the synergistic effect of Ni and Co atoms improved the electronic configuration and made it easier for Ni species to form inside the two-dimensional MOFs. These species were found to be active for the oxidation of aldehydes and alcohols. Using conductive MOFs, Zhang et al. synthesized an electrocatalytic material (M-CATs, M = Co, Ni, Cu, etc.) that indicated excellent catalytic efficiency in HMFOR and had well-precise M-O_4 active sites [44]. For instance, the Co-CAT catalyst was able to achieve an FDCA yield of 99.3% and 97.6% FE at 1.42 V versus RHE. On the other hand, the 98.7% FDCA yield with 86.8% FE was observed using the Ni-CAT catalyst. For Co-CAT, the rate-determining step was the adsorption of the reactant, whereas the desorption of the intermediate was the rate-determining step for Ni-CAT. Ni/Co/Fe sulfide nanoclusters were recently added to MOF nanoarrays by Qi et al. using a simple solvothermal technique [45]. The resulting NiCoFeS-MOF catalyst showed a staggered structure, and more active catalytic sites, active surface area, and electron transfer were all greatly increased by integrating metal sulfide nanoparticles with 2D MOF. Fan et al. [46] recently used an in situ electrochemical improvement technique to facilitate the activation of the Ni-MOF/Ag interface. For HMFOR, the CF-Ni-MOF/Ag (CF = Cu foam) showed outstanding activity at 1.625 V versus RHE with almost full HMF conversion and 98.6% FE. These results imply that the high activity of CF-Ni-MOF/Ag is primarily due to the collaborative interaction between Ni^{3+} and Ag^{+}, with the main active site being the Ni^{3+} and Ag^{+} generated by oxidation.

Innovative organic porous polymer materials known as covalent organic frameworks (COFs) have several advantages, such as low structural density, large surface area, superior crystallinity, high porosity, uniform pore size distribution, exceptional stability, and adaptable structural units. The nickel(II)-modified COF membrane T_pB_{py}-Ni@FTO was created by Cai et al. [47] and utilized in HMFOR. The obtained results showed that the HMF conversion was up to 96%, but the FDCA yield was only 58% due to suboptimal FDCA selectivity. Although this was the first time COFs were used in HMFOR, the catalytic efficiency was still poor.

The creation of non metal catalysts for HMFOR is still in trouble with difficulties. In 2019, Qin et al. developed a porous carbon BNC-2 co-doped with boron and nitrogen, exhibiting polar pore wall structures and enriched point defects [48]. With this, the first non metal electrocatalyst was introduced for HMFOR, and the results explained that over the course of six h, 57% FDCA yield was observed at 1.9 V, and the HMF conversion reached 71%. It provided useful information for the advancement and design of non metal electrocatalysts, even though its catalytic efficiency was not as good as that of metal catalysts. Table 18.2 reviews recent developments in the electro-oxidation of HMF to FDCA through non noble metal catalysts.

18.3 Mechanism of electrochemical oxidation of HMF

Indirect oxidation and direct oxidation are the two familiar processes for HMF oxidation [49]. The most common and well-known technique is direct oxidation, in which reactants are oxidized by applying voltage. The dehydrogenation of the substrate results from the applied potential's facilitation of OH-adsorption onto the electrode and its promotion of electron transfer. Specifically, hydroxymethyl preferentially sticks to the electrode surface and starts the removal of H from C–H and O–H bonds by activating OH^-, which leads to the formation of DFF intermediates. In contrast, the hydration of HMF aldehyde is reduced in an alkaline medium ($pH < 13$). A geminal diol is then created when the aldehyde group reacts with water in a nucleophilic addition reaction. FFCA is produced by deprotonation after OH-induced activation of the C–H and O–H bonds within the geminal diol. FDCA is then created by repeating the nucleophilic addition and deprotonation processes. Aldehyde groups preferentially stick to the catalyst surface in a strongly alkaline environment ($pH > 13$). During alkaline catalysis, they react with H_2O to produce a geminal diol, and then they dehydrogenate to form HMFCA through OH^- activation. After that, hydroxymethyl undergoes proton abstraction, which yields FFCA. Lastly, FDCA is obtained by continuing the two steps: nucleophilic addition and dehydrogenation (Fig. 18.2).

However, another technique that has attracted a lot of attention from researchers is indirect oxidation. Indirect oxidation basically involves applying a weak voltage at first to change the catalyst's low-valence ions into high-valence ions, which then function as oxidants to oxidize HMF molecules. The intermediate is first oxidized from a reduced state (step I: $Mn^+ \rightarrow M^{n+1}$). Then, the produced high valence oxidized intermediate (M^{n+1}) undergoes non electrochemical oxidation of the nucleophilic substrate via hydrogen atom transfer, and M^{n+1} is again reduced to its initial valence state (M^{n+}) (step II: $M^{n+1} \rightarrow M^{n+}$). Heterogeneous electrocatalytic oxidation of HMF usually requires OH^- to be involved in the reaction process. Therefore the heterogeneous catalytic oxidation of HMF is more favorable in an environment that is strongly alkaline ($pH \geq 13$). Furthermore, when designing catalysts, it is crucial to understand that the electrochemical potential affects the conversion of M^{n+1}, whereas the active

TABLE 18.2 HMFOR catalyzed by non noble metal catalysts.

Entry	Catalyst	HMF (mM)	Potential (V versus RHE)	HMF Conversion (%)	FDCA Yield (%)	FE (%)	References
1	Ni-PA	10	1.6	100	99.1	90	16
2	NiP/NF	100	1.49	96	96	96	17
3	Ce-Co_2P@NC	10	1.20	99.5	99.3	98.5	18
4	Mn-5Ni_2P	10	1.43	100	98	97.8	19
5	CoMoP	100	1.36	99.9	95.8	93	20
6	Ni_2S_3/NF	10	1.498	100	98	94	21
7	S-Ni@C-600	10	1.473	100	96	96	22
8	NiCo-S	10	1.45	99.1	97.1	96.4	23
9	$Mn_{0.2}NiS$/GF	100	1.48	99.3	97.6	94.2	24
10	Cu-Ni_3S_2-R	20	1.40	100	100	100	25
11	W_{20}-Ni_3S_2@NF	10	1.42	100	99.2	97.3	26
12	NiB_x	10	1.426	≥ 99	≥ 99	≥ 99	27
13	NiB_x–$P_{0.07}$	10	1.464	> 99	90.6	92.5	28
14	NiCoFe LDHs	10	1.54	95.5	84.9	90	30
15	d-NiFe LDH/CP	10	1.48	97.35	96.8	84.47	31
16	Ni_xSe_y–NiFe LDH@NF	10	1.423	99.6	99.3	98.9	32
17	Ni_3V_1-LDHs	100	-	83.5	79.5	-	33
18	14%Ce-$Ni(OH)_2$	10	0.45	100	-	86.6	34
19	$CuMn_2O_4$	10	1.31	100	-	96	35
20	NiO-Co_3O_4	10	1.45	100	98	96	37
21	CoP-CoOOH	150	1.42	98.3	96.3	96.3	38
22	$Ni(OH)_2$-NiOOH/NiFeP	10	1.435	99.4	99.4	94.62	39
23	Co-Ni_xP@C	10	1.38	100	100	98.9	40
24	NF/Co_4N@CeO_2	10	1.425	91.1	93.6	84.5	41
25	$NiCo_2$@MoO_2-NF	10	1.40	100	99.6	99.4	41
26	NiCoBDC-NF	10	1.55	-	99	78.8	43
27	NiCoFeS-MOF	50	1.39	100	99	99	45
28	CF-Ni-MOF/Ag	10	1.623	100	-	98.6	46
29	T_pB_{py}-Ni@FTO	0.5	1.55	96	58	-	47

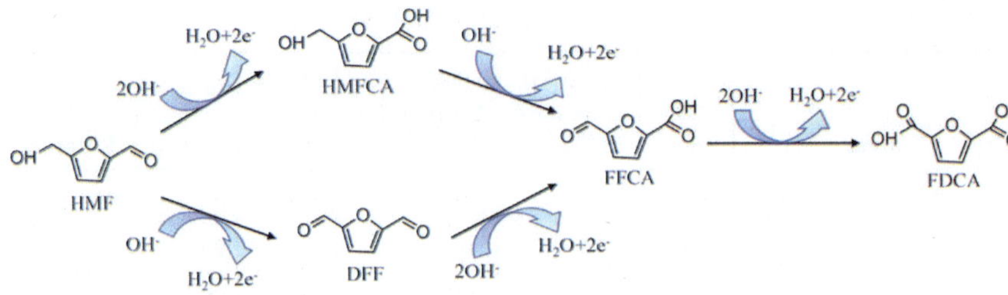

FIGURE 18.2 Reaction mechanism of direct oxidation of HMF.

site concentration or their fundamental activity determines the subsequent oxidation of HMF.

18.4 Conclusion and future prospects

With an importance on electrocatalytic oxidation, this book chapter attempts to methodically investigate the electrochemical conversion of HMF. A variety of efficient electrocatalysts are first introduced in a methodical manner for the electrochemical oxidation of HMF, including nonprecious metal, precious metal, and metal-free catalysts. According to the results, metal-free catalysts display lower activity and have little research available in the electrocatalytic oxidation of HMF, while noble metal catalysts perform exceptionally well. It is clear that the electrocatalytic oxidation of HMF still primarily relies on metal catalysts. This book chapter provides insightful information for the construction of superior materials for the electrocatalytic oxidation of HMF from a mechanistic standpoint. Because of its substantial and long-lasting economic benefits, the process of preparing FDCA from HMF has attracted more and more attention. Notably, there has been an increase in interest in the electrocatalytic method in recent years. Additionally, combining the electrocatalytic anodic oxidation of HMF with cathodic reduction reactions, like CO_2 reduction and H_2 generation, can improve energy efficiency and economic gains. Researchers are very interested in this integration. Numerous dependable materials with exceptional catalytic efficiency have been investigated and documented, and the electrochemical oxidation of HMF for the production of furfural derivatives is thriving; however, a number of technical obstacles and bottlenecks still exist.

It should be noted that most laboratory-scale investigations are limited to lower current densities and mM-level HMF additions, which do not meet the standards of industrial production. Additionally, when exposed to a strong alkaline solution, HMF is prone to ring opening and degradation, which causes the quality of the added HMF to rapidly deteriorate.

Additional work is needed in the following areas if the electrocatalytic method is to be used in industrial production: a greater weight should be placed on understanding and studying the mechanism of the reaction, as well as a thorough understanding of the relationships among catalytic pathways, intermediates, and reaction media; the development of more stable, effective, and eco-friendly catalysts should be based on theoretical calculations; and FDCA is more easily separated and purifiable in acidic environments, whereas electrocatalytic oxidation of HMF currently mostly takes place in alkaline conditions. As a result, developing catalysts that work well in acidic environments has gained immediate attention.

Acknowledgment

Anupam Singha Roy acknowledges Woxsen University, Hyderabad, and the Royal Society of Chemistry for his Newton International Fellowship Alumni Fund. Amit Kumar acknowledges Dev Bhoomi Uttarakhand University, Dehradun.

References

[1] C.H. Zhou, X. Xia, C.X. Lin, D.S. Tong, J. Beltramini, Catalytic conversion of biomass to chemicals and fuels, Chem. Soc. Rev. 40 (2011) 5588–5617.

[2] Q.S. Kong, X.L. Li, H.J. Xu, Y. Fu, Conversion of 5-hydroxymethylfurfural to chemicals: a review of catalytic routes and product applications, Fuel Process. Technol. 209 (2020) 106528.

[3] H. Liu, X. Tang, X. Zeng, Y. Sun, X. Ke, T. Li, J. Zhang, L. Lin, Catalyst design strategy toward the efficient heterogeneously catalyzed selective oxidation of 5-hydroxymethylfurfural, Green Energy Environ 7 (5) (2022) 900–932.

[4] Y. Yu, J. Ma, L.L Zhang, T. Sun, M. Wang, J. Zhu, J. Wang, Selective electrooxidation of 5-hydroxymethylfurfural to value-added 2,5-furanodiformic acid: mechanism, electrolyzer system, and electrocatalyst regulation, Chem. Commun. 61 (2025) 7751–7769.

[5] N.N. Rosman, W.S. Ng, N. Shah, M.S. Masdar, N.A. Karim, N. Ataollahi, R. Mohamad Yunus, Nanostructured transition metal-based electrocatalysts: a promising pathway in anion exchange membrane water electrolysis, Mater. Today Sustain. 31 (2025) 101203.

[6] J.S. Lim, Y.J. Sa, S.H. Joo, Catalyst design, measurement guidelines, and device integration for H_2O_2 electrosynthesis from oxygen reduction, Cell Rep. Phys. Sci. 3 (8) (2022) 100987.

[7] M.A. Belgami, A. Patra, S.M. Jeong, C.S. Rout, Exploring the synergy of magnetism and electrocatalysis: a comprehensive review on mechanisms, recent developments and future perspectives, J. Mater. Chem. A 12 (2024) 24005–24040.

[8] K.R. Vuyyuru, P. Strasser, Oxidation of biomass derived 5-hydroxymethylfurfural using heterogeneous and electrochemical catalysis, Catal. Today 195 (2012) 144–154.

[9] D.J. Chadderdon, L. Xin, J. Qi, Y. Qiu, P. Krishna, K.L. More, W. Li, Electrocatalytic oxidation of 5-hydroxymethylfurfural to 2,5-furandicarboxylic acid on supported Au and Pd bimetallic nanoparticles, Green Chem. 16 (2014) 3778–3786.

[10] M. Park, M. Gu, B.S. Kim, Tailorable electrocatalytic 5-hydroxymethylfurfural oxidation and H_2 production: architecture–performance relationship in bifunctional multilayer electrodes, ACS Nano 14 (2020) 6812–6822.

[11] G. Liu, T. Nie, Z. Song, X. Sun, T. Shen, S. Bai, L. Zheng, Y.F. Song, Pd loaded NiCo hydroxides for biomass electrooxidation: understanding the synergistic effect of proton deintercalation and adsorption kinetics, Angew. Chem. Int. Ed. 62 (2023) e202311696.

[12] W. Gu, A. Pei, S. Zhang, F. Jiang, Y. Jia, Q. Qin, R. Du, et al., Atomic-interface effect of single-atom Ru/CoO_x for selective electrooxidation of 5-hydroxymethylfurfural, ACS Appl. Mater. Interfaces 15 (2023) 28036–28043.

[13] L. Zeng, Y. Chen, M. Sun, Q. Huang, K. Sun, J. Ma, J. Li, H. Tan, M. Li, Y. Pan, et al., Cooperative Rh–O_5/Ni(Fe) site for efficient biomass upgrading coupled with H_2 production, J. Am. Chem. Soc. 145 (2023) 17577–17587.

[14] H. Xu, G. Xin, W. Hu, Z. Zhang, C. Si, J. Chen, L. Lu, Y. Peng, X. Li, Single-atom Ru/NiFe layered double hydroxide electrocatalyst: efficient for selective oxidation of 5-hydroxymethylfurfural and oxygen evolution reaction, Appl. Catal. B 339 (2023) 123157.

[15] N. Jiang, B. You, R. Boonstra, I.M Terrero Rodriguez, Y. Sun, Electrocatalytic upgrading of biomass-derived molecules, ACS Energy Lett. 1 (2016) 386–393.

[16] S. Liu, X. Yuan, X. Huang, Y. Huang, C. Sun, K. Qian, et al., Nickel–phytic acid hybrid for highly efficient electrocatalytic upgrading of HMF, Front. Chem. 11 (2023) 1199921.

[17] B. Dasgupta, J.N. Hausmann, R. Beltrán-Suito, S. Kalra, K. Laun, I. Zebger, et al., A facile molecular approach to amorphous nickel pnictides and their reconstruction to crystalline potassium-intercalated γ-$NiOOH_x$ enabling high-performance electrocatalytic water oxidation and selective oxidation of 5-hydroxymethylfurfural, Small 19 (2023) 2301258.

[18] S. Xie, H. Fu, L. Chen, Y. Li, K. Shen, Carbon-based nanoarrays embedded with Ce-doped ultrasmall Co_2P nanoparticles enable efficient electrooxidation of 5-hydroxymethylfurfural coupled with hydrogen production, Sci. China Chem. 66 (2023) 2141–2152.

[19] H. Xu, J. Bi, T. Sang, W. Wang, J. Hao, Z. Li, Mn-doped Ni_2P: nanocrystal-decorated amorphous nanosheets for efficient electrooxidation of 5-hydroxymethylfurfural, Chem. Commun. 59 (2023) 8440–8443.

[20] H. Wang, C. Niu, W. Liu, S. Tao, d-Electron tuned CoMoP for enhanced 5-hydroxymethylfurfural oxidation and HER, Appl. Catal. B 340 (2024) 123249.

[21] W. Wang, F. Kong, Z. Zhang, L. Yang, M. Wang, Sulfidation of nickel foam with enhanced electrocatalytic oxidation of 5-hydroxymethylfurfural to 2,5-furandicarboxylic acid, Dalton Trans. 50 (2021) 10922–10927.

[22] F. Kong, M. Wang, Preparation of sulfur-modulated nickel/carbon composites from lignosulfonate for the electrocatalytic oxidation of 5-hydroxymethylfurfural to 2,5-furandicarboxylic acid, ACS Appl. Energy Mater. 4 (2021) 1182–1188.

[23] Z. Zhao, T. Guo, X. Luo, X. Qin, L. Zheng, L. Yu, Z. Lv, D. Ma, H. Zheng, Bimetallic sites and coordination effects: electronic structure engineering of NiCo-based sulfide for 5-hydroxymethylfurfural electrooxidation, Catal. Sci. Technol. 12 (2022) 3817–3825.

[24] S. Li, S. Wang, Y. Wang, J. He, K. Li, Y. Xu, et al., Doped Mn enhanced NiS electrooxidation performance of HMF into FDCA at industrial-level current density, Adv. Funct. Mater. 33 (2023) 2214488.

[25] P. Xu, Z. Bao, Y. Zhao, L. Zheng, Z. Lv, X. Shi, et al., Anionic regulation and heteroatom doping of Ni-based electrocatalysts to boost biomass valorization coupled with hydrogen production, Adv. Energy Mater. 14 (2024) 2303557.

[26] S. Wang, G. Yang, Y. Jiao, Y. Liu, C. Tian, A. Wu, H. Yan, Tuning electronic structure of Ni_3S_2 with tungsten doping for high-performance electrooxidation of 5-hydroxymethylfurfural, Sci. China Chem. 66 (2023) 3636–3644.

[27] P. Zhang, X. Sheng, X. Chen, Z. Fang, J. Jiang, M. Wang, F. Li, L. Fan, Y. Ren, B. Zhang, et al., Paired electrocatalytic oxygenation and hydrogenation of organic substrates with water as the oxygen and hydrogen source., Angew. Chem. Int. Ed. 58 (2019) 9155–9159.

[28] X. Song, X. Liu, H. Wang, Y. Guo, Y. Wang, Improved performance of nickel boride by phosphorus doping as an efficient electrocatalyst for the oxidation of 5-hydroxymethylfurfural to 2,5-furandicarboxylic acid, Ind. Eng. Chem. Res. 59 (2020) 17348–17356.

[29] N. Zhang, Y. Zou, L. Tao, W. Chen, L. Zhou, Z. Liu, et al., Electrochemical oxidation of 5-hydroxymethylfurfural on nickel nitride/carbon nanosheets: reaction pathway determined by in situ sum frequency generation vibrational spectroscopy, Angew. Chem. Int. Ed. 58 (2019) 15895–15903.

[30] M. Zhang, Y. Liu, B. Liu, Z. Chen, H. Xu, K. Yan, Trimetallic NiCoFe-layered double hydroxide nanosheets efficient for oxygen evolution and highly selective oxidation of biomass-derived 5-hydroxymethylfurfural, ACS Catal. 10 (2020) 5179–5189.

[31] Y.F Qi, K.Y Wang, Y. Sun, J. Wang, C. Wang, Engineering the electronic structure of NiFe layered double hydroxide nanosheet array by implanting cationic vacancies for efficient electrochemical conversion of 5-hydroxymethylfurfural to 2,5-furandicarboxylic acid, ACS Sustain. Chem. Eng. 10 (2022) 645–654.

[32] Y. Zhong, R.Q. Ren, J.B. Wang, Y.Y. Peng, Q. Li, Y.M. Fan, Grass-like NixSey nanowire arrays shelled with NiFe LDH nanosheets as a 3D hierarchical core–shell electrocatalyst for efficient upgrading of biomass-derived 5-hydroxymethylfurfural and furfural, Catal. Sci. Technol. 12 (2022) 201–211.

[33] B. Liu, Z. Zheng, Y. Liu, M. Zhang, Y. Wang, Y. Wan, et al., Efficient electrooxidation of biomass-derived aldehydes over ultrathin NiV-layered double hydroxide films, J. Energy Chem. 78 (2023) 412–421.

[34] Y. Huang, X. Pang, J. Cui, Z. Huang, G. Wang, H. Zhao, et al., Strengthening the stability of the reconstructed NiOOH phase for 5-hydroxymethylfurfural oxidation, Inorg. Chem. 62 (2023) 6499–6509.

[35] B. Zhu, Y. Qin, J. Du, F. Zhang, X. Lei, Ammonia etching to generate oxygen vacancies on $CuMn_2O_4$ for highly efficient electrocatalytic oxidation of 5-hydroxymethylfurfural, ACS Sustain. Chem. Eng. 9 (2021) 11790–11797.

[36] Y. Lu, C.L. Dong, Y.C. Huang, Y. Zou, Z. Liu, Y. Liu, et al., Identifying the geometric site dependence of spinel oxides for the electrooxidation of 5-hydroxymethylfurfural, Angew. Chem. Int. Ed. 59 (2020) 19215–19221.

[37] Y. Lu, C.L. Dong, Y.C. Huang, Y. Zou, Y. Liu, Y. Li, et al., Hierarchically nanostructured NiO–Co_3O_4 with rich interface defects for the electro-oxidation of 5-hydroxymethylfurfural, Sci. China Chem. 63 (2020) 980–986.

[38] H. Wang, Y. Zhou, S. Tao, CoP–CoOOH heterojunction with modulating interfacial electronic structure: a robust biomass-upgrading electrocatalyst, Appl. Catal. B 315 (2022) 121588.

[39] R. Luo, Y. Li, L. Xing, N. Wang, R. Zhong, Z. Qian, et al., A dynamic $Ni(OH)_2$–NiOOH/NiFeP heterojunction enabling high-performance electro-upgrading of hydroxymethylfurfural, Appl. Catal. B 311 (2022) 121357.

[40] M. Xing, D. Zhang, D. Liu, C. Song, D. Wang, Surface engineering of carbon-coated cobalt-doped nickel phosphides bifunctional electrocatalyst for boosting 5-hydroxymethylfurfural oxidation coupled with hydrogen evolution, J. Colloid Interface Sci. 629 (2023) 451–460.

[41] P. Zhou, G. Hai, G. Zhao, R. Li, X. Huang, Y. Lu, et al., CeO_2 as an electron pump to boost the performance of Co_4N in electrocatalytic hydrogen evolution, oxygen evolution, and biomass oxidation valorization, Appl. Catal. B 325 (2023) 122364.

[42] J. Wu, J. Chen, T. Yu, Z. Zhai, Y. Zhu, X. Wu, et al., Boosting electrochemical kinetics of $NiCo_2$ via MoO_2 modification for biomass upgrading assisted hydrogen evolution, ACS Catal. 13 (2023) 13257–13266.

[43] M. Cai, Y. Zhang, Y. Zhao, Q. Liu, Y. Li, G. Li, Two-dimensional metal–organic framework nanosheets for highly efficient electrocatalytic biomass 5-(hydroxymethyl)furfural valorization, J. Mater. Chem. A 8 (2020) 20386–20392.

[44] Y. Zhang, N. Kornienko, Conductive metal–organic frameworks bearing M–O_4 active sites as highly active biomass valorization electrocatalysts, ChemSusChem 15 (2022) e202101587.

[45] Y. Feng, K. Yang, R.L. Smith, X. Qi, Metal sulfide enhanced metal–organic framework nanoarrays for electrocatalytic oxidation of 5-hydroxymethylfurfural to 2,5-furandicarboxylic acid, J. Mater. Chem. A 11 (2023) 6375–6383.

[46] X. Pang, H. Zhao, Y. Huang, B. Luo, H. Bai, W. Fan, Electrochemically induced NiOOH/Ag^+ active species for efficient oxidation of 5-hydroxymethylfurfural, Appl. Surf. Sci. 608 (2023) 155152.

[47] M. Cai, S. Ding, B. Gibbons, X. Yang, M.C. Kessinger, A.J Morris, Nickel(II)-modified covalent-organic framework film for electrocatalytic oxidation of 5-hydroxymethylfurfural, Chem. Commun. 56 (2020) 14361–14364.

[48] Q. Qin, T. Heil, J. Schmidt, M. Schmallegger, G. Gescheidt, M. Antonietti, et al., Electrochemical fixation of nitrogen and its coupling with biomass valorization with a strongly adsorbing and defect optimized boron–carbon–nitrogen catalyst, ACS Appl. Energy Mater. 2 (2019) 8359–8365.

[49] C. Tang, D. Cui, Z. Li, L. Yu, J. Xue, X. Yin, Electrooxidation degradation of hydroxychloroquine in wastewater using a long-acting Ti-based PbO_2 anode with an arc-sprayed (Ti, Zr) N interlayer, Chemosphere 335 (2023) 139074.

CHAPTER

19

Electrocatalytic depolymerization of lignin

Yeddula Nikhileshwar Reddy[a,b], *Lakshmana Reddy Nagappagari*[c], and *Jayeeta Bhaumik*[a]

[a]Department of Biotechnology (DBT), Laboratory of Bioproduct Chemistry, BRIC-National Agri-Food and Biomanufacturing Institute (BRIC NABI), Government of India, Punjab, India, [b]Department of Biotechnology, School of Bioengineering, SRM Institute of Science and Technology, Kattankulathur, Tamil Nadu, India, [c]Catalysis and Spectrochemistry Laboratory, ENSICAEN, CNRS, LCS, University of Caen Normandy, Caen, France

19.1 Introduction

The utilization of lignocellulosic biomass presents a significant challenge due to its recalcitrant nature, which arises from its structural complexity [1]. While the cellulose and hemicellulose components of biomass are vital to industries such as ethanol production, paper manufacturing, and pulp processing, these processes rely heavily on delignification. This step is projected to produce approximately 225 million tons of lignin annually by 2030, with an estimated market value of USD 700 million [2,3]. The lignin market is expected to grow at a compound annual growth rate (CAGR) of 2.2%, potentially exceeding USD 900 million globally [3].

Despite its abundance and being the only aromatic and energy-rich polymer in biomass, lignin remains underutilized due to its amorphous polyphenolic structure. Agricultural residues such as rice and wheat straw, significant sources of lignin, are often discarded, carbonized, or used as animal feed. However, these residues hold tremendous potential as

Electrochemical Biomass Conversion: Sustainable Processes and Applications for Bioenergy, Biofuels, and Bioproducts
DOI: https://doi.org/10.1016/B978-0-443-26757-4.00011-8

reservoirs for high-value lignin-derived products [4]. In lignocellulosic biomass, lignin comprises 15%–30% of the dry weight, providing high-energy content up to 40%, making it a promising renewable resource for platform chemicals and high-quality fuels [5]. Nevertheless, only 2% of industrially produced lignin is transformed into value-added chemicals, such as vanillin and biofuels, while the majority is incinerated to produce heat [6–9].

Lignin's complexity arises from its heteroaromatic nature during plant biosynthesis. It is primarily composed of three main monomers, H (hydroxyphenyl), G (guaiacyl), and S (syringyl) derived from their corresponding monolignols, along with other minor monomers that contribute to structural heterogeneity. The fundamental lignin building block is typically a phenylpropanoid structure. During lignin biosynthesis, various C–O and C–C linkages are formed between these units, resulting in the native polymer. Recent research indicates that lignin in its natural state is less branched and more tractable than previously thought. Additionally, over the past decade, it has been discovered that other compounds, such as flavonoids and hydroxy stilbenes, can integrate into lignin. Fig. 19.1 highlights several lignin model structures, showcasing key characteristics with different plant origins like hardwood, softwood, and grass [10].

Globally, an estimated 700 million metric tons of lignin are produced annually, with demand set to rise due to the expansion of biorefineries. However, a considerable portion of lignin, particularly black liquor from the paper industry, is discarded in wastewater, leading to both environmental concerns and the loss of valuable aromatic compounds [11,12]. Efficient valorization of lignin for aromatic compound production requires overcoming significant challenges in lignin depolymerization. The polymeric structure of lignin, featuring diverse C–C and C–O–C linkages with high bond dissociation energies, coupled with its poor solubility in conventional organic solvents, limits its susceptibility to degradation [13–15]. Depolymerization often results in the repolymerization of lignin-derived monomers into more resistant polymers, further complicating the selective production of phenolic compounds [16]. Over the past decade, extensive research has explored various lignin depolymerization strategies, including acidic and basic treatments, oxidative and reductive processes, photocatalytic and advanced catalytic methods [17–19]. These approaches necessitate precise control over reaction parameters such as catalysts, solvents, temperature, and pH to achieve high efficiency and selectivity. Among emerging technologies, electrocatalysis stands out for its environmental benefits, leveraging renewable electricity and operating under mild conditions with high energy efficiency [20–23]. Electrocatalytic systems are also characterized by simplicity, modularity, and reduced construction and maintenance costs [24].

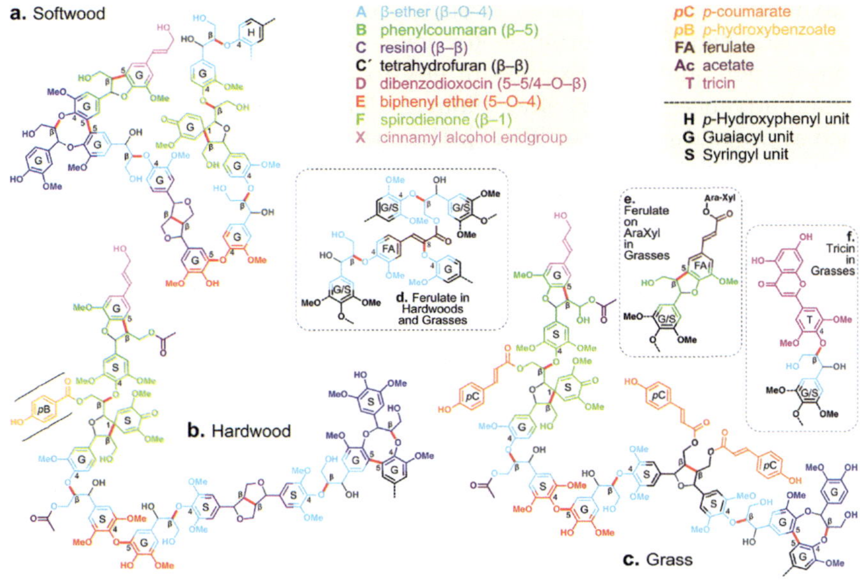

FIGURE 19.1 Model lignin structures containing 11 units for (A) a softwood, (B) a hardwood, and (C) grass, highlighting the major and more intriguing lignin units with "legal" interunit bonding. Key features include: (A) A dibenzodioxin unit (D), which is formed through 5–5 coupling of oligomer units followed by 4-O-β coupling with a coniferyl alcohol monomer, although it does not create a true "Y-type" branch point. It also features a spirodienone (F) formed via β–1 coupling and a 4–O–5 (biphenyl ether) unit (E) from oligomer coupling, which does not lead to a branch point as it exists solely in its free-phenolic form. (B) Similar units to those found in softwood, but incorporating S-units where applicable. This also includes the acylation of a γ-OH with one acetate and one p-hydroxybenzoate, the latter being typical of species like willow, aspen, palm, and poplar. (C) Similar units as in hardwood and softwood but featuring a β–β coupled unit derived from the dimerization of sinapyl p-coumarate, resulting in a tetrahydrofuran (C′) instead of the resinol (C) from unacylated monolignols. It showcases natural lignin acylation by acetate and p-coumarate, which are common features in grasses. Additional notable structures include: (D) A structure involving ferulate units (FA), which arise from lignification with monolignol-ferulate conjugates, typical in grasses and some hardwoods. (E) Units formed from cross-linking between lignin and arabinoxylan via ferulates acylating the latter, a characteristic feature of grasses and commelinid monocots. (F) The tricin (T) is a flavone that acts as a chain starter in nearly all grasses. *From M.M. Abu-Omar, K. Barta, G.T. Beckham, J.S. Luterbacher, J. Ralph, R. Rinaldi, et al., Guidelines for performing lignin-first biorefining, Energy Environ. Sci. 14 (1) (2021) 262–292. https://doi.org/10.1039/D0EE02870C. Copyright @ Royal Society of Chemistry 2021.*

Electrocatalysis offers exceptional tunability in activity and selectivity through variations in applied potentials, electrolytes, and catalyst design. Strategies such as potential adjustment, electrolyte optimization, and advanced catalyst engineering have demonstrated the potential for efficient and selective conversion of lignin into target compounds [25]. Compared with thermocatalysis, photocatalysis, and enzymatic catalysis, electrocatalysis holds significant promise for value-added lignin conversion. Despite its potential, the progress of electrocatalytic lignin upgrading lags behind advancements in other biomass-derived compound conversions, such as polyols and furans [26,27]. The challenges stem from lignin's high molecular weight, hydrophobic functional groups, and poor solubility in aqueous media [28], which reduce its interaction with electrodes and lead to competing reactions such as water splitting. Additionally, the structural diversity of lignin's functional groups and linkages generates a wide array of degradation products, which affects selectivity and, in turn, complicates purification processes and downstream separation.

Effective depolymerization aims to selectively break these bonds to release aromatic monomers and other platform chemicals. At the heart of this method is electrochemical activation, which uses an applied potential to drive redox reactions. Oxidative depolymerization takes place at the anode, where lignin bonds are cleaved through oxidation, while reductive depolymerization involves reducing specific intermediates at the cathode to promote fragmentation. Electrocatalysts are vital in facilitating these reactions. Heterogeneous catalysts, such as transition metal oxides and carbon-based materials, are commonly used to improve selectivity and reaction rates. Additionally, homogeneous catalysts, including soluble redox mediators, can transfer electrons between the electrode and lignin molecules, further enhancing efficiency. The reaction medium and operating conditions play a significant role in the process. The choice of electrolyte—acidic, neutral, or basic—affects lignin solubility and reaction pathways, while adjustments in temperature and pressure can boost reaction rates, although milder conditions are generally preferred to maintain product integrity. Achieving high selectivity for target products, such as vanillin and syringol, is a primary objective. At the same time, careful control of parameters is crucial to minimize unwanted polymerization and char formation. This chapter focuses on lignin degradation via electrocatalysis, highlighting its promise as a sustainable, efficient, and versatile approach to unlock the potential of lignin as a renewable resource for chemical production.

19.2 Role of various components in electrocatalytic depolymerization reactions of lignin

The electrocatalytic depolymerization of lignin is a promising approach to break down this complex biopolymer into valuable aromatic compounds through electrochemical processes. Several components are involved in this process, each playing a critical role in ensuring efficient depolymerization, high selectivity for target products, and overall system performance. These components include the electrode materials, electrolytes, lignin feedstock, applied voltage and current, and various additives or cocatalysts that can enhance reaction rates or selectivity. When multiple electrochemically active species are present in a reaction medium, electrocatalysis occurs at the electrode surface, where substrates are adsorbed. The selectivity of this process is influenced by various factors, such as the redox potentials of the substrates, the morphology of the electrode surface, the solubility and affinity of the substrate for the catalyst surface, and the applied electrode potential. Recent developments highlight the importance of electrode nanostructuring and surface engineering in improving selectivity by adjusting adsorption phenomena and the availability of active sites. Additionally, the application of in situ spectroscopic and computational techniques has offered deeper insights into these interactions, facilitating the rational design of customized catalysts [29–32].

Electrode materials: The electrode materials are crucial as they act as the catalyst for the electrochemical reactions in lignin depolymerization. The choice of electrode material determines the efficiency, selectivity, and stability of the process. Common electrode materials include: metal electrodes, such as rhodium, ruthenium (Ru), palladium (Pd), iridium (Ir), and platinum (Pt), which are examples of noble metals used for their high conductivity and catalytic activity [33–35]. Besides, multimetallic catalysts can also be used for tuning the geometric, electronic, and compositional properties of the products, which are also being explored [36,37]. These metals can promote hydrogenation reactions at the anode surface, which are essential for breaking down the lignin structure. Graphite, carbon cloth, and carbon paper are commonly employed due to their good conductivity, chemical stability, and high surface area [23,33,38]. They can also be modified with different materials to enhance electrocatalytic performance. Other transition earth-abundant metal-based electrode materials such as nickel, copper, and cobalt are also used, either alone or as alloys [39–41]. These metals can be modified with oxide or hydroxide coatings to boost their catalytic activity, particularly in facilitating the cleavage of C–O and C–C bonds in lignin [42]. Besides, there are other nonmetal catalysts that can even catalyze the lignin depolymerization reactions [42].

Electrolytes: The electrolyte is the medium through which ions flow during the electrochemical reaction, enabling the charge transfer between lignin molecules and the electrode. The composition of the electrolyte significantly affects the efficiency and degree of lignin depolymerization. Common electrolyte systems include acidic electrolytes are made up of solutions such as H_2SO_4, HCl, and TFA, which are often used to facilitate the oxidation of lignin and improve the solubility of the products [43,44]. Basic electrolytes such as NaOH, KOH, Na_2CO_3, and $NaHCO_3$ solutions can also be used, particularly when more selective reduction or bond scission. Acidic conditions help break down the complex structure of lignin, resulting in the production of aldehydes, ketones, and various other compounds. On the other hand, basic conditions promote the formation of acids and alcohols [45]. Neutral electrolytes such as KCl and NaCl are used to maintain a more stable pH and to control the reaction rate [46]. In aqueous electrolytes, the electrochemical redox reactions of water, such as the hydrogen evolution reaction (HER) and the oxygen evolution reaction (OER), often compete with the desired organic transformations and also pave the way for hydrogenation reactions [25,47].

Lignin feedstock: The structure of lignin significantly influences its electrodepolymerization process due to its complex and heterogeneous nature. Lignin is composed of three primary phenylpropanoid monomers: coniferyl, coumaryl, and sinapyl alcohol, which can form various linkages, primarily through β-O-4 ether bonds. These structural characteristics affect both the reactivity and the efficiency of the polymerization process impacts of lignin structure on electrodepolymerization [48]. Lignin contains numerous functional groups, including methoxy and hydroxyl groups, which actively participate in electrochemical reactions. These groups enhance the reactivity of lignin during electrodepolymerization, facilitating improved interaction with the surface of the electrode. The molecular weight and the types of linkages present in lignin influence its solubility and the ease with which it can be depolymerized into smaller fragments. Higher molecular weight lignins may require more energy for depolymerization, affecting the overall efficiency of the polymerization process. For instance, lignins with a high β-O-4 linkage content tend to yield better results in terms of monomer production during depolymerization, which is crucial for effective electrodepolymerization. Industrial processes such as pulping and fractionation alter lignin's structure, leading to variations in its polymerization behavior. For example, kraft lignin often exhibits significant cleavage of β-aryl ether linkages, making it more amenable to electrochemical reactions compared with other types that may retain more intact structures. Moreover, the self-assembly behavior of lignin in aqueous environments is driven by noncovalent interactions between its monomers. This property can affect how lignin deposits on electrodes during polymerization, potentially influencing the morphology and conductivity of the resulting polymer films. The structural characteristics of lignin, such as its functional groups, molecular weight, linkage types, and changes induced by processing, play a crucial role in determining its behavior during electrodepolymerization. Understanding these factors is essential for optimizing lignin-based materials for various applications in biofuels and biomaterials. A tabulated summary (Table 19.1) is provided to outline the various types of lignin, highlighting the key and structurally significant characteristics that impact the depolymerization process.

TABLE 19.1 Different types of extracted lignins and their characteristics.

Types of Lignin	Characteristics	References
Kraft lignin	High lignin recovery and purity (>90%) with high 2000–7000 g/mol Mw, β-O-4 bond cleavage. Condensation at C5 in the aromatic ring formation of stilbene and vinyl ethers.	[49]
Soda lignin	High recovery and purity (~93%) of lignin with demethoxylated structure 1000–3000 g/mol Mw. β-O-4 bond cleavage is observed.	[50]
Lignosulfonate	high recovery with a broad molecular weight range of 1000–150,000 g/mol and sulfonate groups attached at the Cα position of lignin.	[51]
Milled wood lignin (MWL)	Close to native lignin with up to 50% recovery.	[52,53]
Cellulolytic enzyme lignin (CEL)	Similar to native lignin, with better recovery than MWL.	[52,53]
Enzymatic mild acidolysis lignin (EMAL)	Breakdown of lignin-carbohydrate bonds by mild acidolysis. High yield and purity are better with close structural similarity to MWL.	[52,53]
Hot-water extracted lignin	low recovery (10%–40%) of lignin with 5000–10,000 g/mol Mw. Mild β-O-4 bond cleavage.	[54]
Ammonia lignin	Good recovery 50%–70% of lignin with ~4000 g/mol Mw. Preserved β-O-4 linkage. Transformation of coumarates and ferulates to corresponding amides.	[55]
Cosolvent-enhanced lignocellulosic fractionation (CELF) lignin	High purity and recovery with substantial β-O-4 bond cleavage resulting in a large condensed structure 1500 – 3000 g/mol. Mw similar to ethanol organosolv lignin.	[56]
Ethanol organosolv lignin	Lignin recovery with high purity with severe β-O-4 cleavage. Large condensed structure 1000–4000 g/mol Mw.	[57]
Cyrene lignin	>60% lignin recovery with 4000–6300 g/mol Mw. Minimal condensation reactions preserving β-O-4 linkages.	[58]
Ionic liquid lignin	Well-preserved linkages of β-O-4 bonds and less condensed structure with a precise 3200– 3400 g/mol Mw with high purity and recovery.	[59]

(continued on next page)

TABLE 19.1 Different types of extracted lignins and their characteristics—cont'd

Types of Lignin	Characteristics	References
Deep eutectic solvents (DES) lignin	Lignin recovery varies depending on HBA and HBD. With residual in the recovered lignin, the degree of condensation varies by conditions. Low Mw lignins (500–2600 g/mol) due to cleavage of β-O-4 linkage.	[60]
γ-valerolactone (GVL) lignin	Good lignin recovery ~80% with preserved β-O-4 linkage and less condensed structure ~5500 g/mol Mw.	[61]
Reductive catalytic fractionated lignin (RCF)	Cleaves β-O-4 linkage and yields low molecular weight aromatic compounds with a high yield of monomers and minimal condensation. Also called the lignin first approach.	[62]
Klason lignin	Residual lignin resists the acid hydrolysis process with a highly condensed nature. High molecular weight lignin.	[63]
Aldehyde-assisted fractionated lignin (AAFL)	Facilitates lignin valorization using industrially relevant waste particle sizes (3–5 mm). Achieves monomer/oligomer yields that are higher compared with those obtained through RCF.	[64]

Applied voltage and current: The rate and extent of lignin depolymerization are influenced by the applied electrical potential (voltage) and current density. When higher voltages are used, they can trigger more aggressive oxidation reactions at the anode, potentially resulting in a wider variety of products. However, if the voltage is too high, it may lead to overoxidation, which can reduce the selectivity for the desired products. The current density plays a crucial role in determining the overall current flow through the system, affecting both the reaction rate and the formation of intermediates. Therefore optimizing these parameters is vital for achieving efficient depolymerization while minimizing side reactions [18].

Additives and cocatalysts: The use of additives or cocatalysts can greatly affect the electrochemical depolymerization process. These substances can serve various functions: Solvents like ethanol, acetonitrile, or dimethyl sulfoxide, 1-ethyl-3-methylimidazolium acetate, and 1-butyl-3-methylimidazolium chloride may be introduced to enhance lignin solubility, stabilize reaction intermediates, or boost product yields [65]. Also, the choice of solvent can influence the electrochemical characteristics of the system, affecting conductivity and product formation. Redox-active mediators, such as quinones, metal ions, N-hydroxyphthalimide (NHPI) [66], TEMPO [67], 2,3-dichloro-5,6-dicyano-1,4-benzoquinone (DDQ) [68], and pyridinium dichromate [69]. can facilitate electron transfer between the electrode and lignin molecules, improving the efficiency of lignin depolymerization. Additionally, these mediators can enhance the selectivity of the reaction by stabilizing certain intermediates or products.

Supporting electrolytes: Additional salts or buffers, like sodium sulfate or potassium phosphate, can be added to help maintain electrolyte conductivity, control pH levels, or stabilize the electrode surface during the process. The operating conditions, such as temperature and pressure, also influence the efficiency of electrocatalytic depolymerization. While these processes are typically carried out at ambient temperature and pressure, slightly raising the temperature can enhance the reaction rate by improving mass transport and lowering reaction barriers. It's also essential to manage temperature to avoid unwanted side reactions, like overoxidation or the creation of byproducts. The long-term stability and durability of electrocatalysts are crucial for their practical use. Many electrocatalysts, particularly those made from noble metals, can degrade over time due to the challenging electrochemical conditions. Catalyst deactivation may occur from fouling, surface oxidation, or the leaching of active species. Consequently, enhancing catalyst stability through surface modifications, alloying, or the use of support materials remains a key focus of ongoing research [70].

19.3 Pathways in electrocatalytic depolymerization of lignin

The electrochemical valorization of lignin has been a topic of research since the 1940s. Early studies by Bailey et al. [71] showed that butanol lignin could be cleaved at a lead cathode, resulting in the production of compounds such as methyl ethyl ketone, acetic acid, and acetone. Following this, Chen et al. [72] performed systematic evaluations of IrO_2-based electrodes for lignin degradation, determining that Ti/RuO_2-IrO_2 was the most effective catalyst. They advanced this research by examining Ti/TiO_2NT/PbO_2 electrodes for lignin valorization in wastewater treatment. Meanwhile, researchers compared various anodes, including Ni, Au, Pt, Cu, dimensionally stable anodes (DSA-O_2), and PbO_2, for the electrooxidation of kraft lignin and spent liquor effluents. Zhu et al. [73] investigated an electrolytic cell setup with a graphite felt cathode and RuO_2-IrO_2/Ti anode [25]. Their findings revealed two primary pathways for lignin depolymerization: direct anodic oxidation and an indirect mechanism facilitated by hydrogen peroxide (H_2O_2) generated at the cathode.

The H_2O_2 decomposed into reactive oxygen species (ROS), which played a vital role in breaking lignin bonds. The study indicated that higher current densities and temperatures increased the production of H_2O_2 and ROS, thereby boosting the yield of LMW products [74]. Furthermore, the addition of external oxygen significantly enhanced lignin depolymerization, a conclusion supported by other studies using similar setups. Much of the research focused on maximizing LMW product yields. Explored a more cost-effective approach by utilizing nickel foam as the working electrode. Their results highlighted the potential of nickel foam for lignin depolymerization, achieving combined yields of vanillin and syringaldehyde (valuable aromatic aldehydes) of up to 17.5%. The study also proposed further advancements by developing hierarchically porous nickel-based electrocatalysts deposited on nickel foam, suggesting promising avenues for future improvements. So far, most reported reactions involving lignin have focused on electrocatalytic oxidation processes, although redox reactions have also been studied separately. This chapter will explore various aspects of lignin depolymerization, which can occur through direct oxidation, reduction, or mediated oxidation.

19.3.1 General mechanism of electrocatalysts

The electrocatalytic conversion of organic molecules depends on the transfer of electrons between the substrate and the electrode surface, which occurs through three main mechanisms. First, direct electrochemical oxidation or reduction happens at the electrode-electrolyte boundary for substrates that are adsorbed on the electrode. Second, at the interface, substrates can undergo a redox reaction, leading to the formation of reactive intermediates that then rearrange or couple to yield the desired products. Third, when larger molecules cannot directly adsorb onto the electrode surface due to steric hindrance, a mediator facilitates electron transfer, allowing the redox reaction to take place away from the electrode [30]. These studies are usually carried out in batch reactors, which can be classified as either undivided cells, where the anode and cathode are situated in the same compartment, or divided cells, where the electrodes are separated by an ion-exchange membrane or a porous barrier. While undivided cells tend to be more cost-effective, divided cells offer benefits when the reaction products are susceptible to degradation or side reactions at the opposite electrode. The selectivity and efficiency of the reaction can be adjusted by varying the applied current, typically expressed as current density ($mA\ cm^{-2}$) [27,75]. However, calculating current density often depends on the geometric area of the electrode, which may not accurately reflect the actual surface area for porous or powdered electrodes. Electrocatalytic oxidation (ECO) at the anode involves a series of reactions. Hydroxyl radicals produced during water oxidation can adsorb

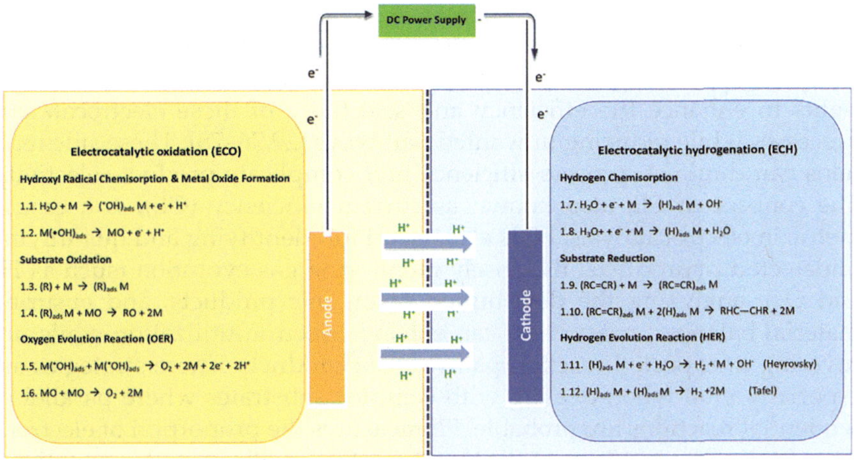

FIGURE 19.2 List of reactions involved in electrocatalytic hydrogenation and electrocatalytic oxidation. *From M. Garedew, C.H. Lam, L. Petitjean, S. Huang, B. Song, F. Lin, et al., Electrochemical upgrading of depolymerized lignin: a review of model compound studies, Green Chem. 23 (8) (2021) 2868–2899. https://doi.org/10.1039/D0GC04127K. Copyright @ Royal Society of Chemistry 2021.*

onto the metal electrode surface and interact with it to form metal oxides. These reactive species, along with free hydroxyl radicals, play a role in the oxidation of organic substrates. At the same time, the OER may take place, producing oxygen gas and diverting electrons from the intended substrate oxidation. The efficiency of these processes is usually assessed through Faradaic efficiency (FE), which indicates the fraction of charge utilized for the desired chemical transformation. The OER often competes with substrate oxidation, which can lower the overall FE of the system. The complete set of reactions involved in the oxidation and reduction has been depicted in Fig. 19.2.

Similarly, electrocatalytic hydrogenation and deoxygenation (ECH) take place at the cathode through a series of steps. Hydrogen is generated by reducing water protons on the catalyst surface, where it becomes chemisorbed. This adsorbed hydrogen then interacts with organic substrates attached to the electrode surface, enabling hydrogenation or deoxygenation. These processes are akin to traditional catalytic hydrogenation, where hydrogen and organic compounds interact on the catalytic surface to yield the desired products. Ultimately, the reduced products and hydrogen gas detach from the electrode, with hydrogen being released via electrochemical reactions such as the Tafel and Heyrovsky mechanisms (discussed in Section 19.3.1). However, the HER can compete with ECH, resulting in a decrease in FE. The properties of the catalyst and the characteristics of the substrate play a crucial role in determining the competition between favorable reactions, such as ECO or ECH, and undesirable side reactions, like OER or HER. Current research is aimed at creating advanced catalysts, optimizing reaction conditions, and improving mechanistic insights to enhance the efficiency and selectivity of these electrocatalytic processes while reducing unwanted pathways [29,76–78]. These side reactions can diminish process efficiency and complicate product selectivity. The concept of FE, also known as current efficiency (CE), is a crucial metric in electrocatalysis. FE is also useful for identifying and quantifying undetected byproducts. By closely monitoring gas evolution (such as H_2 and O_2), analyzing the distribution of organic products, and ensuring material balances, researchers can enhance electron utilization while uncovering unexpected reaction pathways or products. These strategies are especially vital when dealing with complex substrates where parallel or sequential reactions are probable. FE measures the proportion of electrons that effectively contribute to the target reaction, as illustrated in equation:

$$\%\ \text{FE} = \frac{\text{mol} \times \text{F} \times \text{n}}{C_{\text{total}}} \times 100$$

where Faradaic efficiency (%*FE*), *mol* = moles of product formed; *F* = Faraday's constant, 96,485 C mol^{-1}; *n* = number of electrons per reaction; C_{total} = total charge passed.

The HER, oxygen reduction reaction (ORR), and OER are essential electrochemical processes that are fundamental during lignin conversion technologies. These reactions consist of intricate multistep mechanisms that are affected by various factors, including reaction pathways, intermediate species, and the properties of catalysts. Gaining a thorough understanding of the mechanisms behind HER, ORR, and OER is crucial for tackling their inherent challenges, such as slow kinetics and high overpotentials, and for enhancing their performance in real-world applications. In neutral media, the Volmer reaction leads to the creation of a hydrogen intermediate (M-H), which then generates H_2 either through the electrochemical Heyrovsky step or the chemical Tafel step. The mechanism of the HER is depicted schematically in Fig. 19.3A [79]. The ORR can occur through two pathways: a direct four-electron process (the direct $4e^-$ pathway) that is more efficient, or a slower two-step process that involves sequential two-electron transfers ($2e^- + 2e^-$ pathway), resulting in the conversion of oxygen molecules to OH^- ions. As illustrated in Fig. 19.3B, the $2e^- + 2e^-$ pathway starts with a two-electron reduction of oxygen to produce HO_2^-, which is then followed by either another two-electron reduction of HO_2^- or the chemical disproportionation of HO_2^- without direct electron transfer to the electrode. The high bond energy of the O=O bond (498 kJ mol^{-1}) and the complexity of the multielectron transfer process contribute to the inherently slow kinetics of the ORR, necessitating a significant overpotential [80]. To gain insights into

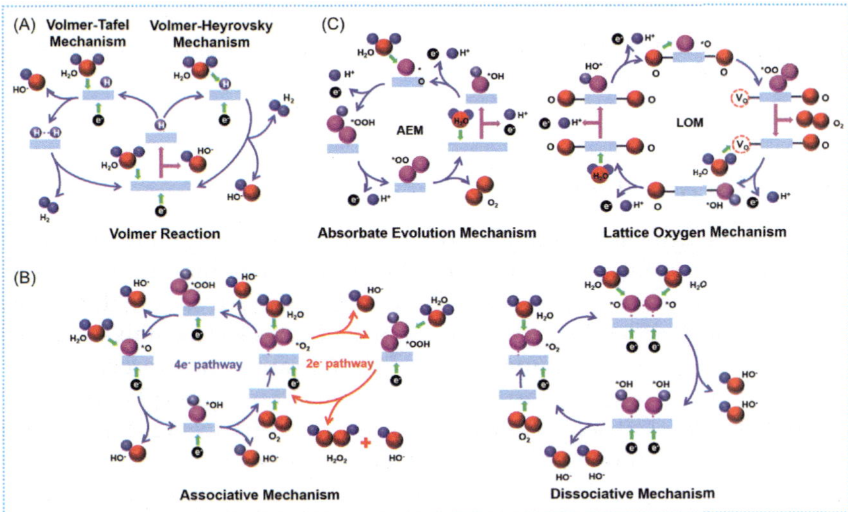

FIGURE 19.3 (A) Schematic diagram of the Volmer–Tafel and Volmer–Heyrovsky mechanisms for HER. (B) Schematic diagram of the associative and dissociative mechanisms for ORR. (C) Schematic diagram of the absorbate evolution mechanism (AEM) and lattice oxygen mechanism (LOM) for OER. *From F. Lai, H. Shang, Y. Jiao, X. Chen, T. Zhang, X. Liu, et al., Recent progress and perspective on electrocatalysis in neutral media: mechanisms, materials, and advanced characterizations, Interdisciplinary Mater. 3 (4) (2024) 492–529. https://doi.org/10.1002/idm2.12172. Copyright @ John Wiley & Sons 2024.*

the sources of this overpotential, Nørskov and colleagues developed a free-energy (ΔG) landscape for the reaction intermediates in the ORR, utilizing density functional theory calculations [81]. They identified two potential ORR mechanisms: the associative mechanism involving the HOO* intermediate and the direct dissociation mechanism of O_2, both of which are elaborated in Fig. 19.3B. The OER, which is the reverse of the ORR, is characterized by a high overpotential and often acts as the rate-limiting step in various energy conversion systems, including rechargeable metal–air batteries and water electrolysis devices. Two main OER mechanisms are proposed: the adsorbate evolution mechanism (AEM) and the lattice oxygen-mediated mechanism (LOM). In neutral media, the AEM operates through four sequential transfers of protons and electrons, involving oxygen intermediates like OH*, O*, and OOH*. The process begins with a water molecule adsorbing onto the catalyst surface, which then undergoes a two-step deprotonation to produce OH* and O*. Next, another water molecule interacts with the O* intermediate via a nucleophilic attack, resulting in the formation of the OOH* intermediate. The final step can either involve the deprotonation of the OOH* intermediate to release oxygen or the coupling of two O* intermediates, as shown in Fig. 19.3C

[82]. The detailed equations can be found elsewhere [46]. Considering these aspects, the electrochemical depolymerization can be categorized into three different pathways.

19.3.2 Electroreduction

Lignin's significant aromatic content makes it a valuable source for producing biochemicals and biofuels. With the rise of biorefineries, lignin is increasingly produced, and its utilization is key to advancing the circular economy. Electrochemical processes, particularly electrocatalysis, play a crucial role in lignin depolymerization. By integrating these processes with other sustainable sources, we can make significant progress [83]. Additionally, renewable electricity sources, such as green hydrogen (H_2), can be used in situ to generate necessary reagents such as oxygenated aromatic monomers during lignin conversion. ECH stands as a promising approach for the valorization of lignin or lignin-derived substrates. For example, ketone-alcohol (KA) oil (a mixture of cyclohexanol and cyclohexanone) obtained by transforming lignin can be used in nylon polymer manufacturing. These products are typically obtained through hydrogenation and hydrodeoxygenation reactions. The electrical reduction of lignin occurs at the cathode, where the process increases the hydrogen and carbon content in the resulting products. The reaction mechanism in electrocatalytic hydrogenolysis involves the reduction of lignin and its hydrogenation at the cathode, along with oxygen formation at the anode.

This pathway is particularly effective for targeting oxygen-containing functional groups, such as methoxy and hydroxyl groups, which are prevalent in lignin's structure. The mechanism involves the generation of adsorbed hydrogen (H_h) on the cathode surface, which reacts with lignin fragments to produce reduced products, such as phenols and hydrocarbons. Nickel-based cathodes have been widely investigated for this purpose due to their high HER activity and low cost. For instance, nickel foam electrodes have demonstrated the ability to achieve selective hydrogenation and deoxygenation of lignin, yielding valuable compounds like vanillin and syringaldehyde [39,44,84,85]. Electroreduction also offers advantages in terms of energy efficiency, as it operates at lower potentials compared with oxidation pathways. However, challenges include competition from HER, which can divert electrons away from lignin reduction. To address this, researchers have explored the use of tailored cathode materials and electrolyte formulations to enhance lignin-substrate interactions and suppress unwanted side reactions. An emerging strategy involves the deposition of hierarchically porous nickel-based electrocatalysts onto conductive substrates, which has been shown to improve both selectivity and yield [86]. By optimizing the surface area and electronic properties of the cathode, these systems can achieve higher Faradaic efficiencies and enable more effective valorization of lignin.

The hydrogenation of phenolic compounds in lignin can proceed via two distinct mechanisms. The first involves adsorbed hydrogen atoms (H_{ads}) in a process known as hydrogen atom transfer (HAT). This process begins with the generation of hydrogen at the metal electrode through the Volmer step, which involves proton-coupled electron transfer (PCET). The organic compounds then adsorb onto the electrode surface, where hydrogenation occurs, followed by desorption of the final product. As discussed previously, factors such as the diffusion kinetics of the electrolyte, hydrogen atom mobility, and reduced FE significantly impact the process. These limitations divert electrons away from the formation of desired products, instead reducing protons and generating hydrogen gas. This phenomenon can be further described by the Volmer–Tafel or Volmer–Heyrovsky mechanisms, where the Volmer step involves hydrogen adsorption, and the Tafel or Heyrovsky steps describe the recombination of adsorbed hydrogen atoms, often without additional electron involvement, as discussed in the previous section [46,87].

In aqueous systems, electron diversion to proton reduction further hinders efficiency, leading to unwanted H_2 gas production. Moreover, if unsaturated substrates must travel long distances to reach the catalyst within the electrode's pores, typically made of a noble metal, the hydrogenation process can be less effective. These challenges highlight critical aspects of optimizing the reduction and hydrogenation processes for lignin conversion [88]. However, electrochemical depolymerization of lignin faces several challenges. The complex structure, high molecular weight, and hydrophobic nature of lignin, which is compounded during extraction, pose significant obstacles. In addition, weak interactions with electrodes, reduced reactivity, and unwanted side reactions such as water splitting in aqueous media, further complicates the process. The most pressing challenge remains the achievement of a diverse range of products and their effective separation, an area where research is still ongoing. To better understand the mechanics of lignin depolymerization and guide the creation of efficient catalysts, research has concentrated on lignin model compounds such as phenol and guaiacol over the past 10 years. These compounds are used to replicate different monomer and dimer configurations. One significant benefit of this lignin deconstruction method is that it does not require external hydrogen gas or hydrogen donors because it is powered by electricity at ambient temperature and pressure. However, transport and kinetic diffusion limits continue to limit reaction speeds even when 3D porous electrodes with huge surface areas are used in ECH. Traditional electrocatalytic methods frequently use conductive polymeric binders to secure active catalyst particles to electrodes. Unfortunately, this approach can result in the loss of active catalytic sites and raises stability concerns for the attached catalysts, particularly under high current densities and prolonged exposure to the reaction environment. Consequently, many electrochemical hydrogenation (ECH) processes are performed at relatively low current densities (approximately 40 mA cm^{-2}) and show limited FE, usually ranging from 20% to 60%[89]. Despite these limitations, the generation of hydrogen gas during the process offers a distinct advantage by eliminating the need for external hydrogen sources or organic hydrogen donors. This also reduces catalyst deactivation caused by coking. Additionally, the byproducts formed during hydrogenation have potential applications in fuel cells, as their utilization can be optimized through controlled current and voltage inputs [83].

Phenol and guaiacol, which come from the complex structure of lignin, have been extensively researched in hydrogenation reactions. Noble metal catalystssuch as Pt, Pd, Ru, and Rh, are frequently utilized in electrode materials because of their outstanding activity in the HER, making them ideal for phenol reduction. Cyclohexanone, a partially hydrogenated product of phenol, is produced more significantly at lower phenol conversion rates. However, single-component noble-metal catalysts encounter challenges such as low current density (<10 mA cm^{-2}) and limited selectivity for cyclohexanone. Mechanistic studies indicate that the ECH of phenol adheres to the Langmuir–Hinshelwood mechanism [84]. The zero-order dependency of the reaction rate on phenol concentration at various currents indicates that the rate-determining step (RDS) entails a thermochemical interaction between adsorbed phenol and Hads. The surface covering of phenol and Hads on the electrode has a significant impact on the FE of the ECH process. Had coverage increases as the Volmer step on the catalyst surface becomes more noticeable at higher negative potentials. This higher concentration of Hads boosts HER, which in turn reduces the FE for desired ECH products such as cyclohexanone and cyclohexanol. Although raising the temperature can speedup the RDS and slightly enhance FE, the overall improvement is still limited. These observations underscore the delicate balance between catalyst activity, reaction conditions, and product selectivity in the hydrogenation of lignin-derived phenol.

Recent advancements in the field of ECH of phenol have emphasized the importance of alloy catalysts and optimized electrocatalyst design. Studies on catalysts supported on nitrogen-doped hierarchically porous carbon (NHPC) have demonstrated significant potential, with Pt-based systems showing remarkable performance and product selectivity [90]. For instance, NiPt/NHPC catalysts were able to control cyclohexanone adsorption and stop further hydrogenation, resulting in over 99% selectivity for cyclohexanone at a production rate of 115.1 mol h^{-1}. Likewise, RuPt/NHPC catalysts produced cyclohexanol at an even greater rate of 213.5 mol h^{-1} while achieving over 99% selectivity. A notable feature of these systems is the simultaneous phenol conversion at both cathode and anode, yielding cyclohexanone and benzoquinone, respectively. Complementing these findings, research on nonnoble catalysts has shown promise. Zhang and colleagues developed Ni nanoparticles supported on molybdenum dioxide (Ni-MoO_2), exploiting strong electronic interactions and oxygen vacancies to enhance ECH efficiency. By tuning the Ni content, they achieved high phenol conversion rates (>90%) and optimized product selectivity, with high oxygen vacancy densities favoring cyclohexanol (95% yield) and lower vacancy densities yielding cyclohexanone (86%) [91]. These results underscore the importance of catalyst design and the regulation of adsorption and desorption processes. In addition, electrocatalyst preparation and material optimization have significantly influenced ECH performance. For example, Pt-supported Vulcan XC-72R catalysts, bound with Teflon, have demonstrated improved current efficiency (85%) and phenol adsorption, as reported [34,35]. Further, Savadogo et al.[34,35] observed that alloying components, such as Pt–Co/C, and metal–support interactions, such as those seen in Pt/C electrodes, enhance activity and selectivity by weakening the Pt–H bond. Mechanistic studies have shown that, instead of phenol adsorption itself, the hydrogenation of phenol to cyclohexanol occurs through a surface interaction between adsorbed species, which is the RDS of the reaction. Song et al. identified that the Langmuir-type mechanism governs this process, wherein surface hydrogen is generated from proton reduction at the catalyst-electrode interface, distinguishing it from the H_2 dissociation mechanism typical of thermal hydrogenation.

In another lignin model, an alkoxy-substituent guaiacol (2-methyl phenol), a simultaneous hydrogenation and hydrodeoxygenation happens, leading to the formation of a different set of products, including 2-methoxycyclohexanol, 2-methoxycyclohexanone, phenol, cyclohexanol, and cyclohexanone. The relationship between the reaction rate and guaiacol concentration follows first or second-order kinetics, indicating that the rate-determining step is governed by the steric hindrance caused by the substituents present on the guaiacol rather than the reaction between adsorbed species, which complicates the electrocatalytic conversion. Thermodynamically, demethoxylation is favored, compromising the product selectivity toward cyclohexane derivatives.

To address these challenges and enhance liquid–solid mass transfer in electrochemical reactions, a stirred slurry electrochemical reactor was developed (Fig. 19.4). This reactor can function as either a moving bed/fluidized bed electrocatalytic system. In neutral–acidic environments, the protons required for ECH diffuse from anolyte to catholyte through the membrane. Ruthenium has proven to be an effective catalyst for hydrogenation, demonstrating exceptional activity in both hydrogenation and hydrolysis reactions. During the ECH reactions of Phenol, Singh et al. observed an unexpected thermal conversions even at low cathode potentials

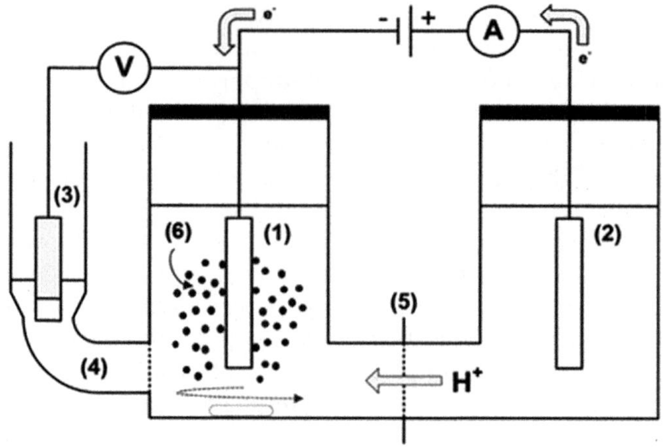

FIGURE 19.4 A sketch of electrochemical cell setup (H-cell) for ECH of guaiacol and phenol: (1) cathode, (2) anode, (3) reference electrode (Ag/AgCl/3 M KCl), (4) fritted side compartment (Luggin probe) for reference electrode filled with 3 M KCl, (5) cation exchange membrane, and (6) catalyst slurry. *From M. Nagy, M. Kosa, H. Theliander, A.J. Ragauskas, Characterization of CO2 precipitated kraft lignin to promote its utilization, Green Chem. 12 (1) (2010) 31–34. https://doi.org/10.1039/B913602A. Copyright @ Royal Society of Chemistry 2010.*

(−0.15 to −0.45 V vs. RHE) using rhodium on carbon (Rh/C) [92]. At higher temperatures reaching nearly 60°C, resulted in selective cyclohexanone formation upon constant cathode superficial current density −182 mA cm^{-2}. In another instance, the conversion of guaiacol was found to be influenced by the cathode potential, as it affects the surface coverage of adsorbed hydrogen radicals produced by proton electroreduction. The pathways for guaiacol conversion are shaped by both temperature and cathode potential, as these factors determine the availability of adsorbed hydrogen radicals crucial for the reaction.

After lignin monomers, lignin dimers that mimic the bonds such as 4-O-5, α-O-4, and β-O-4 have the potential to undergo hydrogenolysis and are rapidly gaining importance. The electronegativity difference between the C–O bond across the dimers can be polarized at the cathode [93]. A report published by using a superacid-noble metal particle dual catalyst suggests that further exploration of electrochemical depolymerization of lignin [94]. A comprehensive study by Xu et al. [95] revealed deeper insights about the hydrogenolysis and hydrolysis rate reaction rates increased in the following order, β-O-4 > 4-O-5 > α-O-4, with specific cleavage at the C–O bond. The reductive cleavage of lignin dimers under electrochemical conditions reveals complex reaction pathways that depend on substrate-specific interactions. Hydrolysis products are observed only under re-ductive conditions, suggesting a mechanism involving the addition of adsorbed hydrogen atoms (H*) followed by water attack. However, side reactions, such as hydrogenation of the phenyl group, hinder the cleavage of the C–O bond, affecting overall efficiency. For diphenyl ether (DPE), the reaction begins with hydrogenolysis but shifts toward hydrolysis as the concentration of DPE decreases. In biphenyl ether (BPE), hydrogenolysis dominates throughout the reaction. In phenoxybenzene (PEB), both hydrogenation and hydrolysis are initially significant, but hydrolysis becomes predominant as the PEB concentration decreases, accompanied by an intensifying recombination pathway. Across all substrates, the FE is adversely affected by the HER, which remains significant despite temperature adjustments, limiting the FE for lignin dimer cleavage to below 55%. The hydrogenolysis pathways of α-O-4 and β-O-4 linkages in lignin dimers have been extensively studied using catalysts such as Pt/C, Ru/C, Rh/C, and Raney Ni [92,96]. Cleavage products are typically phenol and arene, with steric hindrance from substituent groups on aromatic rings significantly reducing efficiency. The delocalized electrons in the phenyl group enhance the reactivity of the C–O bond, making α-O-4 linkages easier to cleave compared with β-O-4. However, hydrogenation of the aromatic rings prior to hydrogenolysis stabilizes the C–O bond, preventing its cleavage. The β-O-4 bond, due to its lower polarity, is harder to break, though introducing carbonyl groups at Cα improves efficiency. Controlled-potential electrolysis in organic electrolytes, demonstrated by Stephensons group, has shown promise in selectively cleaving β-O-4 linkages, producing lignin monomers or coupling products with yields of 59%–92% [97].

Expanding research into raw lignin, with its complex structures, via electrocatalyst reduction is essential for industrial applications. For instance, cathodic ECH catalyzed by Cu/Ni-Mo-Co with PbO_2 anode, achieved both reduction and oxidation of corn stover lignin [98]. These reactions resulted in the production of various valuable chemicals, including vanillin, trans-ferulic acid, syringaldehyde, 3-hydroxy-4-methoxyphenylethanone, 4-methoxy-3-methyl-phenol, and acetosyringone, with yields of 11.09, 22.40, 10.04, 2.37, 38.83, and 6.95 g per kg of lignin, respectively. Commercially available Ni foam was used as a working electrode for the oxidative degradation of ethanol-pretreated raw lignin with Vanillin and syringaldehyde as the primary small-molecule products from aspen and sweetgum lignin, with a maximum yield of 17.5% [99]. Additionally, a bifunctional catalyst was developed by Qin et al. [100], incorporating phosphorus into the $CoMoO_4$ system, enabling the direct conversion of lignin to benzoic acid at the anode with a yield of 56%.

19.3.3 Depolymerization via electrooxidation

As discussed in the earlier sections, the presence of aromatic heterocyclic nature makes it difficult for the lignin molecules to react with any other molecules. The electrochemical oxidation approach is one such approach that can be used for the generation of oxygenated molecules. These have the potential of converting into various derivatives that are important for the active pharmaceutical ingredient synthesis and many other biochemicals. For example, if the KA oil is oxidized, it can be converted into adipic acid by involving the cleavage of the C(O)–C bond, a monomer used for Nylon synthesis [101]. To cite another important example, if a polyhydric alcohol, for example, Glycerol, is electrooxidized, it can be converted into lactic acid, glycolic acid, and formic acid, the important molecules in the chemical and fuel industry. The electrooxidation process can be achieved in two different paths, either by direct oxidation or indirect oxidation. Generally, for lignin, the C–C/C–O bond cleavage and active group functionalization (Cα -Carbonylation) stand as competing reactions. The β-O-4 bond is predominantly observed in lignin structure, with hardwood and softwood comprise about 60% and 45%–50%. This bond plays a crucial role in depolymerization and delignification. Due to the presence of α-carbonyl groups in the β-O-4 bond, the condensation reactions during depolymerization are reduced and also accelerates the bond cleavage. Considering this point, a lot of emphasis has been given to the lignin dimer model compounds to understand the depolymerization via electrooxidation reactions either directly or indirectly. Firstly, we shall look into the aspects of direct oxidation, with a subsequent section dealing with the indirect oxidation.

19.3.3.1 Direct oxidation

In direct electrooxidation reactions, the electrode or the surface functionalized or doped porous electrode directly behaves as a catalyst involved in the conversion of the lignin. Some electrodes, such as RuO_2, PbO_2, and Ni-based catalysts, were among the first few, which were utilized for lignin depolymerization with simultaneous electrolysis happening at the surface [94]. The key challenges in the direct oxidation process are the effectiveness of the catalyst surface and lignin solubility. Direct electrooxidation involves multiple reactions, including depolymerization, functionalization, and various side reactions. During the electrocatalytic process, CV measurements are employed to identify the redox potentials of desired compounds in the electrolyte. For instance, CV analysis (Fig. 19.5A) of the lignin model compound 1G revealed oxidation peaks corresponding to unstable radical cation and dication intermediates: peak (c) at 1.2 V for the single-electron oxidation of 1G to the radical cation 1Ga and peak (d) at 1.8 V for its further oxidation to the dication 1Ga+ (Fig. 19.5B). A reduction peak (c′) at 1.1 V was observed, indicating the reduction of 1Ga back to 1G. Compound 1G undergoes transformation to a carbonyl compound 2G through an ECEC mechanism (E = electron transfer; C = chemical step)

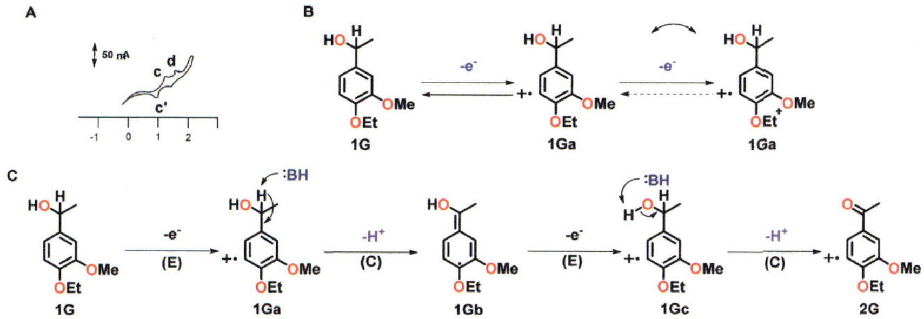

FIGURE 19.5 (A) CV of compound 1G in TFA in 0.1 M $LiClO_4/CH_3CN$; scan rate 0.1 V S^{-1}. (B) Proposed electron transfer and reaction mechanism for the lignin model compound (1G) in the trifluoroacetic anhydride (TFA) system. (C) ECEC (E = electron transfer and C = chemical step) mechanism of compound 1G. *From T. Shiraishi, T. Takano, H. Kamitakahara, F. Nakatsubo, Studies on electrooxidation of lignin and lignin model compounds. Part 1: direct electrooxidation of non-phenolic lignin model compounds, Holzforschung 66 (3) (2012) 303–309. https://doi.org/10.1515/hf.2011.069. Copyright 2011, Walter De Gruyter.*

in four steps (Fig. 19.5C): (E1) a single-electron oxidation, (C1) deprotonation of the Cα-H bond, (E2) a second single-electron oxidation, and (C2) deprotonation of the Cα-OH bond. Direct electrooxidation processes have successfully achieved electrochemical degradation of lignin, involving the cleavage of C–O or C–C bonds. Mechanistic studies demonstrated that the C–O bond in the abundant β-O-4 aryl ether linkage can be cleaved using electrocatalysts. Cleavage of the Cα-Cβ bond is also a common pathway in β-O-4 linkage oxidation.

In some instances, the electrooxidation of β-O-4 dimer models proceeds through Cα-Cβ bond cleavage, resulting in monomer products, confirming competition between Cα-Cβ cleavage and Cα-carbonylation reactions [43]. During the depolymerization process, side reactions, such as polymerization, are also observed. When lignin monomer syringaldehyde was oxidized, due to intermolecular condensation and radical coupling, dimers and corresponding acids as byproducts were formed under mild alkaline conditions and lower-than-usual temperatures. Interestingly, adding lignin to aqueous solutions of anode reduced the cathode potential required for H_2 production from water. Though lignin oxidation was also observed at the anode. The half-cell reactions for lignin-assisted water electrolysis are provided below.

$$\textit{Anode}\text{: } L + 4OH^- \rightarrow [O\text{-}L\text{-}O] + 2H_2O + 4e^-$$

$$\textit{Cathode}\text{: } 4H_2O + 4e^- \rightarrow 4OH^- + 2H_2$$

$$\textit{Overall}\text{: } L + 4H_2O \rightarrow [O\text{-}L\text{-}O] + 2H_2$$

The least thermodynamic voltage (E^0) desired for the described electrolysis process at 25°C can be deduced using the formula E^0 = delta G/(n.F) $\Delta G = 19.4$kcal/mol, n = 4, F = 96,500 C/mol. This calculation yields a minimum voltage of 0.21 V, which is significantly lower than the 1.23 V needed for conventional water electrolysis, indicating the beneficial effects of adding lignin. Additionally, at low overpotentials of the anode, OER is replaced by CO_2 formation [102]. The choice at the anode electrocatalyst significantly influences hydrogen production at the cathode. On the contrary, a β-PbO_2/MWNTs anodic electrocatalyst developed by Bateni et al. [103] showed an increase in the hydrogen production rate of 45.6 mL/h at the cathode with 20% less energy consumption when an anion exchange membrane electrolysis cell was used for electrolysis of alkaline water. This setup also resulted in a substantial increase in lignin oxidation at the anode. Similarly, NiCo/TiO_2 demonstrated comparable effects as an anode electrocatalyst [104]. Polymer electrolyte membrane (PEM)-based reactors proved that, at low electrical potentials ~0.45 V, the continuous-flow lignin electrolysis in alkaline solutions achieved higher H_2 production. Overall, incorporating lignin at the anode lowers the required electrical potential for hydrogen production at the cathode, reducing energy consumption while simultaneously depolymerizing lignin into low molecular weight (LMW) compounds. Direct oxidation of organic compounds involves the transfer of charge from the compound directly to ROS, on the electrocatalyst surface. This process relies on the formation of electron-deficient species such as adsorbed hydroxyl groups (OH*) or bridging oxygen species (μ_2-O) on the catalyst under applied potentials. These species facilitate hydrogen atom abstraction from the organic substrate through a PCET or sequential proton transfer–electron transfer (PT–ET) mechanism. Following this, C–C bonds in the adsorbed organic compounds are cleaved.

For example, Li et al. [105] investigated the selective β-O-4 bond cleavage using a Pt micro disc electrode. Their findings indicated that the hydroxy or carbonyl group at the Cα position of the β-O-4 linkage promotes C–C bond cleavage. Specifically, a benzene radical cation was observed to be involved in the oxidation of 2-phenoxy-1-phenethanol when the phenolic group undergoes single-electron oxidation. The highest occupied molecular orbital (HOMO) of this cation delocalizes across the phenolic group and the β-O-4 linkage, activating the bond. Consequently, cleavage of the Cα-Cβ bond produces phenylacetaldehyde, while C_β-O bond cleavage yields benzaldehyde. In contrast, for 2-phenoxyacetophenone, the carbonyl group's superior electron delocalization ability compared with the hydroxyl group further enhances the cleavage of the Cα-Cβ bond. However, in the electrooxidation of the model compound, devoid of a hydroxyl group on Cα in 2-phenoxy-1-phenylethane, cleavage is of the C_β-O bond, producing phenol and phenethyl alcohol. The presence of a hydroxyl group at Cα and its associated HOMO delocalization are thus critical for enabling C_α-C_β bond cleavage. Nickel-based catalysts demonstrate significant reactivity in electrooxidation and operate via both "indirect oxidation" and "potential-dependent oxidation" mechanisms. These mechanisms involve three key steps: (1) the formation of high-valence Ni active sites from $Ni(OH)_2$, (2) adsorption of organic entities, and (3) reaction of these molecules with high-valence Ni species. For instance, NiOOH has been identified as the active species in the electrooxidation of lignin, where it oxidizes the hydroxyl groups on Cα and facilitates C_β-O bond cleavage in the β-O-4 linkage. In another study, vanillin and syringaldehyde, with a monomer yield of 18% from lignin was achieved using Ni foams [99,106].

Similarly, cobalt-based electrocatalysts exhibit oxidative activity akin to nickel-based systems, but with lower onset potentials. Interestingly, a lignin battery was developed with cobalt sulfide (CoS) as an anodic catalyst. CoS is converted to CoOOH under applied potential, which serves as the active site for β-O-4 bond cleavage, expanding the scope of electrooxidation. However, controlling product selectivity remains a challenge, with side products such as those from ring-opening reactions (acids) and CO_2 [107]. Optimization of reaction conditions, such as controlling pH to mitigate lignin repolymerization or side reactions, is essential for improving lignin degradation efficiency and economic viability. These advancements highlight the importance of mechanistic understanding and tailored processes for effective lignin valorization.

19.3.3.2 *Indirect oxidation*

The indirect oxidation of the electrocatalytic process involves the use of various redox catalytic mediators. Indirect oxidation involves the use of electrocatalysts that mediate the transfer of electrons from lignin or lignin-derived compounds to an electrode. Unlike direct oxidation, where charge transfer occurs directly between the substrate and the electrode, indirect oxidation utilizes high-valence metal oxides or other catalytic intermediates to facilitate bond cleavage in lignin or its dimers. This approach is advantageous for improving reaction selectivity, lowering activation energy, and stabilizing reactive intermediates. Polyoxometalates (POMs) are a class of transition metal-oxygen clusters with strong redox activity, making them effective mediators for indirect lignin depolymerization. POMs undergo redox cycling, abstracting electrons from lignin to oxidize hydroxyl groups and cleave β-O-4 linkages. Their stability and tunable redox potentials allow for controlled lignin degradation with minimal side reactions. Nitroaromatics, such as nitrobenzene, also serve as effective mediators by facilitating electron transfer from lignin to the catalyst surface. These mediators generate reactive intermediates that target specific bonds, including β-O-4 and Cα-Cβ linkages, resulting in depolymerization and the production of aromatic monomers. The mechanism of indirect oxidation typically begins with the activation of the catalyst, often a transition metal-based material such as Ni, Co, or their derivatives. Under an applied potential, the catalyst undergoes a transformation from its base state (e.g., $Ni(OH)_2$ or $Co(OH)_2$ to a high-valence state (e.g., NiOOH or CoOOH), (as mentioned in the above section), which serves as the reactive species. The generalized reaction for catalyst activation is:

$$M(OH)_2 + OH- \rightarrow MOOH + H_2O + e^-$$

where M represents the transition metal.

In the oxidation of lignin or lignin dimers, high-valence species interact with hydroxyl groups or ether linkages in the β-O-4 bond, a key structural unit in lignin. This interaction typically involves hydrogen atom abstraction and electron transfer, leading to bond cleavage. For example, in β-O-4 linkages, the cleavage of C_β-O or Cα -C_β bonds produces smaller aromatic compounds. The reaction mechanism can be summarized as:

$$Ar\text{-}CH(OH)\text{-}CH_2\text{-}Ar'+MOOH \rightarrow Ar\text{-}CHO + Ar'\text{-}CH_2\text{-}OH+MOH$$

Here, Ar/Ar′ are aromatic groups, where M represents the transition metal; the reaction products include aldehydes and alcohols.

Nickel-based electrocatalysts, such as NiOOH, have been widely studied for lignin oxidation. These catalysts preferentially oxidize the Cα-hydroxyl group, facilitating the cleavage of the Cβ -O bond in the β-O-4 structure. Thermodynamic studies indicate that the transformation occurs at potentials of 0.5–0.6 V vs. RHE, with an energy efficiency of ~80%. However, side reactions such as ring-opening or mineralization (ΔG ~ 200 kJ/mol) can lower product selectivity. Similarly, cobalt-based catalysts, such as CoOOH, exhibit high activity at lower onset potentials (~0.45 V vs. RHE) and exhibit similar reactivity, although selectivity challenges remain due to competing reactions, often enabling lignin degradation with reduced energy consumption. A practical example of indirect oxidation involves the conversion of lignin-derived dimers, such as 2-phenoxy-1-phenethanol, into aromatic aldehydes (e.g., benzaldehyde) and alcohols (e.g., phenethyl alcohol). The overall reaction can be expressed as:

$$C_6H_5\text{-}CH(OH)\text{-}CH_2\text{-}O\text{-}C_6H_5 + 2MOOH \rightarrow C_6H_5\text{-}CHO + C_6H_5\text{-}CH_2\text{-}OH + 2MOH$$

Despite its promise, indirect oxidation often produces side reactions such as mineralization and ring-opening , reducing product selectivity. Controlling the reaction conditions, such as pH, potential, and catalyst structure, is critical to minimizing these undesired pathways. Advances in mediator-assisted oxidation, using agents such as TEMPO, have further improved product selectivity and energy efficiency, offering a complementary strategy for lignin valorization compared to direct oxidative cleavage.

HAT-mediated oxidation provides higher selectivity and operates at lower potentials. For example, TEMPO oxidation targets the primary hydroxyl group at Cγ of lignin, converting it to a carboxyl group to form oxidized lignin while avoiding overoxidation. A sequential degradation strategy has been proposed to enhance efficiency and minimize side reactions. This involves a preoxidation step with TEMPO followed by acidification and hydrolysis under mild conditions (100°C, H_2O/HCOOH). Compared with a one-pot procedure, this two-step process significantly increases the monomer yield up to 29%. The HAT mechanism involves a two-step process: oxidation of the mediator at the anode followed by reduction in the presence of lignin or lignin-derived organic compounds [108]. This approach allows for selective cleavage of bonds in lignin's complex aromatic structure, facilitated by specific mediators such as tert-butyl hydroperoxide (tBuOOH), NHPI, TEMPO derivatives, polyoxometalates, and nitroaromatics. Ma et al. demonstrated a one-step strategy for β-O-4 bond cleavage using tBuOOH as a radical initiator. Upon losing an electron at the Pt electrode surface, tBuOOH generates tert-butoxy radicals (tBuO) [109]. Concurrently, at the Cβ position, the dimers with β-O-4 linkages undergo electron loss, forming secondary carbon radicals. These radicals interact with the $tBuO_h$ radical, forming an unstable intermediate through C–O bond coupling. The intermediate B undergoes intramolecular electron transfer, leading to cleavage of the C–C bond and the formation of phenol as a primary product. This method achieved a lignin dimer conversion of 93% and a benzaldehyde yield of 67%. The radical-mediated mechanism offers high selectivity and efficiency for cleaving lignin's β-O-4 linkages.

Stephenson's group employed a photoelectrocatalytic system coupling NHPI with 2,6-lutidine and an iridium complex to cleave β-O-4 bonds. NHPI loses an electron to form the phthalimide N-oxyl radical (PINO), a key electron-deficient species. PINO oxidizes the hydroxyl group on Cα to a carbonyl group, after which it is regenerated to NHPI. This process results in the formation of keto-products, which are subsequently cleaved through photocatalytic processes involving C–O bond cleavage. In neutral or acidic conditions, HAT mediators primarily target secondary benzylic alcohols in lignin via a bimolecular hydride transfer mechanism. Conversely, under alkaline conditions, HAT mediators preferentially oxidize primary alcohols due to reduced steric hindrance, following an inner-sphere electron transfer mechanism [110,111].

Rafiee et al. developed a method using 4-acetamido-TEMPO (ACT) as a mediator for the selective oxidation of primary hydroxyl groups in β-O-4 linkages. Using $NaHCO_3/Na_2CO_3$ electrolyte (pH 10), ACT oxidizes the primary hydroxyl group to a carboxyl group, forming a carboxyl intermediate [112]. This intermediate exhibits enhanced reactivity, enabling hydrolysis of the Cα–Cβ bond to generate depolymerized lignin products. This method also effectively cleaves the Cα–Cβ bonds in β-1–linked lignin dimers. The efficiency of HAT-mediated oxidation depends on the compatibility of mediators with lignin's structural diversity and the reaction conditions (e.g., pH, electrolyte composition, and temperature). Acidic and neutral conditions favor oxidation via hydride transfer mechanisms, whereas alkaline conditions promote inner-sphere electron transfer. The combination of mediators such as POMs and nitroaromatics, with optimized reaction pathways, presents an efficient strategy for lignin depolymerization, enabling the generation of valuable aromatic compounds while minimizing side reactions and energy consumption.

Raw lignin oxidation not only facilitates the production of high-value chemicals but also provides an opportunity to integrate with cathodic reactions at the counter electrode, enhancing overall process efficiency. Deng and colleagues developed an indirect electrooxidation system using POM or ferric chloride ($FeCl_3$) as charge-transfer agents. In this system, raw lignin served as the anodic hydrogen source to enable efficient hydrogen production. The process involves two key steps: in the electrolyte, POM or Fe^{3+} is involved in lignin oxidation, followed by the regeneration. Notably, a significant reduction of 2.54 kWh Nm^{-3} for POM and 2.30 kWh Nm^{-3} for $FeCl_3$ in energy usage by 39.5% for POM and 45.2% for $FeCl_3$ was recorded in comparison with the best-reported proton exchange membrane (PEM) water electrolysis system, which usually requires 4.2 kWh Nm^{-3} [25,113]. This innovative strategy highlights the potential of lignin as a sustainable feedstock for coupled hydrogen production and depolymerization processes.

19.4 Summary and future outlook

The investigation of small organic molecules as catalysts for lignin oxidation offers a promising path for enhancing catalytic systems. This method is particularly appealing because it helps clarify the complex mechanisms involved, unlike the difficulties encountered in enzymatic and surface catalysis. Lignin depolymerization into monomers is challenging due to its complex structure, which features various linkages. The changes occurring within lignin, along with the condensation of fragments into stable forms, add to the complexity of the process. A major obstacle is the separation of monomers that have similar molecular structures, which increases costs and creates a larger disparity between biomass-derived and fossil fuel-based chemical production. Therefore the main challenge in lignin depolymerization is to develop selective and effective catalysts that can efficiently break common C–O and C–C bonds while preserving the integrity of aromatic rings, ultimately resulting in a limited range of products.

The design of electrocatalysts for lignin depolymerization presents several unique challenges: (1) Structural heterogeneity of lignin: Lignin is a complex and diverse biopolymer, with its structure varying significantly based on its biological origin and growth conditions. This variability complicates the design of universal electrocatalysts, as different sources of lignin may necessitate distinct catalytic systems or strategies for effective depolymerization. (2) Selective bond cleavage: Lignin consists of numerous C–O and C–C bonds that need to be selectively cleaved to yield the desired monomeric products. Electrocatalysts must facilitate these selective bond-breaking reactions while maintaining the integrity of essential aromatic structures. Achieving this precision demands a thorough understanding of electrocatalytic reaction pathways and active site engineering. (3) Catalyst stability and fouling: During the electrocatalytic depolymerization process, intermediates, reaction products, or impurities can adsorb onto the catalyst's surface. This can lead to catalyst poisoning, reduced activity, and lower efficiency over time. Developing strategies to reduce surface fouling and ensure long-term catalyst performance is crucial. Consequently, creating mild, energy-efficient, and highly selective electrocatalysts for lignin depolymerization remains a significant challenge, necessitating advancements in catalyst design, reaction engineering, and process optimization. An insightful review highlights the design of electrocatalysts and provides valuable guidance for researchers in optimizing catalyst performance [114].

Sustainability plays a vital role in electrocatalytic lignin depolymerization. By integrating this process with renewable energy sources such as solar or wind, we can enhance its environmental benefits and lower its carbon footprint. Achieving high FE is essential for effective energy use in bond cleavage, making the process both cost-effective and eco-friendly. For large-scale implementation, electrode materials and designs need to be scalable, and the overall economic feasibility of the process should be regularly assessed. Moreover, this process supports the principles of a circular economy by reducing waste and allowing for the use of renewable feedstocks. By valuing byproducts and leftover lignin fractions, we can maximize resource efficiency while fostering the development of sustainable chemical manufacturing methods. Therefore electrocatalytic lignin depolymerization stands out as a promising approach to tackle global energy and material challenges in a renewable and environmentally conscious way.

Acknowledgment

Y.N.R and J.B. thank BRIC NABI for providing funding, infrastructural support, and DBT for the funding of the lignin valorization flagship project. Y.N.R acknowledges the financial support provided by SRMIST through the Seed Grant under the Selective Excellence Research Initiative (SRMIST/R/AR(A)/SERI2025/174/08) for funding.

References

[1] M.E. Himmel, S.Y. Ding, D.K. Johnson, W.S. Adney, M.R. Nimlos, J.W. Brady, et al., Biomass recalcitrance: engineering plants and enzymes for biofuels production, Science 315 (5813) (2007) 804–807, doi:10.1126/SCIENCE.1137016.

[2] Y. Luo, M.E.A. Razzaq, W. Qu, A.A.B A. Mohammed, A. Aui, H. Zobeiri, et al., Introducing thermo-mechanochemistry of lignin enabled the production of high-quality low-cost carbon fiber, Green Chem. 26 (6) (2024) 3281–3300, doi:10.1039/D3GC04288J.

[3] D.S. Bajwa, G. Pourhashem, A.H. Ullah, S.G. Bajwa, A concise review of current lignin production, applications, products and their environmental impact, Ind. Crops Prod. 139 (2019) 111526, doi:10.1016/J.INDCROP.2019.111526.

[4] H. Luo, M.M. Abu-Omar, Chemicals from lignin, Encyclopedia Sustain. Technol. (2017) 573–585, doi:10.1016/B978-0-12-409548-9.10235-0.

[5] S. Gundekari, J. Mitra, M. Varkolu, Classification, characterization, and properties of edible and non-edible biomass feedstocks, Advanced Functional Solid Catalysts for Biomass Valorization, Elsevier, 2020, pp. 89–120, doi:10.1016/B978-0-12-820236-4.00004-0.

[6] J. Zakzeski, P.C.A. Bruijnincx, A.L. Jongerius, B.M. Weckhuysen, The catalytic valorization of lignin for the production of renewable chemicals, Chem. Rev. 110 (6) (2010) 3552–3599, doi:10.1021/cr900354u.

[7] M. Fache, B. Boutevin, S. Caillol, Vanillin production from lignin and its use as a renewable chemical, ACS Sustain. Chem. Eng. 4 (1) (2016) 35–46, doi:10.1021/acssuschemeng.5b01344.

[8] Z. Chen, C. Wan, Biological valorization strategies for converting lignin into fuels and chemicals, Renew. Sustain. Energy Rev. 73 (2017) 610–621, doi:10.1016/J.RSER.2017.01.166.

[9] E. Subbotina, J.S.M. Samec, Cleavage of challenging chemical bonds in lignin enables biofuels, Nature Chem. Eng. 1 (1) (2024) 28–30, doi:10.1038/s44286-023-00012-2.

[10] M.M. Abu-Omar, K. Barta, G.T. Beckham, J.S. Luterbacher, J. Ralph, R. Rinaldi, et al., Guidelines for performing lignin-first biorefining, Energy Environ. Sci. 14 (1) (2021) 262–292, doi:10.1039/D0EE02870C.

[11] M. Mujtaba, L. Fernandes Fraceto, M. Fazeli, S. Mukherjee, S.M. Savassa, G. Araujo de Medeiros, et al., Lignocellulosic biomass from agricultural waste to the circular economy: a review with focus on biofuels, biocomposites and bioplastics, J. Clean Prod. 402 (2023) 136815, doi:10.1016/J.JCLEPRO.2023.136815.

[12] A. Haile, G.G. Gelebo, T. Tesfaye, W. Mengie, M.A. Mebrate, A. Abuhay, et al., Pulp and paper mill wastes: utilizations and prospects for high value-added biomaterials, Bioresour. Bioprocess 8 (1) (2021) 1–22, doi:10.1186/S40643-021-00385-3.

[13] S. Kim, S.C. Chmely, M.R. Nimlos, Y.J. Bomble, T.D. Foust, R.S. Paton, et al., Computational study of bond dissociation enthalpies for a large range of native and modified lignins, J. Phys. Chem. Lett. 2 (22) (2011) 2846–2852, doi:10.1021/JZ201182W.

[14] A. Yamaguchi, N. Mimura, M. Shirai, O. Sato, Bond Cleavage of lignin model compounds into aromatic monomers using supported metal catalysts in supercritical water, Sci. Rep. 7 (1) (2017) 1–7, doi:10.1038/srep46172.

[15] P. Azadi, O.R. Inderwildi, R. Farnood, D.A. King, Liquid Fuels hydrogen and chemicals from lignin: a critical review, Renew. Sustain. Energy Rev. 21 (2013) 506–523, doi:10.1016/J.RSER.2012.12.022.

[16] J. Li, G. Henriksson, G. Gellerstedt, Lignin depolymerization/repolymerization and its critical role for delignification of aspen wood by steam explosion, Bioresour. Technol. 98 (16) (2007) 3061–3068, doi:10.1016/J.BIORTECH.2006.10.018.

[17] X. Liu, F.P. Bouxin, J. Fan, V.L. Budarin, C. Hu, J.H. Clark, Recent advances in the catalytic depolymerization of lignin towards phenolic chemicals: a review, ChemSusChem 13 (17) (2020) 4296–4317, doi:10.1002/CSSC.202001213.

[18] Z. Sun, B. Fridrich, A. De Santi, S. Elangovan, K. Barta, Bright side of lignin depolymerization: toward new platform chemicals, Chem. Rev. 118 (2) (2018) 614–678, doi:10.1021/ACS.CHEMREV.7B00588.

[19] Y.N. Reddy, S. Kirar, N.S. Thakur, M.D. Patil, J. Bhaumik, Sunlight assisted photocatalytic valorization of lignin using recyclable light harvesters, ACS Sustain. Chem. Eng. 11 (12) (2023) 4568–4579, doi:10.1021/ACSSUSCHEMENG.2C05917.

[20] J. Luo, T.L. Liu, Electrochemical valorization of lignin: status, challenges, and prospects, J. Bioresour. Bioprod. 8 (1) (2023) 1–14, doi:10.1016/J.JOBAB.2022.11.003.

[21] M.G.A. da Cruz, R. Gueret, J. Chen, J. Piątek, B. Beele, M.H. Sipponen, et al., Electrochemical depolymerization of lignin in a biomass-based solvent, ChemSusChem 15 (15) (2022) e202200718, doi:10.1002/CSSC.202200718.

[22] R.S. Weber, K.K. Ramasamy, Electrochemical oxidation of lignin and waste plastic, ACS Omega 5 (43) (2020) 27735–27740, doi:10.1021/ACSOMEGA.0C03989.

[23] L.M. Lindenbeck, V.C. Barra, B.B. Beele, B.V.M. Rodrigues, A. Slabon, Revisiting the electrocatalyst role on lignin depolymerization, Adv. Energy Sustain. Res. 5 (10) (2024) 2400130, doi:10.1002/AESR.202400130.

[24] P.C. Sherrell, M. Iesalnieks, Y. Ehrnst, A.R. Rezk, A. Šutka, Electrocatalysis for green(Er) chemistry: limitations and opportunities with traditional and emerging characterization methods for tangible societal impact, Adv. Energy Sustain. Res. 5 (8) (2024) 2400008, doi:10.1002/AESR.202400008.

[25] X. Du, H. Zhang, K.P. Sullivan, P. Gogoi, Y. Deng, Electrochemical lignin conversion, ChemSusChem 13 (17) (2020) 4318–4343, doi:10.1002/CSSC.202001187.

[26] R. Hu, Y. Zhao, C. Tang, Y. Shi, G. Luo, J. Fan, et al., Electrochemical biorefinery toward chemicals synthesis and bio-oil upgrading from lignin, Engineering 27 (2023) 178–198, doi:10.1016/J.ENG.2022.10.013.

[27] M. Garedew, C.H. Lam, L. Petitjean, S. Huang, B. Song, F. Lin, et al., Electrochemical upgrading of depolymerized lignin: a review of model compound studies, Green Chem. 23 (8) (2021) 2868–2899, doi:10.1039/D0GC04127K.

[28] N. Shukry, S.M. Fadel, F.A. Agblevor, S.F. El-Kalyoubi, Some physical properties of acetosolv lignins from bagasse, J. Appl. Polym. Sci. 109 (1) (2008) 434–444, doi:10.1002/APP.28059.

[29] M. Antonietta Casadei, D. Pletcher, The influence of conditions on the electrocatalytic hydrogenation of organic molecules, Electrochim. Acta 33 (1) (1988) 117–120, doi:10.1016/0013-4686(88)80042-2.

[30] J.M. Léger, F. Hahn, Contribution of in-situ infrared reflectance spectroscopy in the study of nanostructured fuel cell electrodes, In-situ Spectroscopic Studies of Adsorption at the Electrode and Electrocatalysis, Elsevier, 2007, pp. 63–98, doi:10.1016/B978-044451870-5/50004-X.

[31] C. Ling, Y. Cui, S. Lu, X. Bai, J. Wang, How computations accelerate electrocatalyst discovery, Chem. 8 (6) (2022) 1575–1610, doi:10.1016/J.CHEMPR.2022.03.015.

[32] H. Hosseini, C.J. Herring, C.F. Nwaokorie, G.A. Sulley, M.M. Montemore, Computational design of catalysts with experimental validation: recent successes, effective strategies, and pitfalls, J. Phys. Chem. C 128 (2024) 18144–18157, doi:10.1021/ACS.JPCC.4C04949.

[33] Z. Li, M. Garedew, C.H. Lam, J.E. Jackson, D.J. Miller, C.M. Saffron, Electrocatalytic hydrogenation and hydrodeoxygenation of bio-oil derived phenolic compounds using ruthenium supported on activated carbon cloth, Green Chem. 14 (9) (2012) 2540–2549, doi:10.1039/C2GC35552C.

[34] K. Amouzegar, O. Savadogo, Electrocatalytic hydrogenation of phenol on highly dispersed Pt electrodes, Electrochim. Acta 39 (4) (1994) 557–559, doi:10.1016/0013-4686(94)80101-0.

[35] K. Amouzegar, O. Savadogo, Electrocatalytic hydrogenation of phenol on dispersed Pt: effect of metal electrochemically active surface area and electrode material, J. Appl. Electrochem. 27 (5) (1997) 539–542, doi:10.1023/A:1018442527348.
[36] C.L. Yang, L.N. Wang, P. Yin, J. Liu, M.X. Chen, Q.Q. Yan, et al., Sulfur-anchoring synthesis of platinum intermetallic nanoparticle catalysts for fuel cells, Science 374 (6566) (2021) 459–464, doi:10.1126/science.abJ9980.
[37] K. Liu, R. Qin, N. Zheng, Insights into the interfacial effects in heterogeneous metal nanocatalysts toward selective hydrogenation, J. Am. Chem. Soc. 143 (12) (2021) 4483–4499, doi:10.1021/JACS.0C13185.
[38] N. Di Fidio, J.W. Timmermans, C. Antonetti, A.M. Raspolli Galletti, R.J.A. Gosselink, R.J.M. Bisselink, et al., Electro-oxidative depolymerisation of technical lignin in water using platinum, nickel oxide hydroxide and graphite electrodes, New J. Chem. 45 (21) (2021) 9647–9657, doi:10.1039/D1NJ01037A.
[39] C.H. Lam, C.B. Lowe, Z. Li, K.N. Longe, J.T. Rayburn, M.A. Caldwell, et al., Electrocatalytic upgrading of model lignin monomers with earth abundant metal electrodes, Green Chem. 17 (1) (2014) 601–609, doi:10.1039/C4GC01632G.
[40] Y. Zhou, G.E. Klinger, E.L. Hegg, C.M. Saffron, J.E. Jackson, Skeletal Ni electrode-catalyzed C-O cleavage of diaryl ethers entails direct elimination via benzyne intermediates, Nature Commun. 13 (1) (2022) 1–13, doi:10.1038/s41467-022-29555-3.
[41] S. Gnaim, A. Bauer, H.J. Zhang, L. Chen, C. Gannett, C.A. Malapit, et al., Cobalt-electrocatalytic HAT for functionalization of unsaturated C–C bonds, Nature 605 (2022) 687–695, doi:10.1038/s41586-022-04595-3.
[42] G. Liu, Z. Zhai, Y. Lu, J. Lu, Y. Wang, S. Liang, et al., Electrocatalytic cleavage of C–C bonds in lignin models using nonmetallic catalysts at ambient conditions, Chem. Bio. Eng. 1 (4) (2024) 357–365, doi:10.1021/CBE.3C00122.
[43] T. Shiraishi, T. Takano, H. Kamitakahara, F. Nakatsubo, Studies on electrooxidation of lignin and lignin model compounds. Part 1: direct electrooxidation of non-phenolic lignin model compounds, Holzforschung 66 (3) (2012) 303–309, doi:10.1515/HF.2011.069.
[44] S. Jung, E.J. Biddinger, Electrocatalytic hydrogenation and hydrogenolysis of furfural and the impact of homogeneous side reactions of Furanic compounds in acidic electrolytes, ACS Sustain. Chem. Eng. 4 (12) (2016) 6500–6508, doi:10.1021/ACSSUSCHEMENG.6B01314.
[45] R. Roy, M.S. Rahman, T.A. Amit, B. Jadhav, Recent Advances in lignin depolymerization techniques: a comparative overview of traditional and greener approaches, Biomass 2 (3) (2022) 130–154, doi:10.3390/BIOMASS2030009.
[46] F. Lai, H. Shang, Y. Jiao, X. Chen, T. Zhang, X. Liu, et al., Recent progress and perspective on electrocatalysis in neutral media: mechanisms, materials, and advanced characterizations, Interdisciplinary Mater. 3 (4) (2024) 492–529, doi:10.1002/IDM2.12172.
[47] Y. He, X. Zeng, Z. Lu, S. Mo, Q. An, Q. Liu, et al., Aqueous electrocatalytic hydrogenation depolymerization of lignin β-O-4 linkage via selective Caryl-O(C) bond cleavage: the regulation of adsorption, J. Am. Chem. Soc. (2024), doi:10.1021/JACS.4C12220.
[48] J. Rumpf, X.T. Do, R. Burger, Y. Monakhova, M. Schulze, Types of lignin, properties, and structural characterization techniques, Lignin-based Materials for Biomedical Applications: Preparation, Characterization, and Implementation, Elsevier, 2021, pp. 105–158, doi:10.1016/B978-0-12-820303-3.00001-1.
[49] M. Nagy, M. Kosa, H. Theliander, A.J. Ragauskas, Characterization of CO_2 precipitated kraft lignin to promote its utilization, Green Chem. 12 (1) (2010) 31–34, doi:10.1039/B913602A.
[50] P. Mousavioun, W.O.S. Doherty, Chemical and thermal properties of fractionated bagasse soda lignin, Ind. Crops Prod. 31 (1) (2010) 52–58, doi:10.1016/J.INDCROP.2009.09.001.

[51] T. Aro, P. Fatehi, Production and application of lignosulfonates and sulfonated lignin, ChemSusChem 10 (9) (2017) 1861–1877, doi:10.1002/CSSC.201700082.

[52] A. Guerra, I. Filpponen, L.A. Lucia, D.S. Argyropoulos, Comparative evaluation of three lignin isolation protocols for various wood species, J. Agric. Food Chem. 54 (26) (2006) 9696–9705, doi:10.1021/JF062433C.

[53] A. Guerra, I. Filpponen, L.A. Lucia, C. Saquing, S. Baumberger, D.S. Argyropoulos, Toward a better understanding of the lignin isolation process from wood, J. Agric. Food Chem. 54 (16) (2006) 5939–5947, doi:10.1021/JF060722V.

[54] K.T. Wang, C. Jing, C. Wood, A. Nagardeolekar, N. Kohan, P. Dongre, et al., Toward complete utilization of miscanthus in a hot-water extraction-based biorefinery, Energies 11 (1) (2017) 39, doi:10.3390/EN11010039.

[55] C.G. Yoo, H. Kim, F. Lu, A. Azarpira, X. Pan, K.K. Oh, et al., Understanding the physicochemical characteristics and the improved enzymatic saccharification of corn stover pretreated with aqueous and gaseous ammonia, Bioenergy Res. 9 (1) (2016) 67–76, doi:10.1007/S12155-015-9662-6/TABLES/3.

[56] X. Meng, A. Parikh, B. Seemala, R. Kumar, Y. Pu, C.E. Wyman, et al., Characterization of fractional cuts of co-solvent enhanced lignocellulosic fractionation lignin isolated by sequential precipitation, Bioresour. Technol. 272 (2019) 202–208, doi:10.1016/J.BIORTECH.2018.09.130.

[57] X. Meng, S. Bhagia, Y. Wang, Y. Zhou, Y. Pu, J.R. Dunlap, et al., Effects of the advanced organosolv pretreatment strategies on structural properties of woody biomass, Ind. Crops Prod. 146 (2020) 112144, doi:10.1016/J.INDCROP.2020.112144.

[58] X. Meng, Y. Pu, M. Li, A.J. Ragauskas, A biomass pretreatment using cellulose-derived solvent cyrene, Green Chem. 22 (9) (2020) 2862–2872, doi:10.1039/D0GC00661K.

[59] E. Liu, M. Li, L. Das, Y. Pu, T. Frazier, B. Zhao, et al., Understanding lignin fractionation and characterization from engineered switchgrass treated by an aqueous ionic liquid, ACS Sustain. Chem. Eng. 6 (5) (2018) 6612–6623, doi:10.1021/ACSSUSCHEMENG.8B00384.

[60] Y. Wang, X. Meng, K. Jeong, S. Li, G. Leem, K.H. Kim, et al., Investigation of a lignin-based deep eutectic solvent using p-hydroxybenzoic acid for efficient woody biomass conversion, ACS Sustain. Chem. Eng. 8 (33) (2020) 12542–12553, doi:10.1021/ACSSUSCHEMENG.0C03533.

[61] L. Shuai, Y.M. Questell-Santiago, J.S.A Luterbacher, mild biomass pretreatment using γ-valerolactone for concentrated sugar production, Green Chem. 18 (4) (2016) 937–943, doi:10.1039/C5GC02489G.

[62] S. Van Den Bosch, W. Schutyser, R. Vanholme, T. Driessen, S.F. Koelewijn, T. Renders, et al., Reductive lignocellulose fractionation into soluble lignin-derived phenolic monomers and dimers and processable carbohydrate pulps, Energy Environ. Sci. 8 (6) (2015) 1748–1763, doi:10.1039/C5EE00204D.

[63] G.S. Lin, W. Peng, J. Gao, A. Wahlen, Z. Tong, Functional naturally derived materials to improve the environment: chemical structures, modifications, applications, and future perspectives, Adv. Bioenergy 8 (2023) 93–144, doi:10.1016/BS.AIBE.2023.02.001.

[64] J. Behaghel De Bueren, F. Héroguel, C. Wegmann, G.R. Dick, R. Buser, J.S. Luterbacher, Aldehyde-assisted fractionation enhances lignin valorization in endocarp waste biomass, ACS Sustain. Chem. Eng. 8 (45) (2020) 16737–16745, doi:10.1021/ACSSUSCHEMENG.0C03360.

[65] M. Fasching, P. Schröder, R.P. Wollboldt, H.K. Weber, H. Sixta, A new and facile method for isolation of lignin from wood based on complete wood dissolution, Holzforschung 62 (1) (2008) 15–23, doi:10.1515/HF.2008.003.

[66] C. Zhang, H. Li, J. Lu, X. Zhang, K.E. Macarthur, M. Heggen, et al., Promoting lignin depolymerization and restraining the condensation via an oxidation-hydrogenation strategy, ACS Catal 7 (5) (2017) 3419–3429, doi:10.1021/ACSCATAL.7B00148.

[67] A. Rahimi, A. Azarpira, H. Kim, J. Ralph, S.S. Stahl, Chemoselective metal-free aerobic alcohol oxidation in lignin, J. Am. Chem. Soc. 135 (17) (2013) 6415–6418, doi:10.1021/JA401793N.

[68] C.S. Lancefield, O.S. Ojo, F. Tran, N.J. Westwood, Isolation of functionalized phenolic monomers through selective oxidation and c?o bond cleavage of the β-O-4 linkages in lignin, Angew. Chem. 127 (1) (2015) 260–264, doi:10.1002/ANGE.201409408.

[69] S. Ljunggren, A. Olsson, The specificity in oxidation ofsome lignin and carbohydrate models and pine wood shavings permanganate and pyridinium dichromate before the kraft pulping process, Holzforschung 38 (2) (1984) 91–99, doi:10.1515/HFSG.1984.38.2.91.

[70] H. Zhang, Z. Li, S. Tang, X. Yang, M. li, J. Yang, et al., The efficient and directional electrocatalytic depolymerization of α-o-4 bond in lignin by auxiliary electrolyte, Ind. Crops Prod. 220 (2024) 119292, doi:10.1016/J.INDCROP.2024.119292.

[71] A. Bailey, H.M. Brooks, Electrolytic oxidation of lignin, J. Am. Chem. Soc. 68 (3) (1946) 445–446, doi:10.1021/JA01207A029.

[72] R. Tolba, M. Tian, J. Wen, Z.H. Jiang, A. Chen, Electrochemical oxidation of lignin at irO2-based oxide electrodes, J. Electroanal. Chem. 649 (1–2) (2010) 9–15, doi:10.1016/J.JELECHEM.2009.12.013.

[73] H. Zhu, Y. Chen, T. Qin, L. Wang, Y. Tang, Y. Sun, et al., Lignin depolymerization via an integrated approach of anode oxidation and electro-generated H_2O_2 oxidation, RSC Adv. 4 (12) (2014) 6232–6238, doi:10.1039/C3RA47516F.

[74] H. Jiang, A. Xue, Z. Wang, R. Xia, L. Wang, Y. Tang, et al., Electrochemical degradation of lignin by ROS, Sustain. Chem. 1 (3) (2020) 345–360, doi:10.3390/SUSCHEM1030023.

[75] M. Garedew, F. Lin, B. Song, T.M. DeWinter, J.E. Jackson, C.M. Saffron, et al., Greener routes to biomass waste valorization: lignin transformation through electrocatalysis for renewable chemicals and fuels production, ChemSusChem 13 (17) (2020) 4214–4237, doi:10.1002/CSSC.202000987.

[76] D. Tarasov, M. Leitch, P. Fatehi, Lignin–carbohydrate complexes: properties, applications, analyses, and methods of extraction: a review, Biotechnol. Biofuels 11 (1) (2018) 1–28, doi:10.1186/S13068-018-1262-1.

[77] D. Robin, M. Comtois, A. Martel, R. Lemieux, A.K. Cheong, G. Belot, et al., The electrocatalytic hydrogenation of fused poly cyclic aromatic compounds at raney nickel electrodes: the influence of catalyst activation and electrolysis conditions, Can. J. Chem. 68 (7) (1990) 1218–1227, doi:10.1139/V90-189.

[78] H. Ilikti, N. Rekik, M. Thomalla, Electrocatalytic hydrogenation of phenol in aqueous solutions at a raney nickel electrode in the presence of cationic surfactants, J. Appl. Electrochem. 32 (6) (2002) 603–609, doi:10.1023/A:1020151429953.

[79] D. Strmcnik, P.P. Lopes, B. Genorio, V.R. Stamenkovic, N.M. Markovic, Design principles for hydrogen evolution reaction catalyst materials, Nano Energy 29 (2016) 29–36, doi:10.1016/J.NANOEN.2016.04.017.

[80] Y. Li, H. Dai, Recent advances in zinc–air batteries, Chem. Soc. Rev. 43 (15) (2014) 5257–5275, doi:10.1039/C4CS00015C.

[81] J.K. Nørskov, J. Rossmeisl, A. Logadottir, L. Lindqvist, J.R. Kitchin, T. Bligaard, et al., Origin of the overpotential for oxygen reduction at a fuel-cell cathode, J. Phys. Chem. B 108 (46) (2004) 17886–17892, doi:10.1021/JP047349J.

[82] J. Song, C. Wei, Z.F. Huang, C. Liu, L. Zeng, X. Wang, et al., Review on fundamentals for designing oxygen evolution electrocatalysts, Chem. Soc. Rev. 49 (7) (2020) 2196–2214, doi:10.1039/C9CS00607A.

[83] Y.P. Wijaya, K.J. Smith, C.S. Kim, E.L. Gyenge, Electrocatalytic hydrogenation and depolymerization pathways for lignin valorization: toward mild synthesis of chemicals and fuels from biomass, Green Chem. 22 (21) (2020) 7233–7264, doi:10.1039/D0GC02782K.

[84] Y. Song, O.Y. Gutiérrez, J. Herranz, J.A. Lercher, Aqueous phase electrocatalysis and thermal catalysis for the hydrogenation of phenol at mild conditions, Appl. Catal. B 182 (2016) 236–246, doi:10.1016/J.APCATB.2015.09.027.

[85] Y. Song, S.H. Chia, U. Sanyal, O.Y. Gutiérrez, J.A. Lercher, Integrated catalytic and electrocatalytic conversion of substituted phenols and diaryl ethers, J. Catal. 344 (2016) 263–272, doi:10.1016/J.JCAT.2016.09.030.

[86] H.Y. Hoang, N.T. Hoang, B.D. Nhi, H. Le-Quoc, H. Van Nam, Nickel foam-based electrode materials for multifunctional electrochemical applications: From substrate to active platform, Journal of Environmental Chemical Engineering 13 (5) (2025) 118695, doi:10.1016/j.jece.2025.118695.

[87] Q. Hu, K. Gao, X. Wang, H. Zheng, J. Cao, L. Mi, et al., Subnanometric Ru clusters with upshifted D band center improve performance for alkaline hydrogen evolution reaction, Nat. Commun. 13 (1) (2022) 1–10, doi:10.1038/s41467-022-31660-2.

[88] A. Cyr, F. Chiltz, P. Jeanson, A. Martel, L. Brossard, J. Lessard, et al., Electrocatalytic hydrogenation of lignin models at Raney nickel and palladium-based electrodes, Can. J. Chem. 78 (3) (2011) 307–315, doi:10.1139/V00-009.

[89] Y. Wang, J. He, Y. Zhang, CeCl3-Promoted Simultaneous Photocatalytic cleavage and amination of Cα–Cβ bond in lignin model compounds and native lignin, CCS Chemistry 2 (3) (2020) 107–117, doi:10.31635/CCSCHEM.020.201900076.

[90] R. Wu, Q. Meng, J. Yan, H. Liu, Q. Zhu, L. Zheng, et al., Electrochemical strategy for the simultaneous production of cyclohexanone and benzoquinone by the reaction of phenol and water, J. Am. Chem. Soc. 144 (4) (2022) 1556–1571, doi:10.1021/JACS.1C09021.

[91] P. Zhou, S.X. Guo, L. Li, T. Ueda, Y. Nishiwaki, L. Huang, et al., Selective electrochemical hydrogenation of phenol with earth-abundant Ni–MoO2 heterostructured catalysts: effect of oxygen vacancy on product selectivity, Angew. Chem. Int. Ed. 62 (8) (2023) e202214881, doi:10.1002/ANIE.202214881.

[92] N. Singh, Y. Song, O.Y. Gutiérrez, D.M. Camaioni, C.T. Campbell, J.A. Lercher, Electrocatalytic hydrogenation of phenol over platinum and rhodium: unexpected temperature effects resolved, ACS Catal 6 (11) (2016) 7466–7470, doi:10.1021/ACSCATAL.6B02296.

[93] X. Liu, Y. Wang, H. Duan, Recent Progress in Electrocatalytic conversion of lignin: from monomers, dimers, to raw lignin, Precision Chem. 2 (2024) 428–446, doi:10.1021/PRECHEM.4C00024.

[94] X. Du, W. iu, Z. Zhang, A. Mulyadi, A. lex Brittain, J. ian Gong, et al., Low-energy catalytic electrolysis for simultaneous hydrogen evolution and lignin depolymerization, ChemSusChem 10 (5) (2017) 847–854, doi:10.1002/CSSC.201601685.

[95] Q. Zhai, S. Han, K. Wang, J. Jiang, J. Xu, Electrocatalytic cleavage of aryl ether C–O linkages in lignin model dimers, Fuel Process. Technol. 235 (2022) 107350, doi:10.1016/J.FUPROC.2022.107350.

[96] B. Mahdavi, A. Lafrance, A. Martel, J. Lessard, H. Ménard, L. Brossard, Electrocatalytic hydrogenolysis of lignin model dimers at raney nickel electrodes, J. Appl. Electrochem. 27 (5) (1997) 605–611, doi:10.1023/A:1018463131891.

[97] C. Yang, G. Magallanes, S. Maldonado, C.R.J. Stephenson, Electro-reductive fragmentation of oxidized lignin models, J. Org. Chem. 86 (22) (2021) 15927–15934, doi:10.1021/ACS.JOC.1C00346.

[98] P. Cai, H. Fan, S. Cao, J. Qi, S. Zhang, G. Li, Electrochemical conversion of corn stover lignin to biomass-based chemicals between Cu/NiMoCo cathode and Pb/PbO2 anode in alkali solution, Electrochim. Acta 264 (2018) 128–139, doi:10.1016/J.ELECTACTA.2018.01.111.

[99] K. Yan, Y. Zhang, M. Tu, Y. Sun, Electrocatalytic valorization of organosolv lignin utilizing a nickel-based electrocatalyst, Energy Fuels 34 (10) (2020) 12703–12709, doi:10.1021/ACS.ENERGYFUELS.0C02284.

[100] Y. Qi, B. Liu, X. Qiu, X. Zeng, Z. Luo, W. Wu, et al., Simultaneous oxidative cleavage of lignin and reduction of furfural via efficient electrocatalysis by P-doped CoMoO4, Adv. Mater. 35 (14) (2023) 2208284, doi:10.1002/ADMA.202208284.
[101] A. Corona, M.J. Biddy, D.R. Vardon, M. Birkved, M.Z. Hauschild, G.T. Beckham, Life cycle assessment of adipic acid production from lignin, Green Chem. 20 (16) (2018) 3857–3866, doi:10.1039/C8GC00868J.
[102] S.B. Lalvani, R. Rajagopal, Hydrogen production from lignin-water solution by electrolysis, Holzforschung 47 (4) (1993) 283–286, doi:10.1515/HFSG.1993.47.4.283.
[103] F. Bateni, M. NaderiNasrabadi, R. Ghahremani, J.A. Staser, Low-cost nanostructured electrocatalysts for hydrogen evolution in an anion exchange membrane lignin electrolysis cell, J. Electrochem. Soc. 166 (14) (2019) F1037–F1046, doi:10.1149/2.0221914JES.
[104] A. Caravaca, W.E. Garcia-Lorefice, S. Gil, A. de Lucas-Consuegra, P. Vernoux, Towards a sustainable technology for H2 production: direct lignin electrolysis in a continuous-flow polymer electrolyte membrane reactor, Electrochem. Commun. 100 (2019) 43–47, doi:10.1016/J.ELECOM.2019.01.016.
[105] J. Chen, H. Yang, H. Fu, H. He, Q. Zeng, X. Li, Electrochemical oxidation mechanisms for selective products due to C–O and C–C Cleavages of β-O-4 kinkages in lignin model compounds, Phys. Chem. Chem. Phys. 22 (20) (2020) 11508–11518, doi:10.1039/D0CP01091J.
[106] M.T. Bender, K.S. Choi, Electrochemical oxidation of HMF via hydrogen atom transfer and hydride transfer on NiOOH and the impact of NiOOH composition, ChemSusChem 15 (13) (2022) e202200675, doi:10.1002/CSSC.202200675.
[107] D. Ouyang, D. Gao, Y. Qiang, X. Zhao, Highly-efficient conversion of lignin to electricity by nickel foam anode loaded with solid electrocatalysts, Appl. Catal. B 328 (2023) 122491, doi:10.1016/J.APCATB.2023.122491.
[108] A. Badalyan, S.S. Stahl, Cooperative electrocatalytic alcohol oxidation with electron-proton-transfer mediators, Nature 535 (7612) (2016) 406–410, doi:10.1038/nature18008.
[109] L. Ma, H. Zhou, X. Kong, Z. Li, H. Duan, An Electrocatalytic strategy for C-C bond cleavage in lignin model compounds and lignin under ambient conditions, ACS Sustain. Chem. Eng. 9 (4) (2021) 1932–1940, doi:10.1021/ACSSUSCHEMENG.0C08612.
[110] T.M. Monos, G. Magallanes, L.J. Sebren, C.R.J. Stephenson, Visible light mediated reductions of ethers, amines and sulfides, J. Photochem. Photobiol A Chem. 328 (2016) 240–248, doi:10.1016/J.JPHOTOCHEM.2016.05.014.
[111] I. Bosque, G. Magallanes, M. Rigoulet, M.D. Kärkäs, C.R.J. Stephenson, Redox catalysis facilitates lignin depolymerization, ACS Cent Sci 3 (6) (2017) 621–628, doi:10.1021/ACSCENTSCI.7B00140.
[112] M. Rafiee, M. Alherech, S.D. Karlen, S.S. Stahl, Electrochemical aminoxyl-mediated oxidation of primary alcohols in lignin to carboxylic acids: polymer modification and depolymerization, J. Am. Chem. Soc. 141 (38) (2019) 15266–15276, doi:10.1021/JACS.9B07243.
[113] M. Carmo, D.L. Fritz, J. Mergel, D. Stolten, A Comprehensive review on PEM water electrolysis, Int. J. Hydrogen Energy 38 (12) (2013) 4901–4934, doi:10.1016/J.IJHYDENE.2013.01.151.
[114] J. Linnemann, K. Kanokkanchana, K. Tschulik, Design strategies for electrocatalysts from an electrochemist's perspective, ACS Catal 11 (9) (2021) 5318–5346, doi:10.1021/ACSCATAL.0C04118.

CHAPTER

20

Biomass electrochemistry: Fundamentals, catalysts and approaches

N. Soujanya[a], Pallavi Saxena[b], Ruchi Pathania[c], Priya Pallavi[b], Kumud Kant Awasthi[d], and Ashish Kapoor[e]

[a] School of Technology, Woxsen University, Hyderabad, Telangana, India, [b] School of Sciences, Woxsen University, Hyderabad, Telangana, India, [c] Department of Medicinal Chemistry, Center for Natural Products, Drug Discovery and Development, University of Florida, Gainesville, FL, United States, [d] School of Basic & Applied Sciences, K.R. Mangalam University, Gurugram, Haryana, India, [e] Department of Chemical Engineering, Harcourt Butler Technical University, Kanpur, Uttar Pradesh, India

20.1 Introduction

20.1.1 Basic electrochemical principles

Electrochemistry is the study of chemical processes that involve the transfer of electrons at electrode–electrolyte interfaces. In an electrochemical cell, oxidation, which means loss of electrons, occurs at the anode, whereas reduction, which is the gain of electrons, occurs at the cathode [1]. The cell potential E_{cell} replicates the energy change that happens due to the redox reaction. Thermodynamically, the maximum (reversible) work from an electrochemical reaction is associated with the change in Gibbs free energy by Eq. (20.1) as below:

$$\Delta G = -nFE_{CELL} \tag{20.1}$$

Electrochemical Biomass Conversion: Sustainable Processes and Applications for Bioenergy, Biofuels, and Bioproducts
DOI: https://doi.org/10.1016/B978-0-443-26757-4.00004-0

where n is the transferred number of electrons, and F is the Faraday constant [2]. Under standard conditions, this becomes as in Eq. (20.2):

$$\Delta G^0 = -nFE^0{}_{cell} \quad (20.2)$$

Thus a spontaneous reaction having a negative ΔG corresponds to a positive standard cell voltage E_{cell}. E_{cell} is calculated as the difference between the standard reduction potentials of the cathode and anode half-reactions as given in Eq. (20.3):

$$E^O_{CELL} = E^o_{CATHODE} - E^O_{ANODE} \quad (20.3)$$

Where the cell voltages can be predicted by opting for known tables of standard potentials, such as the Standard Hydrogen Electrode $E^O = 0$. For nonstandard conditions, the Nernst equation [3] relates electrode potential to reactant activities as in Eq. (20.4).

$$E = E^\circ - \frac{RT}{nF}\ln Q \quad (20.4)$$

Or in base-10 logarithm form as in Eq. (20.5).

$$E = E^O - (2.303RT/nF)\log_{10}Q \quad (20.5)$$

At 298 K, this simplifies to $E = E^O - (0.0592/n)\ log_{10}Q$ [4]. Here Q is the reaction quotient. The Nernst equation thus shows that each tenfold change in activity shifts a one-electron half-cell potential by approximately 59.2 mV (at 298 K) [3]. This relation underpins pH electrodes and sensor responses: for example, Eq. (20.5) gives the equilibrium potential of a hydrogen electrode that depends on the H^+ concentration by

$$E = E^o_{H^+/H_2} - 0.0592\ log^{H^+}_{10} \quad (20.6)$$

These basic principles govern batteries and fuel cells. For example, in a Li-ion battery, the overall cell reaction (e.g., $LiCoO_2 + C \rightleftarrows Li_{1\text{-}x}CoO_2 + Li_x C_6$) has a large n and high E^o, yielding high energy density. Open-circuit voltages in sensors and energy devices are directly predicted by the Nernst equation and tabulated E^o values [2,3]. The same fundamentals apply to corrosion (spontaneous oxidation in air/water) and to electrolysis (nonspontaneous reactions driven by external voltage, that is, $E < 0$ for the reaction as written).

20.1.2 Electrode reactions and redox processes

Electrode processes involve coupled electron and ion transfer at the interface. A redox (reduction–oxidation) half-reaction can be written as, for example, in Eq. (20.6),

$$CeO + ne^- --> R$$

where species O is reduced to R by gaining n electrons. At the electrode surface, electrons flow from the anode to the cathode through the external circuit; meanwhile, ions in the electrolyte migrate to maintain charge balance. The exchange current density j_0 is a key concept: it quantifies the equilibrium rate of electron transfer (per electrode area) when no net current flows. A high j_0 indicates facile (reversible) kinetics [4,5].

The kinetics of electrode reactions are often described by the Butler–Volmer Eq. (20.7). For a single-step, one-electron transfer, the current density j as a function of overpotential $\eta = E - E_{eq}$ is

$$j = j_0\left(\exp\left[\frac{\alpha_a zF\eta}{RT}\right] - \exp\left[-\frac{\alpha_c zF\eta}{RT}\right]\right) \tag{20.7}$$

where α_a and α_c are the anodic and cathodic transfer coefficients, and z is the number of electrons transferred [6]. At small overpotentials, that is, a small η, the above equation simplifies to

$$j = j_0\frac{zF}{RT}\eta \tag{20.8}$$

And at high overpotential region, the Butler–Volmer equation simplifies to the Tafel Eq. (20.8) [7]. When ($E - E_{eq} >> 0$), the first term dominates, and when ($E - E_{eq} << 0$), the second term dominates.where a and b are constants (for a given reaction and temperature) and are called the Tafel equation constants. The theoretical values of the Tafel equation constants are different for the cathodic and anodic processes. However, the Tafel slope s can be defined as in Eq. (20.9):

$$\mathrm{b} = (\partial E \partial ln|IF|)ci, T, p\ s = \left(\frac{\partial E}{\partial ln(I_f)}\right) \tag{20.9}$$

where I_f is the Faradaic current given by $I_f = I_a + I_c$, with I_a and I_c being the anodic and cathodic partial currents, respectively.

In every case, oxidation occurs at the anode and reduction at the cathode [1]. Electrode materials and surface structure strongly affect kinetics—for example, platinum has a much higher exchange current density for the hydrogen electrode reaction than does mercury. In electrosynthesis, control of electrode potential and current allows selective driving of desired redox steps; in corrosion science, the Butler–Volmer form is used to model the rate of metal dissolution under different potentials. Redox reactions are also central to analytical techniques (e.g., cyclic voltammetry), where the peak currents and potentials can be analyzed using these kinetic models.

20.1.3 Electrochemical cell configurations

Electrochemical cells come in various configurations, but two broad categories are galvanic (voltaic) and electrolytic. A galvanic cell spontaneously converts chemical energy to electrical energy (e.g., a battery or fuel cell). For instance, in a Daniell cell (Zn–Cu), Zn is oxidized at the anode and Cu^{2+} is reduced at the cathode, yielding a positive cell voltage [8]. An electrolytic cell, by contrast, drives a nonspontaneous reaction by applying external voltage (e.g., water electrolysis to produce H_2 and O_2) [9]. In both cases, the fundamental roles of anode/cathode (oxidation/reduction) remain the same, but the sign of E_{cell} and the direction of electron flow differ.

Real cells may be two- or three-electrode systems. In a simple two-electrode cell, only a working electrode (WE) and a counter (auxiliary) electrode complete the circuit. In modern experiments, a three-electrode setup is preferred: a WE where the reaction of interest occurs, a reference electrode of fixed and known potential (e.g., saturated calomel or Ag/AgCl), and a counter electrode that carries the current between the cell and the potentiostat [10].

The potentiostat applies a potential between the WE and RE and measures current through WE/CE, while isolating the RE so that its potential remains stable. Placement of these electrodes is crucial: the WE and RE tips are kept very close (often using a Luggin capillary) to minimize solution resistance (iR drop) between them. A separate compartment or frit may isolate the RE/CE to prevent cross-contamination of electrolytes [11].

Common cell variants include:

- Concentration cells, with identical electrodes but differing ion concentrations, produce a small voltage by the Nernst relation.
- Fuel cells, which continuously consume fuel (H_2, CO, etc.) at one electrode and oxidant (O_2) at the other; for instance, proton-exchange membrane fuel cells [12] use H_2 oxidation at the anode and O_2 reduction at the cathode with a Nafion membrane between.
- Batteries, with fixed electrodes and stored reactants; typical Li-ion cells [13] have intercalation materials for electrodes and organic liquid electrolytes.
- Sensors, such as pH or oxygen electrodes, which measure a steady-state potential corresponding to a redox equilibrium.

Quantitatively, cell design is captured by the Nernst equation (for open-circuit voltage) and kinetic models (e.g., Butler–Volmer) under load. For example, the open-circuit voltage of a hydrogen–oxygen fuel cell is 1.23 V [14] (from $2H_2 + O_2 \rightarrow 2H_2O$), but under operating conditions, the effective potential drops according to polarization curves. Instruments (potentiostats and batteries) also use ohmic drop compensation and other controls to account for electrolyte resistance and electrode kinetics.

20.1.4 Electrolytes and membranes

An electrolyte is an ionic conductor that completes the circuit between electrodes. Electrolytes may be liquid (aqueous acids/bases, organic solvents with salts, ionic liquids), gel, polymeric, or solid inorganic [15]. A good electrolyte has high ionic conductivity, a wide electrochemical stability window (so it does not decompose under cell potential), and chemical compatibility with electrodes.

- Liquid electrolytes are common: aqueous KOH or H_2SO_4 in alkaline/fuel cells, or organic carbonate solutions with Li salts ($LiPF_6$, LiTFSI) in lithium batteries [16]. They provide high conductivity (~0.1–1 S/cm) but often require supporting salts to maintain charge neutrality.
- Polymer electrolytes (e.g., polyethylene oxide, PEO) [17] offer mechanical flexibility. PEO solvates Li^+ and can be cast into thin films, but its ionic conductivity at room temperature is modest ($<10^{-4}$ S/cm) due to crystallinity. Gel polymer electrolytes (liquid trapped in polymer matrix) combine some advantages of liquids and solids.
- Solid electrolytes include inorganic ceramics (β-alumina, perovskite oxides, sulfides) [18], which can achieve high conductivities and thermal stability. Recent advances (e.g., $Li_7La_3Zr_2O_{12}$ garnets) are promising for all-solid-state batteries, but interfacial impedance remains a challenge.
- Ionic liquids (molten salts at room temperature) are attractive because they are nonvolatile and electrochemically stable. Their ionic conductivity depends on viscosity; many ILs have lower conductivity than aqueous electrolytes but wider voltage windows [19].

20.1.5 Membranes and separators

Membranes and separators are specialized electrolytes that also act as physical barriers. In fuel cells, proton-exchange membranes (PEMs) [20] such as Nafion conduct H^+ selectively while blocking fuel gases. Nafion, a perfluorosulfonic acid polymer, is the state-of-the-art PEM due to its high proton conductivity and chemical stability [21]. Alkaline exchange membranes [22] similarly conduct OH^-. In batteries, a *separator* is typically a microporous polymer film soaked with liquid electrolyte; it prevents shorting by keeping the anode and cathode apart but allows ion flow. Advanced concepts use polymer or ceramic single-ion conductors to suppress unwanted side reactions (e.g., dendrite growth in Li metal batteries).

For example, vanadium redox-flow batteries [23] use liquid vanadium electrolytes on both sides of an ion-exchange membrane: one tank contains

V(II)/V(III) and the other V(IV)/V(V). The membrane allows H^+ (or other charge carriers) to move to balance charge as electrons flow through the circuit, while preventing vanadium ion cross-mixing. This ensures that each half-cell reaction can proceed independently and reversibly.

20.1.6 Electrocatalysts for biomass electrochemical conversion

To optimize performance under electrochemical conditions, while dealing with complex biomass substrates, an ideal electrocatalyst must possess specific features as specified below [24]. Active sites are the places on the catalyst surface on which reactant molecules adsorb and are converted. The distribution and character of the sites control catalytic activity directly. Through a change in geometry, coordination environment, and electronic structure of active sites, there can be enhanced interaction with biomass-derived molecules, that is, polyols, sugars, and lignin fragments. For facile access of large biomass molecules to active sites, hierarchical porosity and high surface area materials are favorable. Microporous and mesoporous structures provide ease of mass transportation, reduce diffusion constraints, and maximize accessible active sites. Facile electron transfer is critical in electrocatalytic reactions. The catalysts need to be inherently highly conductive or be supported on conductive supports to facilitate fast electron transfer between the catalytic centers and the electrode. Low conductivity results in energy loss and lower overall efficiency. The electrochemical conversion of biomass is more likely in aqueous systems. Hence, the surface wettability of the catalyst controls the interaction with hydrophilic biomass molecules. Hydroxyl, carbonyl, and amino functional groups on the surface can also act as points of anchoring for reactants or additional modification of the electronic environment of active sites. Chemical and structural stability under the harsh and repeated oxidative chemical conditions of biomass electrooxidation (e.g., radical presence, acid presence, high potentials), catalysts should be able to preserve activity and structure in the long term. Leaching, aggregation, and poisoning should be avoided to the maximum possible degree through rational material design [25].

20.1.7 Biomass-derived carbon materials

Biomass-derived carbon materials are beginning to function as low-cost and renewable substitutes for catalyst supports or even electrocatalysts. Biomass is used to produce these materials from renewable feedstocks such as agricultural waste, lignin, sugars, and cellulose through thermal or chemical processing [26]. Common preparation methods include pyrolysis, hydrothermal carbonization, and activation (chemical or physical) [27]. The as-synthesized carbon materials are of variable porosity, high surface area, and varying surface functionalities depending on precursor type and processing conditions. Carbon doping with heteroatoms such as nitrogen (N) [28], sulfur (S) [29], phosphorus (P), or boron (B), creates defect sites and modifies the electronic density, thus becoming catalytically active. N-doped carbon, for example, is highly active in oxidation and reduction reactions because of the presence of graphitic-N and pyridinic-N sites [28,29]. Carbon materials are also suitable supports for metal nanoparticles due to the fact that they are chemically inert, well-conductive, and surface chemistry tunable. They inhibit nanoparticle agglomeration, facilitate dispersion, and promote overall catalytic activity. Hybrid Catalysts Carbon–metal oxide hybrid materials, metal-organic frameworks, or transition metal phosphides add further variety to biomass-based catalysts. Composite materials combine the strength of each individual component and show improved electrocatalytic activity in biomass valorization reactions.

20.1.8 Challenges of catalyst stability and selectivity

Despite significant progress in catalyst design, several challenges need to be overcome to commercialize biomass electroconversion [25]. One of the significant problems is the deactivation of the catalyst through poisoning by highly adsorbed species such as CO, formates, or phenolic species. These byproducts block active sites and reduce long-term performance. Other causes of stability issues include metal dissolution, sintering, or restructuring under reaction conditions. The complex biomass molecules are structurally multifunctional. Electrochemical reduction or oxidation of the molecules will likely produce multiple parallel reaction pathways with a blend of end products. High selectivity to one product is still difficult and requires stringent control of reaction conditions and surface chemistry of the catalyst. Large, water-soluble macromolecules of biomass, such as cellulose derivatives and polyols, diffuse slowly, particularly in low-conductivity electrolytes. These generate concentration gradients and lower reaction efficiency. Open, accessible pore structures and optimization of the composition of the electrolyte can reverse these effects. High-performance catalysts might entail complex synthesis routes or rare elements, posing scalability issues. Earth-abundant, low-cost materials with similar performance need to be discovered for application. To provide a self-sustaining operation, electrochemical systems must be combined with sustainable sources of energy like solar or wind. This puts constraints on the catalyst's response to changes in voltage and intermittent operation, necessitating the creation of resilient, adaptive materials.

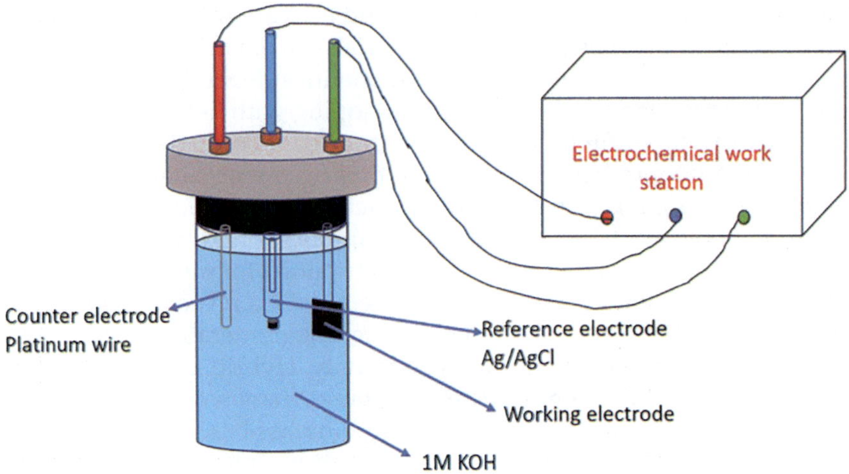

FIGURE 20.1 Schematic showing electrochemical analysis.

20.2 Analytical and characterization techniques

Cyclic voltammetry (CV) and linear sweep voltammetry (LSV) are widely employed electroanalytical techniques that provide critical insights into the redox behavior, electron transfer kinetics, and electrocatalytic efficiency of materials used in biomass electroconversion. EIS is a powerful technique for characterizing interfacial properties and charge transfer processes in electrochemical systems. It involves applying a small sinusoidal voltage perturbation and measuring the current response across a range of frequencies.

20.2.1 Cyclic voltammetry and linear sweep voltammetry

The CV experimental setup is typically a three-electrode electrochemical cell, as in Fig. 20.1. This consists of a working electrode, usually made of conductive materials like glassy carbon, platinum, or gold; a reference electrode, such as Ag/AgCl or saturated calomel electrode, with a fixed potential; and a counter (or auxiliary) electrode, usually platinum wire, completing the circuit and maintaining a balance of current through the working electrode. The potentiostat, another essential instrument in CV, controls the applied potential between the working and reference electrodes and measures the current between the working and counter electrodes. In the course of the experiment, the potential imposed on the WE is scanned from a starting value to a vertex (or switching) potential and then reversed. As the potential is near a value for which a redox

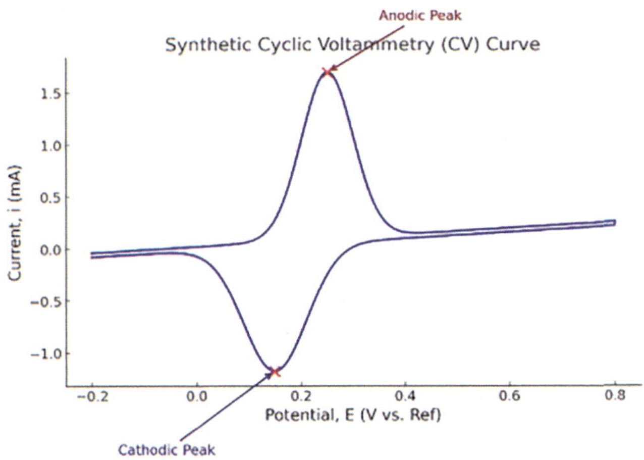

FIGURE 20.2 Cyclic voltammogram.

reaction is thermodynamically favorable, an electrochemical process is invoked—either oxidation (loss of electrons) or reduction (gain of electrons)—and a measurable current is generated.

In a reversible redox couple, two widely separated peaks in the voltammogram are typically observed, as in Fig. 20.2: The anodic peak is the oxidation of the analyte, and the cathodic peak is the reduction of the analyte. The peak-to-peak potential difference provides insight into the reversibility of the redox process. Under ideal conditions, for one-electron transfer at room temperature, the peak-to-peak separation (ΔE_p) is approximately 59 mV [30,31].

Randles–Ševčík equation [32] is a fundamental equation that governs the CV that gives a relation between the square root of the scan rate ($v^{1/2}$) and the observed peak current (i_p), as given in Eq. (20.10):

$$i_p = \left(2.69 \times 10^5\right) \cdot n^{3/2} \cdot A \cdot D^{1/2} C \cdot v^{1/2} \tag{20.10}$$

where i_p is the peak current, n is the number of electrons transferred, i_p A is the surface area of the electrode in cm^2, D is the diffusion coefficient in units cm^2/s, C is represented by the concentration of the electroactive species in mol/cm^3, and v is the scan rate in v/s. This is the proportionality of the peak current to the square root of the scan rate in a mass transport-controlled system by diffusion.

CV can be used to distinguish between different electrochemical systems based on their kinetic response. Reversible systems exhibit symmetric and narrow redox peaks with expected peak separations. Quasireversible systems exhibit broader peaks with higher separations due to more sluggish electron transfer kinetics, while irreversible systems tend to exhibit a single redox peak, that is, the reverse process is not observed on the measurement timescale.

CV is an important technique in various scientific applications. In materials chemistry, CV is used for the characterization of redox-active molecules, the study of the electron-transfer kinetics of the catalysts, and the electrochemical stability evaluation of electrode materials. In energy devices such as batteries and supercapacitors, CV is used for the study of charge storage mechanisms and capacitive behavior. Additionally, in electrocatalysis, particularly in biomass electroconversion, CV is used to study the onset potential of catalytic reactions, reaction reversibility, and catalytic activity and material durability.

Although it has many advantages, a CV has some disadvantages [33]. It performs optimally when reaction kinetics of systems are average and never provides true kinetic information for very rapid or sluggish reactions without the assistance of ancillary methods. Moreover, the interpretation of voltammograms is complex in multistep redox systems or in the event of simultaneously coupled chemical and electrochemical reactions.

In LSV, the potential applied to the WE is varied linearly with time from an initial potential to a final potential, unlike in cyclic voltammetry, where the potential is swept in a triangular waveform. The same experimental setup is followed for LSV, too. The only difference is that the scan direction is unidirectional, meaning that the potential is not reversed as it is in cyclic voltammetry. CV and LSV are both essential in the design and assessment of electrocatalysts for biomass conversion. CV is better applied in mechanistic and fundamental research, providing data regarding reaction intermediates, reversibility, and the electrochemical character of redox processes [34,35]. LSV, however, is mainly used to quantify catalytic performance, determine onset potentials, and evaluate electrochemical stability (Fig. 20.3).

Both CV and LSV are complementary in electrochemical studies of biomass conversion. Combined, they allow the complete characterization of electrocatalysts and reaction pathways, thus enabling the rational design of efficient and selective systems for the sustainable production of chemicals from renewable biomass resources.

20.2.2 Electrochemical impedance spectroscopy

It is a highly sensitive and multifunctional analytical tool for investigating the interfacial and transport processes in electrochemical systems. In contrast to sweeping potential voltametric methods that record current, EIS studies the frequency response of an electrochemical system by applying an alternating current signal with a small amplitude over a set of frequencies. This method allows for the decomposition of complicated electrochemical processes into their basic building blocks, for example,

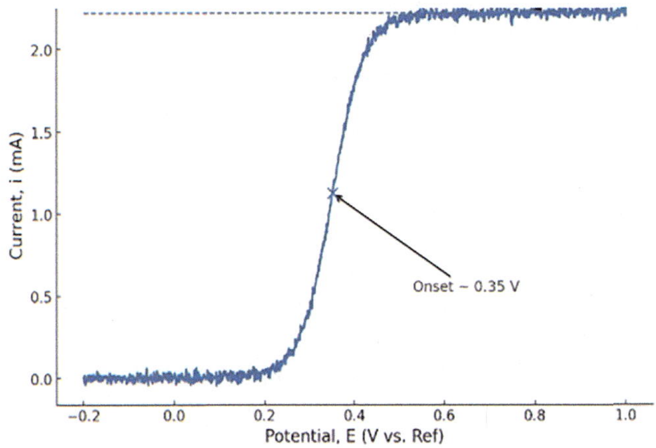

FIGURE 20.3 Basic LSV curve.

charge transfer, mass transport, and double-layer capacitance, and through that, acquire information regarding the kinetics and mechanisms that control electrochemical reactions.

20.2.2.1 Basic principles of electrochemical impedance spectroscopy

EIS measures the current response from applying a sinusoidal voltage signal of small amplitude (usually 5–10 mV). The signal's frequency varies over multiple orders of magnitude, usually between 100 kHz and 0.01 Hz. The system's impedance (Z), a complex number that may be written as follows, is determined by the voltage to current ratio at each frequency (Eq. 20.11):

$$Z(\omega) = V(\omega)/I(\omega) = Z' + Z'' \tag{20.11}$$

Where Z' and Z'' are the real and imaginary resistive components. Nyquist plots, which plot the imaginary part of impedance versus the real part, are frequently used to display the results. Bode plots, on the other hand, display the impedance's phase and magnitude as a function of frequency. The electrochemical system is analyzed by these graphical representations, which show processes such as electrolyte ohmic resistance, charge-transfer resistance, double-layer capacitance, Warburg impedance, etc.

Kinetic and transport parameters pertinent to the system under study can be quantitatively extracted by fitting the EIS data to equivalent electrical circuits (made up of resistors, capacitors, and constant phase elements) [36].

20.2.2.2 Electrochemical impedance spectroscopy in the electrochemical conversion of biomass

EIS is an essential tool in the field of biomass electrochemical conversion for assessing electrocatalyst performance and comprehending the electrode/electrolyte interface during the oxidation of chemicals produced from biomass. With several electron and proton transfers, intermediate species, and competitive side reactions, these reactions are frequently complicated. By measuring the diffusion impedance and charge transfer resistance (R_{ct}), which are essential for improving catalyst design and reaction conditions, EIS allows researchers to analyse these processes.

EIS, for instance, can show how effectively electrons are moved from the substrate to the electrode via the catalyst layer during the electrooxidation of glycerol or 5-hydroxymethylfurfural [37]. Effective catalysis is indicated by a low charge transfer resistance, whereas poor catalytic activity or electrode fouling is suggested by a high resistance. Similarly, Warburg-type diffusion impedance at low frequencies can be used to detect mass transport constraints, which are frequently seen in viscous or complicated biomass electrolytes [38].

EIS is also very helpful for assessing the stability and mechanisms of degradation of catalysts over time. Researchers can track changes in interfacial characteristics and provide early warning signs of catalyst deactivation, surface passivation, or structural alterations by doing time-resolved EIS studies. Although CV and LSV offer information on peak potentials and current densities, EIS offers a more thorough comprehension of the mechanisms governing these reactions, including kinetic, mass transport, and interfacial processes.

In biomass electrochemical conversion, various products—such as organic acids, alcohols, aldehydes, ketones, and other value-added chemicals—are generated as intermediates or final products. To determine the efficiency and selectivity of the conversion process, it is essential to accurately identify and quantify these products. Two major analytical approaches are employed for this purpose: chromatography and spectroscopy. These methods allow detailed product characterization, purity assessment, and yield calculation.

20.2.3 Chromatography

Chromatography is a separation technique that distributes the components of a mixture between a stationary phase and a mobile phase based on differences in physical or chemical properties such as polarity, size, or volatility. In biomass product analysis, the most commonly used chromatographic methods are:

- *Gas chromatography (GC)*: Suitable for volatile and thermally stable products such as methanol, ethanol, acetone, and light aldehydes. GC often uses flame ionization detection (FID) or mass spectrometry (GC–MS) for both quantification and structural identification [39].
- *High-performance liquid chromatography (HPLC)*: Applied for nonvolatile and thermally labile products such as organic acids (formic acid, acetic acid, lactic acid), sugars, and other polar compounds. HPLC can be coupled with UV-Vis or refractive index detectors depending on the product type [40].

Through these techniques, complex mixtures resulting from biomass conversion can be separated into individual components, each of which can then be quantified based on calibration curves with known standards.

Accurate quantification of products is essential to evaluate the efficiency of biomass electrochemical conversion processes. Depending on the reaction pathway, the products can be gaseous, liquid, or solid.

20.2.4 Gas products

Gaseous products such as hydrogen (H_2), carbon dioxide (CO_2), carbon monoxide (CO), methane (CH_4), and light hydrocarbons are typically analyzed using GC.

- Carrier gases (e.g., Ar, N_2, and He) are used to transport the sample through the GC column.
- Thermal conductivity detector or FID may be employed for detection, often coupled with a methanizer for CO and CO_2 quantification.
- Calibration with standard gas mixtures ensures accurate concentration measurement.

The output is generally expressed in terms of mole fraction, which can be converted to molar production rates using the known flow rate of the electrolyte or gas stream.

20.2.5 Liquid products

Liquid-phase products—such as alcohols, acids, ketones, aldehydes, and other oxygenated compounds—are analyzed using:

- HPLC for nonvolatile compounds like formic acid, lactic acid, glycerol, and glucose derivatives.
- Nuclear magnetic resonance (NMR) spectroscopy for qualitative and quantitative analysis without requiring separation.
- MS or GC–MS for volatile or semivolatile compounds.

The Faradaic efficiency (FE) represents the fraction of the total charge passed through the electrode that is utilized for producing a specific product [41].

$$FE = (n.F.M_{product}/Q_{total}) \times 100$$

Where:

- n = number of electrons involved per mole of product.
- F = Faraday's constant (96,485 C/mol).
- $M_{product}$ = moles of the product formed.
- Q_{total} = total charge passed (from the integrated current–time curve).

High FE indicates that the process is selective toward the desired product rather than forming side products or causing parasitic reactions.

20.2.6 Spectroscopy in biomass electrochemical conversion

Spectroscopy refers to a set of analytical techniques that study how matter interacts with electromagnetic radiation. Depending on the wavelength or energy of the radiation, different spectroscopic methods provide information about molecular structures, bonding, and composition. In biomass conversion, spectroscopy is indispensable for product identification, reaction monitoring, and mechanistic studies. Below is a detailed discussion of the most commonly employed techniques.

20.2.6.1 UV—Visible spectroscopy

UV–Vis's spectroscopy measures the absorption of ultraviolet (200–400 nm) and visible (400–700 nm) light by molecules. The method is based on the principle that when photons of suitable energy strike a molecule, electrons in π or nonbonding (n) orbitals can be excited to higher-energy π^* or σ^* orbitals [42].

Applications in biomass electroconversion:

- *Monitoring reaction progress:* Many biomass-derived molecules, such as aromatic aldehydes, conjugated ketones, and phenolic compounds, absorb strongly in the UV–Vis range.
- *Kinetic analysis:* By recording the absorbance at characteristic wavelengths over time, the rate of product formation or reactant consumption can be quantified.
- *Quantification:* Using the Beer–Lambert Law,

$$A = \varepsilon\, c\, l$$

where A is absorbance, ε is the molar absorptivity, c is concentration, and l is path length, the concentration of products can be determined from absorbance measurements.

20.2.6.2 Fourier transform infrared spectroscopy

FTIR spectroscopy examines the absorption of infrared radiation (4000–400 cm^{-1}), which causes molecular vibrations. Each functional group (e.g., O–H, C=O, C–O, C–H) absorbs IR radiation at characteristic frequencies, giving rise to a "molecular fingerprint."

- *Functional group identification:* For example, a strong peak around 1700 cm^{-1} indicates C=O stretching (carbonyl group), confirming the presence of aldehydes, ketones, or acids.
- *Tracking functional transformations:* FTIR can show the disappearance of carbohydrate O–H bands and appearance of carbonyl or aromatic bands during biomass depolymerization and oxidation.
- *In-situ FTIR:* When coupled with an electrochemical cell, in-situ FTIR can monitor intermediate formation in real time, helping to elucidate reaction mechanisms [43].

20.2.6.3 Nuclear magnetic resonance spectroscopy

NMR spectroscopy exploits the magnetic properties of nuclei (commonly ^1H, ^13C, and ^31P) in an external magnetic field. When subjected to radiofrequency radiation, these nuclei resonate at characteristic frequencies depending on their chemical environment.

- *Structural elucidation:* NMR reveals the connectivity of atoms within a molecule, providing a complete molecular structure of complex biomass-derived products.
- *Quantification without separation:* Integration of NMR peaks gives direct molar ratios, allowing quantification even in unpurified mixtures.
- *Tracking intermediates:* ^{13}C NMR can be used to follow carbon atom changes during the electrooxidation of biomass molecules.

20.2.6.4 Mass spectrometry

MS determines the molecular mass and fragmentation pattern of analytes. Molecules are ionized (by electron impact, electrospray ionization, etc.), separated by their mass-to-charge ratio (m/z) in a mass analyzer, and detected.

- *Molecular identification:* MS can distinguish between compounds with similar molecular formulas by analysing their fragmentation patterns.
- *Coupling with chromatography:* GC–MS or LC–MS enables separation of complex mixtures followed by mass analysis of each fraction, which is crucial for identifying minor products in biomass-derived mixtures.
- *Isotope tracking:* By using isotope-labelled biomass precursors (e.g., ^13C-labeled glucose), MS can trace the incorporation of atoms into products, revealing mechanistic pathways.

20.2.7 Importance of combining techniques

In biomass electrochemical conversion, no single spectroscopic method can provide all the necessary information. For instance, FTIR can confirm functional group changes, NMR can reveal full molecular structures, UV–Vis can monitor reaction kinetics, and MS can identify molecular weights and fragmentation pathways. When used together, these techniques provide a comprehensive understanding of product distribution, purity, and reaction mechanisms.

20.3 Conclusion and future outlook

Biomass electrochemistry is one of the most promising technologies we can have for sustainable energy and chemical production by employing renewable feedstocks. This electroconversion coupling not only utilizes waste but also helps in reducing carbon emissions while creating a value-added product. It could be a vital technology to be used in respect of the circular economy, having low carbon emissions prospects. Furthermore, research is required in terms of catalyst innovation, Integration of renewable energy generation, and generation of hybrid systems. To accelerate commercialization, techno-economic analysis and policy making around the globe are much needed. With global efforts and constant advancements, biomass electrochemistry industrial-scale output can be redefined with the evolving demands.

References

[1] C. Lefrou, P. Fabry, J.-C. Poignet, Electrochemistry: The Basics, With Examples, Springer Science & Business Media, 2012, doi:10.1007/978-3-642-30250-3.

[2] S. Gildemyn, R.A. Rozendal, K. Rabaey, A Gibbs free energy-based assessment of microbial electrocatalysis, Trends. Biotechnol. 35 (2017) 393–406.

[3] F.J. Vidal-Iglesias, J. Solla-Gullón, A. Rodes, E. Herrero, A. Aldaz, Understanding the Nernst equation and other electrochemical concepts: an easy experimental approach for students, J. Chem. Educ. 89 (2012) 936–939.

[4] M.R. Wright, Concepts and Ideas: Setting the Stage, An Introduction to Aqueous Electrolyte Solutions, John Wiley & Sons, 2007.

[5] E.J.F. Dickinson, A.J. Wain, The Butler–Volmer equation in electrochemical theory: origins, value, and practical application, J. Electroanal. Chem. 872 (2020) 114145.

[6] E.J.F. Dickinson, G. Hinds, The Butler–Volmer equation for polymer electrolyte membrane fuel cell (PEMFC) electrode kinetics: a critical discussion, J. Electrochem. Soc. 166 (2019) F221.

[7] S. Fletcher, T.S. Varley, Beyond the Butler–Volmer equation. Curved Tafel slopes from steady-state current–voltage curves, Phys. Chem. Chem. Phys. 13 (2011) 5359–5364.

[8] Z. He, J. Guo, F. Xiong, S. Tan, Y. Yang, R. Cao, et al., Re-imagining the Daniell cell: ampere-hour-level rechargeable Zn–Cu batteries, Energy Environ. Sci. 16 (2023) 5832–5841.

[9] K. Scott, Introduction to electrolysis, electrolysers and hydrogen production, in: Electrochemical Methods for Hydrogen Production, 2019, doi:10.1039/9781788016049-00001.

[10] N.A. Ogude, J.D. Bradley, Electrode processes and aspects relating to cell EMF, current, and cell components in operating electrochemical cells, J. Chem. Educ. 73 (1996) 1145–1149.

[11] M.H. Chakrabarti, R.A.W. Dryfe, E.P.L. Roberts, Evaluation of electrolytes for redox flow battery applications, Electrochim. Acta 52 (2007) 2189–2195.

[12] K. Jiao, J. Xuan, Q. Du, Z. Bao, B. Xie, B. Wang, et al., Designing the next generation of proton-exchange membrane fuel cells, Nature 595 (2021) 361–369.

[13] N. Meddings, M. Heinrich, F. Overney, J.-S. Lee, V. Ruiz, E. Napolitano, et al., Application of electrochemical impedance spectroscopy to commercial Li-ion cells: a review, J. Power Sources 480 (2020) 228742.

[14] J. Zhang, Y. Tang, C. Song, J. Zhang, H. Wang, PEM fuel cell open circuit voltage (OCV) in the temperature range of 23 C to 120 C, J. Power Sources 163 (2006) 532–537.

[15] M. Farkhondeh, M. Pritzker, C. Delacourt, S.-W. Liu, M. Fowler, Method of the four-electrode electrochemical cell for the characterization of concentrated binary electrolytes: theory and application, J. Phys. Chem. C 121 (2017) 4112–4129.

[16] S. Wang, L. Zhang, Z. Hu, B. Zhang, N. Li, Y. Tong, et al., Intrinsic structural and coordination chemistry insights of li salts in rechargeable lithium batteries, Adv. Mater. 37 (2025) 2420428.

[17] A. Arya, A.L. Sharma, Insights into the use of polyethylene oxide in energy storage/conversion devices: a critical review, J. Phys. D. Appl. Phys. 50 (2017) 443002.

[18] M.V. Reddy, C.M. Julien, A. Mauger, K. Zaghib, Sulfide and oxide inorganic solid electrolytes for all-solid-state Li batteries: a review, Nanomaterials 10 (2020) 1606.

[19] A. Lewandowski, A. Świderska-Mocek, Ionic liquids as electrolytes for Li-ion batteries—an overview of electrochemical studies, J. Power. Sources 194 (2009) 601–609.

[20] S.J. Peighambardoust, S. Rowshanzamir, M. Amjadi, Review of the proton exchange membranes for fuel cell applications, Int. J. Hydrogen. Energy 35 (2010) 9349–9384.

[21] H. Kawakami, Development of composite electrolyte membranes with functional polymer nanofiber frameworks, Polym. J. 57 (2025) 623–633.

[22] G. Merle, M. Wessling, K. Nijmeijer, Anion exchange membranes for alkaline fuel cells: a review, J. Memb. Sci. 377 (2011) 1–35.

[23] Á. Cunha, J. Martins, N. Rodrigues, F.P. Brito, Vanadium redox flow batteries: a technology review, Int. J. Energy Res. 39 (2015) 889–918.

[24] J.C. Vedrine, Revisiting active sites in heterogeneous catalysis: their structure and their dynamic behaviour, Appl. Catal. A Gen. 474 (2014) 40–50.

[25] Y. Jiao, Y. Zheng, K. Davey, S.-Z. Qiao, Activity origin and catalyst design principles for electrocatalytic hydrogen evolution on heteroatom-doped graphene, Nat. Energy 1 (2016) 1–9.

[26] A. Demirbaş, Biomass resource facilities and biomass conversion processing for fuels and chemicals, Energy Convers. Manag. 42 (2001) 1357–1378.

[27] A. Plavniece, G. Dobele, A. Volperts, A. Zhurinsh, Hydrothermal carbonization vs. pyrolysis: effect on the porosity of the activated carbon materials, Sustainability 14 (2022) 15982.

[28] L. Yan, J. Yu, J. Houston, N. Flores, H. Luo, Biomass-derived porous nitrogen-doped carbon for electrochemical devices, Green. Energy Environ. 2 (2017) 84–99.

[29] G. Xu, J. Han, B. Ding, P. Nie, J. Pan, H. Dou, et al., Biomass-derived porous carbon materials with sulfur and nitrogen dual-doping for energy storage, Green Chem. 17 (2015) 1668–1674.

[30] R. Bhujel, S. Rai, B.P. Swain, Investigation of cyclic voltammetry, impedance spectroscopy and electrical properties of thermally exfoliated biomass-synthesized graphene, Appl. Nanosci. 9 (2019) 1319–1331.

[31] L. Wang, Q. Zhang, S. Chen, F. Xu, S. Chen, J. Jia, et al., Electrochemical sensing and biosensing platform based on biomass-derived macroporous carbon materials, Anal. Chem. 86 (2014) 1414–1421.

[32] O.A. González-Meza, E.R. Larios-Durán, A. Gutiérrez-Becerra, N. Casillas, J.I. Escalante, M. Bárcena-Soto, Development of a Randles–Ševčík-like equation to predict the peak current of cyclic voltammetry for solid metal hexacyanoferrates, J. Solid State Electrochem. 23 (2019) 3123–3133.

[33] F. Marken, A. Neudeck, A.M. Bond, Cyclic voltammetry, in: Electroanalytical Methods: Guide to Experiments and Applications, Springer, 2009, pp. 57–106, doi:10.1007/978-3-642-02915-8_4.

[34] G. Hussain, D.S. Silvester, Comparison of voltammetric techniques for ammonia sensing in ionic liquids, Electroanalysis 30 (2018) 75–83.

[35] K. Sathish-Kumar, O. Solorza-Feria, R. Hernández-Vera, G. Vazquez-Huerta, H.M. Poggi-Varaldo, Comparison of various techniques to characterize a single chamber microbial fuel cell loaded with sulfate reducing biocatalysts, J. New Mater. Electrochem. Syst. 15 (2012) 195–201.

[36] H.S. Magar, R.Y.A. Hassan, A. Mulchandani, Electrochemical impedance spectroscopy (EIS): principles, construction, and biosensing applications, Sensors 21 (2021) 6578.

[37] Y. Yang, T. Mu, Electrochemical oxidation of biomass derived 5-hydroxymethylfurfural (HMF): pathway, mechanism, catalysts and coupling reactions, Green Chem. 23 (12) (2021) 4228–4254.

[38] S.A. Mareev, V.V. Nikonenko, A numerical experiment approach to modeling impedance: application to study a Warburg-type spectrum in a membrane system with diffusion coefficients depending on concentration, Electrochim. Acta 81 (2012) 268–274.

[39] C.P. Long, M.R. Antoniewicz, Quantifying biomass composition by gas chromatography/mass spectrometry, Anal. Chem. 86 (19) (2014) 9423–9427.

[40] G. Bonn, O. Bobleter, HPLC-analyses of plant biomass hydrolysis and fermentation solutions, Chromatographia 18 (8) (1984) 445–448.

[41] I. Garagounis, V. Kyriakou, M. Stoukides, Electrochemical promotion of catalytic reactions: thermodynamic analysis and calculation of the limits in Faradaic efficiency, Solid. State Ion. 231 (2013) 58–62.

[42] M. Zirbes, S.R. Waldvogel, Electro-conversion as sustainable method for the fine chemical production from the biopolymer lignin, Curr. Opin. Green Sustain. Chem. 14 (2018) 19–25.

[43] V. Thirumal, R. Yuvakkumar, B. Saravanakumar, G. Ravi, M. Isacfranklin, M. Shobana, et al., Carbonization and optimization of biomass waste for HER application, Fuel 324 (2022) 124466.

Index

Page numbers followed by "*f*" and "*t*" indicate, figures and tables respectively.

9780443267574